Student Solutions Manual

Trigonometry

TENTH EDITION

Ron Larson

The Pennsylvania State University, The Behrend College

Prepared by

Ron Larson

The Pennsylvania State University, The Behrend College

CENGAGE
Learning·

Australia · Brazil · Mexico · Singapore · United Kingdom · United States

CONTENTS

CONTENTS

C H A P T E R   P
Prerequisites

CHAPTER P
Prerequisites

Section P.1   Review of Real Numbers and Their Properties

1. irrational

3. absolute value

5. terms

7. $-9, -\frac{7}{2}, 5, \frac{2}{3}, \sqrt{2}, 0, 1, -4, 2, -11$

  (a) Natural numbers: 5, 1, 2

  (b) Whole numbers: 0, 5, 1, 2

  (c) Integers: $-9, 5, 0, 1, -4, 2, -11$

  (d) Rational numbers: $-9, -\frac{7}{2}, 5, \frac{2}{3}, 0, 1, -4, 2, -11$

  (e) Irrational numbers: $\sqrt{2}$

9. $2.01, 0.\overline{6}, -13, 0.010110111\ldots, 1, -6$

  (a) Natural numbers: 1

  (b) Whole numbers: 1

  (c) Integers: $-13, 1, -6$

  (d) Rational numbers: $2.01, 0.\overline{6}, -13, 1, -6$

  (e) Irrational numbers: $0.010110111\ldots$

11. (a)

  (b)

  (c)

  (d)

13. $-4 > -8$

15. $\frac{5}{6} > \frac{2}{3}$

17. (a) The inequality $x \le 5$ denotes the set of all real numbers less than or equal to 5.

  (b)

  (c) The interval is unbounded.

19. (a) The inequality $-2 < x < 2$ denotes the set of all real numbers greater than -2 and less than 2.

  (b)

  (c) The interval is bounded.

21. (a) The interval $[4, \infty)$ denotes the set of all real numbers greater than or equal to 4.

  (b)

  (c) The interval is unbounded.

23. (a) The interval $[-5, 2)$ denotes the set of all real numbers greater than or equal to -5 and less than 2.

  (b)

  (c) The interval is bounded.

25. $y \ge 0; [0, \infty)$

27. $10 \le t \le 22; [10, 22]$

29. $|-10| = -(-10) = 10$

31. $|3 - 8| = |-5| = -(-5) = 5$

33. $|-1| - |-2| = 1 - 2 = -1$

35. $5|-5| = 5(5) = 25$

37. If $x < -2$, then $x + 2$ is negative.

  So, $\dfrac{|x + 2|}{x + 2} = \dfrac{-(x + 2)}{x + 2} = -1.$

39. $|-4| = |4|$ because $|-4| = 4$ and $|4| = 4.$

41. $-|-6| < |-6|$ because $|-6| = 6$ and $-|-6| = -(6) = -6.$

43. $d(126, 75) = |75 - 126| = 51$

45. $d\left(-\frac{5}{2}, 0\right) = \left|0 - \left(-\frac{5}{2}\right)\right| = \frac{5}{2}$

47. $d(x, 5) = |x - 5|$ and $d(x, 5) \le 3$, so $|x - 5| \le 3.$

Receipts, R	Expenditures, E	$\lvert R - E \rvert$
49. $2524.0 billion	$2982.5 billion	$\lvert 2524.0 - 2982.5 \rvert = \$458.5$ billion
51. $2450.0 billion	$3537.0 billion	$\lvert 2450.0 - 3537.0 \rvert = \$1087.0$ billion

53. $7x + 4$

    Terms: $7x, 4$

    Coefficient: 7

55. $6x^3 - 5x$

    Terms: $6x^3, -5x$

    Coefficients: 6, −5

57. $3\sqrt{3}x^2 + 1$

    Terms: $3\sqrt{3}x^2, 1$

    Coefficient: $3\sqrt{3}$

59. $4x - 6$

    (a) $4(-1) - 6 = -4 - 6 = -10$

    (b) $4(0) - 6 = 0 - 6 = -6$

67. $x(3y) = (x \cdot 3)y$ Associative Property of Multiplication

          $= (3x)y$ Commutative Property of Multiplication

69. $\dfrac{2x}{3} - \dfrac{x}{4} = \dfrac{8x}{12} - \dfrac{3x}{12} = \dfrac{5x}{12}$

71. $\dfrac{3x}{10} \cdot \dfrac{5}{6} = \dfrac{x}{2} \cdot \dfrac{1}{2} = \dfrac{x}{4}$

73. False. Because zero is nonnegative but not positive, not every nonnegative number is positive.

75. The product of two negative numbers is positive.

61. $x^2 - 3x + 2$

    (a) $(0)^2 - 3(0) + 2 = 2$

    (b) $(-1)^2 - 3(-1) + 2 = 1 + 3 + 2 = 6$

63. $\dfrac{x + 1}{x - 1}$

    (a) $\dfrac{1 + 1}{1 - 1} = \dfrac{2}{0}$

    Division by zero is undefined.

    (b) $\dfrac{-1 + 1}{-1 - 1} = \dfrac{0}{-2} = 0$

65. $\dfrac{1}{(h + 6)}(h + 6) = 1, h \neq -6$

    Multiplicative Inverse Property

77. (a)

n	0.0001	0.01	1	100	10,000
$5/n$	50,000	500	5	0.05	0.0005

    (b) (i) As n approaches 0, the value of $5/n$ increases without bound (approaches infinity).

       (ii) As n increases without bound (approaches infinity), the value of $5/n$ approaches 0.

Section P.2   Solving Equations

1. equation

3. extraneous

5.

$$x + 11 = 15$$
$$x + 11 - 11 = 15 - 11$$
$$x = 4$$

7.

$$7 - 2x = 25$$
$$7 - 7 - 2x = 25 - 7$$
$$-2x = 18$$
$$\frac{-2x}{-2} = \frac{18}{-2}$$
$$x = -9$$

9.

$$3x - 5 = 2x + 7$$
$$3x - 2x - 5 = 2x - 2x + 7$$
$$x - 5 = 7$$
$$x - 5 + 5 = 7 + 5$$
$$x = 12$$

11. $x - 3(2x + 3) = 8 - 5x$

$x - 6x - 9 = 8 - 5x$

$-5x - 9 = 8 - 5x$

$-5x + 5x - 9 = 8 - 5x + 5x$

$-9 \neq 8$

Because $-9 = 8$ is a contradiction, the equation has no solution.

13.
$$\frac{3x}{8} - \frac{4x}{3} = 4$$

$$(24)\frac{3x}{8} - (24)\frac{4x}{3} = (24)4$$

$$9x - 32x = 96$$

$$-23x = 96$$

$$x = -\frac{96}{23}$$

19. $\dfrac{x}{x + 4} + \dfrac{4}{x + 4} = -2$

$$\frac{x + 4}{x + 4} = -2$$

$$1 \neq -2$$

Because $1 = -2$ is a contradiction, the equation has no solution.

21. $\dfrac{2}{(x - 4)(x - 2)} = \dfrac{1}{x - 4} + \dfrac{2}{x - 2}$ Multiply each term by $(x - 4)(x - 2)$.

$$2 = 1(x - 2) + 2(x - 4)$$

$$2 = x - 2 + 2x - 8$$

$$2 = 3x - 10$$

$$12 = 3x$$

$$4 = x$$

A check reveals that $x = 4$ yields a denominator of zero. So, $x = 4$ is an extraneous solution, and the original equation has no real solution.

23. $\dfrac{1}{x - 3} + \dfrac{1}{x + 3} = \dfrac{10}{x^2 - 9}$ Multiply each term by $(x + 3)(x - 3)$.

$$\frac{1}{x - 3} + \frac{1}{x + 3} = \frac{10}{(x + 3)(x - 3)}$$

$$1(x + 3) + 1(x - 3) = 10$$

$$2x = 10$$

$$x = 5$$

15.
$$\frac{5x - 4}{5x + 4} = \frac{2}{3}$$

$$3(5x - 4) = 2(5x + 4)$$

$$15x - 12 = 10x + 8$$

$$5x = 20$$

$$x = 4$$

17. $10 - \dfrac{13}{x} = 4 + \dfrac{5}{x}$

$$\frac{10x - 13}{x} = \frac{4x + 5}{x}$$

$$10x - 13 = 4x + 5$$

$$6x = 18$$

$$x = 3$$

25. $6x^2 + 3x = 0$

$3x(2x + 1) = 0$

$3x = 0$ or $2x + 1 = 0$

$x = 0$ or $\quad\quad x = -\frac{1}{2}$

27. $x^2 + 10x + 25 = 0$

$(x + 5)(x + 5) = 0$

$x + 5 = 0$

$x = -5$

29. $3 + 5x - 2x^2 = 0$

$(3 - x)(1 + 2x) = 0$

$3 - x = 0$ or $1 + 2x = 0$

$x = 3$ or $\quad\quad x = -\frac{1}{2}$

31. $\quad\quad 16x^2 - 9 = 0$

$(4x + 3)(4x + 3) = 0$

$4x + 3 = 0 \Rightarrow x = -\frac{3}{4}$

$4x - 3 = 0 \Rightarrow x = \frac{3}{4}$

33. $\frac{3}{4}x^2 + 8x + 20 = 0$

$4\left(\frac{3}{4}x^2 + 8x + 20\right) = 4(0)$

$3x^2 + 32x + 80 = 0$

$(3x + 20)(x + 4) = 0$

$3x + 20 = 0$ or $x + 4 = 0$

$x = -\frac{20}{3}$ or $\quad x = -4$

35. $x^2 = 49$

$x = \pm 7$

37. $3x^2 = 81$

$x^2 = 27$

$x = \pm 3\sqrt{3}$

$\approx \pm 5.20$

39. $(x - 4)^2 = 49$

$x - 4 = \pm 7$

$x = 4 \pm 7$

$x = 11 \text{ or } x = -3$

41. $(2x - 1)^2 = 18$

$2x - 1 = \pm\sqrt{18}$

$2x = 1 \pm 3\sqrt{2}$

$x = \dfrac{1 \pm 3\sqrt{2}}{2}$

$\approx 2.62, \ -1.62$

43. $x^2 + 4x - 32 = 0$

$x^2 + 4x = 32$

$x^2 + 4x + 2^2 = 32 + 2^2$

$(x + 2)^2 = 36$

$x + 2 = \pm 6$

$x = -2 \pm 6$

$x = 4$ or $x = -8$

45. $x^2 + 4x + 2 = 0$

$x^2 + 4x = -2$

$x^2 + 4x + 2^2 = -2 + 2^2$

$(x + 2)^2 = 2$

$x + 2 = \pm\sqrt{2}$

$x = -2 \pm \sqrt{2}$

47. $6x^2 - 12x = -3$

$x^2 - 2x = -\frac{1}{2}$

$x^2 - 2x + 1^2 = -\frac{1}{2} + 1^2$

$(x - 1)^2 = \frac{1}{2}$

$x - 1 = \pm\sqrt{\dfrac{1}{2}}$

$x = 1 \pm \sqrt{\dfrac{1}{2}}$

$x = 1 \pm \dfrac{\sqrt{2}}{2}$

49.
$$2x^2 + 5x - 8 = 0$$
$$2x^2 + 5x = 8$$
$$x^2 + \frac{5}{2}x = 4$$
$$x^2 + \frac{5}{2}x + \left(\frac{5}{4}\right)^2 = 4 + \left(\frac{5}{4}\right)^2$$
$$\left(x + \frac{5}{4}\right)^2 = \frac{89}{16}$$
$$x + \frac{5}{4} = \pm\frac{\sqrt{89}}{4}$$
$$x = -\frac{5}{4} \pm \frac{\sqrt{89}}{4}$$
$$x = \frac{-5 \pm \sqrt{89}}{4}$$

51.
$$\frac{1}{x^2 - 2x + 5} = \frac{1}{x^2 - 2x + 1^2 + 5}$$
$$= \frac{1}{(x - 1)^2 + 4}$$

53.
$$\frac{1}{\sqrt{3 + 2x - x^2}} = \frac{1}{\sqrt{-1(x^2 - 2x - 3)}}$$
$$= \frac{1}{\sqrt{-1\left[x^2 - 2x + (1)^2 - (1)^2 - 3\right]}}$$
$$= \frac{1}{\sqrt{-1(x^2 - 2x + 1) + 4}}$$
$$= \frac{1}{\sqrt{4 - (x - 1)^2}}$$

55. $2x^2 + x - 1 = 0$
$$x = \frac{-b \pm \sqrt{b^2 - 4ac}}{2a}$$
$$= \frac{-1 \pm \sqrt{1^2 - 4(2)(-1)}}{2(2)}$$
$$= \frac{-1 \pm 3}{4} = \frac{1}{2}, -1$$

57. $9x^2 + 30x + 25 = 0$
$$x = \frac{-b \pm \sqrt{b^2 - 4ac}}{2a}$$
$$= \frac{-30 \pm \sqrt{30^2 - 4(9)(25)}}{2(9)}$$
$$= \frac{-30 \pm 0}{18} = -\frac{5}{3}$$

59. $2x^2 - 7x + 1 = 0$
$$x = \frac{-b \pm \sqrt{b^2 - 4ac}}{2a}$$
$$= \frac{-(-7) \pm \sqrt{(-7)^2 - 4(2)(1)}}{2(2)}$$
$$= \frac{7 \pm \sqrt{49 - 8}}{2(2)}$$
$$= \frac{7 \pm \sqrt{41}}{4}$$
$$= \frac{7}{4} \pm \frac{\sqrt{41}}{4}$$

61.
$$12x - 9x^2 = -3$$
$$-9x^2 + 12x + 3 = 0$$
$$x = \frac{-b \pm \sqrt{b^2 - 4ac}}{2a}$$
$$= \frac{-12 \pm \sqrt{12^2 - 4(-9)(3)}}{2(-9)}$$
$$= \frac{-12 \pm 6\sqrt{7}}{-18} = \frac{2}{3} \pm \frac{\sqrt{7}}{3}$$

63. $2 + 2x - x^2 = 0$
$$-x^2 + 2x + 2 = 0$$
$$x = \frac{-b \pm \sqrt{b^2 - 4ac}}{2a}$$
$$= \frac{-2 \pm \sqrt{2^2 - 4(-1)(2)}}{2(-1)}$$
$$= \frac{-2 \pm 2\sqrt{3}}{-2}$$
$$= 1 \pm \sqrt{3}$$

65.
$$8t = 5 + 2t^2$$
$$-2t^2 + 8t - 5 = 0$$
$$t = \frac{-b \pm \sqrt{b^2 - 4ac}}{2a}$$
$$= \frac{-8 \pm \sqrt{8^2 - 4(-2)(-5)}}{2(-2)}$$
$$= \frac{-8 \pm 2\sqrt{6}}{-4} = 2 \pm \frac{\sqrt{6}}{2}$$

67. $(y - 5)^2 = 2y$

$y^2 - 12y + 25 = 0$

$$y = \frac{-b \pm \sqrt{b^2 - 4ac}}{2a}$$

$$= \frac{-(-12) \pm \sqrt{(-12)^2 - 4(1)(25)}}{2(1)}$$

$$= \frac{12 \pm 2\sqrt{11}}{2} = 6 \pm \sqrt{11}$$

69. $5.1x^2 - 1.7x = 3.2$

$5.1x^2 - 1.7x - 3.2 = 0$

$$x = \frac{1.7 \pm \sqrt{(-1.7)^2 - 4(5.1)(-3.2)}}{2(5.1)}$$

$$\approx 0.976, -0.643$$

71. $-0.005x^2 + 0.101x - 0.193 = 0$

$$x = \frac{-b \pm \sqrt{b^2 - 4ac}}{2a}$$

$$= \frac{-0.101 \pm \sqrt{(0.101)^2 - 4(-0.005)(-0.193)}}{2(-0.005)}$$

$$= \frac{-0.101 \pm \sqrt{0.006341}}{-0.01}$$

$$\approx 2.137, 18.063$$

73. $x^2 - 2x - 1 = 0$ Complete the square.

$x^2 - 2x = 1$

$x^2 - 2x + 1^2 = 1 + 1^2$

$(x - 1)^2 = 2$

$x - 1 = \pm\sqrt{2}$

$x = 1 \pm \sqrt{2}$

75. $(x + 2)^2 = 64$ Extract square roots.

$x + 2 = \pm 8$

$x + 2 = 8$ or $x + 2 = -8$

$x = 6$ or $x = -10$

77. $x^2 - x - \frac{11}{4} = 0$ Complete the square.

$x^2 - x = \frac{11}{4}$

$x^2 - x + \left(\frac{1}{2}\right)^2 = \frac{11}{4} + \left(\frac{1}{2}\right)^2$

$\left(x - \frac{1}{2}\right)^2 = \frac{12}{4}$

$x - \frac{1}{2} = \pm\sqrt{\frac{12}{4}}$

$x = \frac{1}{2} \pm \sqrt{3}$

79. $3x + 4 = 2x^2 - 7$ Quadratic Formula

$0 = 2x^2 - 3x - 11$

$$x = \frac{-(-3) \pm \sqrt{(-3)^2 - 4(2)(-11)}}{2(2)}$$

$$= \frac{3 \pm \sqrt{97}}{4}$$

$$= \frac{3}{4} \pm \frac{\sqrt{97}}{4}$$

81. $\dfrac{1}{x} - \dfrac{1}{x + 1} = 3$

$x(x + 1)\dfrac{1}{x} - x(x + 1)\dfrac{1}{x + 1} = x(x + 1)(3)$

$x + 1 - x = 3x(x + 1)$

$1 = 3x^2 + 3x$

$0 = 3x^2 + 3x - 1$

$$x = \frac{-3 \pm \sqrt{(3)^2 - 4(3)(-1)}}{2(3)} = \frac{-3 \pm \sqrt{21}}{6}$$

83. $\dfrac{x}{x^2 - 4} + \dfrac{1}{x + 2} = 3$

$(x + 2)(x - 2)\dfrac{x}{x^2 - 4} + (x + 2)(x - 2)\dfrac{1}{x + 2} = 3(x + 2)(x - 2)$

$x + x - 2 = 3x^2 - 12$

$3x^2 - 2x - 10 = 0$

$$x = \frac{-(-2) \pm \sqrt{(-2)^2 - 4(3)(-10)}}{2(3)}$$

$$= \frac{2 \pm \sqrt{124}}{6} = \frac{2 \pm 2\sqrt{31}}{6} = \frac{1 \pm \sqrt{31}}{3}$$

85. $6x^4 - 54x^2 = 0$

$6x^2(x^2 - 9) = 0$

$6x^2 = 0 \Rightarrow x = 0$

$x^2 - 9 = 0 \Rightarrow x = \pm 3$

87. $\qquad x^3 + 2x^2 - 8x = 16$

$x^3 + 2x^2 - 8x - 16 = 0$

$x^2(x + 2) - 8(x + 2) = 0$

$(x + 2)(x^2 - 8) = 0$

$x + 2 = 0 \Rightarrow x = -2$

$x^2 - 8 = 0 \Rightarrow \pm\sqrt{8} = \pm 2\sqrt{2}$

89. $\qquad x^4 - 4x^2 + 3 = 0$

$(x^2)^2 - 4(x^2) + 3 = 0$

Let $u = x^2$.

$u^2 - 4u + 3 = 0$

$(u - 3)(u - 1) = 0$

$u - 3 = 0 \Rightarrow u = 3$

$u - 1 = 0 \Rightarrow u = 1$

$\begin{array}{ll} u = 1 & u = 3 \\ x^2 = 1 & x^2 = 3 \\ x = \pm 1 & x = \pm\sqrt{3} \end{array}$

91. $\sqrt{5x} - 10 = 0$

$\sqrt{5x} = 10$

$\left(\sqrt{5x}\right)^2 = (10)^2$

$5x = 100$

$x = 20$

93. $4 + \sqrt[3]{2x - 9} = 0$

$\sqrt[3]{2x - 9} = -4$

$\left(\sqrt[3]{2x - 9}\right)^3 = (-4)^3$

$2x - 9 = -64$

$2x = -55$

$x = -\dfrac{55}{2}$

95. $\qquad \sqrt{x + 8} = 2 + x$

$\left(\sqrt{x + 8}\right)^2 = (2 + x)^2$

$x + 8 = x^2 + 4x + 4$

$0 = x^2 + 3x - 4$

$x^2 + 3x - 4 = 0$

$(x + 4)(x - 1) = 0$

$x + 4 = 0 \Rightarrow x = -4,\ \text{extraneous}$

$x - 1 = 0 \Rightarrow x = 1$

97. $\sqrt{x - 3} + 1 = \sqrt{x}$

$\sqrt{x - 3} = \sqrt{x} - 1$

$\left(\sqrt{x - 3}\right)^2 = \left(\sqrt{x} - 1\right)^2$

$x - 3 = x - 2\sqrt{x} + 1$

$-4 = -2\sqrt{x}$

$2 = \sqrt{x}$

$(2)^2 = \left(\sqrt{x}\right)^2$

$4 = x$

99. $(x - 5)^{3/2} = 8$

$(x - 5)^3 = 8^2$

$x - 5 = \sqrt[3]{64}$

$x = 5 + 4 = 9$

101. $3x(x - 1)^{1/2} + 2(x - 1)^{3/2} = 0$

$(x - 1)^{1/2}[3x + 2(x - 1)] = 0$

$(x - 1)^{1/2}(5x - 2) = 0$

$(x - 1)^{1/2} = 0 \Rightarrow x - 1 = 0 \Rightarrow x = 1$

$5x - 2 = 0 \Rightarrow x = \frac{2}{5},\ \text{extraneous}$

103. $|2x - 5| = 11$

$2x - 5 = 11 \Rightarrow x = 8$

$-(2x - 5) = 11 \Rightarrow x = -3$

105. $\left| x + 1 \right| = x^2 - 5$

First equation:

$$x + 1 = x^2 - 5$$
$$x^2 - x - 6 = 0$$
$$(x - 3)(x + 2) = 0$$
$$x - 3 = 0 \Rightarrow x = 3$$
$$x + 2 = 0 \Rightarrow x = -2$$

Second equation:

$$-(x + 1) = x^2 - 5$$
$$-x - 1 = x^2 - 5$$
$$x^2 + x - 4 = 0$$
$$x = \frac{-1 + \sqrt{17}}{2}$$

Only $x = 3$ and $x = \dfrac{-1 - \sqrt{17}}{2}$ are solutions of the original equation. $x = -2$ and $x = \dfrac{-1 + \sqrt{17}}{2}$ are extraneous.

107. Let $y = 18$.

$$y = 0.514x - 14.75$$
$$18 = 0.514x - 14.75$$
$$32.75 = 0.514x$$
$$\frac{32.75}{0.514} = x$$
$$63.7 = x$$

So, the height of the female is about 63.7 inches.

109. False.

$$\sqrt{2x + 1} = -2 + \sqrt{x + 1}$$
$$2x + 1 = 4 - 4\sqrt{x + 1} + (x + 1)$$
$$x - 4 = -4\sqrt{x + 1}$$
$$x^2 - 8x + 16 = 16(x + 1)$$
$$x^2 - 24x = 0$$
$$x(x - 24) = 0$$
$$x = 0 \quad 1 \neq -2 + 1, \; x = 24 \quad 5 \neq -2 + 5$$

111. $\sqrt{x - 10} - \sqrt{x - 10} = 0$

$$\sqrt{x - 10} = \sqrt{x - 10}$$

False. The equation is an identity, so every real number is a solution.

113. $\dfrac{3x + 2}{5} = 7$

$$3x + 2 = 35 \quad \text{and} \quad x + 9 = 20$$
$$3x = 33 \qquad\qquad x = 11$$
$$x = 11$$

Yes, they are equivalent equations. They both have the solution $x = 11$.

Section P.3   The Cartesian Plane and Graphs of Equations

1. Cartesian

3. Midpoint Formula

5. graph

7. y-axis

9. $A: (2, 6)$, $B: (-6, -2)$, $C: (4, -4)$, $D: (-3, 2)$

11.

13. $(-3, 4)$

15. $x > 0$ and $y < 0$ in Quadrant IV.

17. $x = -4$ and $y > 0$ in Quadrant II.

19. $x + y = 0, x \neq 0, y \neq 0$ means $x = -y$ or $y = -x$. This occurs in Quadrant II or IV.

21.

Year, x	Number of Stores, y
2007	7276
2008	7720
2009	8416
2010	8970
2011	10,130
2012	10,773
2013	10,942
2014	11,453

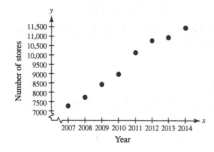

23. $d = \sqrt{(x_2 - x_1)^2 + (y_2 - y_1)^2}$

$\quad = \sqrt{(3 - (-2))^2 + (-6 - 6)^2}$

$\quad = \sqrt{(5)^2 + (-12)^2}$

$\quad = \sqrt{25 + 144}$

$\quad = 13$ units

25. $d = \sqrt{(x_2 - x_1)^2 + (y_2 - y_1)^2}$

$\quad = \sqrt{(-5 - 1)^2 + (-1 - 4)^2}$

$\quad = \sqrt{(-6)^2 + (-5)^2}$

$\quad = \sqrt{36 + 25}$

$\quad = \sqrt{61}$ units

27. (a) $(1, 0), (13, 5)$

$\quad$ Distance $= \sqrt{(13 - 1)^2 + (5 - 0)^2}$

$\quad\quad\quad\quad = \sqrt{12^2 + 5^2} = \sqrt{169} = 13$

$\quad (13, 5), (13, 0)$

$\quad$ Distance $= |5 - 0| = |5| = 5$

$\quad (1, 0), (13, 0)$

$\quad$ Distance $= |1 - 13| = |-12| = 12$

(b) $5^2 + 12^2 = 25 + 144 = 169 = 13^2$

29. $d_1 = \sqrt{(4 - 2)^2 + (0 - 1)^2} = \sqrt{4 + 1} = \sqrt{5}$

$\quad d_2 = \sqrt{(4 + 1)^2 + (0 + 5)^2} = \sqrt{25 + 25} = \sqrt{50}$

$\quad d_3 = \sqrt{(2 + 1)^2 + (1 + 5)^2} = \sqrt{9 + 36} = \sqrt{45}$

$\quad \left(\sqrt{5}\right)^2 + \left(\sqrt{45}\right)^2 = \left(\sqrt{50}\right)^2$

31. $d_1 = \sqrt{(1 - 3)^2 + (-3 - 2)^2} = \sqrt{4 + 25} = \sqrt{29}$

$\quad d_2 = \sqrt{(3 + 2)^2 + (2 - 4)^2} = \sqrt{25 + 4} = \sqrt{29}$

$\quad d_3 = \sqrt{(1 + 2)^2 + (-3 - 4)^2} = \sqrt{9 + 49} = \sqrt{58}$

$\quad d_1 = d_2$

33. (a)

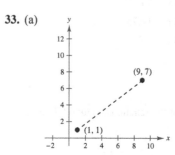

(b) $d = \sqrt{(9 - 1)^2 + (7 - 1)^2} = \sqrt{64 + 36} = 10$

(c) $\left(\dfrac{9 + 1}{2}, \dfrac{7 + 1}{2}\right) = (5, 4)$

35. (a)

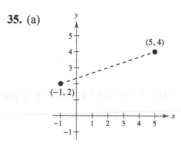

(b) $d = \sqrt{(5 + 1)^2 + (4 - 2)^2}$

$\quad\quad = \sqrt{36 + 4} = 2\sqrt{10}$

(c) $\left(\dfrac{-1 + 5}{2}, \dfrac{2 + 4}{2}\right) = (2, 3)$

37. $d = \sqrt{120^2 + 150^2}$

$\quad = \sqrt{36,900}$

$\quad = 30\sqrt{41}$

$\quad \approx 192.09$

The plane flies about 192 kilometers.

39. midpoint $= \left(\dfrac{x_1 + x_2}{2}, \dfrac{y_1 + y_2}{2} \right)$

$\phantom{\text{midpoint}} = \left(\dfrac{2010 + 2014}{2}, \dfrac{35{,}123 + 45{,}998}{2} \right)$

$\phantom{\text{midpoint}} = (2012, \, 40{,}560.5)$

In 2012, the sales for the Coca-Cola Company were about \$40,560.5 million.

41. (a) $(0, 2)$: $2 \overset{?}{=} \sqrt{0 + 4}$

$ 2 = 2$

Yes, the point *is* on the graph.

(b) $(5, 3)$: $3 \overset{?}{=} \sqrt{5 + 4}$

$ 3 \overset{?}{=} \sqrt{9}$

$ 3 = 3$

Yes, the point *is* on the graph.

43. (a) $(2, 0)$: $(2)^2 - 3(2) + 2 \overset{?}{=} 0$

$ 4 - 6 + 2 \overset{?}{=} 0$

$ 0 = 0$

Yes, the point *is* on the graph.

(b) $(-2, 8)$: $(-2)^2 - 3(-2) + 2 \overset{?}{=} 8$

$ 4 + 6 + 2 \overset{?}{=} 8$

$ 12 \neq 8$

No, the point *is not* on the graph.

45. (a) $(3, -2)$: $(3)^2 + (-2)^2 \overset{?}{=} 20$

$ 9 + 4 \overset{?}{=} 20$

$ 13 \neq 20$

No, the point *is not* on the graph.

(b) $(-4, 2)$: $(-4)^2 + (2)^2 \overset{?}{=} 20$

$ 16 + 4 \overset{?}{=} 20$

$ 20 = 20$

Yes, the point *is* on the graph.

47. $y = -2x + 5$

x	-1	0	1	2	$\frac{5}{2}$
y	7	5	3	1	0
(x, y)	$(-1, 7)$	$(0, 5)$	$(1, 3)$	$(2, 1)$	$\left(\frac{5}{2}, 0\right)$

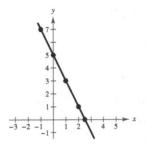

49. $y + 3x = x^2$

x	-1	0	1	2	3
y	4	0	-2	-2	0
(x, y)	$(-1, 4)$	$(0, 0)$	$(1, -2)$	$(2, -2)$	$(3, 0)$

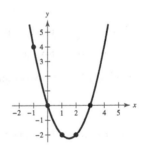

51. $y = 5x - 6$

x-intercept: $0 = 5x - 6$

$\phantom{x\text{-intercept:}} 6 = 5x$

$\phantom{x\text{-intercept:}} \frac{6}{5} = x$

$\phantom{x\text{-intercept:}} \left(\frac{6}{5}, 0 \right)$

y-intercept: $y = 5(0) - 6 = -6$

$\phantom{y\text{-intercept:}} (0, -6)$

53. $y = \sqrt{x + 4}$

    x-intercept: $0 = \sqrt{x + 4}$

                $0 = x + 4$

              $-4 = x$

              $(-4, 0)$

    y-intercept: $y = \sqrt{0 + 4} = 2$

              $(0, 2)$

55. $y = |3x - 7|$

    x-intercept: $0 = |3x - 7|$

                $0 = 3x - 7$

                $\frac{7}{3} = 0$

              $\left(\frac{7}{3}, 0\right)$

    y-intercept: $y = |3(0) - 7| = 7$

              $(0, 7)$

57. $y = 2x^3 - 4x^2$

    x-intercept: $0 = 2x^3 - 4x^2$

                $0 = 2x^2(x - 2)$

              $x = 0$ or $x = 2$

             $(0, 0), (2, 0)$

    y-intercept: $y = 2(0)^3 - 4(0)^2$

                $y = 0$

              $(0, 0)$

59. $y^2 = 6 - x$

    x-intercept: $0 = 6 - x$

                $x = 6$

              $(6, 0)$

    y-intercepts: $y^2 = 6 - 0$

                $y = \pm\sqrt{6}$

         $\left(0, \sqrt{6}\right), \left(0, -\sqrt{6}\right)$

61. $x^2 - y = 0$

    $(-x)^2 - y = 0 \Rightarrow x^2 - y = 0 \Rightarrow y$-axis symmetry

    $x^2 - (-y) = 0 \Rightarrow x^2 + y = 0 \Rightarrow$ No x-axis symmetry

    $(-x)^2 - (-y) = 0 \Rightarrow x^2 + y = 0 \Rightarrow$ No origin symmetry

63. $y = x^3$

    $y = (-x)^3 \Rightarrow y = -x^3 \Rightarrow$ No y-axis symmetry

    $-y = x^3 \Rightarrow y = -x^3 \Rightarrow$ No x-axis symmetry

    $-y = (-x)^3 \Rightarrow -y = -x^3 \Rightarrow y = x^3 \Rightarrow$ Origin symmetry

65. $y = \dfrac{x}{x^2 + 1}$

    $y = \dfrac{-x}{(-x)^2 + 1} \Rightarrow y = \dfrac{-x}{x^2 + 1} \Rightarrow$ No y-axis symmetry

    $-y = \dfrac{x}{x^2 + 1} \Rightarrow y = \dfrac{-x}{x^2 + 1} \Rightarrow$ No x-axis symmetry

    $-y = \dfrac{-x}{(-x)^2 + 1} \Rightarrow -y = \dfrac{-x}{x^2 + 1} \Rightarrow y = \dfrac{x}{x^2 + 1} \Rightarrow$ Origin symmetry

67. $xy^2 + 10 = 0$

    $(-x)y^2 + 10 = 0 \Rightarrow -xy^2 + 10 = 0 \Rightarrow$ No y-axis symmetry

    $x(-y)^2 + 10 = 0 \Rightarrow xy^2 + 10 = 0 \Rightarrow x$-axis symmetry

    $(-x)(-y)^2 + 10 = 0 \Rightarrow -xy^2 + 10 = 0 \Rightarrow$ No origin symmetry

69.

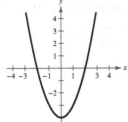

71.

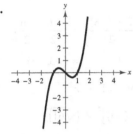

73. $y = -3x + 1$

x-intercept: $\left(\frac{1}{3}, 0\right)$

y-intercept: $(0, 1)$

No symmetry

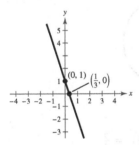

75. $y = x^2 - 2x$

x-intercepts: $(0, 0), (2, 0)$

y-intercept: $(0, 0)$

No symmetry

x	−1	0	1	2	3
y	3	0	−1	0	3

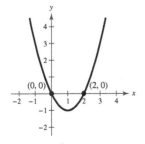

77. $y = x^3 + 3$

x-intercept: $\left(\sqrt[3]{-3}, 0\right)$

y-intercept: $(0, 3)$

No symmetry

x	−2	−1	0	1	2
y	−5	2	3	4	11

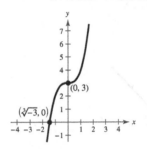

79. $y = \sqrt{x - 3}$

x-intercept: $(3, 0)$

y-intercept: none

No symmetry

x	3	4	7	12
y	0	1	2	3

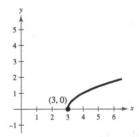

81. $y = |x - 6|$

x-intercept: $(6, 0)$

y-intercept: $(0, 6)$

No symmetry

x	-2	0	2	4	6	8	10
y	8	6	4	2	0	2	4

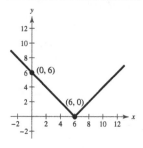

83. Center: $(0, 0)$; Radius: 7

$$(x - 0)^2 + (y - 0)^2 = 7^2$$
$$x^2 + y^2 = 49$$

85. Center: $(3, 8)$; Solution point: $(-9, 13)$

$$r = \sqrt{(x - h)^2 + (y - k)^2}$$
$$= \sqrt{(-9 - 3)^2 + (13 - 8)^2}$$
$$= \sqrt{(-12)^2 + (5)^2}$$
$$= \sqrt{144 + 25}$$
$$= \sqrt{169}$$
$$= 13$$
$$(x - h)^2 + (y - k)^2 = r^2$$
$$(x - 3)^2 + (y - 8)^2 = 13^2$$
$$(x - 3)^2 + (y - 8)^2 = 169$$

87. Endpoints of a diameter: $(3, 2), (-9, -8)$

$$r = \frac{1}{2}\sqrt{(-9 - 3)^2 + (-8 - 2)^2}$$
$$= \frac{1}{2}\sqrt{(-12)^2 + (-10)^2}$$
$$= \frac{1}{2}\sqrt{144 + 100}$$
$$= \frac{1}{2}\sqrt{244} = \frac{1}{2}(2)\sqrt{61} = \sqrt{61}$$
$$(h, k): \left(\frac{3 + (-9)}{2} \cdot \frac{2 + (-8)}{2}\right) = \left(\frac{-6}{2} \cdot \frac{-6}{2}\right)$$
$$= (-3, -3)$$
$$(x - h)^2 + (y - k)^2 = r^2$$
$$[x - (-3)]^2 + [y - (-3)]^2 = (\sqrt{61})^2$$
$$(x + 3)^2 + (y + 3)^2 = 61$$

89. $x^2 + y^2 = 25$

Center: $(0, 0)$, Radius: 5

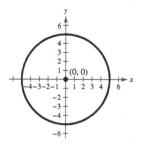

91. $\left(x - \frac{1}{2}\right)^2 + \left(y - \frac{1}{2}\right)^2 = \frac{9}{4}$

Center: $\left(\frac{1}{2}, \frac{1}{2}\right)$, Radius: $\frac{3}{2}$

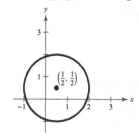

93. $y = 1,200,000 - 80,000t, \ 0 \le t \le 10$

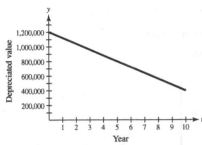

95. (a)

x	5	10	20	30	40	50	60	70	80	90	100
y	414.8	103.7	25.9	11.5	6.5	4.1	2.9	2.1	1.6	1.3	1.0

(b)

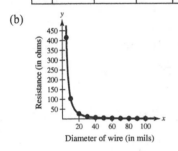

Diameter of wire (in mils)

When $x = 85.5$, the resistance is about 1.4 ohms.

(c) When $x = 85.5$,

$$y = \frac{10,370}{(85.5)^2} = 1.4 \text{ ohms.}$$

(d) As the diameter of the copper wire increases, the resistance decreases.

97. False, you would have to use the Midpoint Formula 15 times.

99. False. The line $y = x$ is symmetric with respect to the origin.

101. Answers will vary. *Sample answer:* When the x-values are much larger or smaller than the y-values, different scales for the coordinate axes should be used.

103. Use the Midpoint Formula to prove the diagonals of the parallelogram bisect each other.

$$\left(\frac{b + a}{2}, \frac{c + 0}{2} \right) = \left(\frac{a + b}{2}, \frac{c}{2} \right)$$

$$\left(\frac{a + b + 0}{2}, \frac{c + 0}{2} \right) = \left(\frac{a + b}{2}, \frac{c}{2} \right)$$

105. Because $x_m = \dfrac{x_1 + x_2}{2}$ and $y_m = \dfrac{y_1 + y_2}{2}$ we have:

$$2x_m = x_1 + x_2 \qquad\qquad 2y_m = y_1 + y_2$$
$$2x_m - x_1 = x_2 \qquad\qquad 2y_m - y_1 = y_2$$

So, $(x_2, y_2) = (2x_m - x_1, 2y_m - y_1)$.

(a) $(x_2, y_2) = (2x_m - x_1, 2y_m - y_1) = (2 \cdot 4 - 1, 2(-1) - (-2)) = (7, 0)$

(b) $(x_2, y_2) = (2x_m - x_1, 2y_m - y_1) = (2 \cdot 2 - (-5), 2 \cdot 4 - 11) = (9, -3)$

Section P.4   Linear Equations in Two Variables

1. linear

3. point-slope

5. perpendicular

7. linear extrapolation

9. (a) $m = \frac{2}{3}$. Because the slope is positive, the line rises. Matches L_2.

(b) m is undefined. The line is vertical. Matches L_3.

(c) $m = -2$. The line falls. Matches L_1.

11.

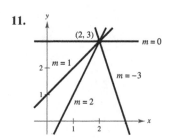

13. Two points on the line: $(0, 0)$ and $(4, 6)$

Slope $= \dfrac{y_2 - y_1}{x_2 - x_1} = \dfrac{6}{4} = \dfrac{3}{2}$

15. $y = 5x + 3$

Slope: $m = 5$

y-intercept: $(0, 3)$

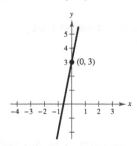

17. $y = -\frac{3}{4}x - 1$

Slope: $m = -\frac{3}{4}$

y-intercept: $(0, -1)$

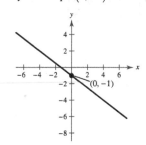

19. $y - 5 = 0$

$y = 5$

Slope: $m = 0$

y-intercept: $(0, 5)$

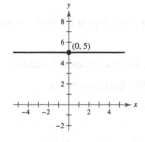

21. $5x - 2 = 0$

$x = \frac{2}{5}$, vertical line

Slope: undefined

y-intercept: none

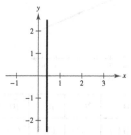

23. $7x - 6y = 30$

$-6y = -7x + 30$

$y = \frac{7}{6}x - 5$

Slope: $m = \frac{7}{6}$

y-intercept: $(0, -5)$

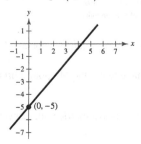

25. $m = \dfrac{0 - 9}{6 - 0} = \dfrac{-9}{6} = -\dfrac{3}{2}$

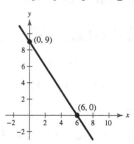

27. $m = \dfrac{6 - (-2)}{1 - (-3)} = \dfrac{8}{4} = 2$

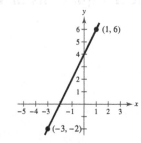

29. $m = \dfrac{-7 - (-7)}{8 - 5} = \dfrac{0}{3} = 0$

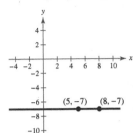

31. $m = \dfrac{4 - (-1)}{-6 - (-6)} = \dfrac{5}{0}$

m is undefined.

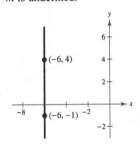

33. $m = \dfrac{1.6 - 3.1}{-5.2 - 4.8} = \dfrac{-1.5}{-10} = 0.15$

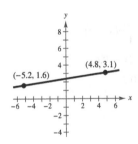

35. Point: $(5, 7)$, Slope: $m = 0$

Because $m = 0$, y does not change. Three other points are $(-1, 7)$, $(0, 7)$, and $(4, 7)$.

37. Point: $(-5, 4)$, Slope: $m = 2$

Because $m = 2 = \frac{2}{1}$, y increases by 2 for every one unit increase in x. Three additional points are $(-4, 6)$, $(-3, 8)$, and $(-2, 10)$.

39. Point: $(4, 5)$, Slope: $m = -\frac{1}{3}$

Because $m = -\frac{1}{3}$, y decreases by 1 unit for every three unit increase in x. Three additional points are $(-2, 7)$, $\left(0, -\frac{19}{4}\right)$, and $(1, 6)$.

41. Point: $(-4, 3)$, Slope is undefined.

Because m is undefined, x does not change. Three points are $(-4, 0)$, $(-4, 5)$, and $(-4, 2)$.

43. Point: $(0, -2)$; $m = 3$

$$y + 2 = 3(x - 0)$$
$$y = 3x - 2$$

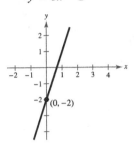

45. Point: $(-3, 6)$; $m = -2$

$$y - 6 = -2(x + 3)$$
$$y = -2x$$

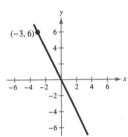

47. Point: $(4, 0)$; $m = -\frac{1}{3}$

$$y - 0 = -\frac{1}{3}(x - 4)$$
$$y = -\frac{1}{3}x + \frac{4}{3}$$

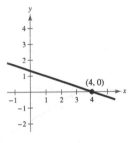

49. Point: $(2, -3)$; $m = -\dfrac{1}{2}$

$$y - (-3) = -\dfrac{1}{2}(x - 2)$$

$$y + 3 = -\dfrac{1}{2}x + 1$$

$$y = -\dfrac{1}{2}x - 2$$

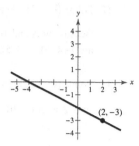

51. Point: $\left(4, \dfrac{5}{2}\right)$; $m = 0$

$$y - \dfrac{5}{2} = 0(x - 4)$$

$$y - \dfrac{5}{2} = 0$$

$$y = \dfrac{5}{2}$$

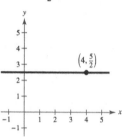

53. Point: $(-5.1, 1.8)$; $m = 5$

$$y - 1.8 = 5(x - (5.1))$$

$$y = 5x + 27.3$$

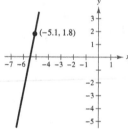

55. $(5, -1)$, $(-5, 5)$

$$y + 1 = \dfrac{5 + 1}{-5 - 5}(x - 5)$$

$$y = -\dfrac{3}{5}(x - 5) - 1$$

$$y = -\dfrac{3}{5}x + 2$$

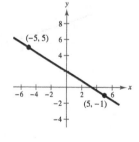

57. $(-7, 2)$, $(-7, 5)$

$$m = \dfrac{5 - 2}{-7 - (-7)} = \dfrac{3}{0}$$

m is undefined.

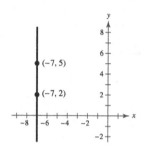

59. $\left(2, \dfrac{1}{2}\right)$, $\left(\dfrac{1}{2}, \dfrac{5}{4}\right)$

$$y - \dfrac{1}{2} = \dfrac{\dfrac{5}{4} - \dfrac{1}{2}}{\dfrac{1}{2} - 2}(x - 2)$$

$$y = -\dfrac{1}{2}(x - 2) + \dfrac{1}{2}$$

$$y = -\dfrac{1}{2}x + \dfrac{3}{2}$$

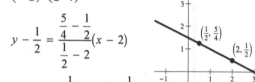
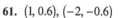

61. $(1, 0.6)$, $(-2, -0.6)$

$$y - 0.6 = \dfrac{-0.6 - 0.6}{-2 - 1}(x - 1)$$

$$y = 0.4(x - 1) + 0.6$$

$$y = 0.4x + 0.2$$

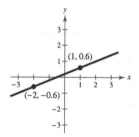

63. $(2, -1)$, $\left(\dfrac{1}{3}, -1\right)$

$$y + 1 = \dfrac{-1 - (-1)}{\dfrac{1}{3} - 2}(x - 2)$$

$$y + 1 = 0$$

$$y = -1$$

The line is horizontal.

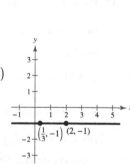

65. $L_1: y = -\frac{2}{3}x - 3$

$\quad m_1 = -\frac{2}{3}$

$\quad L_2: y = -\frac{2}{3}x - 1$

$\quad m_2 = -\frac{2}{3}$

The slopes are equal, so the lines are parallel.

67. $L_1: y = \frac{1}{2}x - 3$

$\quad m_1 = \frac{1}{2}$

$\quad L_2: y = -\frac{1}{2}x + 1$

$\quad m_2 = -\frac{1}{2}$

The lines are neither parallel nor perpendicular.

69. $L_1: (0, -1), (5, 9)$

$\quad m_1 = \dfrac{9 + 1}{5 - 0} = 2$

$\quad L_2: (0, 3), (4, 1)$

$\quad m_2 = \dfrac{1 - 3}{4 - 0} = -\dfrac{1}{2}$

The slopes are negative reciprocals, so the lines are perpendicular.

71. $L_1: (-6, -3), (2, -3)$

$\quad m_1 = \dfrac{-3 - (-3)}{2 - (-6)} = \dfrac{0}{8} = 0$

$\quad L_2: \left(3, -\frac{1}{2}\right), \left(6, -\frac{1}{2}\right)$

$\quad m_2 = \dfrac{-\frac{1}{2} - \left(-\frac{1}{2}\right)}{6 - 3} = \dfrac{0}{3} = 0$

L_1 and L_2 are both horizontal lines, so they are parallel.

73. $4x - 2y = 3$

$\quad y = 2x - \frac{3}{2}$

Slope: $m = 2$

(a) $(2, 1), m = 2$

$\quad y - 1 = 2(x - 2)$

$\quad y = 2x - 3$

(b) $(2, 1), m = -\frac{1}{2}$

$\quad y - 1 = -\frac{1}{2}(x - 2)$

$\quad y = -\frac{1}{2}x + 2$

75. $3x + 4y = 7$

$\quad y = -\frac{3}{4}x + \frac{7}{4}$

Slope: $m = -\frac{3}{4}$

(a) $\left(-\frac{2}{3}, \frac{7}{8}\right), m = -\frac{3}{4}$

$\quad y - \frac{7}{8} = -\frac{3}{4}\left(x - \left(-\frac{2}{3}\right)\right)$

$\quad y = -\frac{3}{4}x + \frac{3}{8}$

(b) $\left(-\frac{2}{3}, \frac{7}{8}\right), m = \frac{4}{3}$

$\quad y - \frac{7}{8} = \frac{4}{3}\left(x - \left(-\frac{2}{3}\right)\right)$

$\quad y = \frac{4}{3}x + \frac{127}{72}$

77. $y + 5 = 0$

$\quad y = -5$

Slope: $m = 0$

(a) $(-2, 4), m = 0$

$\quad y = 4$

(b) $(-2, 4), m$ is undefined.

$\quad x = -2$

79. $x - y = 4$

$\quad y = x - 4$

Slope: $m = 1$

(a) $(2.5, 6.8), m = 1$

$\quad y - 6.8 = 1(x - 2.5)$

$\quad y = x + 4.3$

(b) $(2.5, 6.8), m = -1$

$\quad y - 6.8 = (-1)(x - 2.5)$

$\quad y = -x + 9.3$

81. $\quad \dfrac{x}{3} + \dfrac{y}{5} = 1$

$\quad (15)\left(\dfrac{x}{3} + \dfrac{y}{5}\right) = 1(15)$

$\quad 5x + 3y - 15 = 0$

83. $\quad \dfrac{x}{-1/6} + \dfrac{y}{-2/3} = 1$

$\quad 6x + \dfrac{3}{2}y = -1$

$\quad 12x + 3y + 2 = 0$

85.
$$\frac{x}{c} + \frac{y}{c} = 1, c \neq 0$$
$$x + y = c$$
$$1 + 2 = c$$
$$3 = c$$
$$x + y = 3$$
$$x + y - 3 = 0$$

87. (a) $m = 135$. The sales are increasing 135 units per year.

(b) $m = 0$. There is no change in sales during the year.

(c) $m = -40$. The sales are decreasing 40 units per year.

89. $y = \frac{6}{100}x$

$y = \frac{6}{100}(200) = 12$ feet

91. $(16, 3000)$, $m = -150$

$$V - 3000 = -150(t - 16)$$
$$V - 3000 = -150t + 2400$$
$$V = -150t + 5400, \ 16 \leq t \leq 21$$

93. The C-intercept measures the fixed costs of manufacturing when zero bags are produced.

The slope measures the cost to produce one laptop bag.

95. Using the points $(0, 875)$ and $(5, 0)$, where the first coordinate represents the year t and the second coordinate represents the value V, you have

$$m = \frac{0 - 875}{5 - 0} = -175$$
$$V = -175t + 875, \ 0 \leq t \leq 5.$$

97. Using the points $(0, 32)$ and $(100, 212)$, where the first coordinate represents a temperature in degrees Celsius and the second coordinate represents a temperature in degrees Fahrenheit, you have

$$m = \frac{212 - 32}{100 - 0} = \frac{180}{100} = \frac{9}{5}.$$

Since the point $(0, 32)$ is the F- intercept, $b = 32$, the

equation is $F = 1.8C + 32$ or $C = \frac{5}{9}F - \frac{160}{9}$.

99. (a) Total Cost = cost for fuel and maintainance + cost for operator + purchase cost

$$C = 9.5t + 11.5t + 42,000$$
$$C = 21t + 42,000$$

(b) Revenue = Rate per hour $\cdot$ Hours

$$R = 45t$$

(c) $P = R - C$

$$P = 45t - (21t + 42,000)$$
$$P = 24t - 42,000$$

(d) Let $P = 0$, and solve for t.

$$0 = 24t - 42,000$$
$$42,000 = 24t$$
$$1750 = t$$

The equipment must be used 1750 hours to yield a profit of 0 dollars.

101. False. The slope with the greatest magnitude corresponds to the steepest line.

103. Find the slope of the line segments between the points A and B, and B and C.

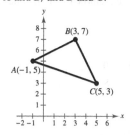

$$m_{AB} = \frac{7 - 5}{3 - (-1)} = \frac{2}{4} = \frac{1}{2}$$

$$m_{BC} = \frac{3 - 7}{5 - 3} = \frac{-4}{2} = -2$$

Since the slopes are negative reciprocals, the line segments are perpendicular and therefore intersect to form a right angle. So, the triangle is a right triangle.

105. Since the scales for the y-axis on each graph is unknown, the slopes of the lines cannot be determined.

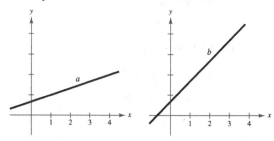

107. No, the slopes of two perpendicular lines have opposite signs. (Assume that neither line is vertical or horizontal.)

109. The line $y = 4x$ rises most quickly.

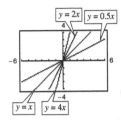

The line $y = -4x$ falls most quickly.

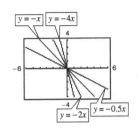

The greater the magnitude of the slope (the absolute value of the slope), the faster the line rises or falls.

111. Set the distance between $(4, -1)$ and (x, y) equal to the distance between $(-2, 3)$ and (x, y).

$$\sqrt{(x-4)^2 + [y-(-1)]^2} = \sqrt{[x-(-2)]^2 + (y-3)^2}$$
$$(x-4)^2 + (y+1)^2 = (x+2)^2 + (y-3)^2$$
$$x^2 - 8x + 16 + y^2 + 2y + 1 = x^2 + 4x + 4 + y^2 - 6y + 9$$
$$-8x + 2y + 17 = 4x - 6y + 13$$
$$-12x + 8y + 4 = 0$$
$$-4(3x - 2y - 1) = 0$$
$$3x - 2y - 1 = 0$$

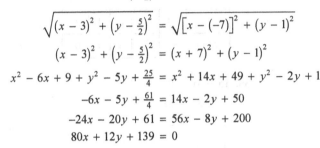

This line is the perpendicular bisector of the line segment connecting $(4, -1)$ and $(-2, 3)$.

113. Set the distance between $\left(3, \frac{5}{2}\right)$ and (x, y) equal to the distance between $(-7, 1)$ and (x, y).

$$\sqrt{(x-3)^2 + \left(y-\frac{5}{2}\right)^2} = \sqrt{[x-(-7)]^2 + (y-1)^2}$$
$$(x-3)^2 + \left(y-\frac{5}{2}\right)^2 = (x+7)^2 + (y-1)^2$$
$$x^2 - 6x + 9 + y^2 - 5y + \frac{25}{4} = x^2 + 14x + 49 + y^2 - 2y + 1$$
$$-6x - 5y + \frac{61}{4} = 14x - 2y + 50$$
$$-24x - 20y + 61 = 56x - 8y + 200$$
$$80x + 12y + 139 = 0$$

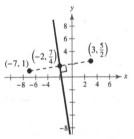

This line is the perpendicular bisector of the line segment connecting $\left(3, \frac{5}{2}\right)$ and $(-7, 1)$.

Section P.5   Functions

1. domain; range; function

3. implied domain

5. Yes, the relationship is a function. Each domain value is matched with exactly one range value.

7. No, it does not represent a function. The input values of 10 and 7 are each matched with two output values.

9. (a) Each element of A is matched with exactly one element of B, so it does represent a function.

 (b) The element 1 in A is matched with two elements, -2 and 1 of B, so it does not represent a function.

 (c) Each element of A is matched with exactly one element of B, so it does represent a function.

 (d) The element 2 in A is not matched with an element of B, so the relation does not represent a function.

11. $x^2 + y^2 = 4 \Rightarrow y = \pm\sqrt{4 - x^2}$

No, y is *not* a function of x.

13. $y = \sqrt{16 - x^2}$

Yes, y *is* a function of x.

15. $y = 4 - |x|$

Yes, y *is* a function of x.

17. $y = -75$ or $y = -75 + 0x$

Yes, y *is* a function of x.

19. $f(x) = 3x - 5$

    (a) $f(1) = 3(1) - 5 = -2$

    (b) $f(-3) = 3(-3) - 5 = -14$

    (c) $f(x + 2) = 3(x + 2) - 5$

               $= 3x + 6 - 5$

               $= 3x + 1$

21. $g(t) = 4t^2 - 3t + 5$

    (a) $g(2) = 4(2)^2 - 3(2) + 5$

             $= 15$

    (b) $g(t - 2) = 4(t - 2)^2 - 3(t - 2) + 5$

                $= 4t^2 - 19t + 27$

    (c) $g(t) - g(2) = 4t^2 - 3t + 5 - 15$

                $= 4t^2 - 3t - 10$

23. $f(y) = 3 - \sqrt{y}$

    (a) $f(4) = 3 - \sqrt{4} = 1$

    (b) $f(0.25) = 3 - \sqrt{0.25} = 2.5$

    (c) $f(4x^2) = 3 - \sqrt{4x^2} = 3 - 2|x|$

25. $q(x) = \dfrac{1}{x^2 - 9}$

    (a) $q(0) = \dfrac{1}{0^2 - 9} = -\dfrac{1}{9}$

    (b) $q(3) = \dfrac{1}{3^2 - 9}$ is undefined.

    (c) $q(y + 3) = \dfrac{1}{(y + 3)^2 - 9} = \dfrac{1}{y^2 + 6y}$

27. $f(x) = \dfrac{|x|}{x}$

    (a) $f(2) = \dfrac{|2|}{2} = 1$

    (b) $f(-2) = \dfrac{|-2|}{-2} = -1$

    (c) $f(x - 1) = \dfrac{|x - 1|}{x - 1} = \begin{cases} -1, & \text{if } x < 1 \\ 1, & \text{if } x > 1 \end{cases}$

29. $f(x) = \begin{cases} 2x + 1, & x < 0 \\ 2x + 2, & x \geq 0 \end{cases}$

    (a) $f(-1) = 2(-1) + 1 = -1$

    (b) $f(0) = 2(0) + 2 = 2$

    (c) $f(2) = 2(2) + 2 = 6$

31. $f(x) = -x^2 + 5$

    $f(-2) = -(-2)^2 + 5 = 1$

    $f(-1) = -(-1)^2 + 5 = 4$

    $f(0) = -(0)^2 + 5 = 5$

    $f(1) = -(1)^2 + 5 = 4$

    $f(2) = -(2)^2 + 5 = 1$

x	-2	-1	0	1	2
$f(x)$	1	4	5	-4	1

33. $f(x) = \begin{cases} -\frac{1}{2}x + 4, & x \leq 0 \\ (x - 2)^2, & x > 0 \end{cases}$

    $f(-2) = -\frac{1}{2}(-2) + 4 = 5$

    $f(-1) = -\frac{1}{2}(-1) + 4 = 4\frac{1}{2} = \frac{9}{2}$

    $f(0) = -\frac{1}{2}(0) + 4 = 4$

    $f(1) = (1 - 2)^2 = 1$

    $f(2) = (2 - 2)^2 = 0$

x	-2	-1	0	1	2
$f(x)$	5	$\frac{9}{2}$	4	1	0

35. $15 - 3x = 0$

    $3x = 15$

    $x = 5$

37. $\dfrac{3x - 4}{5} = 0$

    $3x - 4 = 0$

    $x = \dfrac{4}{3}$

39. $f(x) = x^2 - 81$

    $x^2 - 81 = 0$

    $x^2 = 81$

    $x = \pm 9$

41. $x^3 - x = 0$

    $x(x^2 - 1) = 0$

    $x(x + 1)(x - 1) = 0$

    $x = 0, x = -1, \text{ or } x = 1$

43.
$$f(x) = g(x)$$
$$x^2 = x + 2$$
$$x^2 - x - 2 = 0$$
$$(x - 2)(x + 1) = 0$$
$$x - 2 = 0 \quad x + 1 = 0$$
$$x = 2 \qquad x = -1$$

45.
$$f(x) = g(x)$$
$$x^4 - 2x^2 = 2x^2$$
$$x^4 - 4x^2 = 0$$
$$x^2(x^2 - 4) = 0$$
$$x^2(x + 2)(x - 2) = 0$$
$$x^2 = 0 \Rightarrow x = 0$$
$$x + 2 = 0 \Rightarrow x = -2$$
$$x - 2 = 0 \Rightarrow x = 2$$

47. $f(x) = 5x^2 + 2x - 1$

Because $f(x)$ is a polynomial, the domain is all real numbers x.

49. $g(y) = \sqrt{y + 6}$

Domain: $y + 6 \geq 0$
$$y \geq -6$$

The domain is all real numbers y such that $y \geq -6$.

51. $g(x) = \dfrac{1}{x} - \dfrac{3}{x + 2}$

The domain is all real numbers x except
$x = 0, x = -2$.

53. $f(s) = \dfrac{\sqrt{s - 1}}{s - 4}$

Domain: $s - 1 \geq 0 \Rightarrow s \geq 1$ and $s \neq 4$

The domain consists of all real numbers s, such that
$s \geq 1$ and $s \neq 4$.

55. $f(x) = \dfrac{x - 4}{\sqrt{x}}$

The domain is all real numbers x such that $x > 0$ or
$(0, \infty)$.

57. (a)

Height, x	Volume, V
1	484
2	800
3	972
4	1024
5	980
6	864

The volume is maximum when $x = 4$ and
$V = 1024$ cubic centimeters.

(b)

V is a function of x.

(c) $V = x(24 - 2x)^2$

Domain: $0 < x < 12$

59. $A = s^2$ and $P = 4s \Rightarrow \dfrac{P}{4} = s$

$$A = \left(\dfrac{P}{4}\right)^2 = \dfrac{P^2}{16}$$

61.
$$y = -\tfrac{1}{10}x^2 + 3x + 6$$
$$y(25) = -\tfrac{1}{10}(25)^2 + 3(25) + 6 = 18.5 \text{ feet}$$

If the child holds a glove at a height of 5 feet, then the ball *will* be over the child's head because it will be at a height of 18.5 feet.

63. $A = \dfrac{1}{2}bh = \dfrac{1}{2}xy$

Because $(0, y), (2, 1),$ and $(x, 0)$ all lie on the same line, the slopes between any pair are equal.

$$\frac{1 - y}{2 - 0} = \frac{0 - 1}{x - 2}$$

$$\frac{1 - y}{2} = \frac{-1}{x - 2}$$

$$y = \frac{2}{x - 2} + 1$$

$$y = \frac{x}{x - 2}$$

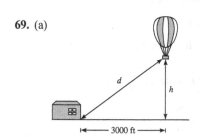

So, $A = \dfrac{1}{2}x\left(\dfrac{x}{x - 2}\right) = \dfrac{x^2}{2(x - 2)}.$

The domain of A includes x-values such that $x^2/[2(x - 2)] > 0.$ By solving this inequality, the domain is $x > 2.$

65. For 2008 through 2011, use

$$p(t) = 2.77t + 45.2.$$

2008: $p(8) = 2.77(8) + 45.2 = 67.36\%$

2009: $p(9) = 2.77(9) + 45.2 = 70.13\%$

2010: $p(10) = 2.77(10) + 45.2 = 72.90\%$

2011: $p(11) = 2.77(11) + 45.2 = 75.67\%$

For 2011 through 2014, use

$$p(t) = 1.95t + 55.9.$$

2012: $p(12) = 1.95(12) + 55.9 = 79.30\%$

2013: $p(13) = 1.95(13) + 55.9 = 81.25\%$

2014: $p(14) = 1.95(14) + 55.9 = 83.20\%$

67. (a) Cost = variable costs + fixed costs

$$C = 12.30x + 98{,}000$$

(b) Revenue = price per unit × number of units

$$R = 17.98x$$

(c) Profit = Revenue − Cost

$$P = 17.98x - (12.30x + 98{,}000)$$

$$P = 5.68x - 98{,}000$$

69. (a)

(b) $(3000)^2 + h^2 = d^2$

$$h = \sqrt{d^2 - (3000)^2}$$

Domain: $d \geq 3000$ (because both $d \geq 0$ and $d^2 - (3000)^2 \geq 0$)

71. (a) $R = n(\text{rate}) = n[8.00 - 0.05(n - 80)], n \geq 80$

$$R = 12.00n - 0.05n^2 = 12n - \frac{n^2}{20} = \frac{240n - n^2}{20}, n \geq 80$$

(b)

n	90	100	110	120	130	140	150
$R(n)$	\$675	\$700	\$715	\$720	\$715	\$700	\$675

The revenue is maximum when 120 people take the trip.

73.

$$f(x) = x^2 - 2x + 4$$

$$f(2 + h) = (2 + h)^2 - 2(2 + h) + 4$$

$$= 4 + 4h + h^2 - 4 - 2h + 4$$

$$= h^2 + 2h + 4$$

$$f(2) = (2)^2 - 2(2) + 4 = 4$$

$$f(2 + h) - f(2) = h^2 + 2h$$

$$\frac{f(2 + h) - f(2)}{h} = \frac{h^2 + 2h}{h} = h + 2, h \neq 0$$

75.
$$f(x) = x^3 + 3x$$
$$f(x + h) = (x + h)^3 + 3(x + h)$$
$$= x^3 + 3x^2h + 3xh^2 + h^3 + 3x + 3h$$
$$\frac{f(x + h) - f(x)}{h} = \frac{(x^3 + 3x^2h + 3xh^2 + h^3 + 3x + 3h) - (x^3 + 3x)}{h}$$
$$= \frac{h(3x^2 + 3xh + h^2 + 3)}{h}$$
$$= 3x^2 + 3xh + h^2 + 3, h \neq 0$$

77.
$$g(x) = \frac{1}{x^2}$$
$$\frac{g(x) - g(3)}{x - 3} = \frac{\frac{1}{x^2} - \frac{1}{9}}{x - 3}$$
$$= \frac{9 - x^2}{9x^2(x - 3)}$$
$$= \frac{-(x + 3)(x - 3)}{9x^2(x - 3)}$$
$$= -\frac{x + 3}{9x^2}, x \neq 3$$

79. $f(x) = \sqrt{5x}$
$$\frac{f(x) - f(5)}{x - 5} = \frac{\sqrt{5x} - 5}{x - 5}, x \neq 5$$

81. By plotting the points, we have a parabola, so $g(x) = cx^2$. Because $(-4, -32)$ is on the graph, you have $-32 = c(-4)^2 \Rightarrow c = -2$. So, $g(x) = -2x^2$.

83. Because the function is undefined at 0, we have $r(x) = c/x$. Because $(-4, -8)$ is on the graph, you have $-8 = c/-4 \Rightarrow c = 32$. So, $r(x) = 32/x$.

85. False. The equation $y^2 = x^2 + 4$ is a relation between x and y. However, $y = \pm\sqrt{x^2 + 4}$ does not represent a function.

87. False. The range is $[-1, \infty)$.

89. The domain of $f(x) = \sqrt{x - 1}$ includes $x = 1, x \geq 1$ and the domain of $g(x) = \frac{1}{\sqrt{x - 1}}$ does not include $x = 1$ because you cannot divide by 0. The domain of $g(x) = \frac{1}{\sqrt{x - 1}}$ is $x > 1$. So, the functions do not have the same domain.

91. No; x is the independent variable, f is the name of the function.

93. (a) Yes. The amount that you pay in sales tax will increase as the price of the item purchased increases.

(b) No. The length of time that you study the night before an exam does not necessarily determine your score on the exam.

Section P.6   Analyzing Graphs of Function

1. Vertical Line Test

3. decreasing

5. average rate of change; secant

7. Domain: $(-2, 2]$; Range: $[-1, 8]$

(a) $f(-1) = -1$

(b) $f(0) = 0$

(c) $f(1) = -1$

(d) $f(2) = 8$

9. Domain: $(-\infty, \infty)$; Range: $(-2, \infty)$

(a) $f(2) = 0$

(b) $f(1) = 1$

(c) $f(3) = 2$

(d) $f(-1) = 3$

11. A vertical line intersects the graph at most once, so y *is* a function of x.

13. A vertical line intersects the graph more than once, so y is *not* a function of x.

15. $f(x) = 3x + 18$

$3x + 18 = 0$

$3x = -18$

$x = -6$

17. $f(x) = 2x^2 - 7x - 30$

$2x^2 - 7x - 30 = 0$

$(2x + 5)(x - 6) = 0$

$2x + 5 = 0$ or $x - 6 = 0$

$x = -\frac{5}{2}$ $x = 6$

19. $f(x) = \dfrac{x + 3}{2x^2 - 6}$

$\dfrac{x + 3}{2x^2 - 6} = 0$

$x + 3 = 0$

$x = -3$

21. $f(x) = \frac{1}{3}x^3 - 2x$

$\frac{1}{3}x^3 - 2x = 0$

$(3)\left(\frac{1}{3}x^3 - 2x\right) = 0(3)$

$x^3 - 6x = 0$

$x\left(x^2 - 6\right) = 0$

$x = 0$ or $x^2 - 6 = 0$

$x^2 = 6$

$x = \pm\sqrt{6}$

23. $f(x) = x^3 - 4x^2 - 9x + 36$

$x^3 - 4x^2 - 9x + 36 = 0$

$x^2(x - 4) - 9(x - 4) = 0$

$(x - 4)(x^2 - 9) = 0$

$x - 4 = 0 \Rightarrow x = 4$

$x^2 - 9 = 0 \Rightarrow x = \pm 3$

25. $f(x) = \sqrt{2x} - 1$

$\sqrt{2x} - 1 = 0$

$\sqrt{2x} = 1$

$2x = 1$

$x = \frac{1}{2}$

27. (a)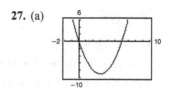

Zeros: $x = 0, 6$

(b) $f(x) = x^2 - 6x$

$x^2 - 6x = 0$

$x(x - 6) = 0$

$x = 0 \Rightarrow x = 0$

$x - 6 = 0 \Rightarrow x = 6$

29. (a)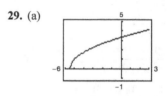

Zero: $x = -5.5$

(b) $f(x) = \sqrt{2x + 11}$

$\sqrt{2x + 11} = 0$

$2x + 11 = 0$

$x = -\frac{11}{2}$

31. (a)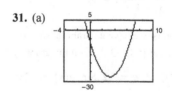

Zero: $x = 0.3333$

(b) $f(x) = \dfrac{3x - 1}{x - 6}$

$\dfrac{3x - 1}{x - 6} = 0$

$3x - 1 = 0$

$x = \dfrac{1}{3}$

33. $f(x) = -\frac{1}{2}x^3$

The function is decreasing on $(-\infty, \infty)$.

35. $f(x) = \sqrt{x^2 - 1}$

The function is decreasing on $(-\infty, -1)$ and increasing on $(1, \infty)$.

37. $f(x) = |x + 1| + |x - 1|$

The function is increasing on $(1, \infty)$.

The function is constant on $(-1, 1)$.

The function is decreasing on $(-\infty, -1)$.

39. $f(x) = \begin{cases} 2x + 1, & x \le -1 \\ x^2 - 2, & x > -1 \end{cases}$

The function is decreasing on $(-1, 0)$ and increasing on $(-\infty, -1)$ and $(0, \infty)$.

41. $f(x) = 3$

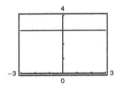

Constant on $(-\infty, \infty)$

x	-2	-1	0	1	2
$f(x)$	3	3	3	3	3

43. $g(x) = \frac{1}{2}x^2 - 3$

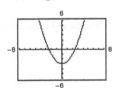

Decreasing on $(-\infty, 0)$.

Increasing on $(0, \infty)$.

x	-2	-1	0	1	2
$g(x)$	-1	$-\frac{5}{2}$	-3	$-\frac{5}{2}$	-1

45. $f(x) = \sqrt{1 - x}$

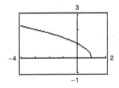

Decreasing on $(-\infty, 1)$

x	-3	-2	-1	0	1
$f(x)$	2	$\sqrt{3}$	$\sqrt{2}$	1	0

47. $f(x) = x^{3/2}$

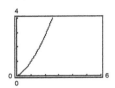

Increasing on $(0, \infty)$

x	0	1	2	3	4
$f(x)$	0	1	2.8	5.2	8

49. $f(x) = x(x + 3)$

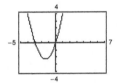

Relative minimum: $(-1.5, -2.25)$

51. $h(x) = x^3 - 6x^2 + 15$

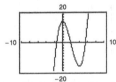

Relative minimum: $(4, -17)$

Relative maximum: $(0, 15)$

53. $h(x) = (x - 1)\sqrt{x}$

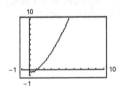

Relative minimum: $(0.33, -0.38)$

55. $f(x) = 4 - x$

$f(x) \ge 0$ on $(-\infty, 4]$

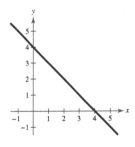

57. $f(x) = 9 - x^2$

$f(x) \geq 0$ on $[-3, 3]$

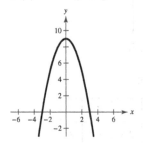

59. $f(x) = \sqrt{x - 1}$

$f(x) \geq 0$ on $[1, \infty)$

$\sqrt{x - 1} \geq 0$

$x - 1 \geq 0$

$x \geq 1$

$[1, \infty)$

61. $f(x) = -2x + 15$

$$\frac{f(3) - f(0)}{3 - 0} = \frac{9 - 15}{3} = -2$$

The average rate of change from $x_1 = 0$ to $x_2 = 3$ is -2.

63. $f(x) = x^3 - 3x^2 - x$

$$\frac{f(2) - f(-1)}{2 - (-1)} = \frac{-6 - (-3)}{3} = \frac{-3}{3} = -1$$

The average rate of change from $x_1 = -1$ to $x_2 = 2$ to is -1.

65. (a)

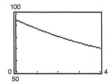

(b) To find the average rate of change of the amount the U.S. Department of Energy spent for research and development from 2010 to 2014, find the average rate of change from $(0, f(0))$ to $(4, f(4))$.

$$\frac{f(4) - f(0)}{4 - 0} = \frac{70.5344 - 95.08}{4} = \frac{-24.5456}{4} = -6.1364$$

The amount the U.S. Department of Energy spent on research and development for defense decreased by about \$6.14 billion each year from 2010 to 2014.

67. $s_0 = 6, v_0 = 64$

(a) $s = -16t^2 + 64t + 6$

(b)

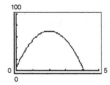

(c) $\dfrac{s(3) - s(0)}{3 - 0} = \dfrac{54 - 6}{3} = 16$

(d) The slope of the secant line is positive.

(e) $s(0) = 6, m = 16$

Secant line: $y - 6 = 16(t - 0)$

$y = 16t + 6$

(f)

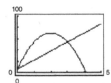

69. $v_0 = 120$, $s_0 = 0$

(a) $s = -16t^2 + 120t$

(b)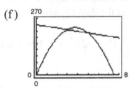

(c) The average rate of change from $t = 3$ to $t = 5$:

$$\frac{s(5) - s(3)}{5 - 3} = \frac{200 - 216}{2} = -\frac{16}{2} = -8 \text{ feet per}$$
second

(d) The slope of the secant line through $\left(3, s(3)\right)$ and $\left(5, s(5)\right)$ is negative.

(e) The equation of the secant line: $m = -8$

Using $\left(5, s(5)\right) = (5, 200)$ we have

$$y - 200 = -8(t - 5)$$
$$y = -8t + 240.$$

(f)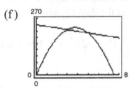

71. $f(x) = x^6 - 2x^2 + 3$

$$f(-x) = (-x)^6 - 2(-x)^2 + 3$$
$$= x^6 - 2x^2 + 3$$
$$= f(x)$$

The function is even. y-axis symmetry.

73. $h(x) = x\sqrt{x + 5}$

$$h(-x) = (-x)\sqrt{-x + 5}$$
$$= -x\sqrt{5 - x}$$
$$\neq h(x)$$
$$\neq -h(x)$$

The function is neither odd nor even. No symmetry.

75. $f(s) = 4s^{3/2}$

$$= 4(-s)^{3/2}$$
$$\neq f(s)$$
$$\neq -f(s)$$

The function is neither odd nor even. No symmetry.

77.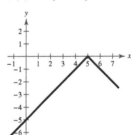

The graph of $f(x) = -9$ is symmetric to the y-axis, which implies $f(x)$ is even.

$$f(-x) = -9$$
$$= f(x)$$

The function is even.

79. $f(x) = -|x - 5|$

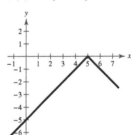

The graph displays no symmetry, which implies $f(x)$ is neither odd nor even.

$$f(x) = -|(-x) - 5|$$
$$= -|-x - 5|$$
$$\neq f(x)$$
$$\neq -f(x)$$

The function is neither even nor odd.

81. $f(x) = \sqrt[3]{4x}$

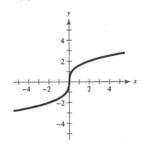

The graph displays origin symmetry, which implies $f(x)$ is odd.

$$f(-x) = \sqrt[3]{4(-x)}$$
$$= \sqrt[3]{-4x}$$
$$= -\sqrt[3]{4x}$$
$$= -f(x)$$

The function is odd.

83. $h = \text{top} - \text{bottom}$
$$= 3 - \left(4x - x^2\right)$$
$$= 3 - 4x + x^2$$

85. $L = \text{right} - \text{left}$
$$= 2 - \sqrt[3]{2y}$$

87. The error is that $-2x^3 - 5 \neq -\left(2x^3 - 5\right)$. The correct process is as follows.

$$f(x) = 2x^3 - 5$$
$$f(-x) = 2(-x)^3 - 5$$
$$= -2x^3 - 5$$
$$= -\left(2x^3 + 5\right)$$

$f(-x) \neq -f(x)$ and $f(-x) \neq f(x)$, so the function $f(x) = 2x^3 - 5$ is neither odd nor even.

89. (a) For the average salary of college professors, a scale of $10,000 would be appropriate.

(b) For the population of the United States, use a scale of 10,000,000.

(c) For the percent of the civilian workforce that is unemployed, use a scale of 10%.

(d) For the number of games a college football team wins in a single season, single digits would be appropriate.

For each of the graphs, using the suggested scale would show yearly changes in the data clearly.

91. False. The function $f(x) = \sqrt{x^2 + 1}$ has a domain of all real numbers.

93. True. A graph that is symmetric with respect to the y-axis cannot be increasing on its entire domain.

95. $\left(-\frac{5}{3}, -7\right)$

(a) If f is even, another point is $\left(\frac{5}{3}, -7\right)$.

(b) If f is odd, another point is $\left(\frac{5}{3}, 7\right)$.

97. (a) $y = x$

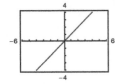

(b) $y = x^2$

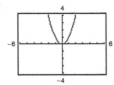

(c) $y = x^3$

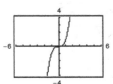

(d) $y = x^4$

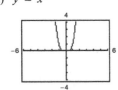

(e) $y = x^5$

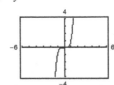

(f) $y = x^6$

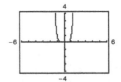

All the graphs pass through the origin. The graphs of the odd powers of x are symmetric with respect to the origin and the graphs of the even powers are symmetric with respect to the y-axis. As the powers increase, the graphs become flatter in the interval $-1 < x < 1$.

99. (a) Even. The graph is a reflection in the *x*-axis.

 (b) Even. The graph is a reflection in the *y*-axis.

 (c) Even. The graph is a vertical translation of *f*.

 (d) Neither. The graph is a horizontal translation of *f*.

Section P.7   A Library of Parent Functions

1. Greatest integer function

3. Reciprocal function

5. Square root function

7. Absolute value function

9. Linear function

11. (a) $f(1) = 4, f(0) = 6$

$(1, 4), (0, 6)$

$$m = \frac{6 - 4}{0 - 1} = -2$$

$$y - 6 = -2(x - 0)$$

$$y = -2x + 6$$

$$f(x) = -2x + 6$$

 (b)

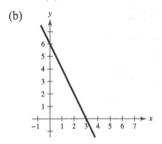

13. (a) $f\left(\frac{1}{2}\right) = -\frac{5}{3}, f(6) = 2$

$\left(\frac{1}{2}, -\frac{5}{3}\right), (6, 2)$

$$m = \frac{2 - \left(-\frac{5}{3}\right)}{6 - \left(\frac{1}{2}\right)}$$

$$= \frac{\frac{11}{3}}{\frac{11}{2}} = \left(\frac{11}{3}\right) \cdot \left(\frac{2}{11}\right) = \frac{2}{3}$$

$$f(x) - 2 = \frac{2}{3}(x - 6)$$

$$f(x) - 2 = \frac{2}{3}x - 4$$

$$f(x) = \frac{2}{3}x - 2$$

 (b)

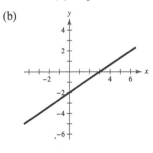

15. $f(x) = 2.5x - 4.25$

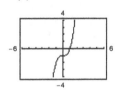

17. $g(x) = x^2 + 3$

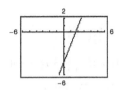

19. $f(x) = x^3 - 1$

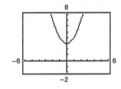

21. $f(x) = \sqrt{x} + 4$

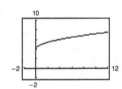

23. $f(x) = \dfrac{1}{x - 2}$

25. $g(x) = |x| - 5$

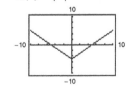

27. $f(x) = [\![x]\!]$

   (a) $f(2.1) = 2$

   (b) $f(2.9) = 2$

   (c) $f(-3.1) = -4$

   (d) $f\left(\frac{7}{2}\right) = 3$

29. $k(x) = [\![2x + 1]\!]$

   (a) $k\left(\frac{1}{3}\right) = \left[\!\left[2\left(\frac{1}{3}\right) + 1\right]\!\right] = \left[\!\left[\frac{5}{3}\right]\!\right] = 1$

   (b) $k(-2.1) = [\![2(-2.1) + 1]\!] = [\![-3.1]\!] = -4$

   (c) $k(1.1) = [\![2(1.1) + 1]\!] = [\![3.2]\!] = 3$

   (d) $k\left(\frac{2}{3}\right) = \left[\!\left[2\left(\frac{2}{3}\right) + 1\right]\!\right] = \left[\!\left[\frac{7}{3}\right]\!\right] = 2$

31. $g(x) = -[\![x]\!]$

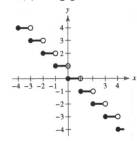

33. $g(x) = [\![x]\!] - 1$

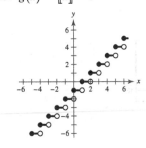

35. $g(x) = \begin{cases} x + 6, & x \leq -4 \\ \frac{1}{2}x - 4, & x > -4 \end{cases}$

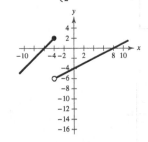

37. $f(x) = \begin{cases} 1 - (x - 1)^2, & x \leq 2 \\ \sqrt{x - 2}, & x > 2 \end{cases}$

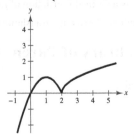

39. $h(x) = \begin{cases} 4 - x^2, & x < -2 \\ 3 + x, & -2 \leq x < 0 \\ x^2 + 1, & x \geq 0 \end{cases}$

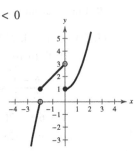

41. $s(x) = 2\left(\frac{1}{4}x - \left[\!\left[\frac{1}{4}x\right]\!\right]\right)$

   (a)

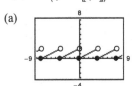

   (b) Domain: $(-\infty, \infty)$; Range: $[0, 2)$

43. (a) $W(30) = 14(30) = 420$

       $W(40) = 14(40) = 560$

       $W(45) = 21(45 - 40) + 560 = 665$

       $W(50) = 21(50 - 40) + 560 = 770$

   (b) $W(h) = \begin{cases} 14h, & 0 < h \leq 36 \\ 21(h - 36) + 504, & h > 36 \end{cases}$

   (c) $W(h) = \begin{cases} 16h, & 0 < h \leq 40 \\ 24(h - 40) + 640, & h > 40 \end{cases}$

45. Answers will vary. *Sample answer:*

Interval	Input Pipe	Drain Pipe 1	Drain Pipe 2
[0, 5]	Open	Closed	Closed
[5, 10]	Open	Open	Closed
[10, 20]	Closed	Closed	Closed
[20, 30]	Closed	Closed	Open
[30, 40]	Open	Open	Open
[40, 45]	Open	Closed	Open
[45, 50]	Open	Open	Open
[50, 60]	Open	Open	Closed

47. For the first two hours, the slope is 1. For the next six hours, the slope is 2. For the final hour, the slope is $\frac{1}{2}$.

$$f(t) = \begin{cases} t, & 0 \le t \le 2 \\ 2t - 2, & 2 < t \le 8 \\ \frac{1}{2}t + 10, & 8 < t \le 9 \end{cases}$$

To find $f(t) = 2t - 2$, use $m = 2$ and $(2, 2)$.

$$y - 2 = 2(t - 2) \Rightarrow y = 2t - 2$$

To find $f(t) = \frac{1}{2}t + 10$, use $m = \frac{1}{2}$ and $(8, 14)$.

$$y - 14 = \frac{1}{2}(t - 8) \Rightarrow y = \frac{1}{2}t + 10$$

Total accumulation = 14.5 inches.

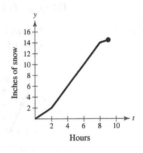

49. False. A piecewise-defined function is a function that is defined by two or more equations over a specified domain. That domain may or may not include x- and y-intercepts.

Section P.8   Transformations of Functions

1. rigid

3. vertical stretch; vertical shrink

5. (a) $f(x) = |x| + c$ Vertical shifts

     $c = -2$: $f(x) = |x| - 2$ 2 units down

     $c = -1$: $f(x) = |x| - 1$ 1 unit down

     $c = 1$: $f(x) = |x| + 1$ 1 unit up

     $c = 2$: $f(x) = |x| + 2$ 2 units up

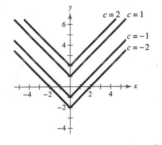

(b) $f(x) = |x - c|$ Horizontal shifts

     $c = -2$: $f(x) = |x - (-2)| = |x + 2|$ 2 units left

     $c = -1$: $f(x) = |x - (-1)| = |x + 1|$ 1 unit left

     $c = 1$: $f(x) = |x - (1)| = |x - 1|$ 1 unit right

     $c = 2$: $f(x) = |x - (2)| = |x - 2|$ 2 units right

7. (a) $f(x) = [\![x]\!] + c$ Vertical shifts

$c = -4$: $f(x) = [\![x]\!] - 4$ 4 units down

$c = -1$: $f(x) = [\![x]\!] - 1$ 1 unit down

$c = 2$: $f(x) = [\![x]\!] + 2$ 2 units up

$c = 5$: $f(x) = [\![x]\!] + 5$ 5 units up

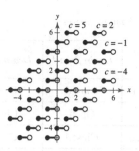

(b) $f(x) = [\![x + c]\!]$ Horizontal shifts

$c = -4$: $f(x) = [\![x - (-4)]\!] = [\![x + 4]\!]$ 4 units left

$c = -1$: $f(x) = [\![x - (-1)]\!] = [\![x + 1]\!]$ 1 unit left

$c = 2$: $f(x) = [\![x - (2)]\!] = [\![x - 2]\!]$ 2 units right

$c = 5$: $f(x) = [\![x - (5)]\!] = [\![x - 5]\!]$ 5 units right

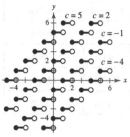

9. (a) $y = f(-x)$

Reflection in the y-axis

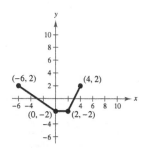

(b) $y = f(x) + 4$

Vertical shift 4 units upward

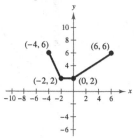

(c) $y = 2f(x)$

Vertical stretch (each y-value is multiplied by 2)

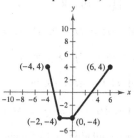

(d) $y = -f(x - 4)$

Reflection in the x-axis and a horizontal shift 4 units to the right

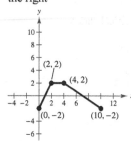

(e) $y = f(x) - 3$

Vertical shift 3 units downward

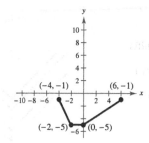

(f) $y = -f(x) - 1$

Reflection in the x-axis and a vertical shift 1 unit downward

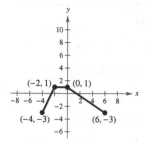

(g) $y = f(2x)$

Horizontal shrink (each x-value is divided by 2)

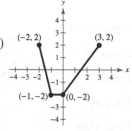

11. Parent function: $f(x) = x^2$

   (a) Vertical shift 1 unit downward

     $g(x) = x^2 - 1$

   (b) Reflection in the *x*-axis, horizontal shift 1 unit to the left, and a vertical shift 1 unit upward

     $g(x) = -(x + 1)^2 + 1$

13. Parent function: $f(x) = |x|$

   (a) Reflection in the *x*-axis and a horizontal shift 3 units to the left

     $g(x) = -|x + 3|$

   (b) Horizontal shift 2 units to the right and a vertical shift 4 units downward

     $g(x) = |x - 2| - 4$

15. Parent function: $f(x) = x^3$

   Horizontal shift 2 units to the right

   $y = (x - 2)^3$

17. Parent function: $f(x) = x^2$

   Reflection in the *x*-axis

   $y = -x^2$

19. Parent function: $f(x) = \sqrt{x}$

   Reflection in the *x*-axis and a vertical shift 1 unit upward

   $y = -\sqrt{x} + 1$

21. $g(x) = x^2 + 6$

   (a) Parent function: $f(x) = x^2$

   (b) A vertical shift 6 units upward

   (c)

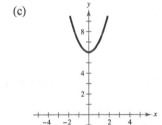

   (d) $g(x) = f(x) + 6$

23. $g(x) = -(x - 2)^3$

   (a) Parent function: $f(x) = x^3$

   (b) Horizontal shift of 2 units to the right and a reflection in the *x*-axis

   (c)

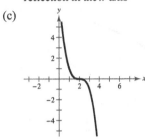

   (d) $g(x) = -f(x - 2)$

25. $g(x) = -3 - (x + 1)^2$

   (a) Parent function: $f(x) = x^2$

   (b) Reflection in the *x*-axis, a vertical shift 3 units downward and a horizontal shift 1 unit left

   (c)

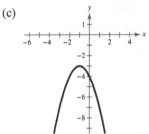

   (d) $g(x) = -f(x + 1) - 3$

27. $g(x) = |x - 1| + 2$

   (a) Parent function: $f(x) = |x|$

   (b) A horizontal shift 1 unit right and a vertical shift 2 units upward

   (c)

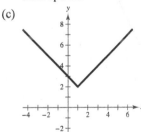

   (d) $g(x) = f(x - 1) + 2$

29. $g(x) = 2\sqrt{x}$

    (a) Parent function: $f(x) = \sqrt{x}$

    (b) A vertical stretch (each y value is multiplied by 2)

    (c)

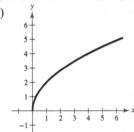

    (d) $g(x) = 2f(x)$

31. $g(x) = 2[\![x]\!] - 1$

    (a) Parent function: $f(x) = [\![x]\!]$

    (b) A vertical shift of 1 unit downward and a vertical stretch (each y value is multiplied by 2)

    (c)

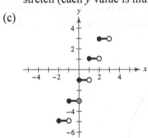

    (d) $g(x) = 2f(x) - 1$

33. $g(x) = |2x|$

    (a) Parent function: $f(x) = |x|$

    (b) A horizontal shrink

    (c)

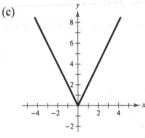

    (d) $g(x) = f(2x)$

35. $g(x) = -2x^2 + 1$

    (a) Parent function: $f(x) = x^2$

    (b) A vertical stretch, reflection in the x-axis and a vertical shift 1 unit upward

    (c)

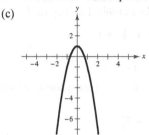

    (d) $g(x) = -2f(x) + 1$

37. $g(x) = 3|x - 1| + 2$

    (a) Parent function: $f(x) = |x|$

    (b) A horizontal shift of 1 unit to the right, a vertical stretch, and a vertical shift 2 units upward

    (c)

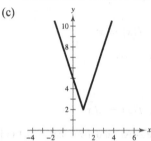

    (d) $g(x) = 3f(x - 1) + 2$

39. $g(x) = (x - 3)^2 - 7$

41. $f(x) = x^3$ moved 13 units to the right

    $g(x) = (x - 13)^3$

43. $g(x) = -|x| + 12$

45. $f(x) = \sqrt{x}$ moved 6 units to the left and reflected in both the x- and y-axes

    $g(x) = -\sqrt{-x + 6}$

47. $f(x) = x^2$

   (a) Reflection in the *x*-axis and a vertical stretch (each
     y-value is multiplied by 3)

$$g(x) = -3x^2$$

   (b) Vertical shift 3 units upward and a vertical stretch
     (each *y*-value is multiplied by 4)

$$g(x) = 4x^2 + 3$$

49. $f(x) = |x|$

   (a) Reflection in the *x*-axis and a vertical shrink
     $\left(\text{each } y\text{-value is multiplied by } \frac{1}{2}\right)$

$$g(x) = -\tfrac{1}{2}|x|$$

   (b) Vertical stretch (each *y*-value is multiplied by 3) and
     a vertical shift 3 units downward

$$g(x) = 3|x| - 3$$

51. Parent function: $f(x) = x^3$

   Vertical stretch (each *y*-value is multiplied by 2)

$$g(x) = 2x^3$$

53. Parent function: $f(x) = x^2$

   Reflection in the *x*-axis, vertical shrink
   $\left(\text{each } y\text{-value is multiplied by } \frac{1}{2}\right)$

$$g(x) = -\tfrac{1}{2}x^2$$

55. Parent function: $f(x) = \sqrt{x}$

   Reflection in the *y*-axis, vertical shrink
   $\left(\text{each } y\text{-value is multiplied by } \frac{1}{2}\right)$

$$g(x) = \tfrac{1}{2}\sqrt{-x}$$

57. Parent function: $f(x) = x^3$

   Reflection in the *x*-axis, horizontal shift 2 units to the
   right and a vertical shift 2 units upward

$$g(x) = -(x - 2)^3 + 2$$

59. Parent function: $f(x) = \sqrt{x}$

   Reflection in the *x*-axis and a vertical shift 3 units
   downward

$$g(x) = -\sqrt{x} - 3$$

61. (a)

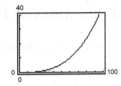

   (b) $H(x) = 0.00004636x^3$

$$H\!\left(\frac{x}{1.6}\right) = 0.00004636\left(\frac{x}{1.6}\right)^3$$

$$= 0.00004636\left(\frac{x^3}{4.096}\right)$$

$$= 0.0000113184x^3 = 0.00001132x^3$$

   The graph of $H\!\left(\dfrac{x}{1.6}\right)$ is a horizontal stretch of the

   graph of $H(x)$.

63. False. $y = f(-x)$ is a reflection in the *y*-axis.

65. True. Because $|x| = |-x|$, the graphs of
   $f(x) = |x| + 6$ and $f(x) = |-x| + 6$ are identical.

67. $y = f(x + 2) - 1$

   Horizontal shift 2 units to the left and a vertical shift
   1 unit downward

$$(0, 1) \to (0 - 2, 1 - 1) = (-2, 0)$$

$$(1, 2) \to (1 - 2, 2 - 1) = (-1, 1)$$

$$(2, 3) \to (2 - 2, 3 - 1) = (0, 2)$$

69. Since the graph of $g(x)$ is a horizontal shift one unit to
   the right of $f(x) = x^3$, the equation should be
   $g(x) = (x - 1)^3$ and not $g(x) = (x + 1)^3$.

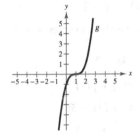

71. (a) The profits were only $\frac{3}{4}$ as large as expected:

$$g(t) = \tfrac{3}{4}f(t)$$

   (b) The profits were \$10,000 greater than predicted:

$$g(t) = f(t) + 10,000$$

   (c) There was a two-year delay: $g(t) = f(t - 2)$

Section P.9 Combinations of Functions: Composite Functions

1. addition; subtraction; multiplication; division

3.

x	0	1	2	3
f	2	3	1	2
g	-1	0	$\frac{1}{2}$	0
$f + g$	1	3	$\frac{3}{2}$	2

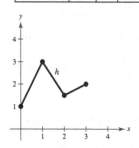

5. $f(x) = x + 2$, $g(x) = x - 2$

  (a) $(f + g)(x) = f(x) + g(x)$
$$= (x + 2) + (x - 2)$$
$$= 2x$$

  (b) $(f - g)(x) = f(x) - g(x)$
$$= (x + 2) - (x - 2)$$
$$= 4$$

  (c) $(fg)(x) = f(x) \cdot g(x)$
$$= (x + 2)(x - 2)$$
$$= x^2 - 4$$

  (d) $\left(\dfrac{f}{x}\right)(x) = \dfrac{f(x)}{g(x)} = \dfrac{x + 2}{x - 2}$

    Domain: all real numbers x except $x = 2$

7. $f(x) = x^2$, $g(x) = 4x - 5$

  (a) $(f + g)(x) = f(x) + g(x)$
$$= x^2 + (4x - 5)$$
$$= x^2 + 4x - 5$$

  (b) $(f - g)(x) = f(x) - g(x)$
$$= x^2 - (4x - 5)$$
$$= x^2 - 4x + 5$$

  (c) $(fg)(x) = f(x) \cdot g(x)$
$$= x^2(4x - 5)$$
$$= 4x^3 - 5x^2$$

  (d) $\left(\dfrac{f}{g}\right)(x) = \dfrac{f(x)}{g(x)}$
$$= \dfrac{x^2}{4x - 5}$$

    Domain: all real numbers x except $x = \dfrac{5}{4}$

9. $f(x) = x^2 + 6$, $g(x) = \sqrt{1 - x}$

  (a) $(f + g)(x) = f(x) + g(x) = x^2 + 6 + \sqrt{1 - x}$

  (b) $(f - g)(x) = f(x) - g(x) = x^2 + 6 - \sqrt{1 - x}$

  (c) $(fg)(x) = f(x) \cdot g(x) = \left(x^2 + 6\right)\sqrt{1 - x}$

  (d) $\left(\dfrac{f}{g}\right)(x) = \dfrac{f(x)}{g(x)} = \dfrac{x^2 + 6}{\sqrt{1 - x}} = \dfrac{\left(x^2 + 6\right)\sqrt{1 - x}}{1 - x}$

    Domain: $x < 1$

11. $f(x) = \dfrac{x}{x + 1}$, $g(x) = x^3$

  (a) $(f + g)(x) = \dfrac{x}{x + 1} + x^3 = \dfrac{x + x^4 + x^3}{x + 1}$

  (b) $(f - g)(x) = \dfrac{x}{x + 1} - x^3 = \dfrac{x - x^4 - x^3}{x + 1}$

  (c) $(fg)(x) = \dfrac{x}{x + 1} \cdot x^3 = \dfrac{x^4}{x + 1}$

  (d) $\left(\dfrac{f}{g}\right)(x) = \dfrac{x}{x + 1} + x^3 = \dfrac{x}{x + 1} \cdot \dfrac{1}{x^3} = \dfrac{1}{x^2(x + 1)}$

    Domain: all real numbers x except $x = 0$ and $x = -1$

For Exercises 13–23, $f(x) = x + 3$ and $g(x) = x^2 - 2$.

13. $(f + g)(2) = f(2) + g(2)$
$$= (2 + 3) + \left(2^2 - 2\right)$$
$$= 7$$

15. $(f - g)(0) = f(0) - g(0)$
$$= (0 + 3) - \left((0)^2 - 2\right)$$
$$= 5$$

17. $(f - g)(3t) = f(3t) - g(3t)$
$$= \left((3t) + 3\right) - \left((3t)^2 - 2\right)$$
$$= 3t + 3 - \left(9t^2 - 2\right)$$
$$= -9t^2 + 3t + 5$$

19. $(fg)(6) = f(6)g(6)$

$\qquad = \big((6) + 3\big)\big((6)^2 - 2\big)$

$\qquad = (9)(34)$

$\qquad = 306$

21. $(f/g)(5) = f(5) / g(5)$

$\qquad = \big((5) + 3\big) / \big((5)^2 - 2\big)$

$\qquad = \dfrac{8}{23}$

23. $(f/g)(-1) - g(3) = f(-1) / g(-1) - g(3)$

$\qquad = \big((-1) + 3\big) / \big((-1)^2 - 2\big) - \big((3)^2 - 2\big)$

$\qquad = (2/-1) - 7$

$\qquad = -2 - 7 = -9$

25. $f(x) = 3x,\ g(x) = -\dfrac{x^3}{10}$

$(f + g)(x) = 3x - \dfrac{x^3}{10}$

For $0 \le x \le 2$, $f(x)$ contributes most to the magnitude.

For $x > 6$, $g(x)$ contributes most to the magnitude.

27. $f(x) = 3x + 2,\ g(x) = -\sqrt{x + 5}$

$(f + g)x = 3x - \sqrt{x + 5} + 2$

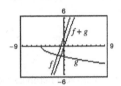

For $0 \le x \le 2$, $f(x)$ contributes most to the magnitude.

For $x > 6$, $f(x)$ contributes most to the magnitude.

29. $f(x) = x + 8,\ g(x) = x - 3$

(a) $(f \circ g)(x) = f\big(g(x)\big) = f(x - 3) = (x - 3) + 8 = x + 5$

(b) $(g \circ f)(x) = g\big(f(x)\big) = g(x + 8) = (x + 8) - 3 = x + 5$

(c) $(g \circ g)(x) = g\big(g(x)\big) = g(x - 3) = (x - 3) - 3 = x - 6$

31. $f(x) = x^2,\ g(x) = x - 1$

(a) $(f \circ g)(x) = f\big(g(x)\big) = f(x - 1) = (x - 1)^2$

(b) $(g \circ f)(x) = g\big(f(x)\big) = g(x^2) = x^2 - 1$

(c) $(g \circ g)(x) = g\big(g(x)\big) = g(x - 1) = x - 2$

33. $f(x) = \sqrt[3]{x - 1},\ g(x) = x^3 + 1$

(a) $(f \circ g)(x) = f\big(g(x)\big)$

$\qquad = f(x^3 + 1)$

$\qquad = \sqrt[3]{(x^3 + 1) - 1}$

$\qquad = \sqrt[3]{x^3} = x$

(b) $(g \circ f)(x) = g\big(f(x)\big)$

$\qquad = g\big(\sqrt[3]{x - 1}\big)$

$\qquad = \big(\sqrt[3]{x - 1}\big)^3 + 1$

$\qquad = (x - 1) + 1 = x$

(c) $(g \circ g)(x) = g\big(g(x)\big)$

$\qquad = g(x^3 + 1)$

$\qquad = (x^3 + 1)^3 + 1$

$\qquad = x^9 + 3x^6 + 3x^3 + 2$

35. $f(x) = \sqrt{x + 4}$ Domain: $x \geq -4$

$g(x) = x^2$ Domain: all real numbers x

(a) $(f \circ g)(x) = f(g(x)) = f(x^2) = \sqrt{x^2 + 4}$

Domain: all real numbers x

(b) $(g \circ f)(x) = g(f(x))$

$= g(\sqrt{x + 4}) = (\sqrt{x + 4})^2 = x + 4$

Domain: $x \geq -4$

37. $f(x) = x^3$ Domain: all real numbers x

$g(x) = x^{2/3}$ Domain: all real numbers x

(a) $(f \circ g)(x) = f(g(x)) = f(x^{2/3}) = (x^{2/3})^3 = x^2$

Domain: all real numbers x.

(b) $(g \circ f)(x) = g(f(x)) = g(x^3) = (x^3)^{2/3} = x^2$

Domain: all real numbers x.

39. $f(x) = |x|$ Domain: all real numbers x

$g(x) = x + 6$ Domain: all real numbers x

(a) $(f \circ g)(x) = f(g(x)) = f(x + 6) = |x + 6|$

Domain: all real numbers x

(b) $(g \circ f)(x) = g(f(x)) = g(|x|) = |x| + 6$

Domain: all real numbers x

41. $f(x) = \dfrac{1}{x}$ Domain: all real numbers x except $x = 0$

$g(x) = x + 3$ Domain: all real numbers x

(a) $(f \circ g)(x) = f(g(x)) = f(x + 3) = \dfrac{1}{x + 3}$

Domain: all real numbers x except $x = -3$

(b) $(g \circ f)(x) = g(f(x)) = g\left(\dfrac{1}{x}\right) = \dfrac{1}{x} + 3$

Domain: all real numbers x except $x = 0$

43. $f(x) = \frac{1}{2}x, \ g(x) = x - 4$

(a)

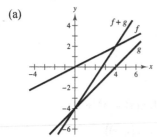

(b)

45. (a) $(f + g)(3) = f(3) + g(3) = 2 + 1 = 3$

(b) $\left(\dfrac{f}{g}\right)(2) = \dfrac{f(2)}{g(2)} = \dfrac{0}{2} = 0$

47. (a) $(f \circ g)(2) = f(g(2)) = f(2) = 0$

(b) $(g \circ f)(2) = g(f(2)) = g(0) = 4$

49. $h(x) = (2x^2 + 1)^2$

One possibility: Let $f(x) = x^2$ and $g(x) = 2x + 1$,

then $(f \circ g)(x) = h(x)$.

51. $h(x) = \sqrt[3]{x^2 - 4}$

One possibility: Let $f(x) = \sqrt[3]{x}$ and $g(x) = x^2 - 4$,

then $(f \circ g)(x) = h(x)$.

53. $h(x) = \dfrac{1}{x + 2}$

One possibility: Let $f(x) = 1/x$ and $g(x) = x + 2$,

then $(f \circ g)(x) = h(x)$.

55. $h(x) = \dfrac{-x^2 + 3}{4 - x^2}$

One possibility: Let $f(x) = \dfrac{x + 3}{4 + x}$ and $g(x) = -x^2$,

then $(f \circ g)(x) = h(x)$.

57. (a) $T(x) = R(x) + B(x) = \frac{3}{4}x + \frac{1}{15}x^2$

(b)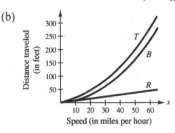

(c) $B(x)$; As x increases, $B(x)$ increases at a faster rate.

59. (a) $c(t) = \dfrac{b(t) - d(t)}{p(t)} \times 100$

(b) $c(16)$ represents the percent change in the population due to births and deaths in the year 2016.

61. (a) $r(x) = \dfrac{x}{2}$

(b) $A(r) = \pi r^2$

(c) $(A \circ r)(x) = A(r(x)) = A\!\left(\dfrac{x}{2}\right) = \pi\!\left(\dfrac{x}{2}\right)^2$

 $(A \circ r)(x)$ represents the area of the circular base of the tank on the square foundation with side length x.

63. (a) $f(g(x)) = f(0.03x) = 0.03x - 500{,}000$

(b) $g(f(x)) = g(x - 500{,}000) = 0.03(x - 500{,}000)$

 $g(f(x))$ represents your bonus of 3% of an amount over \$500,000.

65. False. $(f \circ g)(x) = 6x + 1$ and $(g \circ f)(x) = 6x + 6$

67. Let O = oldest sibling, M = middle sibling, Y = youngest sibling.

Then the ages of each sibling can be found using the equations:

 $O = 2M$

 $M = \frac{1}{2}Y + 6$

(a) $O(M(Y)) = 2\!\left(\frac{1}{2}(Y) + 6\right) = 12 + Y$; Answers will vary.

(b) Oldest sibling is 16: $O = 16$

 Middle sibling: $O = 2M$

 $16 = 2M$

 $M = 8$ years old

 Youngest sibling: $M = \frac{1}{2}Y + 6$

 $8 = \frac{1}{2}Y + 6$

 $2 = \frac{1}{2}Y$

 $Y = 4$ years old

69. Let $f(x)$ and $g(x)$ be two odd functions and define $h(x) = f(x)g(x)$. Then

 $\begin{aligned} h(-x) &= f(-x)g(-x) \\ &= \left[-f(x)\right]\!\left[-g(x)\right] \quad \text{because } f \text{ and } g \text{ are odd} \\ &= f(x)g(x) \\ &= h(x). \end{aligned}$

So, $h(x)$ is even.

Let $f(x)$ and $g(x)$ be two even functions and define $h(x) = f(x)g(x)$. Then

 $\begin{aligned} h(-x) &= f(-x)g(-x) \\ &= f(x)g(x) \quad \text{because } f \text{ and } g \text{ are even} \\ &= h(x). \end{aligned}$

So, $h(x)$ is even.

71. (a) Answer not unique. *Sample answer*:

 $f(x) = x + 3, \; g(x) = x + 2$

 $(f \circ g)(x) = f(g(x)) = (x + 2) + 3 = x + 5$

 $(g \circ f)(x) = g(f(x)) = (x + 3) + 2 = x + 5$

(b) Answer not unique. *Sample answer*: $f(x) = x^2, \; g(x) = x^3$

 $(f \circ g)(x) = f(g(x)) = \left(x^3\right)^2 = x^6$

 $(g \circ f)(x) = g(f(x)) = \left(x^2\right)^3 = x^6$

73. (a) $g(x) = \frac{1}{2}[f(x) + f(-x)]$

To determine if $g(x)$ is even, show $g(-x) = g(x)$.

$g(-x) = \frac{1}{2}[f(-x) + f(-(-x))] = \frac{1}{2}[f(-x) + f(x)] = \frac{1}{2}[f(x) + f(-x)] = g(x)$ ✓

$h(x) = \frac{1}{2}[f(x) - f(-x)]$

To determine if $h(x)$ is odd show $h(-x) = -h(x)$.

$h(-x) = \frac{1}{2}[f(-x) - f(-(-x))] = \frac{1}{2}[f(-x) - f(x)] = -\frac{1}{2}[f(x) - f(-x)] = -h(x)$ ✓

(b) Let $f(x) = a$ function

$f(x) =$ even function + odd function.

Using the result from part (a) $g(x)$ is an even function and $h(x)$ is an odd function.

$f(x) = g(x) + h(x) = \frac{1}{2}[f(x) + f(-x)] + \frac{1}{2}[f(x) - f(-x)] = \frac{1}{2}f(x) + \frac{1}{2}f(-x) + \frac{1}{2}f(x) - \frac{1}{2}f(-x) = f(x)$ ✓

(c) $f(x) = x^2 - 2x + 1$

$f(x) = g(x) + h(x)$

$g(x) = \frac{1}{2}[f(x) + f(-x)] = \frac{1}{2}[x^2 - 2x + 1 + (-x)^2 - 2(-x) + 1]$

$\quad = \frac{1}{2}[x^2 - 2x + 1 + x^2 + 2x + 1] = \frac{1}{2}[2x^2 + 2] = x^2 + 1$

$h(x) = \frac{1}{2}[f(x) - f(-x)] = \frac{1}{2}[x^2 - 2x + 1 - ((-x)^2 - 2(-x) + 1)]$

$\quad = \frac{1}{2}[x^2 - 2x + 1 - x^2 - 2x - 1] = \frac{1}{2}[-4x] = -2x$

$f(x) = (x^2 + 1) + (-2x)$

$k(x) = \dfrac{1}{x + 1}$

$k(x) = g(x) + h(x)$

$g(x) = \frac{1}{2}[k(x) + k(-x)] = \frac{1}{2}\left[\dfrac{1}{x + 1} + \dfrac{1}{-x + 1}\right]$

$\quad = \frac{1}{2}\left[\dfrac{1 - x + x + 1}{(x + 1)(1 - x)}\right] = \frac{1}{2}\left[\dfrac{2}{(x + 1)(1 - x)}\right]$

$\quad = \dfrac{1}{(x + 1)(1 - x)} = \dfrac{-1}{(x + 1)(x - 1)}$

$h(x) = \frac{1}{2}[k(x) - k(-x)] = \frac{1}{2}\left[\dfrac{1}{x + 1} - \dfrac{1}{1 - x}\right]$

$\quad = \frac{1}{2}\left[\dfrac{1 - x - (x + 1)}{(x + 1)(1 - x)}\right] = \frac{1}{2}\left[\dfrac{-2x}{(x + 1)(1 - x)}\right]$

$\quad = \dfrac{-x}{(x + 1)(1 - x)} = \dfrac{x}{(x + 1)(x - 1)}$

$k(x) = \left(\dfrac{-1}{(x + 1)(x - 1)}\right) + \left(\dfrac{x}{(x + 1)(x - 1)}\right)$

Section P.10   Inverse Functions

1. inverse

3. range; domain

5. one-to-one

7. $f(x) = 6x$

$$f^{-1}(x) = \frac{x}{6} = \frac{1}{6}x$$

$$f(f^{-1}(x)) = f\left(\frac{x}{6}\right) = 6\left(\frac{x}{6}\right) = x$$

$$f^{-1}(f(x)) = f^{-1}(6x) = \frac{6x}{6} = x$$

11. $f(x) = x^2 - 4, x \geq 0$

$$f^{-1}(x) = \sqrt{x + 4}$$

$$f(f^{-1}(x)) = f(\sqrt{x + 4}) = (\sqrt{x + 4})^2 - 4 = (x + 4) - 4 = x$$

$$f^{-1}(f(x)) = f^{-1}(x^2 - 4) = \sqrt{(x^2 - 4) + 4} = \sqrt{x^2} = x$$

13. $f(x) = x^3 + 1$

$$f^{-1}(x) = \sqrt[3]{x - 1}$$

$$f(f^{-1}(x)) = f(\sqrt[3]{x - 1}) = (\sqrt[3]{x - 1})^3 + 1 = (x - 1) + 1 = x$$

$$f^{-1}(f(x)) = f^{-1}(x^3 + 1) = \sqrt[3]{(x^3 + 1) - 1} = \sqrt[3]{x^3} = x$$

15. $(f \circ g)(x) = f(g(x)) = f(4x + 9) = \dfrac{4x + 9 - 9}{4} = \dfrac{4x}{4} = x$

$(g \circ f)(x) = g(f(x)) = g\left(\dfrac{x - 9}{4}\right) = 4\left(\dfrac{x - 9}{4}\right) + 9 = x - 9 + 9 = x$

17. $f(g(x)) = f(\sqrt[3]{4x}) = \dfrac{(\sqrt[3]{4x})^3}{4} = \dfrac{4x}{4} = x$

$g(f(x)) = g\left(\dfrac{x^3}{4}\right) = \sqrt[3]{4\left(\dfrac{x^3}{4}\right)} = \sqrt[3]{x^3} = x$

19.

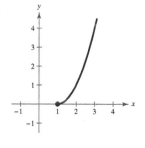

9. $f(x) = 3x + 1$

$$f^{-1}(x) = \frac{x - 1}{3}$$

$$f(f^{-1}(x)) = f\left(\frac{x - 1}{3}\right) = 3\left(\frac{x - 1}{3}\right) + 1 = x$$

$$f^{-1}(f(x)) = f^{-1}(3x + 1) = \frac{(3x + 1) - 1}{3} = x$$

21. $f(x) = x - 5, g(x) = x + 5$

(a) $f(g(x)) = f(x + 5) = (x + 5) - 5 = x$

$g(f(x)) = g(x - 5) = (x - 5) + 5 = x$

(b)

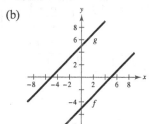

23. $f(x) = 7x + 1$, $g(x) = \dfrac{x-1}{7}$

(a) $f(g(x)) = f\left(\dfrac{x-1}{7}\right) = 7\left(\dfrac{x-1}{7}\right) + 1 = x$

$g(f(x)) = g(7x + 1) = \dfrac{(7x+1) - 1}{7} = x$

(b)

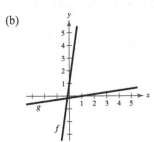

25. $f(x) = x^3$, $g(x) = \sqrt[3]{x}$

(a) $f(g(x)) = f(\sqrt[3]{x}) = (\sqrt[3]{x})^3 = x$

$g(f(x)) = g(x^3) = \sqrt[3]{(x^3)} = x$

(b)

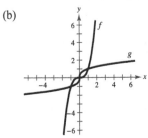

27. $f(x) = \sqrt{x+5}$, $g(x) = x^2 - 5$, $x \geq 0$

(a) $f(g(x)) = f(x^2 - 5)$, $x \geq 0$

$= \sqrt{(x^2 - 5) + 5} = x$

$g(f(x)) = g(\sqrt{x+5})$

$= (\sqrt{x+5})^2 - 5 = x$

(b)

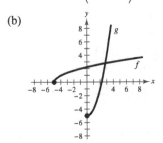

29. $f(x) = \dfrac{1}{x}$, $g(x) = \dfrac{1}{x}$

(a) $f(g(x)) = f\left(\dfrac{1}{x}\right) = \dfrac{1}{1/x} = 1 \div \dfrac{1}{x} = 1 \cdot \dfrac{x}{1} = x$

$g(f(x)) = g\left(\dfrac{1}{x}\right) = \dfrac{1}{1/x} = 1 \div \dfrac{1}{x} = 1 \cdot \dfrac{x}{1} = x$

(b)

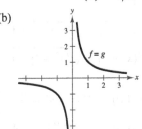

31. $f(x) = \dfrac{x-1}{x+5}$, $g(x) = -\dfrac{5x+1}{x-1}$

(a) $f(g(x)) = f\left(-\dfrac{5x+1}{x-1}\right) = \dfrac{\left(-\dfrac{5x+1}{x-1} - 1\right)}{\left(-\dfrac{5x+1}{x-1} + 5\right)} \cdot \dfrac{x-1}{x-1} = \dfrac{-(5x+1) - (x-1)}{-(5x+1) + 5(x-1)} = \dfrac{-6x}{-6} = x$

$g(f(x)) = g\left(\dfrac{x-1}{x+5}\right) = -\dfrac{\left[5\left(\dfrac{x-1}{x+5}\right) + 1\right]}{\left[\dfrac{x-1}{x+5} - 1\right]} \cdot \dfrac{x+5}{x+5} = -\dfrac{5(x-1) + (x+5)}{(x-1) - (x+5)} = -\dfrac{6x}{-6} = x$

(b)

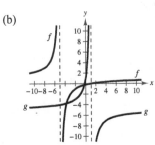

33. No, $\{(-2, -1), (1, 0), (2, 1), (1, 2), (-2, 3), (-6, 4)\}$ does not represent a function. -2 and 1 are paired with two different values.

35.

x	3	5	7	9	11	13
$f^{-1}(x)$	-1	0	1	2	3	4

37. Yes, because no horizontal line crosses the graph of f at more than one point, f has an inverse.

39. No, because some horizontal lines cross the graph of f twice, f *does not* have an inverse.

41. $g(x) = (x + 3)^2 + 2$

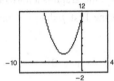

g does not pass the Horizontal Line Test, so g *does not* have an inverse.

43. $f(x) = x\sqrt{9 - x^2}$

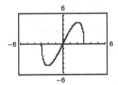

f does not pass the Horizontal Line Test, so f *does not* have an inverse.

45. (a) $\quad f(x) = x^5 - 2$ (b)

$$y = x^5 - 2$$
$$x = y^5 - 2$$
$$y = \sqrt[5]{x + 2}$$
$$f^{-1}(x) = \sqrt[5]{x + 2}$$

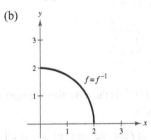

(c) The graph of f^{-1} is the reflection of the graph of f in the line $y = x$.

(d) The domains and ranges of f and f^{-1} are all real numbers.

47. (a) $\quad f(x) = \sqrt{4 - x^2}, 0 \le x \le 2$

$$y = \sqrt{4 - x^2}$$
$$x = \sqrt{4 - y^2}$$
$$x^2 = 4 - y^2$$
$$y^2 = 4 - x^2$$
$$y = \sqrt{4 - x^2}$$
$$f^{-1}(x) = \sqrt{4 - x^2}, 0 \le x \le 2$$

(b)

(c) The graph of f^{-1} is the same as the graph of f.

(d) The domains and ranges of f and f^{-1} are all real numbers x such that $0 \le x \le 2$.

49. (a) $\quad f(x) = \dfrac{4}{x}$ (b)

$$y = \frac{4}{x}$$
$$x = \frac{4}{y}$$
$$xy = 4$$
$$y = \frac{4}{x}$$
$$f^{-1}(x) = \frac{4}{x}$$

(c) The graph of f^{-1} is the same as the graph of f.

(d) The domains and ranges of f and f^{-1} are all real numbers except for 0.

51. (a) $f(x) = \dfrac{x+1}{x-2}$ **(b)**

$y = \dfrac{x+1}{x-2}$

$x = \dfrac{y+1}{y-2}$

$x(y-2) = y+1$

$xy - 2x = y+1$

$xy - y = 2x+1$

$y(x-1) = 2x+1$

$y = \dfrac{2x+1}{x-1}$

$f^{-1}(x) = \dfrac{2x+1}{x-1}$

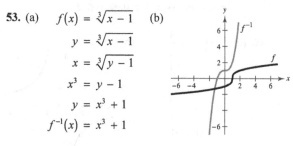

(c) The graph of f^{-1} is the reflection of graph of f in the line $y = x$.

(d) The domain of f and the range of f^{-1} is all real numbers except 2.

The range of f and the domain of f^{-1} is all real numbers except 1.

53. (a) $f(x) = \sqrt[3]{x-1}$ **(b)**

$y = \sqrt[3]{x-1}$

$x = \sqrt[3]{y-1}$

$x^3 = y-1$

$y = x^3 + 1$

$f^{-1}(x) = x^3 + 1$

(c) The graph of f^{-1} is the reflection of the graph of f in the line $y = x$.

(d) The domains and ranges of f and f^{-1} are all real numbers.

55. $f(x) = x^4$

$y = x^4$

$x = y^4$

$y = \pm\sqrt[4]{x}$

This does not represent y as a function of x. f does not have an inverse.

57. $g(x) = \dfrac{x+1}{6}$

$y = \dfrac{x+1}{6}$

$x = \dfrac{y+1}{6}$

$6x = y+1$

$y = 6x - 1$

This is a function of x, so g has an inverse.

$g^{-1}(x) = 6x - 1$

59. $p(x) = -4$

$y = -4$

Because $y = -4$ for all x, the graph is a horizontal line and fails the Horizontal Line Test. p does not have an inverse.

61. $f(x) = (x+3)^2, \ x \geq -3 \Rightarrow y \geq 0$

$y = (x+3)^2, \ x \geq -3, \ y \geq 0$

$x = (y+3)^2, \ y \geq -3, \ x \geq 0$

$\sqrt{x} = y+3, \ y \geq -3, \ x \geq 0$

$y = \sqrt{x} - 3, \ x \geq 0, \ y \geq -3$

This is a function of x, so f has an inverse.

$f^{-1}(x) = \sqrt{x} - 3, \ x \geq 0$

63. $f(x) = \begin{cases} x+3, & x < 0 \\ 6-x, & x \geq 0 \end{cases}$

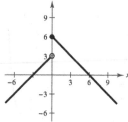

This graph fails the Horizontal Line Test, so f does not have an inverse.

65. $h(x) = |x+1| - 1$

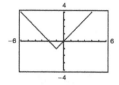

The graph fails the Horizontal Line Test, so h does not have an inverse.

67. $f(x) = \sqrt{2x + 3} \Rightarrow x \geq -\dfrac{3}{2}, y \geq 0$

$\quad y = \sqrt{2x + 3}, x \geq -\dfrac{3}{2}, y \geq 0$

$\quad x = \sqrt{2y + 3}, y \geq -\dfrac{3}{2}, x \geq 0$

$\quad x^2 = 2y + 3, x \geq 0, y \geq -\dfrac{3}{2}$

$\quad y = \dfrac{x^2 - 3}{2}, x \geq 0, y \geq -\dfrac{3}{2}$

This is a function of x, so f has an inverse.

$f^{-1}(x) = \dfrac{x^2 - 3}{2}, x \geq 0$

69. $\qquad f(x) = \dfrac{6x + 4}{4x + 5}$

$\qquad\quad y = \dfrac{6x + 4}{4x + 5}$

$\qquad\quad x = \dfrac{6y + 4}{4y + 5}$

$\quad x(4y + 5) = 6y + 4$

$\quad 4xy + 5x = 6y + 4$

$\quad 4xy - 6y = -5x + 4$

$\quad y(4x - 6) = -5x + 4$

$\qquad\quad y = \dfrac{-5x + 4}{4x - 6}$

$\qquad\qquad = \dfrac{5x - 4}{6 - 4x}$

This is a function of x, so f has an inverse.

$f^{-1}(x) = \dfrac{5x - 4}{6 - 4x}$

71. $f(x) = |x + 2|$

domain of $f: x \geq -2$, range of $f: y \geq 0$

$\quad f(x) = |x + 2|$

$\quad\quad y = |x + 2|$

$\quad\quad x = y + 2$

$\quad x - 2 = y$

So, $f^{-1}(x) = x - 2$.

domain of $f^{-1}: x \geq 0$, range of $f^{-1}: y \geq -2$

73. $f(x) = (x + 6)^2$

domain of $f: x \geq -6$, range of $f: y \geq 0$

$\quad f(x) = (x + 6)^2$

$\quad\quad y = (x + 6)^2$

$\quad\quad x = (y + 6)^2$

$\quad \sqrt{x} = y + 6$

$\quad \sqrt{x} - 6 = y$

So, $f^{-1}(x) = \sqrt{x} - 6$.

domain of $f^{-1}: x \geq 0$, range of $f^{-1}: y \geq -6$

75. $f(x) = -2x^2 + 5$

domain of $f: x \geq 0$, range of $f: y \leq 5$

$\qquad\quad f(x) = -2x^2 + 5$

$\qquad\qquad y = -2x^2 + 5$

$\qquad\qquad x = -2y^2 + 5$

$\qquad x - 5 = -2y^2$

$\qquad 5 - x = 2y^2$

$\qquad \sqrt{\dfrac{5 - x}{2}} = y$

$\qquad \dfrac{\sqrt{5 - x}}{\sqrt{2}} \cdot \dfrac{\sqrt{2}}{\sqrt{2}} = y$

$\qquad \dfrac{\sqrt{2(5 - x)}}{2} = y$

So, $f^{-1}(x) = \dfrac{\sqrt{-2(x - 5)}}{2}$.

domain of $f^{-1}(x): x \leq 5$, range of $f^{-1}(x): y \geq 0$

77. $f(x) = |x - 4| + 1$

domain of $f: x \geq 4$, range of $f: y \geq 1$

$\quad f(x) = |x - 4| + 1$

$\quad\quad y = x - 3$

$\quad\quad x = y - 3$

$\quad x + 3 = y$

So, $f^{-1}(x) = x + 3$.

domain of $f^{-1}: x \geq 1$, range of $f^{-1}: y \geq 4$

In Exercises 79–83, $f(x) = \frac{1}{8}x - 3$, $f^{-1}(x) = 8(x + 3)$, $g(x) = x^3$, $g^{-1}(x) = \sqrt[3]{x}$.

79. $\left(f^{-1} \circ g^{-1}\right)(1) = f^{-1}\left(g^{-1}(1)\right)$

$\qquad = f^{-1}\left(\sqrt[3]{1}\right)$

$\qquad = 8\left(\sqrt[3]{1} + 3\right) = 32$

81. $\left(f^{-1} \circ f^{-1}\right)(4) = f^{-1}\left(f^{-1}(4)\right)$

$\qquad = f^{-1}(8[4 + 3])$

$\qquad = 8\big[8(4 + 3) + 3\big]$

$\qquad = 8\big[8(7) + 3\big]$

$\qquad = 8(59) = 472$

83. $(f \circ g)(x) = f(g(x)) = f(x^3) = \frac{1}{8}x^3 - 3$

$\qquad y = \frac{1}{8}x^3 - 3$

$\qquad x = \frac{1}{8}y^3 - 3$

$\qquad x + 3 = \frac{1}{8}y^3$

$\qquad 8(x + 3) = y^3$

$\qquad \sqrt[3]{8(x + 3)} = y$

$\qquad (f \circ g)^{-1}(x) = 2\sqrt[3]{x + 3}$

In Exercises 85–88, $f(x) = x + 4$, $f^{-1}(x) = x - 4$, $g(x) = 2x - 5$, $g^{-1}(x) = \dfrac{x + 5}{2}$.

85. $\left(g^{-1} \circ f^{-1}\right)(x) = g^{-1}\left(f^{-1}(x)\right)$

$\qquad = g^{-1}(x - 4)$

$\qquad = \dfrac{(x - 4) + 5}{2}$

$\qquad = \dfrac{x + 1}{2}$

87. $(f \circ g)(x) = f(g(x))$

$\qquad = f(2x - 5)$

$\qquad = (2x - 5) + 4$

$\qquad = 2x - 1$

$\qquad (f \circ g)^{-1}(x) = \dfrac{x + 1}{2}$

Note: Comparing Exercises 85 and 87,

$(f \circ g)^{-1}(x) = \left(g^{-1} \circ f^{-1}\right)(x)$.

89. (a) $\qquad y = 10 + 0.75x$

$\qquad\qquad x = 10 + 0.75y$

$\qquad\qquad x - 10 = 0.75y$

$\qquad\qquad \dfrac{x - 10}{0.75} = y$

$\qquad$ So, $f^{-1}(x) = \dfrac{x - 10}{0.75}$.

$\qquad x = $ hourly wage, $y = $ number of units produced

(b) $\quad y = \dfrac{24.25 - 10}{0.75} = 19$

$\qquad$ So, 19 units are produced.

91. False. $f(x) = x^2$ is even and does not have an inverse.

93.

x	1	3	4	6
f	1	2	6	7

x	1	2	6	7
$f^{-1}(x)$	1	3	4	6

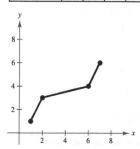

95. Let $(f \circ g)(x) = y$. Then $x = (f \circ g)^{-1}(y)$. Also,

$\qquad (f \circ g)(x) = y \Rightarrow f(g(x)) = y$

$\qquad\qquad\qquad\qquad g(x) = f^{-1}(y)$

$\qquad\qquad\qquad\qquad x = g^{-1}\left(f^{-1}(y)\right)$

$\qquad\qquad\qquad\qquad x = \left(g^{-1} \circ f^{-1}\right)(y)$.

Because f and g are both one-to-one functions,

$(f \circ g)^{-1} = g^{-1} \circ f^{-1}$.

97. If $f(x) = k(2 - x - x^3)$ has an inverse and $f^{-1}(3) = -2$, then $f(-2) = 3$. So,

$\qquad f(-2) = k\left(2 - (-2) - (-2)^3\right) = 3$

$\qquad k(2 + 2 + 8) = 3$

$\qquad\qquad 12k = 3$

$\qquad\qquad k = \frac{3}{12} = \frac{1}{4}$.

So, $k = \frac{1}{4}$.

99.

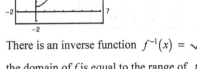

There is an inverse function $f^{-1}(x) = \sqrt{x-1}$ because the domain of f is equal to the range of f^{-1} and the range of f is equal to the domain of f^{-1}.

101. This situation could be represented by a one-to-one function if the runner does not stop to rest. The inverse function would represent the time in hours for a given number of miles completed.

Review Exercises for Chapter P

1. $\left\{11, -14, -\frac{8}{9}, \frac{5}{2}, \sqrt{6}, 0.4\right\}$

(a) Natural numbers: 11

(b) Whole numbers: 0, 11

(c) Integers: 11, −14

(d) Rational numbers: $11, -14, -\frac{8}{9}, \frac{5}{2}, 0.4$

(e) Irrational numbers: $\sqrt{6}$

3.

$\frac{5}{4} > \frac{7}{8}$

5. $d(-74, 48) = |48 - (-74)| = 122$

7. $d(x, 7) = |x - 7|$ and $d(x, 7) \geq 4$, thus $|x - 7| \geq 4$.

9. $-x^2 + x - 1$

(a) $-(1)^2 + 1 - 1 = -1$

(b) $-(-1)^2 + (-1) - 1 = -3$

11. $0 + (a - 5) = a - 5$

Illustrates the Additive Identity Property

13. $2x + (3x - 10) = (2x + 3x) - 10$

Illustrates the Associative Property of Addition

15. $(t^2 + 1) + 3 = 3 + (t^2 + 1)$

Illustrates the Commutative Property of Addition

17. $-6 + 6 = 0$

19. $(-8)(-4) = 32$

21. $\frac{x}{2} - \frac{2x}{5} = \frac{5x}{10} - \frac{4x}{10} = \frac{x}{10}$

23. $2(x + 5) - 7 = x + 9$

$2x + 10 - 7 = x + 9$

$2x + 3 = x + 9$

$x = 6$

25. $\frac{x}{5} - 3 = \frac{x}{3} + 1$

$15\left(\frac{x}{5} - 3\right) = \left(\frac{x}{3} + 1\right)15$

$3x - 45 = 5x + 15$

$-2x = 60$

$x = -30$

27. $2x^2 - x - 28 = 0$

$(2x + 7)(x - 4) = 0$

$2x + 7 = 0 \Rightarrow x = -\frac{7}{2}$

$x - 4 = 0 \Rightarrow x = 4$

29. $(x + 13)^2 = 25$

$x + 13 = \pm 5$

$x = -13 \pm 5$

$x = -18 \text{ or } x = -8$

31. $5x^4 - 12x^3 = 0$

$x^3(5x - 12) = 0$

$x^3 = 0 \text{ or } 5x - 12 = 0$

$x = 0 \text{ or } \qquad x = \frac{12}{5}$

33. $\sqrt{x + 4} = 3$

$\left(\sqrt{x + 4}\right)^2 = (3)^2$

$x + 4 = 9$

$x = 5$

35. $(x - 1)^{2/3} - 25 = 0$

$$(x - 1)^{2/3} = 25$$
$$(x - 1)^2 = 25^3$$
$$x - 1 = \pm\sqrt{25^3}$$
$$x = 1 \pm 125$$
$$x = 126 \text{ or } x = -124$$

37. $|x - 5| = 10$

$$x - 5 = -10 \text{ or } x - 5 = 10$$
$$x = -5 \qquad\qquad x = 15$$

39. (a)

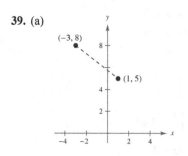

(b) $d = \sqrt{(-3 - 1)^2 + (8 - 5)^2} = \sqrt{16 + 9} = 5$

(c) Midpoint: $\left(\dfrac{-3 + 1}{2}, \dfrac{8 + 5}{2}\right) = \left(-1, \dfrac{13}{2}\right)$

41.

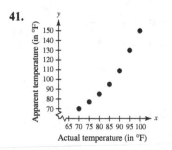

43. $y = 2x - 6$

x	1	2	3	4	5
y	−4	−2	0	2	4

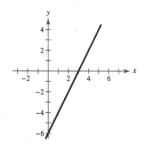

45. $y = x^2 + 2x$

x	−3	−2	−1	0	1
y	3	0	−1	0	3

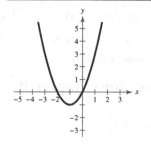

47. x-intercepts: $(1, 0), (5, 0)$

y-intercept: $(0, 5)$

49. $y = -3x + 7$

Intercepts: $\left(\dfrac{7}{3}, 0\right), (0, 7)$

$y = -3(-x) + 7 \Rightarrow y = 3x + 7 \Rightarrow$ No y-axis symmetry

$-y = -3x + 7 \Rightarrow y = 3x - 7 \Rightarrow$ No x-axis symmetry

$-y = -3(-x) + 7 \Rightarrow y = -3x - 7 \Rightarrow$ No origin symmetry

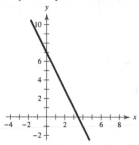

51. $y = -x^4 + 6x^2$

Intercept: $(0, 0)$

$y = -(-x)^4 + 6(-x)^2 \Rightarrow y = -x^4 + 6x^2 \Rightarrow y$-axis
   symmetry

$-y = -x^4 + 6x^2 \Rightarrow y = x^4 - 6x^2 \Rightarrow$ No x-axis
   symmetry

$-y = -(-x)^4 + 6(-x)^2 \Rightarrow y = x^4 - 6x^2 \Rightarrow$ No
   origin symmetry

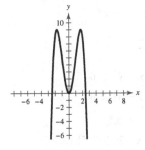

53. Endpoints of a diameter: $(0, 0)$ and $(4, -6)$

Center: $\left(\dfrac{0 + 4}{2}, \dfrac{0 + (-6)}{2} \right) = (2, -3)$

Radius: $r = \sqrt{(2 - 0)^2 + (-3 - 0)^2} = \sqrt{4 + 9} = \sqrt{13}$

Standard form: $(x - 2)^2 + (y - (-3))^2 = \left(\sqrt{13} \right)^2$

$\qquad\qquad (x - 2)^2 + (y + 3)^2 = 13$

55. $x^2 + y^2 = 9$

Center: $(0, 0)$

Radius: 3

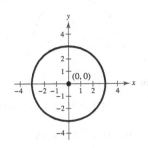

57. $y = -\frac{1}{2}x + 1$

Slope: $m = -\frac{1}{2}$

y-intercept: $(0, 1)$

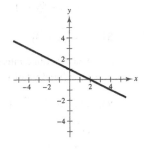

59. $y = 1$

Slope: $m = 0$

y-intercept: $(0, 1)$

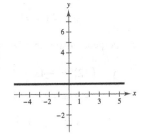

61. $(5, -2), (-1, 4)$

$m = \dfrac{4 - (-2)}{-1 - 5} = \dfrac{6}{-6} = -1$

63. $(6, -5), m = \frac{1}{3}$

$y - (-5) = \frac{1}{3}(x - 6)$

$y + 5 = \frac{1}{3}x - 2$

$y = \frac{1}{3}x - 7$

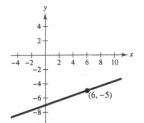

65. $(-6, 4)$, $(4, 9)$

$$m = \frac{9 - 4}{4 - (-6)} = \frac{5}{10} = \frac{1}{2}$$

$$y - 4 = \frac{1}{2}(x - (-6))$$

$$y - 4 = \frac{1}{2}(x + 6)$$

$$y - 4 = \frac{1}{2}x + 3$$

$$y = \frac{1}{2}x + 7$$

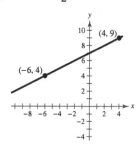

67. Point: $(3, -2)$

$$5x - 4y = 8$$

$$y = \frac{5}{4}x - 2$$

(a) Parallel slope: $m = \frac{5}{4}$

$$y - (-2) = \frac{5}{4}(x - 3)$$

$$y + 2 = \frac{5}{4}x - \frac{15}{4}$$

$$y = \frac{5}{4}x - \frac{23}{4}$$

(b) Perpendicular slope: $m = -\frac{4}{5}$

$$y - (-2) = -\frac{4}{5}(x - 3)$$

$$y + 2 = -\frac{4}{5}x + \frac{12}{5}$$

$$y = -\frac{4}{5}x + \frac{2}{5}$$

69. *Verbal Model:* Sale price $=$ (List price) $-$ (Discount)

Labels: Sale price $= S$

           List price $= L$

           Discount $= 20\%$ of $L = 0.2L$

Equation: $S = L - 0.2L$

            $S = 0.8L$

71. $16x - y^4 = 0$

$$y^4 = 16x$$

$$y = \pm 2\sqrt[4]{x}$$

No, y is not a function of x. Some x-values correspond to two y-values.

73. $y = \sqrt{1 - x}$

Yes, the equation represents y as a function of x. Each x-value, $x \leq 1$, corresponds to only one y-value.

75. $g(x) = x^{4/3}$

(a) $g(8) = 8^{4/3} = 2^4 = 16$

(b) $g(t + 1) = (t + 1)^{4/3}$

(c) $(-27)^{4/3} = (-3)^4 = 81$

(d) $g(-x) = (-x)^{4/3} = x^{4/3}$

77. $f(x) = \sqrt{25 - x^2}$

Domain: $25 - x^2 \geq 0$

              $(5 + x)(5 - x) \geq 0$

Critical numbers: $x = \pm 5$

Test intervals: $(-\infty, -5)$, $(-5, 5)$, $(5, \infty)$

Test: Is $25 - x^2 \geq 0$?

Solution set: $-5 \leq x \leq 5$

Domain: all real numbers x such that $-5 \leq x \leq 5$, or $[-5, 5]$

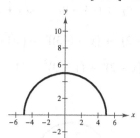

79. (a) $v(t) = -32t + 48$

      $v(1) = 16$ feet per second

(b) $0 = -32t + 48$

      $t = \frac{48}{32} = 1.5$ seconds

81. $y = (x - 3)^2$

A vertical line intersects the graph no more than once, so y *is* a function of x.

83. $f(x) = \sqrt{2x + 1}$

$$\sqrt{2x + 1} = 0$$

$$2x + 1 = 0$$

$$2x = -1$$

$$x = -\frac{1}{2}$$

85. $f(x) = |x| + |x + 1|$

f is increasing on $(0, \infty)$.

f is decreasing on $(-\infty, -1)$.

f is constant on $(-1, 0)$.

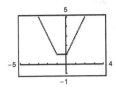

(b)

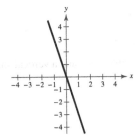

97. $g(x) = [\![x]\!] - 2$

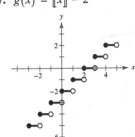

87. $f(x) = -x^2 + 2x + 1$

Relative maximum: $(1, 2)$

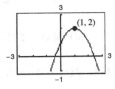

89. $f(x) = -x^2 + 8x - 4$

$$\frac{f(4) - f(0)}{4 - 0} = \frac{12 - (-4)}{4} = 4$$

The average rate of change of f from $x_1 = 0$ to $x_2 = 4$ is 4.

99. $f(x) = \begin{cases} 5x - 3, & x \geq -1 \\ -4x + 5, & x < -1 \end{cases}$

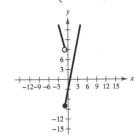

91. $f(x) = x^5 + 4x - 7$

$f(-x) = (-x)^5 + 4(-x) - 7$

$\quad = -x^5 - 4x - 7$

$\quad \neq f(x)$

$\quad \neq -f(x)$

The function is neither even nor odd, so the graph has no symmetry.

93. $f(x) = 2x\sqrt{x^2 + 3}$

$f(-x) = 2(-x)\sqrt{(-x)^2 + 3}$

$\quad = -2x\sqrt{x^2 + 3}$

$\quad = -f(x)$

The function is odd, so the graph has origin symmetry.

101. (a) $f(x) = x^2$

(b) $h(x) = x^2 - 9$

Vertical shift 9 units downward

(c)

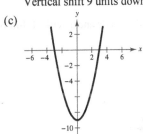

95. (a) $f(2) = -6, f(-1) = 3$

Points: $(2, -6), (-1, 3)$

$$m = \frac{3 - (-6)}{-1 - 2} = \frac{9}{-3} = -3$$

$y - (-6) = -3(x - 2)$

$y + 6 = -3x + 6$

$\quad\quad y = -3x$

$\quad f(x) = -3x$

(d) $h(x) = f(x) - 9$

103. (a) $f(x) = \sqrt{x}$

(b) $h(x) = -\sqrt{x} + 4$

Reflection in the x-axis and a vertical shift 4 units upward

(c)

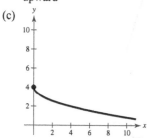

(d) $h(x) = -f(x) + 4$

105. (a) $f(x) = x^2$

(b) $h(x) = -(x + 2)^2 + 3$

Reflection in the x-axis, a horizontal shift 2 units to the left, and a vertical shift 3 units upward

(c)

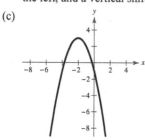

(d) $h(x) = -f(x + 2) + 3$

107. (a) $f(x) = [\![x]\!]$

(b) $h(x) = -[\![x]\!] + 6$

Reflection in the x-axis and a vertical shift 6 units upward

(c)

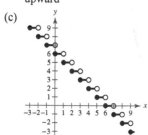

(d) $h(x) = -f(x) + 6$

109. (a) $f(x) = [\![x]\!]$

(b) $h(x) = 5[\![x - 9]\!]$

Horizontal shift 9 units to the right and a vertical stretch (each y-value is multiplied by 5)

(c)

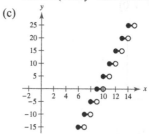

(d) $h(x) = 5f(x - 9)$

111. $f(x) = x^2 + 3,\ g(x) = 2x - 1$

(a) $(f + g)(x) = (x^2 + 3) + (2x - 1) = x^2 + 2x + 2$

(b) $(f - g)(x) = (x^2 + 3) - (2x - 1) = x^2 - 2x + 4$

(c) $(fg)(x) = (x^2 + 3)(2x - 1) = 2x^3 - x^2 + 6x - 3$

(d) $\left(\dfrac{f}{g}\right)(x) = \dfrac{x^2 + 3}{2x - 1},\ \text{Domain: } x \neq \dfrac{1}{2}$

113. $f(x) = \frac{1}{3}x - 3,\ g(x) = 3x + 1$

The domains of f and g are all real numbers.

(a) $(f \circ g)(x) = f(g(x))$

$= f(3x + 1)$

$= \frac{1}{3}(3x + 1) - 3$

$= x + \frac{1}{3} - 3$

$= x - \frac{8}{3}$

Domain: all real numbers

(b) $(g \circ f)(x) = g(f(x))$

$= g\left(\frac{1}{3}x - 3\right)$

$= 3\left(\frac{1}{3}x - 3\right) + 1$

$= x - 9 + 1$

$= x - 8$

Domain: all real numbers

In Exercise 115 use the following functions. $f(x) = x - 100,\ g(x) = 0.95x$

115. $(f \circ g)(x) = f(0.95x) = 0.95x - 100$ represents the sale price if first the 5% discount is applied and then the $100 rebate.

117. $f(x) = \dfrac{x-4}{5}$

$$y = \dfrac{x-4}{5}$$

$$x = \dfrac{y-4}{5}$$

$$5x = y - 4$$

$$y = 5x + 4$$

So, $f^{-1}(x) = 5x + 4$.

$$f\left(f^{-1}(x)\right) = f(5x+4) = \dfrac{5x+4-4}{5} = \dfrac{5x}{5} = x$$

$$f^{-1}\left(f(x)\right) = f^{-1}\left(\dfrac{x-4}{5}\right) = 5\left(\dfrac{x-4}{5}\right) + 4 = x - 4 + 4 = x$$

119. $f(x) = (x-1)^2$

No, the function does not have an inverse because the horizontal line test fails.

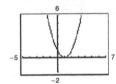

121. (a) $\quad f(x) = \frac{1}{2}x - 3$ (b)

$$y = \tfrac{1}{2}x - 3$$

$$x = \tfrac{1}{2}y - 3$$

$$x + 3 = \tfrac{1}{2}y$$

$$2(x+3) = y$$

$$f^{-1}(x) = 2x + 6$$

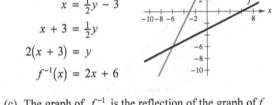

(c) The graph of f^{-1} is the reflection of the graph of f in the line $y = x$.

(d) The domains and ranges of f and f^{-1} are the set of all real numbers.

123. $f(x) = 2(x-4)^2$ is increasing on $(4, \infty)$.

Let $f(x) = 2(x-4)^2$, $x > 4$ and $y > 0$.

$$y = 2(x-4)^2$$

$$x = 2(y-4)^2, \ x > 0, \ y > 4$$

$$\dfrac{x}{2} = (y-4)^2$$

$$\sqrt{\dfrac{x}{2}} = y - 4$$

$$\sqrt{\dfrac{x}{2}} + 4 = y$$

$$f^{-1}(x) = \sqrt{\dfrac{x}{2}} + 4, \ x > 0$$

125. False. The graph is reflected in the x-axis, shifted 9 units to the left, then shifted 13 units downward.

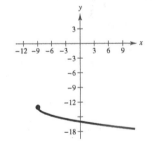

Problem Solving for Chapter P

1. (a) $W_1 = 0.07S + 2000$

(b) $W_2 = 0.05S + 2300$

(c)

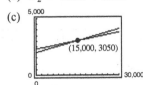

Point of intersection: $(15{,}000, 3050)$

Both jobs pay the same, \$3050, if you sell \$15,000 per month.

(d) No. If you think you can sell \$20,000 per month, keep your current job with the higher commission rate. For sales over \$15,000 it pays more than the other job.

3. (a) Let $f(x)$ and $g(x)$ be two even functions.

Then define $h(x) = f(x) \pm g(x)$.

$h(-x) = f(-x) \pm g(-x)$

$\quad = f(x) \pm g(x)$ because f and g are even

$\quad = h(x)$

So, $h(x)$ is also even.

(b) Let $f(x)$ and $g(x)$ be two odd functions.

Then define $h(x) = f(x) \pm g(x)$.

$h(-x) = f(-x) \pm g(-x)$

$\quad = -f(x) \pm g(x)$ because f and g are odd

$\quad = -h(x)$

So, $h(x)$ is also odd. $\left(\text{If } f(x) \neq g(x) \right)$

(c) Let $f(x)$ be odd and $g(x)$ be even. Then define $h(x) = f(x) \pm g(x)$.

$h(-x) = f(-x) \pm g(-x)$

$\quad = -f(x) \pm g(x)$ because f is odd and g is even

$\quad \neq h(x)$

$\quad \neq -h(x)$

So, $h(x)$ is neither odd nor even.

5. $f(x) = a_{2n}x^{2n} + a_{2n-2}x^{2n-2} + \cdots + a_2 x^2 + a_0$

$f(-x) = a_{2n}(-x)^{2n} + a_{2n-2}(-x)^{2n-2} + \cdots + a_2(-x)^2 + a_0 = a_{2n}x^{2n} + a_{2n-2}x^{2n-2} + \cdots + a_2 x^2 + a_0 = f(x)$

So, $f(x)$ is even.

7. (a) April 11: 10 hours

April 12: 24 hours

April 13: 24 hours

April 14: $23\frac{2}{3}$ hours

Total: $81\frac{2}{3}$ hours

(b) Speed $= \dfrac{\text{distance}}{\text{time}} = \dfrac{2100}{81\frac{2}{3}} = \dfrac{180}{7} = 25\frac{5}{7}$ mph

(c) $D = -\dfrac{180}{7}t + 3400$

Domain: $0 \leq t \leq \dfrac{1190}{9}$

Range: $0 \leq D \leq 3400$

(d)

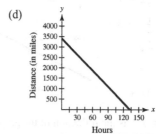

9. (a)–(d) Use $f(x) = 4x$ and $g(x) = x + 6$.

(a) $(f \circ g)(x) = f(x + 6) = 4(x + 6) = 4x + 24$

(b) $(f \circ g)^{-1}(x) = \dfrac{x - 24}{4} = \dfrac{1}{4}x - 6$

(c) $f^{-1}(x) = \dfrac{1}{4}x$

$g^{-1}(x) = x - 6$

(d) $(g^{-1} \circ f^{-1})(x) = g^{-1}\left(\dfrac{1}{4}x\right) = \dfrac{1}{4}x - 6$

(e) $f(x) = x^3 + 1$ and $g(x) = 2x$

$(f \circ g)(x) = f(2x) = (2x)^3 + 1 = 8x^3 + 1$

$(f \circ g)^{-1}(x) = \sqrt[3]{\dfrac{x - 1}{8}} = \dfrac{1}{2}\sqrt[3]{x - 1}$

$f^{-1}(x) = \sqrt[3]{x - 1}$

$g^{-1}(x) = \dfrac{1}{2}x$

$(g^{-1} \circ f^{-1})(x) = g^{-1}\left(\sqrt[3]{x - 1}\right) = \dfrac{1}{2}\sqrt[3]{x - 1}$

(f) Answers will vary.

(g) Conjecture: $(f \circ g)^{-1}(x) = (g^{-1} \circ f^{-1})(x)$

11. $H(x) = \begin{cases} 1, & x \geq 0 \\ 0, & x < 0 \end{cases}$

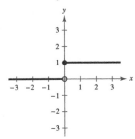

(a) $H(x) - 2$

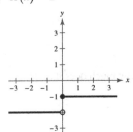

(b) $H(x - 2)$

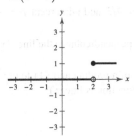

(c) $-H(x)$

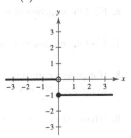

(d) $H(-x)$

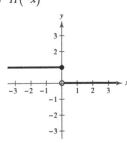

(e) $\frac{1}{2}H(x)$

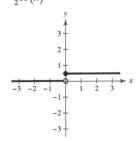

(f) $-H(x - 2) + 2$

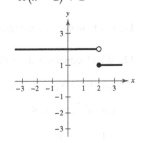

13. $\left(f \circ (g \circ h)\right)(x) = f\left((g \circ h)(x)\right) = f\left(g(h(x))\right) = (f \circ g \circ h)(x)$

$\left((f \circ g) \circ h\right)(x) = (f \circ g)(h(x)) = f\left(g(h(x))\right) = (f \circ g \circ h)(x)$

15.

x	$f(x)$	$f^{-1}(x)$
-4	—	2
-3	4	1
-2	1	0
-1	0	—
0	-2	-1
1	-3	-2
2	-4	—
3	—	—
4	—	-3

(a)

x	$f\left(f^{-1}(x)\right)$
-4	$f\left(f^{-1}(-4)\right) = f(2) = -4$
-2	$f\left(f^{-1}(-2)\right) = f(0) = -2$
0	$f\left(f^{-1}(0)\right) = f(-1) = 0$
4	$f\left(f^{-1}(4)\right) = f(-3) = 4$

(b)

x	$(f + f^{-1})(x)$
-3	$f(-3) + f^{-1}(-3) = 4 + 1 = 5$
-2	$f(-2) + f^{-1}(-2) = 1 + 0 = 1$
0	$f(0) + f^{-1}(0) = -2 + (-1) = -3$
1	$f(1) + f^{-1}(1) = -3 + (-2) = -5$

(c)

x	$(f \cdot f^{-1})(x)$
-3	$f(-3)f^{-1}(-3) = (4)(1) = 4$
-2	$f(-2)f^{-1}(-2) = (1)(0) = 0$
0	$f(0)f^{-1}(0) = (-2)(-1) = 2$
1	$f(1)f^{-1}(1) = (-3)(-2) = 6$

(d)

x	$\left|f^{-1}(x)\right|$
-4	$\left|f^{-1}(-4)\right| = \left|2\right| = 2$
-3	$\left|f^{-1}(-3)\right| = \left|1\right| = 1$
0	$\left|f^{-1}(0)\right| = \left|-1\right| = 1$
4	$\left|f^{-1}(4)\right| = \left|-3\right| = 3$

Practice Test for Chapter P

1. Solve for x: $5(x - 2) - 4 = 3x + 8$

2. Solve for x: $x^2 - 16x + 25 = 0$

3. Solve for x: $|3x + 5| = 8$

4. Write the standard equation of the circle with center $(-3, 5)$ and radius 6.

5. Find the distance between the points $(2, 4)$ and $(3, -1)$.

6. Find the equation of the line with slope $m = 4/3$ and y-intercept $b = -3$.

7. Find the equation of the line through $(4, 1)$ perpendicular to the line $2x + 3y = 0$.

8. If it costs a company \$32 to produce 5 units of a product and \$44 to produce 9 units, how much does it cost to produce 20 units? (Assume that the cost function is linear.)

9. Given $f(x) = x^2 - 2x + 1$, find $f(x - 3)$.

10. Given $f(x) = 4x - 11$, find $\dfrac{f(x) - f(3)}{x - 3}$.

11. Find the domain and range of $f(x) = \sqrt{36 - x^2}$.

12. Which equations determine y as a function of x?

    (a) $6x - 5y + 4 = 0$

    (b) $x^2 + y^2 = 9$

    (c) $y^3 = x^2 + 6$

13. Sketch the graph of $f(x) = x^2 - 5$.

14. Sketch the graph of $f(x) = |x + 3|$.

15. Sketch the graph of $f(x) = \begin{cases} 2x + 1, & \text{if } x \geq 0 \\ x^2 - x, & \text{if } x < 0 \end{cases}$.

16. Use the graph of $f(x) = |x|$ to graph the following:

    (a) $f(x + 2)$

    (b) $-f(x) + 2$

17. Given $f(x) = 3x + 7$ and $g(x) = 2x^2 - 5$, find the following:

    (a) $(g - f)(x)$

    (b) $(fg)(x)$

18. Given $f(x) = x^2 - 2x + 16$ and $g(x) = 2x + 3$, find $f(g(x))$.

19. Given $f(x) = x^3 + 7$, find $f^{-1}(x)$.

20. Which of the following functions have inverses?

   (a) $f(x) = |x - 6|$

   (b) $f(x) = ax + b, a \neq 0$

   (c) $f(x) = x^3 - 19$

21. Given $f(x) = \sqrt{\dfrac{3 - x}{x}}, 0 < x \leq 3$, find $f^{-1}(x)$.

Exercises 22–24, true or false?

22. $y = 3x + 7$ and $y = \frac{1}{3}x - 4$ are perpendicular.

23. $(f \circ g)^{-1} = g^{-1} \circ f^{-1}$

24. If a function has an inverse, then it must pass both the Vertical Line Test and the Horizontal Line Test.

25. Use your calculator to find the least square regression line for the data.

x	-2	-1	0	1	2	3
y	1	2.4	3	3.1	4	4.7

CHAPTER 1
Trigonometry

CHAPTER 1
Trigonometry

Section 1.1   Radian and Degree Measure

1. coterminal

3. complementary; supplementary

5. angular

7.

The angle shown is approximately 1 radian.

9.

The angle shown is approximately −3 radians.

11. (a) Because $0 < \dfrac{\pi}{4} < \dfrac{\pi}{2}, \dfrac{\pi}{4}$ lies in Quadrant I.

(b) Because $-\dfrac{5\pi}{4}$ is coterminal with $\dfrac{3\pi}{4}$ and

$\dfrac{\pi}{2} < \dfrac{3\pi}{4} < \pi, -\dfrac{5\pi}{4}$ lies in Quadrant II.

13. (a) $\dfrac{\pi}{3}$

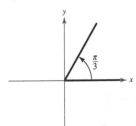

(b) $-\dfrac{2\pi}{3}$

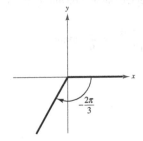

15. *Sample answers:*

(a) $\dfrac{\pi}{6} + 2\pi = \dfrac{13\pi}{6}$

$\dfrac{\pi}{6} - 2\pi = -\dfrac{11\pi}{6}$

(b) $-\dfrac{5\pi}{6} + 2\pi = \dfrac{7\pi}{6}$

$-\dfrac{5\pi}{6} - 2\pi = -\dfrac{17\pi}{6}$

17. (a) Complement: $\dfrac{\pi}{2} - \dfrac{\pi}{12} = \dfrac{5\pi}{12}$

Supplement: $\pi - \dfrac{\pi}{12} = \dfrac{11\pi}{12}$

(b) Complement: Not possible, $\dfrac{11\pi}{12}$ is greater than $\dfrac{\pi}{2}$.

Supplement: $\pi - \dfrac{11\pi}{12} = \dfrac{\pi}{12}$

19. (a) Complement: $\dfrac{\pi}{2} - 1 \approx 0.57$

Supplement: $\pi - 1 \approx 2.14$

(b) Complement: Not possible, 2 is greater than $\dfrac{\pi}{2}$.

Supplement: $\pi - 2 \approx 1.14$

21.

The angle shown is approximately 210°.

23.

The angle shown is approximately −60°.

25. (a) Because $90° < 130° < 180°, 130°$ lies in Quadrant II.

(b) Because $-8.3°$ is coterminal with $351.7°$ and $270° < 351.7° < 360°, -8.3°$ lies in Quadrant IV.

27. (a) $-270°$

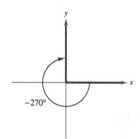

(b) $-120°$

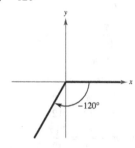

29. (a) Coterminal angles for $120°$

$120° + 360° = 480°$

$120° - 360° = -240°$

(b) Coterminal angles for $-210°$

$-210° + 360° = 150°$

$-210° - 360° = -570°$

31. (a) Complement: $90° - 18° = 72°$

Supplement: $180° - 18° = 162°$

(b) Complement: $90° - 85° = 5°$

Supplement: $180° - 85° = 95°$

33. (a) Complement: $90° - 24° = 66°$

Supplement: $180° - 24° = 156°$

(b) Complement: Not possible. $126°$ is greater than $90°$.

Supplement: $180° - 126° = 54°$

35. (a) $120° = 120°\left(\dfrac{\pi}{180°}\right) = -\dfrac{2\pi}{3}$

(b) $-20° = -20°\left(\dfrac{\pi}{180°}\right) = -\dfrac{\pi}{9}$

37. (a) $\dfrac{3\pi}{2} = \dfrac{3\pi}{2}\left(\dfrac{180°}{\pi}\right) = 270°$

(b) $-\dfrac{7\pi}{6} = -\dfrac{7\pi}{6}\left(\dfrac{180°}{\pi}\right) = -210°$

39. $45° = 45\left(\dfrac{\pi}{180°}\right) \approx 0.785$ radian

41. $-0.54° = -0.54°\left(\dfrac{\pi}{180°}\right) \approx -0.009$ radian

43. $\dfrac{5\pi}{11} = \dfrac{5\pi}{11}\left(\dfrac{180°}{\pi}\right) \approx 81.818°$

45. $-4.2\pi = -4.2\pi\left(\dfrac{180°}{\pi}\right) = -756°$

47. (a) $54°45' = 54° + \left(\dfrac{45}{60}\right)° = 54.75°$

(b) $-128°30' = -128° - \left(\dfrac{30}{60}\right)° = -128.5°$

49. (a) $240.6° = 240° + 0.6(60)' = 240°36'$

(b) $-145.8° = -\left[145° + 0.8(60')\right] = -145°48'$

51. $r = 15$ inches, $\theta = 120°$

$s = r\theta$

$s = 15(120°)\left(\dfrac{\pi}{180°}\right) = 10\pi$ inches

≈ 31.42 inches

53. $r = 80$ kilometers, $s = 150$ kilometers

$s = r\theta$

$150 = 80\theta$

$\theta = \dfrac{150}{80} = \dfrac{15}{8}$ radians

55. $s = r\theta$

$28 = 7\theta$

$\theta = 4$ radians

57. $r = 6$ inches, $\theta = \dfrac{\pi}{3}$

$A = \dfrac{1}{2}r^2\theta = \dfrac{1}{2}(6)^2\left(\dfrac{\pi}{3}\right) = 6\pi$ in.$^2 \approx 18.85$ in.^{2}

59. The angle in degrees should be multiplied by $\dfrac{\pi}{180°}$.

$20° = (20°)\left(\dfrac{\pi \text{ rad}}{180°}\right) = \dfrac{\pi}{9}$ radians.

61. $\theta = 41° 15' 50'' - 32° 47' 9'' \approx 8.47806° \approx 0.14797$ radian

$\quad s = r\theta \approx 4000(0.14782) \approx 592$ miles

63. $\theta = \dfrac{s}{r} = \dfrac{2.5}{6} = \dfrac{25}{60} = \dfrac{5}{12}$ radian $\approx 23.87°$

65. (a) $4 \text{ rpm} = 4(2\pi) \text{ radians/minute} = 8\pi \approx 25$ radians/minute

$\quad$ (b) $r = 25$ ft

$\quad\quad \dfrac{r\theta}{t} = 200\pi$ ft/minute

$\quad\quad$ Linear speed $\approx 25(25.13)$ ft/minute ≈ 628.3 ft/minute

67. (a) Road speed (linear speed) $= \dfrac{\left(\dfrac{25}{2} \text{ in.}\right)\left(\dfrac{1 \text{ ft}}{12 \text{ in.}}\right)\left(\dfrac{1 \text{ mi}}{5280 \text{ ft}}\right)(480)(2\pi)}{1 \text{ minute} \left(\dfrac{1 \text{ hour}}{60 \text{ minutes}}\right)} \approx 35.70$ mi/h

$\quad$ (b) $\dfrac{55 \text{ mi}}{1 \text{ h}} \times \dfrac{5280 \text{ ft}}{1 \text{ mi}} \times \dfrac{12 \text{ in.}}{1 \text{ ft}} \times \dfrac{1 \text{ h}}{60 \text{ min}} = \dfrac{58,080 \text{ in.}}{1 \text{ min}}$

$\quad\quad$ The circumference of the machine is $C = 2\pi\left(\dfrac{25}{2}\right) = 25\pi$ inches.

$\quad\quad$ The number of revolutions per minute is

$\quad\quad r = 58,080/25\pi \approx 739.50$ revolutions/min.

69. $A = \dfrac{1}{2}r^2\theta$

$\quad = \dfrac{1}{2}(15)^2(150°)\left(\dfrac{\pi}{180°}\right)$

$\quad = 93.75\pi$ m^2

$\quad \approx 294.52$ m^2

71. False. $\dfrac{180°}{\pi}$ is in degree measure.

73. True. If α and β are coterminal angles, then

$\quad \alpha = \beta + n(360°)$ or $\alpha = \beta + n(2\pi)$, where n is

$\quad$ an integer. The difference between α and β is

$\quad \alpha - \beta = n(360°)$, or $\alpha - \beta = n(2\pi)$ if expressed

$\quad$ in radians.

75. Since the arc length s is given by $s = r\theta$, if the central angle θ is fixed while the radius r increases, then s increases in proportion to r.

77. The speed increases. The linear speed is proportional to the radius.

79. Area of circle $= \pi r^2$

$\quad \dfrac{\text{Area of sector}}{\text{Area of circle}} = \dfrac{\text{Measure of central angle of sector}}{\text{Measure of central angle of circle}}$

$\quad \dfrac{\text{Area of sector}}{\pi r^2} = \dfrac{\theta}{2\pi}$

$\quad$ Area of sector $= \left(\pi r^2\right)\left(\dfrac{\theta}{2\pi}\right) = \dfrac{1}{2}r^2\theta$

Section 1.2   Trigonometric Functions: The Unit Circle

1. unit circle

3. period

5. $x = \dfrac{12}{13}, y = \dfrac{5}{13}$

$\sin t = y = \dfrac{5}{13}$ $\qquad \csc t = \dfrac{1}{y} = \dfrac{13}{5}$

$\cos t = x = \dfrac{12}{13}$ $\qquad \sec t = \dfrac{1}{x} = \dfrac{13}{12}$

$\tan t = \dfrac{y}{x} = \dfrac{5}{12}$ $\qquad \cot t = \dfrac{x}{y} = \dfrac{12}{5}$

7. $x = -\dfrac{4}{5}, y = -\dfrac{3}{5}$

$\sin t = y = -\dfrac{3}{5}$ $\qquad \csc t = \dfrac{1}{y} = -\dfrac{5}{3}$

$\cos t = x = -\dfrac{4}{5}$ $\qquad \sec t = \dfrac{1}{x} = -\dfrac{5}{4}$

$\tan t = \dfrac{y}{x} = \dfrac{3}{4}$ $\qquad \cot t = \dfrac{x}{y} = \dfrac{4}{3}$

9. $t = \dfrac{\pi}{2}$ corresponds to the point $(x, y) = (0, 1)$.

11. $t = \dfrac{5\pi}{6}$ corresponds to the point $(x, y) = \left(-\dfrac{\sqrt{3}}{2}, \dfrac{1}{2}\right)$.

13. $t = \dfrac{\pi}{4}$ corresponds to the point $(x, y) = \left(\dfrac{\sqrt{2}}{2}, \dfrac{\sqrt{2}}{2}\right)$.

$\sin \dfrac{\pi}{4} = y = \dfrac{\sqrt{2}}{2}$

$\cos \dfrac{\pi}{4} = x = \dfrac{\sqrt{2}}{2}$

$\tan \dfrac{\pi}{4} = \dfrac{y}{x} = 1$

15. $t = -\dfrac{\pi}{6}$ corresponds to $\left(\dfrac{\sqrt{3}}{2}, -\dfrac{1}{2}\right)$.

$\sin -\dfrac{\pi}{6} = y = -\dfrac{1}{2}$

$\cos -\dfrac{\pi}{6} = x = \dfrac{\sqrt{3}}{2}$

$\tan -\dfrac{\pi}{6} = \dfrac{y}{x} = -\dfrac{1}{\sqrt{3}} = -\dfrac{\sqrt{3}}{3}$

17. $t = -\dfrac{7\pi}{4}$ corresponds to the point

$(x, y) = \left(\dfrac{\sqrt{2}}{2}, \dfrac{\sqrt{2}}{2}\right)$.

$\sin\left(-\dfrac{7\pi}{4}\right) = y = \dfrac{\sqrt{2}}{2}$

$\cos\left(-\dfrac{7\pi}{4}\right) = x = \dfrac{\sqrt{2}}{2}$

$\tan\left(-\dfrac{7\pi}{4}\right) = \dfrac{y}{x} = 1$

19. $t = \dfrac{11\pi}{6}$ corresponds to the point $(x, y) = \left(\dfrac{\sqrt{3}}{2}, -\dfrac{1}{2}\right)$.

$\sin \dfrac{11\pi}{6} = y = -\dfrac{1}{2}$

$\cos \dfrac{11\pi}{6} = x = \dfrac{\sqrt{3}}{2}$

$\tan \dfrac{11\pi}{6} = \dfrac{y}{x} = -\dfrac{1}{\sqrt{3}} = -\dfrac{\sqrt{3}}{3}$

21. $t = -\dfrac{3\pi}{2}$ corresponds to the point $(x, y) = (0, 1)$.

$\sin\left(-\dfrac{3\pi}{2}\right) = y = 1$

$\cos\left(-\dfrac{3\pi}{2}\right) = x = 0$

$\tan\left(-\dfrac{3\pi}{2}\right) = \dfrac{y}{x}$ is undefined.

23. $t = \dfrac{2\pi}{3}$ corresponds to the point $(x, y) = \left(-\dfrac{1}{2}, \dfrac{\sqrt{3}}{2}\right)$.

$\sin \dfrac{2\pi}{3} = y = \dfrac{\sqrt{3}}{2}$ $\qquad \csc \dfrac{2\pi}{3} = \dfrac{1}{y} = \dfrac{2\sqrt{3}}{3}$

$\cos \dfrac{2\pi}{3} = x = -\dfrac{1}{2}$ $\qquad \sec \dfrac{2\pi}{3} = \dfrac{1}{x} = -2$

$\tan \dfrac{2\pi}{3} = \dfrac{y}{x} = \dfrac{\frac{\sqrt{3}}{2}}{-\frac{1}{2}} = -\sqrt{3}$ $\qquad \cot \dfrac{2\pi}{3} = \dfrac{x}{y} = \dfrac{-\frac{1}{2}}{\frac{\sqrt{3}}{2}} = -\dfrac{\sqrt{3}}{3}$

25. $t = \dfrac{4\pi}{3}$ corresponds to the point $(x, y) = \left(-\dfrac{1}{2}, -\dfrac{\sqrt{3}}{2}\right)$.

$$\sin \dfrac{4\pi}{3} = y = -\dfrac{\sqrt{3}}{2} \qquad \csc \dfrac{4\pi}{3} = \dfrac{1}{y} = -\dfrac{2\sqrt{3}}{3}$$

$$\cos \dfrac{4\pi}{3} = x = -\dfrac{1}{2} \qquad \sec \dfrac{4\pi}{3} = \dfrac{1}{x} = -2$$

$$\tan \dfrac{4\pi}{3} = \dfrac{y}{x} = \sqrt{3} \qquad \cot \dfrac{4\pi}{3} = \dfrac{x}{y} = \dfrac{\sqrt{3}}{3}$$

27. $t = -\dfrac{5\pi}{3}$ corresponds to the point $\left(\dfrac{1}{2}, \dfrac{\sqrt{3}}{2}\right)$.

$$\sin\left(-\dfrac{5\pi}{3}\right) = y = \dfrac{\sqrt{3}}{2} \qquad \csc\left(-\dfrac{5\pi}{3}\right) = \dfrac{1}{y} = \dfrac{2}{\sqrt{3}}$$

$$\cos\left(-\dfrac{5\pi}{3}\right) = x = \dfrac{1}{2} \qquad \sec\left(-\dfrac{5\pi}{3}\right) = \dfrac{1}{x} = 2$$

$$\tan\left(-\dfrac{5\pi}{3}\right) = \dfrac{y}{x} = \sqrt{3} \qquad \cot\left(-\dfrac{5\pi}{3}\right) = \dfrac{x}{y} = \dfrac{\sqrt{3}}{3}$$

29. $t = -\dfrac{\pi}{2}$ corresponds to the point $(x, y) = (0, -1)$.

$$\sin\left(-\dfrac{\pi}{2}\right) = y = -1 \qquad \csc\left(-\dfrac{\pi}{2}\right) = \dfrac{1}{y} = -1$$

$$\cos\left(-\dfrac{\pi}{2}\right) = x = 0 \qquad \sec\left(-\dfrac{\pi}{2}\right) = \dfrac{1}{x} \text{ is undefined.}$$

$$\tan\left(-\dfrac{\pi}{2}\right) = \dfrac{y}{x} \text{ is undefined.} \qquad \cot\left(-\dfrac{\pi}{2}\right) = \dfrac{x}{y} = 0$$

31. $\sin 4\pi = \sin 0 = 0$

33. $\cos \dfrac{7\pi}{3} = \cos \dfrac{\pi}{3} = \dfrac{1}{2}$

35. $\sin \dfrac{19\pi}{6} = \sin \dfrac{7\pi}{6} = -\dfrac{1}{2}$

37. $\sin t = \dfrac{1}{2}$

   (a) $\sin(-t) = -\sin t = -\dfrac{1}{2}$

   (b) $\csc(-t) = -\csc t = -2$

39. $\cos(-t) = -\dfrac{1}{5}$

   (a) $\cos t = \cos(-t) = -\dfrac{1}{5}$

   (b) $\sec(-t) = \dfrac{1}{\cos(-t)} = -5$

41. $\sin t = \dfrac{4}{5}$

   (a) $\sin(\pi - t) = \sin t = \dfrac{4}{5}$

   (b) $\sin(t + \pi) = -\sin t = -\dfrac{4}{5}$

43. $\sin 0.6 \approx 0.5646$

45. $\tan \dfrac{\pi}{8} \approx 0.4142$

47. $\sec 3.1 = \dfrac{1}{\cos 3.1} \approx -1.0009$

49. $y(t) = \dfrac{1}{2} \cos 6t$

   (a) $y(0) = \dfrac{1}{2} \cos 0 = 0.5$ foot

   (b) $y\left(\dfrac{1}{4}\right) = \dfrac{1}{2} \cos \dfrac{3}{2} \approx 0.04$ foot

   (c) $y\left(\dfrac{1}{2}\right) = \dfrac{1}{2} \cos 3 \approx -0.49$ foot

51. False. $\sin(-t) = -\sin t$ means the function is odd, not that the sine of a negative angle is a negative number.

For example: $\sin\left(-\dfrac{3\pi}{2}\right) = -\sin\left(\dfrac{3\pi}{2}\right) = -(-1) = 1$.

Even though the angle is negative, the sine value is positive.

53. True. $\tan a = \tan(a - 6\pi)$ because the period of the tangent function is π.

55. (a) The points have y-axis symmetry.

(b) $\sin t_1 = \sin(\pi - t_1)$ because they have the same y-value.

(c) $\cos(\pi - t_1) = -\cos t_1$ because the x-values have the opposite signs.

57. The calculator was in degree mode instead of radian mode. $\tan(\pi/2)$ is undefined.

59. (a)

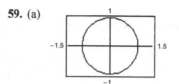

Circle of radius 1 centered at $(0, 0)$

(b) The t-values represent the central angle in radians. The x- and y-values represent the location in the coordinate plane.

(c) $-1 \le x \le 1, -1 \le y \le 1$

61. Let $h(t) = f(t)g(t) = \sin t \cos t$.

Then, $h(-t) = \sin(-t) \cos(-t)$

$$= -\sin t \cos t$$

$$= -h(t).$$

So, $h(t)$ is odd..

Section 1.3   Right Triangle Trigonometry

1. (a) $\dfrac{\text{opposite}}{\text{hypotenuse}} = \sin \theta$ (v)   (b) $\dfrac{\text{adjacent}}{\text{hypotenuse}} = \cos \theta$ (iv)   (c) $\dfrac{\text{opposite}}{\text{adjacent}} = \tan \theta$ (vi)

(d) $\dfrac{\text{hypotenuse}}{\text{opposite}} = \csc \theta$ (iii)   (e) $\dfrac{\text{hypotenuse}}{\text{adjacent}} = \sec \theta$ (i)   (f) $\dfrac{\text{adjacent}}{\text{opposite}} = \cot \theta$ (ii)

3. Complementary

5. $\text{hyp} = \sqrt{6^2 + 8^2} = \sqrt{36 + 64} = \sqrt{100} = 10$

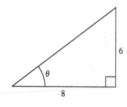

$\sin \theta = \dfrac{\text{opp}}{\text{hyp}} = \dfrac{6}{10} = \dfrac{3}{5}$ $\csc \theta = \dfrac{\text{hyp}}{\text{opp}} = \dfrac{10}{6} = \dfrac{5}{3}$

$\cos \theta = \dfrac{\text{adj}}{\text{hyp}} = \dfrac{8}{10} = \dfrac{4}{5}$ $\sec \theta = \dfrac{\text{hyp}}{\text{adj}} = \dfrac{10}{8} = \dfrac{5}{4}$

$\tan \theta = \dfrac{\text{opp}}{\text{adj}} = \dfrac{6}{8} = \dfrac{3}{4}$ $\cot \theta = \dfrac{\text{adj}}{\text{opp}} = \dfrac{8}{6} = \dfrac{4}{3}$

7. $\text{adj} = \sqrt{41^2 - 9^2} = \sqrt{1681 - 81} = \sqrt{1600} = 40$

$\sin \theta = \dfrac{\text{opp}}{\text{hyp}} = \dfrac{9}{41}$ $\csc \theta = \dfrac{\text{hyp}}{\text{opp}} = \dfrac{41}{9}$

$\cos \theta = \dfrac{\text{adj}}{\text{hyp}} = \dfrac{40}{41}$ $\sec \theta = \dfrac{\text{hyp}}{\text{adj}} = \dfrac{41}{40}$

$\tan \theta = \dfrac{\text{opp}}{\text{adj}} = \dfrac{9}{40}$ $\cot \theta = \dfrac{\text{adj}}{\text{opp}} = \dfrac{40}{9}$

9. hyp $= \sqrt{4^2 + 4^2} = \sqrt{32} = 4\sqrt{2}$

$\sin \theta = \dfrac{\text{opp}}{\text{hyp}} = \dfrac{4}{4\sqrt{2}} = \dfrac{1}{\sqrt{2}} = \dfrac{\sqrt{2}}{2}$ $\csc \theta = \dfrac{\text{hyp}}{\text{opp}} = \dfrac{4\sqrt{2}}{4} = \sqrt{2}$

$\cos \theta = \dfrac{\text{adj}}{\text{hyp}} = \dfrac{4}{4\sqrt{2}} = \dfrac{1}{\sqrt{2}} = \dfrac{\sqrt{2}}{2}$ $\sec \theta = \dfrac{\text{hyp}}{\text{adj}} = \dfrac{4\sqrt{2}}{4} = \sqrt{2}$

$\tan \theta = \dfrac{\text{opp}}{\text{adj}} = \dfrac{4}{4} = 1$ $\cot \theta = \dfrac{\text{adj}}{\text{opp}} = \dfrac{4}{4} = 1$

11.

hyp $= \sqrt{15^2 + 8^2} = \sqrt{289} = 17$

$\sin \theta = \dfrac{\text{opp}}{\text{hyp}} = \dfrac{8}{17}$ $\csc \theta = \dfrac{\text{hyp}}{\text{opp}} = \dfrac{17}{8}$

$\cos \theta = \dfrac{\text{adj}}{\text{hyp}} = \dfrac{15}{17}$ $\sec \theta = \dfrac{\text{hyp}}{\text{adj}} = \dfrac{17}{15}$

$\tan \theta = \dfrac{\text{opp}}{\text{adj}} = \dfrac{8}{15}$ $\cot \theta = \dfrac{\text{adj}}{\text{opp}} = \dfrac{15}{8}$

hyp $= \sqrt{7.5^2 + 4^2} = \dfrac{17}{2}$

$\sin \theta = \dfrac{\text{opp}}{\text{hyp}} = \dfrac{4}{(17/2)} = \dfrac{8}{17}$ $\csc \theta = \dfrac{\text{hyp}}{\text{opp}} = \dfrac{(17/2)}{4} = \dfrac{17}{8}$

$\cos \theta = \dfrac{\text{adj}}{\text{hyp}} = \dfrac{7.5}{(17/2)} = \dfrac{15}{17}$ $\sec \theta = \dfrac{\text{hyp}}{\text{adj}} = \dfrac{(17/2)}{7.5} = \dfrac{17}{15}$

$\tan \theta = \dfrac{\text{opp}}{\text{adj}} = \dfrac{4}{7.5} = \dfrac{8}{15}$ $\cot \theta = \dfrac{\text{adj}}{\text{opp}} = \dfrac{7.5}{4} = \dfrac{15}{8}$

The function values are the same because the triangles are similar, and corresponding sides are proportional.

13. adj $= \sqrt{3^2 - 1^2} = \sqrt{8} = 2\sqrt{2}$

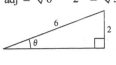

$\sin \theta = \dfrac{\text{opp}}{\text{hyp}} = \dfrac{1}{3}$ $\csc \theta = \dfrac{\text{hyp}}{\text{opp}} = 3$

$\cos \theta = \dfrac{\text{adj}}{\text{hyp}} = \dfrac{2\sqrt{2}}{3}$ $\sec \theta = \dfrac{\text{hyp}}{\text{adj}} = \dfrac{3}{2\sqrt{2}} = \dfrac{3\sqrt{2}}{4}$

$\tan \theta = \dfrac{\text{opp}}{\text{adj}} = \dfrac{1}{2\sqrt{2}} = \dfrac{\sqrt{2}}{4}$ $\cot \theta = \dfrac{\text{adj}}{\text{opp}} = 2\sqrt{2}$

adj $= \sqrt{6^2 - 2^2} = \sqrt{32} = 4\sqrt{2}$

$\sin \theta = \dfrac{\text{opp}}{\text{hyp}} = \dfrac{2}{6} = \dfrac{1}{3}$ $\csc \theta = \dfrac{\text{hyp}}{\text{opp}} = \dfrac{6}{2} = 3$

$\cos \theta = \dfrac{\text{adj}}{\text{hyp}} = \dfrac{4\sqrt{2}}{6} = \dfrac{2\sqrt{2}}{3}$ $\sec \theta = \dfrac{\text{hyp}}{\text{adj}} = \dfrac{6}{4\sqrt{2}} = \dfrac{3}{2\sqrt{2}} = \dfrac{3\sqrt{2}}{4}$

$\tan \theta = \dfrac{\text{opp}}{\text{adj}} = \dfrac{5\sqrt{2}}{4} = \dfrac{1}{2\sqrt{2}} = \dfrac{\sqrt{2}}{4}$ $\cot \theta = \dfrac{\text{adj}}{\text{opp}} = \dfrac{4\sqrt{2}}{2} = 2\sqrt{2}$

The function values are the same since the triangles are similar and the corresponding sides are proportional.

15. Given: $\cos\theta = \dfrac{15}{17} = \dfrac{\text{adj}}{\text{hyp}}$

$(\text{opp})^2 + 15^2 = 17^2$

$\qquad\qquad \text{opp} = \sqrt{289 - 225}$

$\qquad\qquad \text{opp} = \sqrt{64} = 8$

$\sin\theta = \dfrac{\text{opp}}{\text{hyp}} = \dfrac{8}{17}$

$\tan\theta = \dfrac{\text{opp}}{\text{adj}} = \dfrac{8}{15}$

$\csc\theta = \dfrac{\text{hyp}}{\text{opp}} = \dfrac{17}{8}$

$\sec\theta = \dfrac{\text{hyp}}{\text{adj}} = \dfrac{17}{15}$

$\cot\theta = \dfrac{\text{adj}}{\text{opp}} = \dfrac{15}{8}$

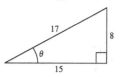

17. Given: $\sec\theta = \dfrac{6}{5} = \dfrac{\text{hyp}}{\text{adj}}$

$(\text{opp})^2 + 5^2 = 6^2$

$\qquad\qquad \text{opp} = \sqrt{36 - 25} = \sqrt{11}$

$\sin\theta = \dfrac{\text{opp}}{\text{hyp}} = \dfrac{\sqrt{11}}{6}$

$\cos\theta = \dfrac{\text{adj}}{\text{hyp}} = \dfrac{5}{6}$

$\tan\theta = \dfrac{\text{opp}}{\text{adj}} = \dfrac{\sqrt{11}}{5}$

$\csc\theta = \dfrac{\text{hyp}}{\text{opp}} = \dfrac{6\sqrt{11}}{11}$

$\cot\theta = \dfrac{\text{adj}}{\text{opp}} = \dfrac{5\sqrt{11}}{11}$

19. Given: $\sin\theta = \dfrac{1}{5} = \dfrac{\text{opp}}{\text{hyp}}$

$1^2 + (\text{adj})^2 = 5^2$

$\qquad\qquad \text{adj} = \sqrt{24} = 2\sqrt{6}$

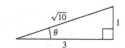

$\cos\theta = \dfrac{\text{adj}}{\text{hyp}} = \dfrac{2\sqrt{6}}{5}$

$\tan\theta = \dfrac{\text{opp}}{\text{adj}} = \dfrac{\sqrt{6}}{12}$

$\csc\theta = \dfrac{\text{hyp}}{\text{opp}} = 5$

$\sec\theta = \dfrac{\text{hyp}}{\text{adj}} = \dfrac{5\sqrt{6}}{12}$

$\cot\theta = \dfrac{\text{adj}}{\text{opp}} = 2\sqrt{6}$

21. Given: $\cot\theta = 3 = \dfrac{3}{1} = \dfrac{\text{adj}}{\text{opp}}$

$1^2 + 3^2 = (\text{hyp})^2$

$\qquad\qquad \text{hyp} = \sqrt{10}$

$\sin\theta = \dfrac{\text{opp}}{\text{hyp}} = \dfrac{\sqrt{10}}{10}$

$\cos\theta = \dfrac{\text{adj}}{\text{hyp}} = \dfrac{3\sqrt{10}}{10}$

$\tan\theta = \dfrac{\text{opp}}{\text{adj}} = \dfrac{1}{3}$

$\csc\theta = \dfrac{\text{hyp}}{\text{opp}} = \sqrt{10}$

$\sec\theta = \dfrac{\text{hyp}}{\text{adj}} = \dfrac{\sqrt{10}}{3}$

23.

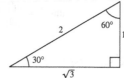

$30° = 30°\left(\dfrac{\pi}{180°}\right) = \dfrac{\pi}{6}$ radian

$\tan 30° = \dfrac{\text{opp}}{\text{adj}} = \dfrac{1}{\sqrt{3}} = \dfrac{\sqrt{3}}{3}$

25.

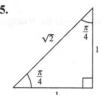

$\dfrac{\pi}{4} = \dfrac{\pi}{4}\left(\dfrac{180°}{\pi}\right) = 45°$

$\sin\dfrac{\pi}{4} = \dfrac{\text{opp}}{\text{hyp}} = \dfrac{1}{\sqrt{2}} = \dfrac{\sqrt{2}}{2}$

27.

$$\frac{\pi}{4} = \frac{\pi}{4}\left(\frac{180°}{\pi}\right) = 45°$$

$$\sec\frac{\pi}{4} = \frac{\text{hyp}}{\text{adj}} = \frac{\sqrt{2}}{1} = \sqrt{2}$$

29. (a) $\sin 20° \approx 0.3420$

  (b) $\cos 70° \approx 0.3420$

35. (a) $\cot 17° 15' = \cot\left(17 + \dfrac{15}{60}\right)° = \dfrac{1}{\tan 17.25°} \approx 3.2205$

  (b) $\tan 17° 15' = \tan\left(17 + \dfrac{15}{60}\right)° = \tan 17.25° \approx 0.3105$

37. $\sin 60° = \dfrac{\sqrt{3}}{2}$, $\cos 60° = \dfrac{1}{2}$

  (a) $\sin 30° = \cos 60° = \dfrac{1}{2}$

  (b) $\cos 30° = \sin 60° = \dfrac{\sqrt{3}}{2}$

  (c) $\tan 60° = \dfrac{\sin 60°}{\cos 60°} = \sqrt{3}$

  (d) $\cot 60° = \dfrac{\cos 60°}{\sin 60°} = \dfrac{1}{\sqrt{3}} = \dfrac{\sqrt{3}}{3}$

39. $\cos \theta = \dfrac{1}{3}$

  (a) $\sin^2 \theta + \cos^2 \theta = 1$

  $$\sin^2 \theta + \left(\frac{1}{3}\right)^2 = 1$$

  $$\sin^2 \theta = \frac{8}{9}$$

  $$\sin \theta = \frac{2\sqrt{2}}{3}$$

  (b) $\tan \theta = \dfrac{\sin \theta}{\cos \theta} = \dfrac{\frac{2\sqrt{2}}{3}}{\frac{1}{3}} = 2\sqrt{2}$

  (c) $\sec \theta = \dfrac{1}{\cos \theta} = 3$

  (d) $\csc\left(90° - \theta\right) = \sec \theta = 3$

31. (a) $\sin 14.21° \approx 0.2455$

  (b) $\csc 14.21° = \dfrac{1}{\sin 14.21°} \approx 4.0737$

33. (a) $\cos 4° 50' 15'' = \cos\left(4 + \dfrac{50}{60} + \dfrac{15}{3600}\right)° \approx 0.9964$

  (b) $\sec 4° 50' 15'' = \dfrac{1}{\cos 4° 50' 15''} \approx 1.0036$

41. $\cot \alpha = 3$

  (a) $\tan \alpha = \dfrac{1}{\cot \alpha} = \dfrac{1}{3}$

  (b) $\csc^2 \alpha = 1 + \cot^2 \alpha$

  $\csc^2 \alpha = 1 + 3^2$

  $\csc^2 \alpha = 10$

  $\csc \alpha = \sqrt{10}$

  (c) $\cot\left(90° - \alpha\right) = \tan \alpha = \dfrac{1}{3}$

  (d) $\csc \alpha = \sqrt{10}$

  $\sin \alpha = \dfrac{1}{\csc \alpha} = \dfrac{1}{\sqrt{10}} = \dfrac{\sqrt{10}}{10}$

43. $\tan \theta \cot \theta = \tan \theta\left(\dfrac{1}{\tan \theta}\right) = 1$

45. $\tan \alpha \cos \alpha = \left(\dfrac{\sin \alpha}{\cos \alpha}\right)\cos \alpha = \sin \alpha$

47. $(1 + \sin \theta)(1 - \sin \theta) = 1 - \sin^2 \theta = \cos^2 \theta$

49. $(\sec \theta + \tan \theta)(\sec \theta - \tan \theta) = \sec^2 \theta - \tan^2 \theta$
$$= \left(1 + \tan^2 \theta\right) - \tan^2 \theta$$
$$= 1$$

51. $\dfrac{\sin \theta}{\cos \theta} + \dfrac{\cos \theta}{\sin \theta} = \dfrac{\sin^2 \theta + \cos^2 \theta}{\sin \theta \cos \theta}$

$$= \dfrac{1}{\sin \theta \cos \theta}$$

$$= \dfrac{1}{\sin \theta} \cdot \dfrac{1}{\cos \theta}$$

$$= \csc \theta \sec \theta$$

53. (a) $\sin \theta = \dfrac{1}{2} \Rightarrow \theta = 30° = \dfrac{\pi}{6}$

   (b) $\csc \theta = 2 \Rightarrow \theta = 30° = \dfrac{\pi}{6}$

55. (a) $\sec \theta = 2 \Rightarrow \theta = 60° = \dfrac{\pi}{3}$

   (b) $\cot \theta = 1 \Rightarrow \theta = 45° = \dfrac{\pi}{4}$

57. (a) $\csc \theta = \dfrac{2\sqrt{3}}{3} \Rightarrow \theta = 60° = \dfrac{\pi}{3}$

   (b) $\sin \theta = \dfrac{\sqrt{2}}{2} \Rightarrow \theta = 45° = \dfrac{\pi}{4}$

59. $\cos 60° = \dfrac{x}{18}$

   $x = 18 \cos 60° = 18\left(\dfrac{1}{2}\right) = 9$

   $\sin 60° = \dfrac{y}{18}$

   $y = 18 \sin 60° = 18\dfrac{\sqrt{3}}{2} = 9\sqrt{3}$

61. $\tan 60° = \dfrac{32}{x}$

   $\sqrt{3} = \dfrac{32}{x}$

   $\sqrt{3}x = 32$

   $x = \dfrac{32}{\sqrt{3}} = \dfrac{32\sqrt{3}}{3}$

   $\sin 60° = \dfrac{32}{r}$

   $r = \dfrac{32}{\sin 60°}$

   $r = \dfrac{32}{\dfrac{\sqrt{3}}{2}} = \dfrac{64\sqrt{3}}{3}$

63.

$\tan 82° = \dfrac{x}{45}$

$x = 45 \tan 82°$

Height of the building:

$123 + 45 \tan 82° \approx 443.2$ meters

Distance between friends:

$\cos 82° = \dfrac{45}{y} \Rightarrow y = \dfrac{45}{\cos 82°}$

≈ 323.34 meters

65. $\sin \theta = \dfrac{1250}{2500} = \dfrac{1}{2}$

   $\theta = 30° = \dfrac{\pi}{6}$

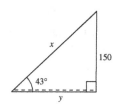

67. (a) $\sin 43° = \dfrac{150}{x}$

   $x = \dfrac{150}{\sin 43°} \approx 219.9$ ft

   (b) $\tan 43° = \dfrac{150}{y}$

   $y = \dfrac{150}{\tan 43°} \approx 160.9$ ft

69.

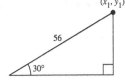

$$\sin 30° = \frac{y_1}{56}$$

$$y_1 = (\sin 30°)(56) = \left(\frac{1}{2}\right)(56) = 28$$

$$\cos 30° = \frac{x_1}{56}$$

$$x_1 = \cos 30°(56) = \frac{\sqrt{3}}{2}(56) = 28\sqrt{3}$$

$$(x_1, y_1) = \left(28\sqrt{3}, 28\right)$$

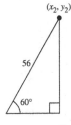

$$\sin 60° = \frac{y_2}{56}$$

$$y_2 = \sin 60°(56) = \left(\frac{\sqrt{3}}{2}\right)(56) = 28\sqrt{3}$$

$$\cos 60° = \frac{x_2}{56}$$

$$x_2 = (\cos 60°)(56) = \left(\frac{1}{2}\right)(56) = 28$$

$$(x_2, y_2) = \left(28, 28\sqrt{3}\right)$$

71. $x \approx 9.397$, $y \approx 3.420$

$$\sin 20° = \frac{y}{10} \approx 0.34$$

$$\cos 20° = \frac{x}{10} \approx 0.94$$

$$\tan 20° = \frac{y}{x} \approx 0.36$$

$$\cot 20° = \frac{x}{y} \approx 2.75$$

$$\sec 20° = \frac{10}{x} \approx 1.06$$

$$\csc 20° = \frac{10}{y} \approx 2.92$$

73. (a)

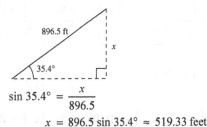

$$\sin 35.4° = \frac{x}{896.5}$$

$$x = 896.5 \sin 35.4° \approx 519.33 \text{ feet}$$

(b) Because the top of the incline is 1693.5 feet above sea level and the vertical rise of the inclined plane is 519.33 feet, the elevation of the lower end of the inclined plan is about

1693.5 − 519.33 = 1174.17 feet.

(c) Ascent time: $d = rt$

$$896.5 = 300t$$

$$3 \approx t$$

It takes about 3 minutes for the cars to get from the bottom to the top.

Vertical rate: $d = rt$

$$519.33 \approx r(3)$$

$$r \approx 173.11 \text{ ft/min}$$

75. $\sin 60° \csc 60° = 1$

True.

$$\csc x = \frac{1}{\sin x} \Rightarrow \sin 60° \csc 60° = \sin 60°\left(\frac{1}{\sin 60°}\right)$$

$$= 1$$

77. False, $\dfrac{\sqrt{2}}{2} + \dfrac{\sqrt{2}}{2} = \sqrt{2} \neq 1$

79. False, $\dfrac{\sin 60°}{\sin 30°} = \dfrac{\cos 30°}{\sin 30°} = \cot 30° \approx 1.7321$

$$\sin 2° \approx 0.0349$$

81. Yes. Given $\tan \theta$, $\sec \theta$ can be found from the identity $1 + \tan^2 \theta = \sec^2 \theta$.

83.

θ	0.1	0.2	0.3	0.4	0.5
$\sin \theta$	0.0998	0.1987	0.2955	0.3894	0.4794

(a) In the interval $(0, 0.5]$, $\theta > \sin \theta$.

(b) As $\theta \to 0$, $\sin \theta \to 0$, and $\dfrac{\theta}{\sin \theta} \to 1$.

Section 1.4   Trigonometric Functions of Any Angle

1. $\dfrac{y}{r}$

3. $\dfrac{y}{x}$

5. $\cos \theta$

7. zero; defined

9. (a) $(x, y) = (4, 3)$

$r = \sqrt{16 + 9} = 5$

$$\sin \theta = \frac{y}{r} = \frac{3}{5} \qquad \csc \theta = \frac{r}{y} = \frac{5}{3}$$

$$\cos \theta = \frac{x}{r} = \frac{4}{5} \qquad \sec \theta = \frac{r}{x} = \frac{5}{4}$$

$$\tan \theta = \frac{y}{x} = \frac{3}{4} \qquad \cot \theta = \frac{x}{y} = \frac{4}{3}$$

(b) $(x, y) = (-8, 15)$

$r = \sqrt{64 + 225} = 17$

$$\sin \theta = \frac{y}{r} = \frac{15}{17} \qquad \csc \theta = \frac{r}{y} = \frac{17}{15}$$

$$\cos \theta = \frac{x}{r} = -\frac{8}{17} \qquad \sec \theta = \frac{r}{x} = -\frac{17}{8}$$

$$\tan \theta = \frac{y}{x} = -\frac{15}{8} \qquad \cot \theta = \frac{x}{y} = -\frac{8}{15}$$

11. (a) $(x, y) = \left(-\sqrt{3}, -1\right)$

$r = \sqrt{3 + 1} = 2$

$$\sin \theta = \frac{y}{r} = -\frac{1}{2} \qquad \csc \theta = \frac{r}{y} = -2$$

$$\cos \theta = \frac{x}{r} = -\frac{\sqrt{3}}{2} \qquad \sec \theta = \frac{r}{x} = -\frac{2\sqrt{3}}{3}$$

$$\tan \theta = \frac{y}{x} = \frac{\sqrt{3}}{3} \qquad \cot \theta = \frac{x}{y} = \sqrt{3}$$

(b) $(x, y) = (4, -1)$

$r = \sqrt{16 + 1} = \sqrt{17}$

$$\sin \theta = \frac{y}{r} = -\frac{1}{\sqrt{17}} = -\frac{\sqrt{17}}{17} \qquad \csc \theta = \frac{r}{y} = -\sqrt{17}$$

$$\cos \theta = \frac{x}{r} = \frac{4}{\sqrt{17}} = \frac{4\sqrt{17}}{17} \qquad \sec \theta = \frac{r}{x} = \frac{\sqrt{17}}{4}$$

$$\tan \theta = \frac{y}{x} = -\frac{1}{4} \qquad \cot \theta = \frac{x}{y} = -4$$

13. $(x, y) = (5, 12)$

$r = \sqrt{25 + 144} = 13$

$$\sin \theta = \frac{y}{r} = \frac{12}{13} \qquad \csc \theta = \frac{r}{y} = \frac{13}{12}$$

$$\cos \theta = \frac{x}{r} = \frac{5}{13} \qquad \sec \theta = \frac{r}{x} = \frac{13}{5}$$

$$\tan \theta = \frac{y}{x} = \frac{12}{5} \qquad \cot \theta = \frac{x}{y} = \frac{5}{12}$$

15. $x = -5$, $y = -2$

$r = \sqrt{(-5)^2 + (-2)^2} = \sqrt{29}$

$$\sin \theta = \frac{y}{r} = \frac{-2}{\sqrt{29}} = -\frac{2\sqrt{29}}{29}$$

$$\cos \theta = \frac{x}{r} = \frac{-5}{\sqrt{29}} = -\frac{5\sqrt{29}}{29}$$

$$\tan \theta = \frac{y}{x} = \frac{-2}{-5} = \frac{2}{5}$$

$$\csc \theta = \frac{r}{y} = \frac{\sqrt{29}}{-2} = -\frac{\sqrt{29}}{2}$$

$$\sec \theta = \frac{r}{x} = \frac{\sqrt{29}}{-5} = -\frac{\sqrt{29}}{5}$$

$$\cot \theta = \frac{x}{y} = \frac{-5}{-2} = \frac{5}{2}$$

17. $(x, y) = (-5.4, 7.2)$

$r = \sqrt{29.16 + 51.84} = 9$

$\sin \theta = \dfrac{y}{r} = \dfrac{7.2}{9} = \dfrac{4}{5}$ $\csc \theta = \dfrac{r}{y} = \dfrac{9}{7.2} = \dfrac{5}{4}$

$\cos \theta = \dfrac{x}{r} = -\dfrac{5.4}{9} = -\dfrac{3}{5}$ $\sec \theta = \dfrac{r}{x} = -\dfrac{9}{5.4} = -\dfrac{5}{3}$

$\tan \theta = \dfrac{y}{x} = -\dfrac{7.2}{5.4} = -\dfrac{4}{3}$ $\tan \theta = \dfrac{x}{y} = -\dfrac{5.4}{7.2} = -\dfrac{3}{4}$

19. $\sin \theta > 0 \Rightarrow \theta$ lies in Quadrant I or in Quadrant II.

$\cos \theta > 0 \Rightarrow \theta$ lies in Quadrant I or in Quadrant IV.

$\sin \theta > 0$ and $\cos \theta > 0 \Rightarrow \theta$ lies in Quadrant I.

21. $\sin \theta > 0 \Rightarrow \theta$ lies in Quadrant I or in Quadrant II.

$\cos \theta < 0 \Rightarrow \theta$ lies in Quadrant II or in Quadrant III.

$\sin \theta > 0$ and $\cos \theta < 0 \Rightarrow \theta$ lies in Quadrant II.

23. $\tan \theta > 0$ and $\sin \theta > 0 \Rightarrow \theta$ is in

Quadrant I $\Rightarrow x > 0$ and $y > 0$.

$\tan \theta = \dfrac{y}{x} = \dfrac{15}{8} \Rightarrow r = 17$

$\sin \theta = \dfrac{y}{r} = \dfrac{15}{7}$ $\sec \theta = \dfrac{r}{x} = \dfrac{17}{8}$

$\cos \theta = \dfrac{x}{r} = \dfrac{8}{17}$ $\cot \theta = \dfrac{x}{y} = \dfrac{8}{15}$

$\csc \theta = \dfrac{r}{y} = \dfrac{17}{15}$

25. $\sin \theta = 0.6 = \dfrac{3}{5}$ and θ in Quadrant II

$x^2 + y^2 = r^2$

$\sin \theta = \dfrac{3}{5},$ $x = -\sqrt{r^2 - y^2}$

$x = -\sqrt{5^2 - 3^2} = -4$

$\cos \theta = \dfrac{x}{r} = -\dfrac{4}{5}$ $\sec \theta = \dfrac{r}{x} = -\dfrac{5}{4}$

$\tan \theta = \dfrac{y}{x} = -\dfrac{3}{4}$ $\cot \theta = \dfrac{x}{y} = -\dfrac{4}{3}$

$\csc \theta = \dfrac{r}{y} = \dfrac{5}{3}$

27. $\cot \theta = \dfrac{x}{y} = -\dfrac{3}{1} = \dfrac{3}{-1}$

$\cos \theta > 0 \Rightarrow \theta$ is in Quadrant IV $\Rightarrow x$ is positive;

$x = 3, y = -1, r = \sqrt{10}$

$\sin \theta = \dfrac{y}{r} = -\dfrac{\sqrt{10}}{10}$ $\csc \theta = \dfrac{r}{y} = -\sqrt{10}$

$\cos \theta = \dfrac{x}{r} = \dfrac{3\sqrt{10}}{10}$ $\sec \theta = \dfrac{r}{x} = \dfrac{\sqrt{10}}{3}$

$\tan \theta = \dfrac{y}{x} = -\dfrac{1}{3}$

29. $\cos \theta = 0 \Rightarrow \theta = \dfrac{\pi}{2} + \pi n$

$\csc \theta = 1 \Rightarrow \theta = \dfrac{\pi}{2} + 2\pi n$

$y = 1, x = 0, r = 1$

$\sin \theta = \dfrac{y}{r} = 1$

$\tan \theta = \dfrac{y}{x}$ is undefined

$\sec \theta$ is undefined

$\cot \theta = 0.$

31. $\cot \theta$ is undefined,

$\dfrac{\pi}{2} \le \theta \le \dfrac{3\pi}{2} \Rightarrow y = 0 \Rightarrow \theta = \pi$

$\sin \theta = 0$ $\csc \theta$ is undefined.

$\cos \theta = -1$ $\sec \theta = -1$

$\tan \theta = 0$ $\cot \theta$ is undefined.

33. To find a point on the terminal side of θ, use any point on the line $y = -x$ that lies in Quadrant II. $(-1, 1)$ is one such point.

$x = -1, y = 1, r = \sqrt{2}$

$\sin \theta = \dfrac{1}{\sqrt{2}} = \dfrac{\sqrt{2}}{2}$ $\csc \theta = \sqrt{2}$

$\cos \theta = -\dfrac{1}{\sqrt{2}} = -\dfrac{\sqrt{2}}{2}$ $\sec \theta = -\sqrt{2}$

$\tan \theta = -1$ $\cot \theta = -1$

35. To find a point on the terminal side of θ, use any point on the line $y = 2x$ that lies in Quadrant I. $(1, 2)$ is one such point.

$$x = 1,\ y = 2,\ y = \sqrt{5}$$

$$\sin \theta = \frac{2}{\sqrt{5}} = \frac{2\sqrt{5}}{5} \qquad\qquad \csc \theta = \frac{\sqrt{5}}{2} = \frac{\sqrt{5}}{2}$$

$$\cos \theta = \frac{1}{\sqrt{5}} = \frac{\sqrt{5}}{5} \qquad\qquad \sec \theta = \frac{\sqrt{5}}{1} = \sqrt{5}$$

$$\tan \theta = \frac{2}{1} = 2 \qquad\qquad \cot \theta = \frac{1}{2}$$

37. $(x, y) = (-1, 0),\ r = 1$

$$\sin \pi = \frac{y}{r} = \frac{0}{1} = 0$$

39. $(x, y) = (0, -1),\ r = 1$

$$\sec \frac{3\pi}{2} = \frac{r}{x} = \frac{1}{0} \Rightarrow \text{undefined}$$

41. $(x, y) = (0, 1),\ r = 1$

$$\sin \frac{\pi}{2} = \frac{y}{r} = \frac{1}{1} = 1$$

43. $(x, y) = (-1, 0),\ r = 1$

$$\csc \pi = \frac{r}{y} = \frac{1}{0} \Rightarrow \text{undefined}$$

45. $(x, y) = (0, 1)$

$$\cot \frac{9\pi}{2} = \frac{x}{y} = \frac{0}{1} = 0$$

47. $\theta = 160°$

$$\theta' = 180° - 160° = 20°$$

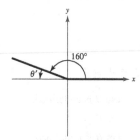

49. $\theta = -125°$

$$360° - 125° = 235° \text{ (coterminal angle)}$$

$$\theta' = 235° - 180° = 55°$$

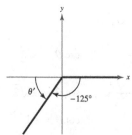

51. $\theta = \dfrac{2\pi}{3}$

$$\theta' = \pi - \frac{2\pi}{3} = \frac{\pi}{3}$$

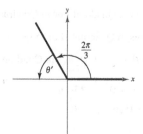

53. $\theta = 4.8$

$$\theta' = 2\pi - 4.8 \approx 1.48$$

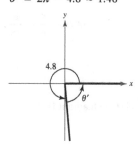

55. $\theta = 225°,\ \theta' = 45°,\ $ Quadrant III

$$\sin 225° = -\sin 45° = -\frac{\sqrt{2}}{2}$$

$$\cos 225° = -\cos 45° = -\frac{\sqrt{2}}{2}$$

$$\tan 225° = \tan 45° = 1$$

57. $\theta = 750°,\ \theta' = 30°,\ $ Quadrant I

$$\sin 750° = \sin 30° = \frac{1}{2}$$

$$\cos 750° = \cos 30° = \frac{\sqrt{3}}{2}$$

$$\tan 750° = \tan 30° = \frac{\sqrt{3}}{3}$$

59. $\theta = -120°$, $\theta' = 60°$, Quadrant III

$$\sin(-120°) = -\sin 60° = -\frac{\sqrt{3}}{2}$$

$$\cos(-120°) = -\cos 60° = -\frac{1}{2}$$

$$\tan(-120°) = \tan 60° = \sqrt{3}$$

61. $\theta = \frac{2\pi}{3}$, $\theta' = \frac{\pi}{3}$ in Quadrant II

$$\sin \frac{2\pi}{3} = \sin \frac{\pi}{3} = \frac{\sqrt{3}}{2}$$

$$\cos \frac{2\pi}{3} = -\cos \frac{\pi}{3} = -\frac{1}{2}$$

$$\tan \frac{2\pi}{3} = -\tan \frac{\pi}{3} = -\sqrt{3}$$

63. $\theta = -\frac{\pi}{6}$, $\theta' = \frac{\pi}{6}$, Quadrant IV

$$\sin\left(-\frac{\pi}{6}\right) = -\sin \frac{\pi}{6} = -\frac{1}{2}$$

$$\cos\left(-\frac{\pi}{6}\right) = \cos \frac{\pi}{6} = \frac{\sqrt{3}}{2}$$

$$\tan\left(-\frac{\pi}{6}\right) = -\tan \frac{\pi}{6} = -\frac{\sqrt{3}}{3}$$

65. $\theta = \frac{11\pi}{4}$, $\theta' = \frac{\pi}{4}$, Quadrant II

$$\sin \frac{11\pi}{4} = \sin \frac{\pi}{4} = \frac{\sqrt{2}}{2}$$

$$\cos \frac{11\pi}{4} = -\cos \frac{\pi}{4} = -\frac{\sqrt{2}}{2}$$

$$\tan \frac{11\pi}{4} = -\tan \frac{\pi}{4} = -1$$

67. $\theta = -\frac{17\pi}{6}$, $\theta' = \frac{\pi}{6}$ in Quadrant III

$$\sin\left(-\frac{17\pi}{6}\right) = -\sin \frac{\pi}{6} = -\frac{1}{2}$$

$$\cos\left(-\frac{17\pi}{6}\right) = -\cos \frac{\pi}{6} = -\frac{\sqrt{3}}{2}$$

$$\tan\left(-\frac{17\pi}{6}\right) = \tan \frac{\pi}{6} = \frac{\sqrt{3}}{3}$$

69.
$$\sin \theta = -\frac{3}{5}$$

$$\sin^2 \theta + \cos^2 \theta = 1$$

$$\cos^2 \theta = 1 - \sin^2 \theta$$

$$\cos^2 \theta = 1 - \left(-\frac{3}{5}\right)^2$$

$$\cos^2 \theta = 1 - \frac{9}{25}$$

$$\cos^2 \theta = \frac{16}{25}$$

$\cos \theta > 0$ in Quadrant IV.

$$\cos \theta = \frac{4}{5}$$

71. $\tan \theta = \frac{3}{2}$

$$\sec^2 \theta = 1 + \tan^2 \theta$$

$$\sec^2 \theta = 1 + \left(\frac{3}{2}\right)^2$$

$$\sec^2 \theta = 1 + \frac{9}{4}$$

$$\sec^2 \theta = \frac{13}{4}$$

$\sec \theta < 0$ in Quadrant III.

$$\sec \theta = -\frac{\sqrt{13}}{2}$$

73.
$$\cos \theta = \frac{5}{8}$$

$$\sin^2 \theta + \cos^2 \theta = 1$$

$$\sin^2 \theta + \left(\frac{5}{8}\right)^2 = 1$$

$$\sin^2 \theta = 1 - \frac{25}{64}$$

$$\sin^2 \theta = \frac{39}{64}$$

$\sin \theta > 0$ in Quadrant I.

$$\sin \theta = \frac{\sqrt{39}}{8}$$

$$\csc \theta = \frac{1}{\sin \theta}$$

$$\csc \theta = \frac{8\sqrt{39}}{39}$$

75. $\sin 10° \approx 0.1736$

77. $\cos(-110°) \approx -0.3420$

79. $\cot 178° \approx -28.6363$

81. $\csc 405° = \dfrac{1}{\sin 405°} \approx 1.4142$

83. $\tan\left(\dfrac{\pi}{9}\right) \approx 0.3640$

85. $\sec\dfrac{11\pi}{8} = \dfrac{1}{\cos\dfrac{11\pi}{8}} \approx -2.6131$

87. $\sin(-0.65) \approx -0.6052$

89. $\csc(-10) = \dfrac{1}{\sin(-10)} \approx 1.8382$

91. (a) $\sin\theta = \dfrac{1}{2} \Rightarrow$ reference angle is $30°$ or $\dfrac{\pi}{6}$ and θ is in Quadrant I or Quadrant II.

Values in degrees: $30°, 150°$

Values in radian: $\dfrac{\pi}{6}, \dfrac{5\pi}{6}$

(b) $\sin\theta = \dfrac{1}{2} \Rightarrow$ reference angle is $30°$ or $\dfrac{\pi}{6}$ and θ is in Quadrant III or Quadrant IV.

Values in degrees: $210°, 330°$

Values in radians: $\dfrac{7\pi}{6}, \dfrac{11\pi}{6}$

93. (a) $\cos\theta = \dfrac{1}{2} \Rightarrow$ reference angle is $60°$ or $\dfrac{\pi}{3}$ and θ is in Quadrant I or IV.

Values in degrees: $60°, 300°$

Values in radians: $\dfrac{\pi}{3}, \dfrac{5\pi}{3}$

(b) $\sec\theta = 2 \Rightarrow \cos\theta = \dfrac{1}{2} \Rightarrow$ reference angle is $60°$ or $\dfrac{\pi}{3}$ and θ is in Quadrant I or IV.

Values in degrees: $60°, 300°$

Values in radians: $\dfrac{\pi}{3}, \dfrac{5\pi}{3}$

95. (a) $\tan\theta = 1 \Rightarrow$ reference angle is $45°$ or $\dfrac{\pi}{4}$ and θ is in Quadrant I or Quadrant III.

Values in degrees: $45°, 225°$

Values in radians: $\dfrac{\pi}{4}, \dfrac{5\pi}{4}$

(b) $\cot\theta = -\sqrt{3} \Rightarrow$ reference angle is $30°$ or $\dfrac{\pi}{6}$ and θ is in Quadrant II or Quadrant IV.

Values in degrees: $150°, 330°$

Values in radians: $\dfrac{5\pi}{6}, \dfrac{11\pi}{6}$

97. $\sin\theta = \dfrac{6}{d} \Rightarrow d = \dfrac{6}{\sin\theta}$

(a) $\theta = 30°$

$d = \dfrac{6}{\sin 30°}$

$= \dfrac{6}{1/2} = 12$ miles

(b) $\theta = 90°$

$d = \dfrac{6}{\sin 90°} = \dfrac{6}{1} = 6$ miles

(c) $\theta = 120°$

$d = \dfrac{6}{\sin 120°}$

$= \dfrac{6}{\sqrt{3}/2} \approx 6.9$ miles

99. (a) Boston: $B \approx 24.593 \sin(0.495t - 2.262) + 57.387$

Fairbanks:

$F \approx 39.071 \sin(0.448t - 1.366) + 32.204$

(b)

Month	Boston, B	Fairbanks, F
February	33.9°	14.5°
April	50.5°	48.3°
May	62.6°	62.2°
July	80.3°	70.5°
September	77.4°	50.1°
October	68.2°	33.3°
December	44.8°	2.4°

(c)

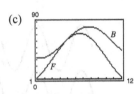

Answers will vary.

101. $d = 312.5 \sin 2\theta$

(a) $d = 312.5 \sin\left(2(30°)\right) \approx 270.63$ feet

(b) $d = 312.5 \sin\left(2(50°)\right) \approx 307.75$ feet

(c) $d = 312.5 \sin\left(2(60°)\right) \approx 270.63$ feet

103. False. In each of the four quadrants, the sign of the secant function and the cosine function will be the same because they are reciprocals of each other.

105. Answers will vary.

Section 1.5   Graphs of Sine and Cosine Functions

1. cycle

3. phase shift

5. $y = 2 \sin 5x$

Period: $\dfrac{2\pi}{5}$

Amplitude: $|2| = 2$

7. $y = \dfrac{3}{4} \cos \dfrac{\pi x}{2}$

Period: $\dfrac{2\pi}{b} = \dfrac{2\pi}{\pi/2} = 4$

Amplitude: $\left|\dfrac{3}{4}\right| = \dfrac{3}{4}$

9. $y = -\dfrac{1}{2} \sin \dfrac{5x}{4}$

Period: $\dfrac{2\pi}{b} = \dfrac{2\pi}{5/4} = \dfrac{8\pi}{5}$

Amplitude: $\left|-\dfrac{1}{2}\right| = \dfrac{1}{2}$

11. $y = -\dfrac{5}{3} \sin \dfrac{\pi x}{12}$

Period: $\dfrac{2\pi}{b} = \dfrac{2\pi}{\pi/12} = 24$

Amplitude: $\left|-\dfrac{5}{3}\right| = \dfrac{5}{3}$

13. $f(x) = \cos x$

$g(x) = \cos 5\,x$

The period of g is one-fifth the period of f.

15. $f(x) = \cos 2x$

$g(x) = -\cos 2x$

g is a reflection of the graph of f in the x-axis.

17. $f(x) = \sin x$

$g(x) = \sin(x - \pi)$

g is a horizontal shift to the right π units of the graph of f.

19. $f(x) = \sin 2x$

$f(x) = 3 + \sin 2x$

g is a vertical shift three units upward of the graph of f.

21. The graph of *g* has twice the amplitude as the graph of *f*. The period is the same.

23. The graph of *g* is a horizontal shift π units to the right of the graph of *f*.

25. $f(x) = \sin x$

Period: $\dfrac{2\pi}{b} = \dfrac{2\pi}{1} = 2\pi$

Amplitude: 1

Symmetry: origin

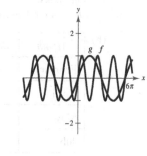

Key points: Intercept   Maximum   Intercept   Minimum   Intercept

$(0, 0)$ $\left(\dfrac{\pi}{2}, 1\right)$ $(\pi, 0)$ $\left(\dfrac{3\pi}{2}, -1\right)$ $(2\pi, 0)$

Because $g(x) = \sin\left(\dfrac{x}{3}\right) = f\left(\dfrac{x}{3}\right)$, the graph of $g(x)$ is the graph of $f(x)$, but stretched horizontally by a factor of 3.

Generate key points for the graph of $g(x)$ by multiplying the *x*-coordinate of each key point of $f(x)$ by 3.

27. $f(x) = \cos x$

Period: $\dfrac{2\pi}{b} = \dfrac{2\pi}{1} = 2\pi$

Amplitude: $|1| = 1$

Symmetry: *y*-axis

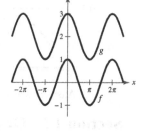

Key points: Maximum   Intercept   Minimum   Intercept   Maximum

$(0, 1)$ $\left(\dfrac{\pi}{2}, 0\right)$ $(\pi, -1)$ $\left(\dfrac{3\pi}{2}, 0\right)$ $(2\pi, 1)$

Because $g(x) = 2 + \cos x = f(x) + 2$, the graph of $g(x)$ is the graph of $f(x)$, but translated upward by two units.

Generate key points of $g(x)$ by adding 2 to the *y*-coordinate of each key point of $f(x)$.

29. $f(x) = -\cos x$

Period: $\dfrac{2\pi}{b} = \dfrac{2\pi}{1} = 2\pi$

Amplitude: 1

Symmetry: *y*-axis

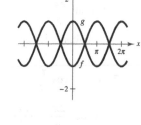

Key points: Minimum   Intercept   Maximum   Intercept   Minimum

$(0, -1)$ $\left(\dfrac{\pi}{2}, 0\right)$ $(\pi, 1)$ $\left(\dfrac{3\pi}{2}, 0\right)$ $(2\pi, -1)$

Because $g(x) = -\cos(x - \pi) = f(x - \pi)$, the graph of $g(x)$ is the graph of $f(x)$, but with a phase shift (horizontal translation) of π. Generate key points for the graph of $g(x)$ by shifting each key point of $f(x)$ π units to the right.

31. $y = 5 \sin x$

Period: 2π

Amplitude: 5

Key points:

$(0, 0), \left(\dfrac{\pi}{2}, 5\right), (\pi, 0),$

$\left(\dfrac{3\pi}{2}, -5\right), (2\pi, 0)$

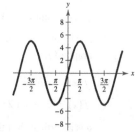

33. $y = \dfrac{1}{3} \cos x$

    Period: 2π

    Amplitude: $\dfrac{1}{3}$

    Key points:

    $\left(0, \dfrac{1}{3}\right), \left(\dfrac{\pi}{2}, 0\right), \left(\pi, -\dfrac{1}{3}\right),$

    $\left(\dfrac{3\pi}{2}, 0\right), \left(2\pi, \dfrac{1}{3}\right)$

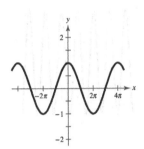

35. $y = \cos \dfrac{x}{2}$

    Period $\dfrac{2\pi}{1/2} = 4\pi$

    Amplitude: 1

    Key points:

    $(0, 1), (\pi, 0), (2\pi, -1),$

    $(3\pi, 0), (4\pi, 1)$

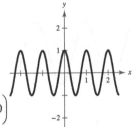

37. $y = \cos 2\pi x$

    Period: $\dfrac{2\pi}{2\pi} = 1$

    Amplitude: 1

    Key points:

    $(0, 1), \left(\dfrac{1}{4}, 0\right), \left(\dfrac{1}{2}, -1\right), \left(\dfrac{3}{4}, 0\right)$

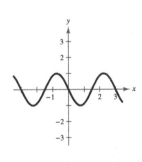

39. $y = -\sin \dfrac{2\pi x}{3}$

    Period: $\dfrac{2\pi}{2\pi/3} = 3$

    Amplitude: 1

    Key points:

    $(0, 0), \left(\dfrac{3}{4}, -1\right), \left(\dfrac{3}{2}, 0\right),$

    $\left(\dfrac{9}{4}, 1\right), (3, 0)$

41. $y = \cos\left(x - \dfrac{\pi}{2}\right)$

    Period: 2π

    Amplitude: 1

    Shift: Set $x - \dfrac{\pi}{2} = 0$ and $x - \dfrac{\pi}{2} = 2\pi$

            $x = \dfrac{\pi}{2}$ $x = \dfrac{5\pi}{2}$

    Key points: $\left(\dfrac{\pi}{2}, 1\right), (\pi, 0), \left(\dfrac{3\pi}{2}, -1\right), (2\pi, 0), \left(\dfrac{5\pi}{2}, 1\right)$

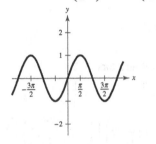

43. $y = 3 \sin(x + \pi)$

    Period: 2π

    Amplitude: 3

    Shift: Set $x + \pi = 0$ and $x + \pi = 2\pi$

           $x = -\pi$ $x = \pi$

    Key points: $(-\pi, 0), \left(-\dfrac{\pi}{2}, 3\right), (0, 0), \left(\dfrac{\pi}{2}, -3\right), (\pi, 0)$

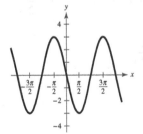

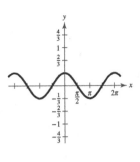

45. $y = 2 - \sin \dfrac{2\pi x}{3}$

Period: $\dfrac{2\pi}{2\pi/3} = 3$

Amplitude: 1

Key points:

$(0, 2), \left(\dfrac{3}{4}, 1\right), \left(\dfrac{3}{2}, 2\right), \left(\dfrac{9}{4}, 3\right), (3, 2)$

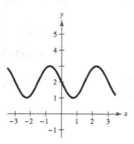

47. $y = 2 + 5\cos 6\pi x$

Period: $\dfrac{2\pi}{6\pi} = \dfrac{1}{3}$

Amplitude: 5

Shift: Set $6\pi x = 0$ and $6\pi x = 2\pi$

$x = 0 \qquad x = \dfrac{1}{3}$

Key points: $\left(\dfrac{1}{3}, 7\right), \left(\dfrac{5}{12}, 2\right), \left(\dfrac{1}{2}, -3\right), \left(\dfrac{7}{12}, 2\right), \left(\dfrac{1}{3}, 7\right)$

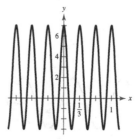

49. $y = 3\sin(x + \pi) - 3$

Period: 2π

Amplitude: 3

Shift: Set $x + \pi = 0$ and $x + \pi = 2\pi$

$x = -\pi \qquad x = \pi$

Key points: $(-\pi, -3), \left(-\dfrac{\pi}{2}, 0\right), (0, -3), \left(\dfrac{\pi}{2}, -6\right), (\pi, -3)$

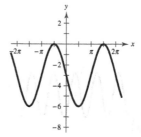

51. $y = \dfrac{2}{3}\cos\left(\dfrac{x}{2} - \dfrac{\pi}{4}\right)$

Period: $\dfrac{2\pi}{1/2} = 4\pi$

Amplitude: $\dfrac{2}{3}$

Shift: $\dfrac{x}{2} - \dfrac{\pi}{4} = 0$ and $\dfrac{\pi}{2} - \dfrac{\pi}{4} = 2\pi$

$x = \dfrac{\pi}{2} \qquad x = \dfrac{9\pi}{2}$

Key points:

$\left(\dfrac{\pi}{2}, \dfrac{2}{3}\right), \left(\dfrac{3\pi}{2}, 0\right), \left(\dfrac{5\pi}{2}, \dfrac{-2}{3}\right), \left(\dfrac{7\pi}{2}, 0\right), \left(\dfrac{9\pi}{2}, \dfrac{2}{3}\right)$

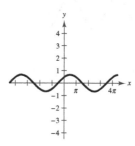

53. $g(x) = \sin(4x - \pi)$

   (a) $g(x)$ is obtained by a horizontal shrink and a phase

      shift of $\dfrac{\pi}{4}$. One cycle of $g(x)$ corresponds to the

      interval $\left[\dfrac{\pi}{4}, \dfrac{3\pi}{4}\right]$.

   (b)

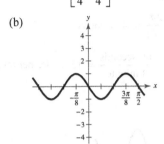

   (c) $g(x) = f(4x - \pi)$ where $f(x) = \sin x$.

55. $g(x) = \cos\left(x - \dfrac{\pi}{2}\right) + 2$

   (a) $g(x)$ is obtained by shifting $f(x)$ two units upward

      and a phase shift of $\dfrac{\pi}{2}$. One cycle of $g(x)$

      corresponds to the interval $[\pi, 3\pi]$.

   (b)

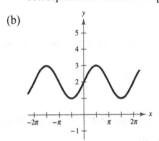

   (c) $g(x) = f\left(x - \dfrac{\pi}{2}\right) + 2$ where $f(x) = \cos x$

57. $g(x) = 2\sin(4x - \pi) - 3$

   (a) $g(x)$ is obtained by a vertical stretch, a horizontal

      shrink, a phase shift of $\dfrac{\pi}{4}$, and shifting $f(x)$ three

      units downward. One cycle of $g(x)$ corresponds to

      the interval $\left[\dfrac{\pi}{4}, \dfrac{3\pi}{4}\right]$.

   (b)

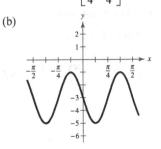

   (c) $g(x) = 2f(4x - \pi) - 3$ where $f(x) = \sin x$

59. $y = -2\sin(4x + \pi)$

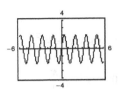

61. $y = \cos\left(2\pi x - \dfrac{\pi}{2}\right) + 1$

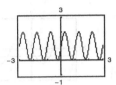

63. $y = -0.1\sin\left(\dfrac{\pi x}{10} + \pi\right)$

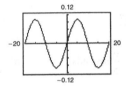

65. $f(x) = a\cos x + d$

     Amplitude: $\dfrac{1}{2}\left[3 - (-1)\right] = 2 \Rightarrow a = 2$

        $3 = 2\cos 0 + d$

        $d = 3 - 2 = 1$

        $a = 2, d = 1$

67. $f(x) = a\cos x + d$

     Amplitude: $\dfrac{1}{2}[8 - 0] = 4$

     Reflected in the x-axis: $a = -4$

        $0 = -4\cos 0 + d$

        $d = 4$

        $a = -4, d = 4$

69. $y = a\sin(bx - c)$

     Amplitude: $|a| = |3|$

     Since the graph is reflected in the x-axis, we have
     $a = -3$.

     Period: $\dfrac{2\pi}{b} = \pi \Rightarrow b = 2$

     Phase shift: $c = 0$

     $a = -3, b = 2, c = 0$

71. $y = a \sin(bx - c)$

Amplitude: $a = 2$

Period: $2\pi \Rightarrow b = 1$

Phase shift: $bx - c = 0$ when $x = -\dfrac{\pi}{4}$

$$(1)\left(-\frac{\pi}{4}\right) - c = 0 \Rightarrow c = -\frac{\pi}{4}$$

$a = 2, b = 1, c = -\dfrac{\pi}{4}$

73. $y_1 = \sin x$

$y_2 = -\dfrac{1}{2}$

In the interval $[-2\pi, 2\pi]$,

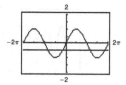

$y_1 = y_2$ when $x = -\dfrac{5\pi}{6}, -\dfrac{\pi}{6}, \dfrac{7\pi}{6}, \dfrac{11\pi}{6}$.

81. $P = 100 - 20 \cos \dfrac{5\pi t}{3}$

(a) Period: $\dfrac{2\pi}{(5\pi)/3} = \dfrac{6}{5}$ seconds

(b) $\dfrac{1 \text{ heartbeat}}{6/5 \text{ seconds}} \cdot \dfrac{60 \text{ seconds}}{1 \text{ minute}} = 50$ heartbeats per minute

83. (a) and (c)

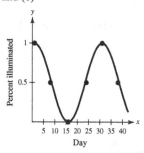

The model fits the data well.

(d) The period of the model is $p = 30$ days.

(e) Because March 12, 2018 is the 76th day of the year, $x = 76$. So, the percent of the moon's face illuminated is $y = 0.5 \cos\left(\dfrac{\pi(71)}{15} - \dfrac{\pi}{15}\right) + 0.5 = 0.25 = 25\%$.

Answers for 75–77 are sample answers.

75. $f(x) = 2 \sin(2x - \pi) + 1$

77. $f(x) = \cos(2x + 2\pi) - \dfrac{3}{2}$

79. $v = 1.75 \sin \dfrac{\pi t}{2}$

(a) Period $= \dfrac{2\pi}{\pi/2} = 4$ seconds

(b) $\dfrac{1 \text{ cycle}}{4 \text{ seconds}} \cdot \dfrac{60 \text{ seconds}}{1 \text{ minute}} = 15$ cycles per minute

(c)

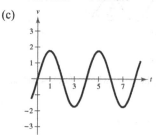

(b) Amplitude $a = \dfrac{1}{2}[\text{max. percent} - \text{min. percent}]$

$$= \frac{1}{2}[1.0 - 0] = 0.5$$

Period:

$p = (2\text{nd day of 1.0 percent} - 1\text{st day of 1.0 percent})$

$= 31 - 1 = 30$

$b = \dfrac{2\pi}{p} = \dfrac{2\pi}{30} = \dfrac{\pi}{15}$

Because the zero percent occurs at day 16, $x = 16$.

$bx - c = \pi$

$\left(\dfrac{\pi}{15}\right)(16) - c = \pi$

$\dfrac{16\pi}{15} - \pi = c$

$\dfrac{\pi}{15} = c$

The average percent of illumination is

$\dfrac{1}{2}(1.0 - 0) = 0.5.$ So, $d = 0.5$.

So the model is $y = 0.5 \cos\left(\dfrac{\pi x}{15} - \dfrac{\pi}{15}\right) + 0.5$.

85. (a) Period $= \dfrac{2\pi}{\left(\dfrac{\pi}{10}\right)} = 20$ seconds

The wheel takes 20 seconds to revolve once.

(b) Amplitude: 50 feet

The radius of the wheel is 50 feet.

(c)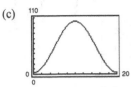

87. False. The graph of $g(x) = \sin(x + 2\pi)$ is the graph of $f(x) = \sin(x)$ translated to the *left* by one period, and the graphs are identical.

89. True. $y = -\cos x$ is a reflection of $y = \sin\left(x + \dfrac{\pi}{2}\right)$ in the x-axis. So, $-\cos x = -\sin\left(x + \dfrac{\pi}{2}\right)$.

91.

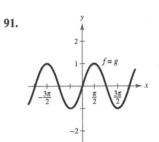

Because the graphs are the same, the conjecture is that

$$\sin(x) = \cos\left(x - \dfrac{\pi}{2}\right).$$

93.

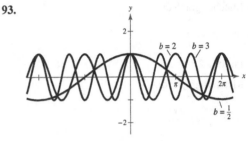

As the value of b increases, the period decreases.

$b = \dfrac{1}{2} \rightarrow \dfrac{1}{2}$ cycle

$b = 2 \rightarrow 2$ cycles

$b = 3 \rightarrow 3$ cycles

95. (a) $\sin \dfrac{1}{2} \approx \dfrac{1}{2} - \dfrac{(1/2)^3}{3!} + \dfrac{(1/2)^5}{5!} \approx 0.4794$

$\sin \dfrac{1}{2} \approx 0.4794$ (by calculator)

(b) $\sin 1 \approx 1 - \dfrac{1}{3!} + \dfrac{1}{5!} \approx 0.8417$

$\sin 1 \approx 0.8415$ (by calculator)

(c) $\sin \dfrac{\pi}{6} \approx 1 - \dfrac{(\pi/6)^3}{3!} + \dfrac{(\pi/6)^5}{5!} \approx 0.5000$

$\sin \dfrac{\pi}{6} = 0.5$ (by calculator)

(d) $\cos(-0.5) \approx 1 - \dfrac{(-0.5)^2}{2!} + \dfrac{(-0.5)^4}{4!} \approx 0.8776$

$\cos(-0.5) \approx 0.8776$ (by calculator)

(e) $\cos 1 \approx 1 - \dfrac{1}{2!} + \dfrac{1}{4!} \approx 0.5417$

$\cos 1 \approx 0.5403$ (by calculator)

(f) $\cos \dfrac{\pi}{4} \approx 1 - \dfrac{(\pi/4)^2}{2!} + \dfrac{(\pi/4)^2}{4!} = 0.7074$

$\cos \dfrac{\pi}{4} \approx 0.7071$ (by calculator)

The error in the approximation is not the same in each case. The error appears to increase as x moves farther away from 0.

Section 1.6   Graphs of Other Trigonometric Functions

1. odd; origin

3. reciprocal

5. π

7. $(-\infty, -1] \cup [1, \infty)$

9. $y = \sec 2x$

Period: $\dfrac{2\pi}{2} = \pi$

Matches graph (e).

10. $y = \tan \dfrac{x}{2}$

Period: $\dfrac{\pi}{b} = \dfrac{\pi}{1/2} = 2\pi$

Asymptotes: $x = -\pi$, $x = \pi$

Matches graph (c).

11. $y = \dfrac{1}{2} \cot \pi x$

Period: $\dfrac{\pi}{\pi} = 1$

Matches graph (a).

12. $y = -\csc x$

Period: 2π

Matches graph (d).

13. $y = \dfrac{1}{2} \sec \dfrac{\pi x}{2}$

Period: $\dfrac{2\pi}{b} = \dfrac{2\pi}{\pi/2} = 4$

Asymptotes: $x = -1$, $x = 1$

Matches graph (f).

14. $y = -2 \sec \dfrac{\pi x}{2}$

Period: $\dfrac{2\pi}{b} = \dfrac{2\pi}{\pi/2} = 4$

Asymptotes: $x = -1$, $x = 1$

Reflected in x-axis

Matches graph (b).

15. $y = \dfrac{1}{3} \tan x$

Period: π

Two consecutive asymptotes:

$x = -\dfrac{\pi}{2}$ and $x = \dfrac{\pi}{2}$

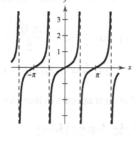

x	$-\dfrac{\pi}{4}$	0	$\dfrac{\pi}{4}$
y	$-\dfrac{1}{3}$	0	$\dfrac{1}{3}$

17. $y = -\dfrac{1}{2} \sec x$

Period: 2π

Two consecutive asymptotes:

$x = -\dfrac{\pi}{2}$, $x = \dfrac{\pi}{2}$

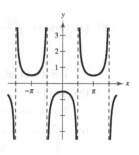

x	$-\dfrac{\pi}{3}$	0	$\dfrac{\pi}{3}$
y	-1	$-\dfrac{1}{2}$	-1

19. $y = -2 \tan 3x$

Period: $\dfrac{\pi}{3}$

Two consecutive asymptotes:

$x = -\dfrac{\pi}{6}$, $x = \dfrac{\pi}{6}$

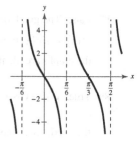

x	$-\dfrac{\pi}{3}$	0	$\dfrac{\pi}{3}$
y	0	0	0

21. $y = \csc \pi x$

Period: $\dfrac{2\pi}{\pi} = 2$

Two consecutive asymptotes:

$x = 0$, $x = 1$

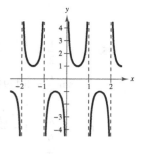

x	$\dfrac{1}{6}$	$\dfrac{1}{2}$	$\dfrac{5}{6}$
y	2	1	2

23. $y = \dfrac{1}{2} \sec \pi x$

Period: 2

Two consecutive asymptotes:

$x = -\dfrac{1}{2}$, $x = \dfrac{1}{2}$

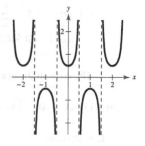

x	-1	0	1
y	$-\dfrac{1}{2}$	$\dfrac{1}{2}$	$-\dfrac{1}{2}$

25. $y = \csc \dfrac{x}{2}$

Period: $\dfrac{2\pi}{1/2} = 4\pi$

Two consecutive asymptotes:

$x = 0,\ x = 2\pi$

x	$\dfrac{\pi}{3}$	π	$\dfrac{5\pi}{3}$
y	2	1	2

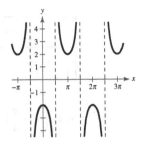

27. $y = 3\cot 2x$

Period: $\dfrac{\pi}{2}$

Two consecutive asymptotes:

$x = -\dfrac{\pi}{2},\ x = \dfrac{\pi}{2}$

x	$-\dfrac{\pi}{6}$	$-\dfrac{\pi}{8}$	$\dfrac{\pi}{8}$	$\dfrac{\pi}{6}$
y	$-3\sqrt{3}$	-3	3	$3\sqrt{3}$

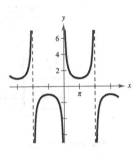

29. $y = \tan \dfrac{\pi x}{4}$

Period: $\dfrac{\pi}{\pi/4} = 4$

Two consecutive asymptotes:

$\dfrac{\pi x}{4} = -\dfrac{\pi}{2} \Rightarrow x = -2$

$\dfrac{\pi x}{4} = \dfrac{\pi}{2} \Rightarrow x = 2$

x	-1	0	1
y	-1	0	1

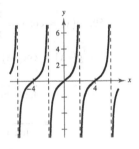

31. $y = 2\csc(x - \pi)$

Period: 2π

Two consecutive asymptotes:

$x = -\pi,\ x = \pi$

x	$-\dfrac{\pi}{2}$	$\dfrac{\pi}{2}$	$\dfrac{3\pi}{2}$
y	2	-2	-2

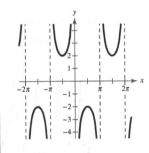

33. $y = 2\sec(x + \pi)$

Period: 2π

Two consecutive asymptotes:

$x = -\dfrac{\pi}{2},\ x = \dfrac{\pi}{2}$

x	$-\dfrac{\pi}{3}$	0	$\dfrac{\pi}{3}$
y	-4	-2	-4

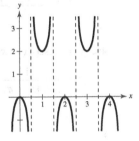

35. $y = -\sec \pi x + 1$

Period: $\dfrac{2\pi}{\pi} = 2$

Two consecutive asymptotes:

$x = -\dfrac{1}{2},\ x = \dfrac{1}{2}$

x	$-\dfrac{1}{3}$	0	$\dfrac{1}{3}$
y	-1	0	1

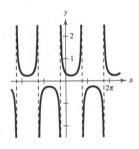

37. $y = \dfrac{1}{4}\csc\left(x + \dfrac{\pi}{4}\right)$

Period: 2π

Two consecutive asymptotes:

$x = -\dfrac{\pi}{4},\ x = \dfrac{3\pi}{4}$

x	$-\dfrac{\pi}{12}$	$\dfrac{\pi}{4}$	$\dfrac{7\pi}{12}$
y	$\dfrac{1}{2}$	$\dfrac{1}{4}$	$\dfrac{1}{2}$

39. $y = \tan \dfrac{x}{3}$

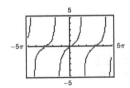

41. $y = -2 \sec 4x = \dfrac{-2}{\cos 4x}$

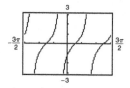

43. $y = \tan\left(x - \dfrac{\pi}{4}\right)$

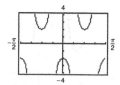

45. $y = -\csc(4x - \pi)$

$y = \dfrac{-1}{\sin(4x - \pi)}$

47. $y = 0.1 \tan\left(\dfrac{\pi x}{4} + \dfrac{\pi}{4}\right)$

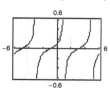

49. $\tan x = 1$

$x = -\dfrac{7\pi}{4}, -\dfrac{3\pi}{4}, \dfrac{\pi}{4}, \dfrac{5\pi}{4}$

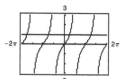

51. $\cot x = -\sqrt{3}$

$x = -\dfrac{7\pi}{6}, -\dfrac{\pi}{6}, \dfrac{5\pi}{6}, \dfrac{11\pi}{6}$

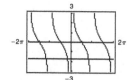

53. $\sec x = -2$

$x = \dfrac{2\pi}{3}, \dfrac{4\pi}{3}, -\dfrac{2\pi}{3}, -\dfrac{4\pi}{3}$

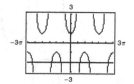

55. $\csc x = \sqrt{2}$

$x = -\dfrac{7\pi}{4}, -\dfrac{5\pi}{4}, \dfrac{\pi}{4}, \dfrac{3\pi}{4}$

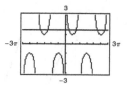

57. $f(x) = \sec x = \dfrac{1}{\cos x}$

$f(-x) = \sec(-x)$

$\quad = \dfrac{1}{\cos(-x)}$

$\quad = \dfrac{1}{\cos x}$

$\quad = f(x)$

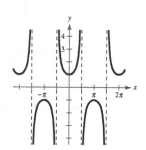

So, $f(x) = \sec x$ is an even function and the graph has y-axis symmetry.

59. $g(x) = \cot x = \dfrac{1}{\tan x}$

$g(-x) = \cot(-x)$

$\quad = \dfrac{1}{\tan(-x)}$

$\quad = -\dfrac{1}{\tan x}$

$\quad = -g(x)$

So, $g(x) = \cot x$ is an odd function and the graph has origin symmetry.

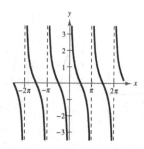

61. $f(x) = x + \tan x$

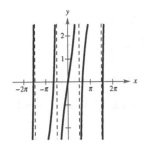

$f(-x) = (-x) + \tan(-x)$

$\qquad = -x - \tan x$

$\qquad = -(x + \tan x)$

$\qquad = -f(x)$

So, $f(x) = x + \tan x$ is an odd function and the graph has origin symmetry.

63. $g(x) = x \csc x = \dfrac{x}{\sin x}$

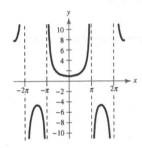

$g(-x) = (-x)\csc(-x)$

$\qquad = \dfrac{-x}{\sin(-x)}$

$\qquad = \dfrac{-x}{-\sin x}$

$\qquad = \dfrac{x}{\sin x}$

$\qquad = x \csc x$

$\qquad = g(x)$

So, $g(x) = x \csc x$ is an even function and the graph has y-axis symmetry.

65. $f(x) = |x \cos x|$

Matches graph (d).

As $x \to 0$, $f(x) \to 0$.

66. $f(x) = x \sin x$

Matches graph (a)

As $x \to 0$, $f(x) \to 0$.

67. $g(x) = |x| \sin x$

Matches graph (b).

As $x \to 0$, $g(x) \to 0$.

68. $g(x) = |x| \cos x$

Matches graph (c).

As $x \to 0$, $g(x) \to 0$.

69. $f(x) = \sin x + \cos\left(x + \dfrac{\pi}{2}\right)$

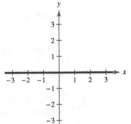

$g(x) = 0$

$f(x) = g(x)$

The functions are equal.

71. $f(x) = \sin^2 x$

$g(x) = \frac{1}{2}(1 - \cos 2x)$

$f(x) = g(x)$

The functions are equal.

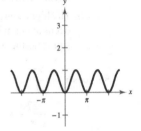

73. $g(x) = x \cos \pi x$

Damping factor: x

As $x \to \infty$, $g(x)$ oscillates and approaches $-\infty$ and ∞.

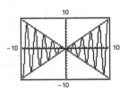

75. $f(x) = x^3 \sin x$

Damping factor: x^3

As $x \to \infty$, $f(x)$ oscillates and approaches $-\infty$ and ∞.

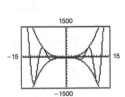

77. $y = \dfrac{6}{x} + \cos x$, $x > 0$

As $x \to 0$, $y \to \infty$.

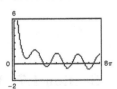

79. $g(x) = \dfrac{\sin x}{x}$

As $x \to 0$, $g(x) \to 1$.

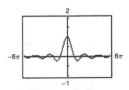

81. $f(x) = \sin \dfrac{1}{x}$

As $x \to 0$, $f(x)$ oscillates between -1 and 1.

83. (a) Period of $\cos \dfrac{\pi t}{6} = \dfrac{2\pi}{\pi/6} = 12$

Period of $\sin \dfrac{\pi t}{6} = \dfrac{2\pi}{\pi/6} = 12$

The period of $H(t)$ is 12 months.

The period of $L(t)$ is 12 months.

(b) From the graph, it appears that the greatest difference between high and low temperatures occurs in the summer. The smallest difference occurs in the winter.

(c) The highest high and low temperatures appear to occur about half of a month after the time when the sun is northernmost in the sky.

85. $\cos x = \dfrac{27}{d}$

$d = \dfrac{27}{\cos x} = 27 \sec x, \ -\dfrac{\pi}{2} < x < \dfrac{\pi}{2}$

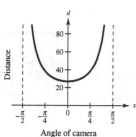

Angle of camera

87. True. Because

$y = \csc x = \dfrac{1}{\sin x}$,

for a given value of x, the y-coordinate of $\csc x$ is the reciprocal of the y-coordinate of $\sin x$.

89. $f(x) = x - \cos x$

(a)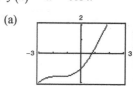

The zero between 0 and 1 occurs at $x \approx 0.7391$.

(b) $x_n = \cos(x_{n-1})$

$x_0 = 1$

$x_1 = \cos 1 \approx 0.5403$

$x_2 = \cos 0.5403 \approx 0.8576$

$x_3 = \cos 0.8576 \approx 0.6543$

$x_4 = \cos 0.6543 \approx 0.7935$

$x_5 = \cos 0.7935 \approx 0.7014$

$x_6 = \cos 0.7014 \approx 0.7640$

$x_7 = \cos 0.7640 \approx 0.7221$

$x_8 = \cos 0.7221 \approx 0.7504$

$x_9 = \cos 0.7504 \approx 0.7314$

$\vdots$

This sequence appears to be approaching the zero of f: $x \approx 0.7391$.

91. $f(x) = \cot x$

(a) $x \to 0^+, f(x) \to \infty$

(b) $x \to 0^-, f(x) \to -\infty$

(c) $x \to \pi^+, f(x) \to \infty$

(d) $x \to \pi^-, f(x) \to -\infty$

93. $f(x) = \tan x$

(a) $x \to \dfrac{\pi}{2}^+, f(x) \to -\infty$

(b) $x \to \dfrac{\pi}{2}^-, f(x) \to \infty$

(c) $x \to -\dfrac{\pi}{2}^+, f(x) \to -\infty$

(d) $x \to -\dfrac{\pi}{2}^-, f(x) \to \infty$

Section 1.7 Inverse Trigonometric Functions

Function	Alternative Notation	Domain	Range
1. $y = \arcsin x$	$y = \sin^{-1} x$	$-1 \le x \le 1$	$-\dfrac{\pi}{2} \le y \le \dfrac{\pi}{2}$
3. $y = \arctan x$	$y = \tan^{-1} x$	$-\infty < x < \infty$	$-\dfrac{\pi}{2} < y < \dfrac{\pi}{2}$

5. $y = \arcsin \dfrac{1}{2} \Rightarrow \sin y = \dfrac{1}{2}$ for $-\dfrac{\pi}{2} \le y \le \dfrac{\pi}{2} \Rightarrow y = \dfrac{\pi}{6}$

7. $y = \arccos \dfrac{1}{2} \Rightarrow \cos y = \dfrac{1}{2}$ for $0 \le y \le \pi \Rightarrow y = \dfrac{\pi}{3}$

9. $y = \arctan \dfrac{\sqrt{3}}{3} \Rightarrow \tan y = \dfrac{\sqrt{3}}{3}$ for $-\dfrac{\pi}{2} < y < \dfrac{\pi}{2} \Rightarrow y = \dfrac{\pi}{6}$

11. It is not possible to evaluate arcsin 3. The domain of the inverse sine function is $[-1, 1]$.

13. $y = \arctan\left(-\sqrt{3}\right) \Rightarrow \tan y = -\sqrt{3}$ for $-\dfrac{\pi}{2} < y < \dfrac{\pi}{2} \Rightarrow y = -\dfrac{\pi}{3}$

15. $y = \arccos\left(-\dfrac{1}{2}\right) \Rightarrow \cos y = -\dfrac{1}{2}$ for $0 \le y \le \pi \Rightarrow y = \dfrac{2\pi}{3}$

17. $y = \sin^{-1}\left(-\dfrac{\sqrt{3}}{2}\right) \Rightarrow \sin y = -\dfrac{\sqrt{3}}{2}$ for $-\dfrac{\pi}{2} \le y \le \dfrac{\pi}{2} \Rightarrow y = -\dfrac{\pi}{3}$

19. $f(x) = \cos x$
$g(x) = \arccos x$
$y = x$

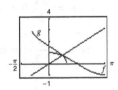

21. $\arccos 0.37 = \cos^{-1}(0.37) \approx 1.19$

23. $\arcsin(-0.75) = \sin^{-1}(-0.75) \approx -0.85$

25. $\arctan(-3) = \tan^{-1}(-3) \approx -1.25$

27. It is not possible to evaluate $\sin^{-1} 1.36$. The domain of the inverse sine function is $[-1, 1]$.

29. $\arccos(-0.41) = \cos^{-1}(-0.41) \approx 1.99$

31. $\arctan 0.92 = \tan^{-1} 0.92 \approx 0.74$

33. $\arcsin \frac{7}{8} = \sin^{-1}\left(\frac{7}{8}\right) \approx 1.07$

35. $\tan^{-1}\left(-\frac{95}{7}\right) \approx -1.50$

37. $\arctan\left(-\sqrt{3}\right) = -\dfrac{\pi}{3}$

$\tan\left(-\dfrac{\pi}{6}\right) = -\dfrac{\sqrt{3}}{3}$

$\tan\left(\dfrac{\pi}{4}\right) = 1$

39. $\tan \theta = \dfrac{x}{4}$

$\theta = \arctan \dfrac{x}{4}$

41. $\sin \theta = \dfrac{x + 2}{5}$

$\theta = \arcsin\left(\dfrac{x + 2}{5}\right)$

43. $\cos \theta = \dfrac{x + 3}{2x}$

$\theta = \arccos \dfrac{x + 3}{2x}$

45. $\sin(\arcsin 0.3) = 0.3$

47. It is not possible to evaluate $\cos\left[\arccos\left(-\sqrt{3}\right)\right]$. The domain of the inverse cosine function is $[-1, 1]$.

49. $\arcsin\left[\sin\left(\dfrac{9\pi}{4}\right)\right] = \arcsin\left(\dfrac{\sqrt{2}}{2}\right) = \dfrac{\pi}{4}$.

Note: $\dfrac{9\pi}{4}$ is not in the range of the arcsin function.

51. Let $u = \arctan \dfrac{3}{4}$.

$\tan u = \dfrac{3}{4}, 0 < u < \dfrac{\pi}{2},$

$\sin\left(\arctan \dfrac{3}{4}\right) = \sin u = \dfrac{3}{5}$

53. Let $u = \tan^{-1} 2$,

$$\tan u = 2 = \frac{2}{1}, 0 < u < \frac{\pi}{2},$$

$$\cos(\tan^{-1} 2) = \cos u = \frac{1}{\sqrt{5}} = \frac{\sqrt{5}}{5}.$$

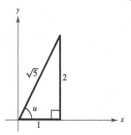

55. Let $u = \arcsin \frac{5}{13}$,

$$\sin u = \frac{5}{13}, 0 < u < \frac{\pi}{2},$$

$$\sec\left(\arcsin \frac{5}{13}\right) = \sec u = \frac{13}{12}.$$

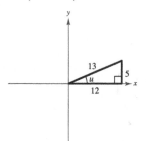

57. Let $u = \arctan\left(-\frac{3}{5}\right)$,

$$\tan u = -\frac{3}{5}, -\frac{\pi}{2} < y < 0,$$

$$\cot\left[\arctan\left(-\frac{3}{5}\right)\right] = \cot u = -\frac{5}{3}.$$

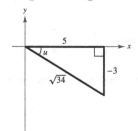

59. Let $u = \arccos\left(-\frac{2}{3}\right)$.

$$\cos u = -\frac{2}{3}, \frac{\pi}{2} < u < \pi,$$

$$\tan\left[\arccos\left(-\frac{2}{3}\right)\right] = \tan u = -\frac{\sqrt{5}}{2}$$

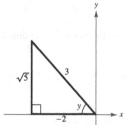

61. Let $u = \cos^{-1} \frac{\sqrt{3}}{2}$.

$$\cos u = \frac{\sqrt{3}}{2}, 0 < u < \frac{\pi}{2},$$

$$\csc\left[\cos^{-1} \frac{\sqrt{3}}{2}\right] = \csc u = 2.$$

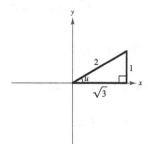

63. Let $u = \arcsin(2x)$.

$$\sin u = 2x = \frac{2x}{1},$$

$$\cos(\arcsin 2x) = \cos u = \sqrt{1 - 4x^2}$$

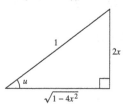

65. Let $u = \arctan x$.

$$\tan u = x = \frac{x}{1},$$

$$\cot(\arctan x) = \cot u = \frac{1}{x}$$

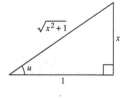

67. Let $u = \arccos x$.

$$\cos u = x = \frac{x}{1},$$

$$\sin(\arccos x) = \sin u = \sqrt{1 - x^2}$$

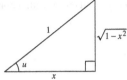

69. Let $u = \arccos\left(\dfrac{x}{3}\right)$.

$$\cos u = \frac{x}{3},$$

$$\tan\left(\arccos\frac{x}{3}\right) = \tan u = \frac{\sqrt{9 - x^2}}{x}$$

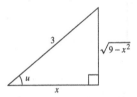

71. Let $u = \arctan\dfrac{x}{a}$.

$$\tan u = \frac{x}{a},$$

$$\csc\left(\arctan\frac{x}{a}\right) = \csc u = \frac{\sqrt{x^2 + a^2}}{a}$$

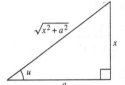

73. $f(x) = \sin(\arctan 2x)$, $g(x) = \dfrac{2x}{\sqrt{1 + 4x^2}}$

They are equal. Let $u = \arctan 2x$,

$$\tan u = 2x = \frac{2x}{1},$$

and $\sin u = \dfrac{2x}{\sqrt{1 + 4x^2}}$.

$$g(x) = \frac{2x}{\sqrt{1 + 4x^2}} = f(x)$$

The graph has horizontal asymptotes at $y = \pm 1$.

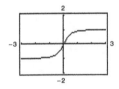

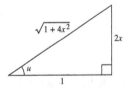

75. Let $u = \arctan\dfrac{9}{x}$.

$$\tan u = \frac{9}{x} \text{ and } \sin u = \frac{9}{\sqrt{x^2 + 81}}, x > 0$$

So,

$$\arctan\frac{9}{x} = \arcsin\frac{9}{\sqrt{x^2 + 81}}, x > 0.$$

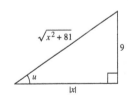

77. Let $u = \arccos \dfrac{3}{\sqrt{x^2 - 2x + 10}}$. Then,

$$\cos u = \dfrac{3}{\sqrt{x^2 - 2x + 10}} = \dfrac{3}{\sqrt{(x-1)^2 + 9}}$$

and $\sin u = \dfrac{|x - 1|}{\sqrt{(x-1)^2 + 9}}$.

So, $u = \arcsin \dfrac{|x - 1|}{\sqrt{x^2 - 2x + 10}}$.

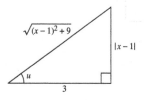

79. $g(x) = 2 \arcsin x$

Domain: $-1 \le x \le 1$

Range: $-\pi \le y \le \pi$

This is the graph of

$f(x) = \arcsin(x)$ with a vertical stretch.

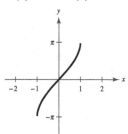

81. $f(x) = \dfrac{\pi}{2} + \arctan x$

Domain: all real numbers

Range: $0 < y \le \pi$

This is the graph of $y = \arctan x$ shifted upward $\pi/2$ units.

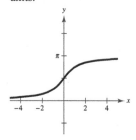

83. $h(v) = \arccos \dfrac{v}{2}$

Domain: $-2 \le v \le 2$

Range: $0 \le y \le \pi$

This is the graph of $h(v) = \arccos v$ with a horizontal stretch.

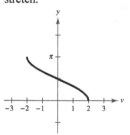

85. $f(x) = 2 \arccos(2x)$

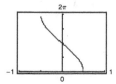

87. $f(x) = \arctan(2x - 3)$

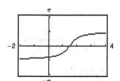

89. $f(x) = \pi - \sin^{-1}\left(\dfrac{2}{3}\right) \approx 2.412$

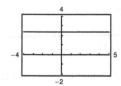

91. $f(t) = 3 \cos 2t + 3 \sin 2t = \sqrt{3^2 + 3^2} \sin\left(2t + \arctan\dfrac{3}{3}\right)$

$\qquad\qquad = 3\sqrt{2} \sin(2t + \arctan 1)$

$\qquad\qquad = 3\sqrt{2} \sin\left(2t + \dfrac{\pi}{4}\right)$

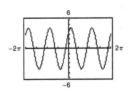

The graph implies that the identity is true.

93. $\dfrac{\pi}{2}$

95. $\dfrac{\pi}{2}$

97. π

99. (a) $\sin\theta = \dfrac{5}{s}$

$\qquad \theta = \arcsin\dfrac{5}{s}$

(b) $s = 40$: $\theta = \arcsin\dfrac{5}{40} \approx 0.13$

$\qquad s = 20$: $\theta = \arcsin\dfrac{5}{20} \approx 0.25$

101. (a) $\tan\theta = \dfrac{5.5}{8.5}$

$\qquad \theta \approx 32.9°$

(b) $\tan 32.9° = \dfrac{h}{10}$

$\qquad h = 10 \tan 32.9°$

$\qquad h \approx 6.49$ meters

The height is about 6.5 meters.

103. $\beta = \arctan\dfrac{3x}{x^2 + 4}$

(a)

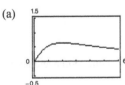

(b) β is maximum when $x = 2$ feet.

(c) The graph has a horizontal asymptote at $\beta = 0$.
As x increases, β decreases.

105. (a) $\tan\theta = \dfrac{x}{20}$

$\qquad \theta = \arctan\dfrac{x}{20}$

(b) $x = 5$: $\theta = \arctan\dfrac{5}{20} \approx 14.0°$

$\qquad x = 12$: $\theta = \arctan\dfrac{12}{20} \approx 31.0°$

107. True. $-\dfrac{\pi}{4}$ is in the range of the arctangent function.

$\tan\left(-\dfrac{\pi}{4}\right) = -1 \Leftrightarrow \arctan(-1) = -\dfrac{\pi}{4}.$

109. False. $\sin^{-1} x \neq \dfrac{1}{\sin x}$

The function $\sin^{-1} x$ is equivalent to arcsin x, which is the inverse sine function. The expression, $\dfrac{1}{\sin x}$ is the reciprocal of the sine function and is equivalent to csc x.

111. $y = \text{arccot } x$ if and only if $\cot y = x$.

Domain: $(-\infty, \infty)$

Range: $(0, \pi)$

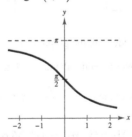

113. $y = \text{arccsc } x$ if and only if $\csc y = x$.

Domain: $(-\infty, -1] \cup [1, \infty)$

Range: $\left[-\dfrac{\pi}{2}, 0\right) \cup \left(0, \dfrac{\pi}{2}\right]$

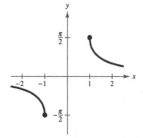

115. $y = \text{arcsec } \sqrt{2} \Rightarrow \sec y = \sqrt{2}$ and

$$0 \le y < \dfrac{\pi}{2} \cup \dfrac{\pi}{2} < y \le \pi \Rightarrow y = \dfrac{\pi}{4}$$

117. $y = \text{arccot}(-1) \Rightarrow \cot y = -1$ and

$$0 < y < \pi \Rightarrow y = \dfrac{3\pi}{4}$$

119. $y = \text{arccsc}(-1) \Rightarrow \csc y = -1$ and

$$-\dfrac{\pi}{2} \le y < 0 \cup 0 < y \le \dfrac{\pi}{2} \Rightarrow y = -\dfrac{\pi}{2}$$

121. $\text{arcsec } 2.54 = \arccos\left(\dfrac{1}{2.54}\right) \approx 1.17$

123. $\text{arccsc}\left(-\dfrac{25}{3}\right) = \arcsin\left(-\dfrac{3}{25}\right) \approx -0.12$

125. $\text{arccot } 5.25 = \arctan\left(\dfrac{1}{5.25}\right) \approx 0.19$

127. Area $= \arctan b - \arctan a$

(a) $a = 0, b = 1$

$$\text{Area} = \arctan 1 - \arctan 0 = \dfrac{\pi}{4} - 0 = \dfrac{\pi}{4}$$

(b) $a = -1, b = 1$

$$\text{Area} = \arctan 1 - \arctan(-1)$$

$$= \dfrac{\pi}{4} - \left(-\dfrac{\pi}{4}\right) = \dfrac{\pi}{2}$$

(c) $a = 0, b = 3$

$$\text{Area} = \arctan 3 - \arctan 0$$

$$\approx 1.25 - 0 = 1.25$$

(d) $a = -1, b = 3$

$$\text{Area} = \arctan 3 - \arctan(-1)$$

$$\approx 1.25 - \left(-\dfrac{\pi}{4}\right) \approx 2.03$$

129. $f(x) = \sin(x)$, $f^{-1}(x) = \arcsin(x)$

(a) $f \circ f^{-1} = \sin(\arcsin x)$ $f^{-1} \circ f = \arcsin(\sin x)$

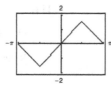

(b) The graphs coincide with the graph of $y = x$ only for certain values of x.

$f \circ f^{-1} = x$ over its entire domain, $-1 \le x \le 1$.

$f^{-1} \circ f = x$ over the region $-\dfrac{\pi}{2} \le x \le \dfrac{\pi}{2}$, corresponding to the region where $\sin x$ is one-to-one and has an inverse.

Section 1.8 Applications and Models

1. bearing

3. period

5. Given: $A = 60°, c = 12$

$$\sin A = \dfrac{a}{c} \Rightarrow a = c \sin A = 12 \sin 60° = \dfrac{12\sqrt{3}}{2} = 6\sqrt{3} \approx 10.39$$

$$\cos A = \dfrac{b}{c} \Rightarrow b = c \cos A = 12 \cos 60° = 12\left(\dfrac{1}{2}\right) = 6$$

$$B = 90° - 60° = 30°$$

7. Given: $B = 72.8°$, $a = 4.4$

$$\cos B = \frac{a}{c} \Rightarrow c = \frac{a}{\cos B} = \frac{4.4}{\cos 72.8°} \approx 14.88$$

$$\tan B = \frac{b}{a} \Rightarrow b = a \tan B = 4.4 \tan 72.8° \approx 14.21$$

$$A = 90° - 72.8° = 17.2°$$

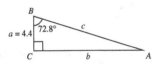

9. Given: $a = 3$, $b = 4$

$$a^2 + b^2 = c^2 \Rightarrow c^2 = (3)^2 + (4)^2 \Rightarrow c = 5$$

$$\tan A = \frac{a}{b} \Rightarrow A = \tan^{-1}\left(\frac{a}{b}\right) = \tan^{-1}\left(\frac{3}{4}\right) \approx 36.87°$$

$$B = 90° - 36.87° = 53.13°$$

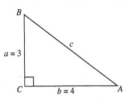

11. Given: $b = 15.70$, $c = 55.16$

$$a = \sqrt{55.16^2 - 15.7^2} \approx 52.88$$

$$\cos A = \frac{b}{c}$$

$$\cos A = \frac{15.7}{55.15}$$

$$A = \arccos \frac{15.7}{55.15} \approx 73.46°$$

$$B = 90° - 73.46° \approx 16.54°$$

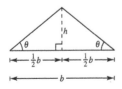

13. $\theta = 45°$, $b = 6$

$$\tan \theta = \frac{h}{(1/2)b} \Rightarrow h = \frac{1}{2}b \tan \theta$$

$$h = \frac{1}{2}(6) \tan 45° = 3.00 \text{ units}$$

15. $\tan \theta = \dfrac{h}{(1/2)b} \Rightarrow h = \dfrac{1}{2}b \tan \theta$

$$h = \frac{1}{2}(8) \tan 32° \approx 2.50 \text{ units}$$

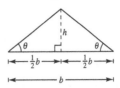

17. $\tan 25° = \dfrac{100}{x}$

$$x = \frac{100}{\tan 25°}$$

$$\approx 214.45 \text{ feet}$$

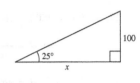

19. $\sin 80° = \dfrac{h}{20}$

$$20 \sin 80° = h$$

$$h \approx 19.7 \text{ feet}$$

21. Let the height of the church $= x$ and the height of the church and steeple $= y$.

$$\tan 35° = \frac{x}{50} \text{ and } \tan 48° = \frac{y}{50}$$

$$x = 50 \tan 35° \approx 35.01 \text{ and } y = 50 \tan 48° \approx 55.53$$

$$h = y - x = 55.53 - 35.01 = 20.52.$$

$$h \approx 20.5 \text{ feet}$$

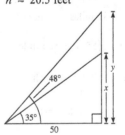

23. $\cot 55 = \dfrac{d}{10} \Rightarrow d \approx 7$ kilometers

$$\cot 28° = \frac{D}{10} \Rightarrow D \approx 18.8 \text{ kilometers}$$

Distance between towns:

$$D - d = 18.8 - 7 = 11.8 \text{ kilometers}$$

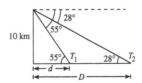

25. $\tan \theta = \frac{75}{50}$

$\theta = \arctan \frac{3}{2} \approx 56.3°$

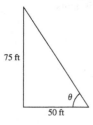

27. $12,500 + 4000 = 16,500$

$\sin \theta = \frac{4000}{16,500}$

$\theta = \arcsin\left(\frac{4000}{16,500}\right)$

$\theta \approx 14.03°$

Angle of depression $= \alpha \approx 90° - 14.03° = 75.97°$

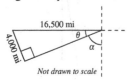

31. (a) $l^2 = (200)^2 + (150)^2$

$l = 250$ feet

$\tan A = \frac{150}{200} \Rightarrow A = \arctan\left(\frac{150}{200}\right) \approx 36.87°$

$\tan B = \frac{200}{150} \Rightarrow B = \arctan\left(\frac{200}{150}\right) \approx 53.13°$

(b) 250 ft $\times \dfrac{\text{mile}}{5280 \text{ ft}} \times \dfrac{\text{hour}}{35 \text{ miles}} \times \dfrac{3600 \text{ sec}}{\text{hour}} \approx 4.87$ seconds

33. The plane has traveled $1.5(550) = 825$ miles.

$\sin 38° = \dfrac{a}{825} \Rightarrow a \approx 508$ miles north

$\cos 38° = \dfrac{b}{825} \Rightarrow b \approx 650$ miles east

29. $\tan 57° = \dfrac{a}{x} \Rightarrow x = a \cot 57°$

$\tan 16° = \dfrac{a}{x + (55/6)}$

$\tan 16° = \dfrac{a}{a \cot 57° + (55/6)}$

$\cot 16° = \dfrac{a \cot 57° + (55/6)}{a}$

$a \cot 16° - a \cot 57° = \dfrac{55}{6} \Rightarrow a \approx 3.23$ miles

$\approx 17,054$ feet

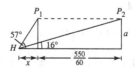

35.

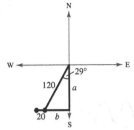

(a) $\cos 29° = \dfrac{a}{120} \Rightarrow a \approx 104.95$ nautical miles south

   $\sin 29° = \dfrac{b}{120} \Rightarrow b \approx 58.18$ nautical miles west

(b) $\tan \theta = \dfrac{20 + b}{a} \approx \dfrac{78.18}{104.95} \Rightarrow \theta \approx 36.7°$

   Bearing: S 36.7° W

   Distance: $d \approx \sqrt{104.95^2 + 78.18^2}$

≈ 130.9 nautical miles from port

37. $\tan \theta = \frac{45}{30} \Rightarrow \theta \approx 56.3°$

   Bearing: N 56.31°

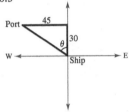

39. $\theta = 32°, \phi = 68°$

(a) $\alpha = 90° - 32° = 58°$

   Bearing from
   A to C: N 58° E

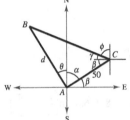

(b) $\beta = \theta = 32°$

   $\gamma = 90° - \phi = 22°$

   $C = \beta + \gamma = 54°$

   $\tan C = \dfrac{d}{50} \Rightarrow \tan 54°$

   $= \dfrac{d}{50} \Rightarrow d \approx 68.82$ meters

41. The diagonal of the base has a length of
   $\sqrt{a^2 + a^2} = \sqrt{2}a$. Now, you have

   $\tan \theta = \dfrac{a}{\sqrt{2}a} = \dfrac{1}{\sqrt{2}}$

   $\theta = \arctan \dfrac{1}{\sqrt{2}}$

   $\theta \approx 35.3°$.

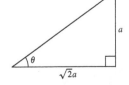

43. $\sin 36° = \dfrac{d}{25} \Rightarrow d \approx 14.69$

   Length of side: $2d \approx 29.4$ inches

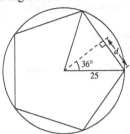

45. Use $d = a \sin \omega t$ because $d = 0$ when $t = 0$.

   Period: $\dfrac{2\pi}{\omega} = 2 \Rightarrow \omega = \pi$

   So, $d = 4 \sin(\pi t)$.

47. Use $d = a \cos \omega t$ because $d = 3$ when $t = 0$.

   Period: $\dfrac{2\pi}{\omega} = 1.5 \Rightarrow \omega = \dfrac{4\pi}{3}$

   So, $d = 3 \cos\left(\dfrac{4\pi}{3}t\right) = 3 \cos\left(\dfrac{4\pi t}{3}\right)$.

49. $d = a \sin \omega t$

   Frequency $= \dfrac{\omega}{2\pi}$

   $262 = \dfrac{\omega}{2\pi}$

   $\omega = 2\pi(262) = 524\pi$

51. $d = 9 \cos \dfrac{6\pi}{5}t$

(a)   Maximum displacement = amplitude = 9

(b)   Frequency $= \dfrac{\omega}{2\pi} = \dfrac{\frac{6\pi}{5}}{2\pi}$

   $= \dfrac{3}{5}$ cycle per unit of time

(c) $d = 9 \cos \dfrac{6\pi}{5}(5) = 9$

(d) $9 \cos \dfrac{6\pi}{5}t = 0$

   $\cos \dfrac{6\pi}{5}t = 0$

   $\dfrac{6\pi}{5}t = \arccos 0$

   $\dfrac{6\pi}{5}t = \dfrac{\pi}{2}$

   $t = \dfrac{5}{12}$

53. $d = \dfrac{1}{4} \sin 6\pi t$

    (a) Maximum displacement = amplitude = $\dfrac{1}{4}$

    (b) Frequency = $\dfrac{\omega}{2\pi} = \dfrac{6\pi}{2\pi}$

                = 3 cycles per unit of time

    (c) $d = \dfrac{1}{4} \sin 30\pi \approx 0$

    (d) $\dfrac{1}{4} \sin 6\pi t = 0$

        $\sin 6\pi t = 0$

          $6\pi t = \arcsin 0$

          $6\pi t = \pi$

             $t = \dfrac{1}{6}$

55. $y = \dfrac{1}{4} \cos 16t, \; t > 0$

    (a)

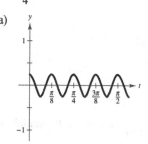

    (b) Period: $\dfrac{2\pi}{16} = \dfrac{\pi}{8}$

    (c) $\dfrac{1}{4} \cos 16t = 0$ when $16t = \dfrac{\pi}{2} \Rightarrow t = \dfrac{\pi}{32}$

57. (a)

    (b) $a = \dfrac{1}{2}(14.3 - 1.7) = 6.3$

        $\dfrac{2\pi}{b} = 12 \Rightarrow b = \dfrac{\pi}{6}$

        Shift: $d = 14.3 - 6.3 = 8$

        $S = d + a \cos bt$

        $S = 8 + 6.3 \cos\left(\dfrac{\pi t}{6}\right)$

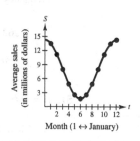

    Note: Another model is $S = 8 + 6.3 \sin\left(\dfrac{\pi t}{6} + \dfrac{\pi}{2}\right)$.

    The model is a good fit.

    (c) The period is $\dfrac{2\pi}{(\pi/6)} = 12$. Yes, sales of outwear are seasonal.

    (d) The amplitude is the maximum displacement from average sales of $8 million.

59. False. The tower isn't vertical and so the triangle formed is not a right triangle.

Review Exercises for Chapter 1

1. $\theta = \dfrac{15\pi}{4}$

(a)

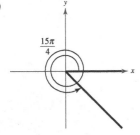

(b) Quadrant IV

(c) $\dfrac{15\pi}{4} - 2\pi = \dfrac{7\pi}{4}$

$\dfrac{7\pi}{4} - 2\pi = -\dfrac{\pi}{4}$

3. $\theta = -110°$

(a)

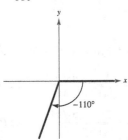

(b) Quadrant III

(c) Coterminal angles:

$-110° + 360° = 250°$

$-110° - 360° = -470°$

5. $450° = 450° \cdot \dfrac{\pi \text{ rad}}{180°} = \dfrac{5\pi}{2} \approx 7.854$ radians

7. $-16° = -16° \cdot \dfrac{\pi \text{ rad}}{180°} \approx -0.279$ radians

9. $\dfrac{3\pi}{10} = \dfrac{3\pi}{10} \cdot \dfrac{180°}{\pi \text{ rad}} = 54°$

11. $-3.5 \text{ rad} = -3.5 \text{ rad} \cdot \dfrac{180°}{\pi \text{ rad}} \approx -200.535°$

13. $198.4° = 198° + 0.4(60)' = 198°24'$

15. $138° = \dfrac{138\pi}{180} = \dfrac{23\pi}{30}$ radians

$s = r\theta = 20\left(\dfrac{23\pi}{30}\right) \approx 48.17$ inches

17. $150° = \dfrac{150\pi}{180} = \dfrac{5\pi}{6}$ radians

$A = \dfrac{1}{2}r^2\theta = \dfrac{1}{2}(20)^2\left(\dfrac{5\pi}{6}\right) = \dfrac{500\pi}{3} \approx 523.6$ square inches

19. $t = \dfrac{2\pi}{3}$ corresponds to the point $\left(-\dfrac{1}{2}, \dfrac{\sqrt{3}}{2}\right)$.

21. $t = \dfrac{7\pi}{6}$ corresponds to the point

$(x, y) = \left(-\dfrac{\sqrt{3}}{2}, -\dfrac{1}{2}\right)$.

23. $t = \dfrac{3\pi}{4}$ corresponds to the point $(x, y) = \left(-\dfrac{\sqrt{2}}{2}, \dfrac{\sqrt{2}}{2}\right)$.

$\sin\dfrac{3\pi}{4} = y = \dfrac{\sqrt{2}}{2}$ $\qquad$ $\csc\dfrac{3\pi}{4} = \dfrac{1}{y} = \sqrt{2}$

$\cos\dfrac{3\pi}{4} = x = -\dfrac{\sqrt{2}}{2}$ $\qquad$ $\sec\dfrac{3\pi}{4} = \dfrac{1}{x} = -\sqrt{2}$

$\tan\dfrac{3\pi}{4} = \dfrac{y}{x} = -1$ $\qquad$ $\cot\dfrac{3\pi}{4} = \dfrac{x}{y} = -1$

25. $\sin \dfrac{11\pi}{4} = \sin \dfrac{3\pi}{4} = \dfrac{\sqrt{2}}{2}$

27. $\cos\left(-\dfrac{17\pi}{6}\right) = \cos \dfrac{7\pi}{6} = -\dfrac{\sqrt{3}}{2}$

29. $\sec\left(\dfrac{12\pi}{5}\right) = \dfrac{1}{\cos\left(\dfrac{12\pi}{5}\right)} \approx 3.2361$

31. $\tan 33 \approx -75.3130$

33. $\text{opp} = 4$, $\text{adj} = 5$, $\text{hyp} = \sqrt{4^2 + 5^2} = \sqrt{41}$

$\sin \theta = \dfrac{\text{opp}}{\text{hyp}} = \dfrac{4}{\sqrt{41}} = \dfrac{4\sqrt{41}}{41}$
$\qquad$
$\csc \theta = \dfrac{\text{hyp}}{\text{opp}} = \dfrac{\sqrt{41}}{4}$

$\cos \theta = \dfrac{\text{adj}}{\text{hyp}} = \dfrac{5}{\sqrt{41}} = \dfrac{5\sqrt{41}}{41}$
$\qquad$
$\sec \theta = \dfrac{\text{hyp}}{\text{adj}} = \dfrac{\sqrt{41}}{5}$

$\tan \theta = \dfrac{\text{opp}}{\text{adj}} = \dfrac{4}{5}$
$\qquad$
$\cot \theta = \dfrac{\text{adj}}{\text{opp}} = \dfrac{5}{4}$

35. $\tan 33° \approx 0.6494$

37. $\cot 15° \, 14' = \dfrac{1}{\tan\left(15 + \dfrac{14}{60}\right)}$

≈ 3.6722

39. $\sin \theta = \dfrac{1}{3}$

(a) $\csc \theta = \dfrac{1}{\sin \theta} = 3$

(b) $\sin^2 \theta + \cos^2 \theta = 1$

$\left(\dfrac{1}{3}\right)^2 + \cos^2 \theta = 1$

$\cos^2 \theta = 1 - \dfrac{1}{9}$

$\cos^2 \theta = \dfrac{8}{9}$

$\cos \theta = \sqrt{\dfrac{8}{9}}$

$\cos \theta = \dfrac{2\sqrt{2}}{3}$

(c) $\sec \theta = \dfrac{1}{\cos \theta} = \dfrac{3}{2\sqrt{2}} = \dfrac{3\sqrt{2}}{4}$

(d) $\tan \theta = \dfrac{\sin \theta}{\cos \theta} = \dfrac{1/3}{\left(2\sqrt{2}\right)/3} = \dfrac{1}{2\sqrt{2}} = \dfrac{\sqrt{2}}{4}$

41. $\sin 1.2° = \dfrac{x}{3.5}$

$x = 3.5 \sin 1.2° \approx 0.0733$ kilometer or 73.3 meters

Not drawn to scale

43. $x = 12$, $y = 16$, $r = \sqrt{144 + 256} = \sqrt{400} = 20$

$\sin \theta = \dfrac{y}{r} = \dfrac{4}{5}$
$\qquad$
$\csc \theta = \dfrac{r}{y} = \dfrac{5}{4}$

$\cos \theta = \dfrac{x}{r} = \dfrac{3}{5}$
$\qquad$
$\sec \theta = \dfrac{r}{x} = \dfrac{5}{3}$

$\tan \theta = \dfrac{y}{x} = \dfrac{4}{3}$
$\qquad$
$\cot \theta = \dfrac{x}{y} = \dfrac{3}{4}$

45. $x = 0.3$, $y = 0.4$

$r = \sqrt{(0.3)^2 + (0.4)^2} = 0.5$

$\sin \theta = \dfrac{y}{r} = \dfrac{0.4}{0.5} = \dfrac{4}{5} = 0.8$
$\qquad$
$\csc \theta = \dfrac{r}{y} = \dfrac{0.5}{0.4} = \dfrac{5}{4} = 1.25$

$\cos \theta = \dfrac{x}{r} = \dfrac{0.3}{0.5} = \dfrac{3}{5} = 0.6$
$\qquad$
$\sec \theta = \dfrac{r}{x} = \dfrac{0.5}{0.3} = \dfrac{5}{3} \approx 1.67$

$\tan \theta = \dfrac{y}{x} = \dfrac{0.4}{0.3} = \dfrac{4}{3} \approx 1.33$
$\qquad$
$\cot \theta = \dfrac{x}{y} = \dfrac{0.3}{0.4} = \dfrac{3}{4} = 0.75$

47. $\sec\theta = \dfrac{6}{5}$, $\tan\theta < 0 \Rightarrow \theta$ is in Quadrant IV.

$r = 6$, $x = 5$, $y = -\sqrt{36 - 25} = -\sqrt{11}$

$\sin\theta = \dfrac{y}{r} = -\dfrac{\sqrt{11}}{6}$

$\cos\theta = \dfrac{x}{r} = \dfrac{5}{6}$

$\tan\theta = \dfrac{y}{x} = -\dfrac{\sqrt{11}}{5}$

$\csc\theta = \dfrac{r}{y} = -\dfrac{6\sqrt{11}}{11}$

$\cot\theta = -\dfrac{5\sqrt{11}}{11}$

49. $\cos\theta = \dfrac{x}{r} = \dfrac{-2}{5} \Rightarrow y^2 = 21$

$\sin\theta > 0 \Rightarrow \theta$ is in Quadrant II $\Rightarrow y = \sqrt{21}$

$\sin\theta = \dfrac{y}{r} = \dfrac{\sqrt{21}}{5}$

$\tan\theta = \dfrac{y}{x} = -\dfrac{\sqrt{21}}{2}$

$\csc\theta = \dfrac{r}{y} = \dfrac{5}{\sqrt{21}} = \dfrac{5\sqrt{21}}{21}$

$\sec\theta = \dfrac{r}{x} = \dfrac{5}{-2} = -\dfrac{5}{2}$

$\cot\theta = \dfrac{x}{y} = \dfrac{-2}{\sqrt{21}} = -\dfrac{2\sqrt{21}}{21}$

51. $\theta = 264°$

$\quad\; = 264° - 180° = 84°$

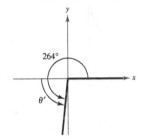

53. $\theta = -\dfrac{6\pi}{5}$

$\quad -\dfrac{6\pi}{5} + 2\pi = \dfrac{4\pi}{5}$

$\quad\quad \theta' = \pi - \dfrac{4\pi}{5} = \dfrac{\pi}{5}$

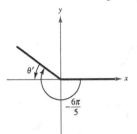

55. $\sin(-150°) = -\dfrac{1}{2}$

$\cos(-150°) = -\dfrac{\sqrt{3}}{2}$

$\tan(-150°) = \dfrac{-1/2}{-\sqrt{3}/2} = \dfrac{\sqrt{3}}{3}$

57. $\sin\dfrac{\pi}{3} = \dfrac{\sqrt{3}}{2}$

$\cos\dfrac{\pi}{3} = \dfrac{1}{2}$

$\tan\dfrac{\pi}{3} = \sqrt{3}$

59. $\sin 106° \approx 0.9613$

61. $\tan\left(-\dfrac{17\pi}{15}\right) \approx -0.4452$

63. $y = \sin 6x$

Amplitude: 1

Period: $\dfrac{2\pi}{6} = \dfrac{\pi}{3}$

65. $y = 5 + \sin \pi x$

Amplitude: 1

Period: $\dfrac{2\pi}{\pi} = 2$

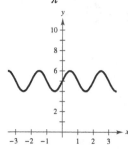

67. $g(t) = \frac{5}{2} \sin(t - \pi)$

Amplitude: $\frac{5}{2}$

Period: 2π

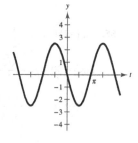

69. $y = a \sin bx$

(a) $a = 2,$

$\dfrac{2\pi}{b} = \dfrac{1}{264} \Rightarrow b = 528\pi$

$y = 2 \sin 528\pi x$

(b) $f = \dfrac{1}{1/264} = 264$ cycles per second

71. $f(t) = \tan\!\left(t + \dfrac{\pi}{2}\right)$

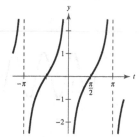

73. $f(x) = \dfrac{1}{2} \csc \dfrac{x}{2}$

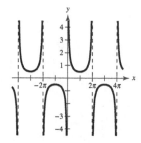

75. $f(x) = x \cos x$

Damping factor: x

As $x \to \infty$, $f(x)$ oscillates.

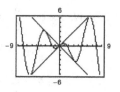

77. $\arcsin(-1) = -\dfrac{\pi}{2}$

79. $\operatorname{arccot} \sqrt{3} = \dfrac{\pi}{6}$

81. $\tan^{-1}(-1.3) \approx -0.92$ radian

83. $\operatorname{arccot} 15.5 = \arctan \dfrac{1}{15.5} \approx 0.06$

85. $f(x) = \arctan\!\left(\dfrac{x}{2}\right) = \tan^{-1}\!\left(\dfrac{x}{2}\right)$

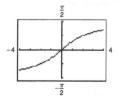

87. Let $u = \arctan \frac{3}{4}$ then $\tan u = \frac{3}{4}$.

$\cos\!\left(\arctan \frac{3}{4}\right) = \frac{4}{5}$

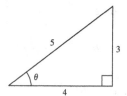

89. Let $u = \arctan \frac{12}{5}$

then $\tan u = \frac{12}{5}$.

$\sec\!\left(\arctan \frac{12}{5}\right) = \frac{13}{5}$

91. Let $y = \arccos\!\left(\dfrac{x}{2}\right)$. Then

$\cos y = \dfrac{x}{2}$ and $\tan y = \tan\!\left(\arccos\!\left(\dfrac{x}{2}\right)\right) = \dfrac{\sqrt{4 - x^2}}{x}$.

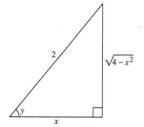

93. $\tan \theta = \frac{70}{30}$

$\theta = \arctan\left(\frac{70}{30}\right) \approx 66.8°$

70 m

30 m

95. $\sin 48° = \dfrac{d_1}{650} \Rightarrow d_1 \approx 483$

$\cos 25° = \dfrac{d_2}{810} \Rightarrow d_2 \approx 734$

$\left.\phantom{\begin{array}{c}a\\b\end{array}}\right\}$ $d_1 + d_2 \approx 1217$

$\cos 48° = \dfrac{d_3}{650} \Rightarrow d_3 \approx 435$

$\sin 25° = \dfrac{d_4}{810} \Rightarrow d_4 \approx 342$

$\left.\phantom{\begin{array}{c}a\\b\end{array}}\right\}$ $d_3 - d_4 \approx 93$

$\tan \theta \approx \dfrac{93}{1217} \Rightarrow \theta \approx 4.4°$

$\sec 4.4° \approx \dfrac{D}{1217} \Rightarrow D \approx 1217 \sec 4.4° \approx 1221$

The distance is 1221 miles and the bearing is 85.6°.

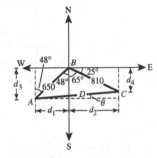

97. False. For each θ there corresponds exactly one value of y.

99. $f(\theta) = \sec \theta$ is undefined at the zeros of $g(\theta) = \cos \theta$ because $\sec \theta = \dfrac{1}{\cos \theta}$.

101. The ranges of the other four trigonometric functions are not bounded.

For $y = \tan x$ and $y = \cot x$, the range is $(-\infty, \infty)$.

For $y = \sec x$ and $y = \csc x$, the range is $(-\infty, -1] \cup [1, \infty)$.

103. Answers will vary.

Problem Solving for Chapter 1

1. (a) $8{:}57 - 6{:}45 = 2$ hours 12 minutes $= 132$ minutes

$\dfrac{132}{48} = \dfrac{11}{4}$ revolutions

$\theta = \left(\dfrac{11}{4}\right)(2\pi) = \dfrac{11\pi}{2}$ radians or 990°

(b) $s = r\theta = 47.25(5.5\pi) \approx 816.42$ feet

3. If you alter the model so that $h = 1$ when $t = 0$, you can use either a sine or a cosine model.

$a = \dfrac{1}{2}[\text{max} - \text{min}] = \dfrac{1}{2}[101 - 1] = 50$

$d = \dfrac{1}{2}[\text{max} + \text{min}] = \dfrac{1}{2}[101 + 1] = 51$

$b = 8\pi$

Cosine model: $h = 51 - 50 \cos(8\pi t)$

Sine model: $h = 51 - 50 \sin\left(8\pi t + \dfrac{\pi}{2}\right)$

Notice that you needed the horizontal shift so that the sine value was one when $t = 0$.

Another model would be: $h = 51 + 50 \sin\left(8\pi t + \dfrac{3\pi}{2}\right)$

Here you wanted the sine value to be 1 when $t = 0$.

5. (a) $\sin 39° = \dfrac{3000}{d}$

$d = \dfrac{3000}{\sin 39°} \approx 4767$ feet

(b) $\tan 39° = \dfrac{3000}{x}$

$x = \dfrac{3000}{\tan 39°} \approx 3705$ feet

(c) $\tan 63° = \dfrac{w + 3705}{3000}$

$3000 \tan 63° = w + 3705$

$w = 3000 \tan 63° - 3705 \approx 2183$ feet

7. (a) $h(x) = \cos^2 x$

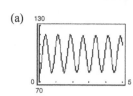

h is even.

(b) $h(x) = \sin^2 x$

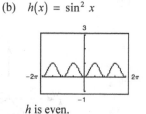

h is even.

9. $P = 100 - 20 \cos\left(\dfrac{8\pi}{3}t\right)$

(a)

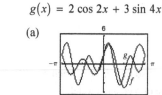

(b) Period $= \dfrac{2\pi}{8\pi/3} = \dfrac{6}{8} = \dfrac{3}{4}$ sec

This is the time between heartbeats.

(c) Amplitude: 20

The blood pressure ranges between $100 - 20 = 80$ and $100 + 20 = 120$.

(d) Pulse rate $= \dfrac{60 \text{ sec/min}}{\dfrac{3}{4} \text{ sec/beat}} = 80$ beats/min

(e) Period $= \dfrac{60}{64} = \dfrac{15}{16}$ sec

$64 = \dfrac{60}{2\pi/b} \Rightarrow b = \dfrac{64}{60} \cdot 2\pi = \dfrac{32}{15}\pi$

11. $f(x) = 2 \cos 2x + 3 \sin 3x$

$g(x) = 2 \cos 2x + 3 \sin 4x$

(a)

(b) The period of $f(x)$ is 2π.

The period of $g(x)$ is π.

(c) $h(x) = A \cos \alpha x + B \sin \beta x$ is periodic because the sine and cosine functions are periodic.

13.

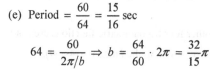

(a) $\dfrac{\sin \theta_1}{\sin \theta_2} = 1.333$

$\sin \theta_2 = \dfrac{\sin \theta_1}{1.333} = \dfrac{\sin 60°}{1.333} \approx 0.6497$

$\theta_2 = 40.5°$

(b) $\tan \theta_2 = \dfrac{x}{2} \Rightarrow x = 2 \tan 40.52° \approx 1.71$ feet

$\tan \theta_1 = \dfrac{y}{2} \Rightarrow y = 2 \tan 60° \approx 3.46$ feet

(c) $d = y - x = 3.46 - 1.71 = 1.75$ feet

(d) As you move closer to the rock, θ_1 decreases, which causes y to decrease, which in turn causes d to decrease.

Practice Test for Chapter 1

1. Express 350° in radian measure.

2. Express $(5\pi)/9$ in degree measure.

3. Convert $135°\,14'\,12''$ to decimal form.

4. Convert $-22.569°$ to $D°\,M'\,S''$ form.

5. If $\cos\theta = \frac{2}{3}$, use the trigonometric identities to find $\tan\theta$.

6. Find θ given $\sin\theta = 0.9063$.

7. Solve for x in the figure below.

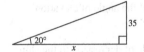

8. Find the magnitude of the reference angle for $\theta = (6\pi)/5$.

9. Evaluate $\csc 3.92$.

10. Find $\sec\theta$ given that θ lies in Quadrant III and $\tan\theta = 6$.

11. Graph $y = 3\sin\dfrac{x}{2}$.

12. Graph $y = -2\cos(x - \pi)$.

13. Graph $y = \tan 2x$.

14. Graph $y = -\csc\left(x + \dfrac{\pi}{4}\right)$.

15. Graph $y = 2x + \sin x$, using a graphing calculator.

16. Graph $y = 3x\cos x$, using a graphing calculator.

17. Evaluate $\arcsin 1$.

18. Evaluate $\arctan(-3)$.

19. Evaluate $\sin\left(\arccos\dfrac{4}{\sqrt{35}}\right)$.

20. Write an algebraic expression for $\cos\left(\arcsin\dfrac{x}{4}\right)$.

For Exercises 21–23, solve the right triangle.

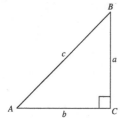

21. $A = 40°, c = 12$

22. $B = 6.84°, a = 21.3$

23. $a = 5, b = 9$

24. A 20-foot ladder leans against the side of a barn. Find the height of the top of the ladder if the angle of elevation of the ladder is 67°.

25. An observer in a lighthouse 250 feet above sea level spots a ship off the shore. If the angle of depression to the ship is 5°, how far out is the ship?

CHAPTER 2
Analytic Trigonometry

C H A P T E R 2
Analytic Trigonometry

Section 2.1 Using Fundamental Identities

1. $\tan u$

3. $\cot u$

5. 1

7. $\sec x = -\dfrac{5}{2}$, $\tan x < 0 \Rightarrow x$ is in Quadrant II.

$\cos x = \dfrac{1}{\sec x} = \dfrac{1}{-\dfrac{5}{2}} = -\dfrac{2}{5}$

$\sin x = \sqrt{1 - \left(-\dfrac{2}{5}\right)^2} = \sqrt{1 - \dfrac{4}{25}} = \dfrac{\sqrt{21}}{5}$

$\tan x = \dfrac{\sin x}{\cos x} = \dfrac{\dfrac{\sqrt{21}}{5}}{-\dfrac{2}{5}} = -\dfrac{\sqrt{21}}{2}$

$\csc x = \dfrac{1}{\sin x} = \dfrac{5}{\sqrt{21}} = \dfrac{5\sqrt{21}}{21}$

$\cot x = \dfrac{1}{\tan x} = -\dfrac{2}{\sqrt{21}} = -\dfrac{2\sqrt{21}}{21}$

9. $\sin \theta = -\dfrac{3}{4}$, $\cos \theta > 0 \Rightarrow \theta$ is in Quadrant IV.

$\cos \theta = \sqrt{1 - \left(-\dfrac{3}{4}\right)^2} = \sqrt{1 - \dfrac{9}{16}} = \dfrac{\sqrt{7}}{4}$

$\tan \theta = \dfrac{\sin \theta}{\cos \theta} = \dfrac{-\dfrac{3}{4}}{\dfrac{\sqrt{7}}{4}} = -\dfrac{3}{\sqrt{7}} = -\dfrac{3\sqrt{7}}{7}$

$\sec \theta = \dfrac{1}{\cos \theta} = \dfrac{1}{\dfrac{\sqrt{7}}{4}} = \dfrac{4}{\sqrt{7}} = \dfrac{4\sqrt{7}}{7}$

$\cot \theta = \dfrac{1}{\tan \theta} = \dfrac{1}{-\dfrac{3}{\sqrt{7}}} = -\dfrac{\sqrt{7}}{3}$

$\csc \theta = \dfrac{1}{\sin \theta} = \dfrac{1}{-\dfrac{3}{4}} = -\dfrac{4}{3}$

11. $\tan x = \dfrac{2}{3}$, $\cos x > 0 \Rightarrow x$ is in Quadrant I.

$\cot x = \dfrac{1}{\tan x} = \dfrac{1}{\dfrac{2}{3}} = \dfrac{3}{2}$

$\sec x = \sqrt{1 + \left(\dfrac{2}{3}\right)^2} = \sqrt{1 + \dfrac{4}{9}} = \dfrac{\sqrt{13}}{3}$

$\csc x = \sqrt{1 + \left(\dfrac{3}{2}\right)^2} = \sqrt{1 + \dfrac{9}{4}} = \dfrac{\sqrt{13}}{2}$

$\sin x = \dfrac{1}{\csc x} = \dfrac{1}{\dfrac{\sqrt{13}}{2}} = \dfrac{2}{\sqrt{13}} = \dfrac{2\sqrt{13}}{13}$

$\cos x = \dfrac{1}{\sec x} = \dfrac{1}{\dfrac{\sqrt{13}}{3}} = \dfrac{3}{\sqrt{13}} = \dfrac{3\sqrt{13}}{13}$

$\cot x = \dfrac{1}{\tan x} = \dfrac{1}{\dfrac{2}{3}} = \dfrac{3}{2}$

13. $\sec x \cos x = \left(\dfrac{1}{\cos x}\right)\cos x$
$= 1$
Matches (c).

14. $\cot^2 x - \csc^2 x = \left(\csc^2 x - 1\right) - \csc^2 x$
$= -1$
Matches (b).

15. $\cos x\left(1 + \tan^2 x\right) = \cos x\left(\sec^2 x\right)$
$= \cos x\left(\dfrac{1}{\cos^2 x}\right)$
$= \dfrac{1}{\cos x}$
$= \sec x$

Matches (f).

16. $\cot x \sec x = \dfrac{\cos x}{\sin x} \cdot \dfrac{1}{\cos x} = \dfrac{1}{\sin x} = \csc x$

Matches (a).

17. $\dfrac{\sec^2 x - 1}{\sin^2 x} = \dfrac{\tan^2 x}{\sin^2 x} = \dfrac{\sin^2 x}{\cos^2 x} \cdot \dfrac{1}{\sin^2 x} = \sec^2 x$

Matches (e).

18. $\dfrac{\cos^2\left[(\pi/2) - x\right]}{\cos x} = \dfrac{\sin^2 x}{\cos x} = \dfrac{\sin x}{\cos x}\sin x = \tan x \sin x$

Matches (d).

19. $\dfrac{\tan \theta \cot \theta}{\sec \theta} = \dfrac{\tan \theta \left(\dfrac{1}{\tan \theta}\right)}{\dfrac{1}{\cos \theta}}$

$= \dfrac{1}{\dfrac{1}{\cos \theta}}$

$= \cos \theta$

21. $\tan^2 x - \tan^2 x \sin^2 x = \tan^2 x(1 - \sin^2 x)$

$= \tan^2 x \cos^2 x$

$= \dfrac{\sin^2 x}{\cos^2 x} \cdot \cos^2 x$

$= \sin^2 x$

27. $\cot^3 x + \cot^2 x + \cot x + 1 = \cot^2 x(\cot x + 1) + (\cot x + 1)$

$= (\cot x + 1)(\cot^2 x + 1)$

$= (\cot x + 1)\csc^2 x$

29. $3 \sin^2 x - 5 \sin x - 2 = (3 \sin x + 1)(\sin x - 2)$

31. $\cot^2 x + \csc x - 1 = \left(\csc^2 x - 1\right) + \csc x - 1$

$= \csc^2 x + \csc x - 2$

$= (\csc x - 1)(\csc x + 2)$

33. $\tan \theta \csc \theta = \dfrac{\sin \theta}{\cos \theta} \cdot \dfrac{1}{\sin \theta} = \dfrac{1}{\cos \theta} = \sec \theta$

35. $\sin \phi(\csc \phi - \sin \phi) = (\sin \phi)\dfrac{1}{\sin \phi} - \sin^2 \phi$

$= 1 - \sin^2 \phi = \cos^2 \phi$

37. $\sin \beta \tan \beta + \cos \beta = (\sin \beta)\dfrac{\sin \beta}{\cos \beta} + \cos \beta$

$= \dfrac{\sin^2 \beta}{\cos \beta} + \dfrac{\cos^2 \beta}{\cos \beta}$

$= \dfrac{\sin^2 \beta + \cos^2 \beta}{\cos \beta}$

$= \dfrac{1}{\cos \beta}$

$= \sec \beta$

45. $\dfrac{\cos x}{1 + \sin x} - \dfrac{\cos x}{1 - \sin x} = \dfrac{\cos x(1 - \sin x) - \cos x(1 + \sin x)}{(1 + \sin x)(1 - \sin x)}$

$= \dfrac{\cos x - \sin x \cos x - \cos x - \sin x \cos x}{(1 + \sin x)(1 - \sin x)}$

$= \dfrac{-2 \sin x \cos x}{1 - \sin^2 x}$

$= \dfrac{-2 \sin x \cos x}{\cos^2 x}$

$= \dfrac{-2 \sin x}{\cos x}$

$= -2 \tan x$

23. $\dfrac{\sec^2 x - 1}{\sec x - 1} = \dfrac{(\sec x + 1)(\sec x - 1)}{\sec x - 1}$

$= \sec x + 1$

25. $1 - 2 \cos^2 x + \cos^4 x = \left(1 - \cos^2 x\right)^2$

$= \left(\sin^2 x\right)^2$

$= \sin^4 x$

39. $\dfrac{1 - \sin^2 x}{\csc^2 x - 1} = \dfrac{\cos^2 x}{\cot^2 x} = \cos^2 x \tan^2 x = \left(\cos^2 x\right)\dfrac{\sin^2 x}{\cos^2 x}$

$= \sin^2 x$

41. $(\sin x + \cos x)^2 = \sin^2 x + 2 \sin x \cos x + \cos^2 x$

$= \left(\sin^2 x + \cos^2 x\right) + 2 \sin x \cos x$

$= 1 + 2 \sin x \cos x$

43. $\dfrac{1}{1 + \cos x} + \dfrac{1}{1 - \cos x} = \dfrac{1 - \cos x + 1 + \cos x}{(1 + \cos x)(1 - \cos x)}$

$= \dfrac{2}{1 - \cos^2 x}$

$= \dfrac{2}{\sin^2 x}$

$= 2 \csc^2 x$

47. $\tan x - \dfrac{\sec^2 x}{\tan x} = \dfrac{\tan^2 x - \sec^2 x}{\tan x}$

$\qquad\qquad = \dfrac{-1}{\tan x} = -\cot x$

49. $\dfrac{\sin^2 y}{1 - \cos y} = \dfrac{1 - \cos^2 y}{1 - \cos y}$

$\qquad\qquad = \dfrac{(1 + \cos y)(1 - \cos y)}{1 - \cos y} = 1 + \cos y$

51. $y_1 = \dfrac{1}{2}(\sin x \cot x + \cos x)$

$\qquad = \dfrac{1}{2}\left(\sin x \left(\dfrac{\cos x}{\sin x}\right) + \cos x\right)$

$\qquad = \dfrac{1}{2}(\cos x + \cos x)$

$\qquad = \cos x$

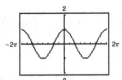

53. $y_1 = \dfrac{\tan x + 1}{\sec x + \csc x}$

$\qquad = \dfrac{\dfrac{\sin x}{\cos x} + 1}{\dfrac{1}{\cos x} + \dfrac{1}{\sin x}}$

$\qquad = \dfrac{\dfrac{\sin x + \cos x}{\cos x}}{\dfrac{\sin x + \cos x}{\sin x \cos x}}$

$\qquad = \left(\dfrac{\sin x + \cos x}{\cos x}\right)\left(\dfrac{\sin x \cos x}{\sin x + \cos x}\right)$

$\qquad = \sin x$

55. Let $x = 3 \cos \theta$.

$\sqrt{9 - x^2} = \sqrt{9 - (3 \cos \theta)^2}$

$\qquad\quad = \sqrt{9 - 9 \cos^2 \theta}$

$\qquad\quad = \sqrt{9(1 - \cos^2 \theta)}$

$\qquad\quad = \sqrt{9 \sin^2 \theta} = 3 \sin \theta$

57. Let $x = 2 \sec \theta$.

$\sqrt{x^2 - 4} = \sqrt{(2 \sec \theta)^2 - 4}$

$\qquad\quad = \sqrt{4(\sec^2 \theta - 1)}$

$\qquad\quad = \sqrt{4 \tan^2 \theta}$

$\qquad\quad = 2 \tan \theta$

59. Let $x = 2 \sin \theta$.

$\sqrt{4 - x^2} = \sqrt{2}$

$\sqrt{4 - (2 \sin \theta)^2} = \sqrt{2}$

$\sqrt{4 - 4 \sin^2 \theta} = \sqrt{2}$

$\sqrt{4(1 - \sin^2 \theta)} = \sqrt{2}$

$\sqrt{4 \cos^2 \theta} = \sqrt{2}$

$2 \cos \theta = \sqrt{2}$

$\cos \theta = \dfrac{\sqrt{2}}{2}$

$\sin \theta = \sqrt{1 - \cos^2 \theta} = \sqrt{1 - \left(\dfrac{\sqrt{2}}{2}\right)^2} = \pm\dfrac{\sqrt{2}}{2}$

61. $x = 6 \sin \theta$

$$3 = \sqrt{36 - x^2}$$
$$= \sqrt{36 - (6 \sin \theta)^2}$$
$$= \sqrt{36(1 - \sin^2 \theta)}$$
$$= \sqrt{36 \cos^2 \theta}$$
$$= 6 \cos \theta$$

$$\cos \theta = \frac{3}{6} = \frac{1}{2}$$
$$\sin \theta = \pm\sqrt{1 - \cos^2 \theta}$$
$$= \pm\sqrt{1 - \left(\frac{1}{2}\right)^2}$$
$$= \pm\sqrt{\frac{3}{4}}$$
$$= \pm\frac{\sqrt{3}}{2}$$

63. $\sin \theta = \sqrt{1 - \cos^2 \theta}$

Let $y_1 = \sin x$ and $y_2 = \sqrt{1 - \cos^2 x}$, $0 \le x \le 2\pi$.

$y_1 = y_2$ for $0 \le x \le \pi$.

So, $\sin \theta = \sqrt{1 - \cos^2 \theta}$ for $0 \le \theta \le \pi$.

65. $\sec \theta = \sqrt{1 + \tan^2 \theta}$

Let $y_1 = \dfrac{1}{\cos x}$ and $y_2 = \sqrt{1 + \tan^2 x}$, $0 \le x \le 2\pi$.

$y_1 = y_2$ for $0 \le x < \dfrac{\pi}{2}$ and $\dfrac{3\pi}{2} < x \le 2\pi$.

So, $\sec \theta = \sqrt{1 + \tan^2 \theta}$ for $0 \le \theta < \dfrac{\pi}{2}$ and

$\dfrac{3\pi}{2} < \theta < 2\pi$.

67. $\mu W \cos \theta = W \sin \theta$

$$\mu = \frac{W \sin \theta}{W \cos \theta} = \tan \theta$$

69. True.

$$\tan u = \frac{\sin u}{\cos u}$$
$$\cot u = \frac{\cos u}{\sin u}$$
$$\sec u = \frac{1}{\cos u}$$
$$\csc u = \frac{1}{\sin u}$$

71. As $x \to \dfrac{\pi^-}{2}$, $\tan x \to \infty$ and $\cot x \to 0$.

73. $\cos(-\theta) \ne -\cos \theta$

$\cos(-\theta) = \cos \theta$

The correct identity is $\dfrac{\sin \theta}{\cos(-\theta)} = \dfrac{\sin \theta}{\cos \theta}$

$$= \tan \theta$$

75. Because $\sin^2 \theta + \cos^2 \theta = 1$, then $\cos^2 \theta = 1 - \sin^2 \theta$.

$\cos \theta = \pm\sqrt{1 - \sin \theta}$

$$\tan \theta = \frac{\sin \theta}{\cos \theta} = \frac{\sin \theta}{\pm\sqrt{1 - \sin^2 \theta}}$$

$$\cot \theta = \frac{\cos \theta}{\sin \theta} = \frac{\pm\sqrt{1 - \sin^2 \theta}}{\sin \theta}$$

$$\sec \theta = \frac{1}{\cos \theta} = \frac{1}{\pm\sqrt{1 - \sin^2 \theta}}$$

$$\csc \theta = \frac{1}{\sin \theta}$$

77. $\dfrac{\sec \theta(1 + \tan \theta)}{\sec \theta + \csc \theta} = \dfrac{\left(\dfrac{1}{\cos \theta}\right)\left(1 + \dfrac{\sin \theta}{\cos \theta}\right)}{\dfrac{1}{\cos \theta} + \dfrac{1}{\sin \theta}}$

$$= \dfrac{\dfrac{\cos \theta + \sin \theta}{\cos^2 \theta}}{\dfrac{\sin \theta + \cos \theta}{\sin \theta \cos \theta}}$$

$$= \left(\dfrac{\sin \theta + \cos \theta}{\cos^2 \theta}\right)\left(\dfrac{\sin \theta \cos \theta}{\sin \theta + \cos \theta}\right)$$

$$= \dfrac{\sin \theta}{\cos \theta}$$

Section 2.2   Verifying Trigonometric Identities

1. identity

3. $\tan u$

5. $\sin u$

7. $-\csc u$

9. $\tan t \cot t = \dfrac{\sin t}{\cos t} \cdot \dfrac{\cos t}{\sin t} = 1$

11. $(1 + \sin \alpha)(1 - \sin \alpha) = 1 - \sin^2 \alpha = \cos^2 \alpha$

13. $\cos^2 \beta - \sin^2 \beta = (1 - \sin^2 \beta) - \sin^2 \beta$

$$= 1 - 2 \sin^2 \beta$$

15. $\tan\left(\dfrac{\pi}{2} - \theta\right) \tan \theta = \cot \theta \tan \theta$

$$= \left(\dfrac{1}{\tan \theta}\right) \tan \theta$$

$$= 1$$

17. $\sin t \csc\left(\dfrac{\pi}{2} - t\right) = \sin t \sec t = \sin t\left(\dfrac{1}{\cos t}\right)$

$$= \dfrac{\sin t}{\cos t} = \tan t$$

19. $\dfrac{1}{\tan x} + \dfrac{1}{\cot x} = \dfrac{\cot x + \tan x}{\tan x \cot x}$

$$= \dfrac{\cot x + \tan x}{1}$$

$$= \tan x + \cot x$$

21. $\dfrac{1 + \sin \theta}{\cos \theta} + \dfrac{\cos \theta}{1 + \sin \theta} = \dfrac{(1 + \sin \theta)^2 + \cos^2 \theta}{\cos \theta(1 + \sin \theta)}$

$$= \dfrac{1 + 2 \sin \theta + \sin^2 \theta + \cos^2 \theta}{\cos \theta(1 + \sin \theta)}$$

$$= \dfrac{2 + 2 \sin \theta}{\cos \theta(1 + \sin \theta)}$$

$$= \dfrac{2(1 + \sin \theta)}{\cos \theta(1 + \sin \theta)}$$

$$= \dfrac{2}{\cos \theta}$$

$$= 2 \sec \theta$$

23. $\dfrac{1}{\cos x + 1} + \dfrac{1}{\cos x - 1} = \dfrac{\cos x - 1 + \cos x + 1}{(\cos x + 1)(\cos x - 1)}$

$$= \dfrac{2 \cos x}{\cos^2 x - 1}$$

$$= \dfrac{2 \cos x}{-\sin^2 x}$$

$$= -2 \cdot \dfrac{1}{\sin x} \cdot \dfrac{\cos x}{\sin x}$$

$$= -2 \csc x \cot x$$

25. $\sec y \cos y = \left(\dfrac{1}{\cos y}\right) \cos y = 1$

27. $\dfrac{\tan^2 \theta}{\sec \theta} = \dfrac{(\sin \theta/\cos \theta)\tan \theta}{1/\cos \theta} = \sin \theta \tan \theta$

29. $\dfrac{1}{\tan \beta} + \tan \beta = \dfrac{1 + \tan^2 \beta}{\tan \beta}$

$$= \dfrac{\sec^2 \beta}{\tan \beta}$$

31. $\dfrac{\cot^2 t}{\csc t} = \dfrac{\cos^2 t/\sin^2 t}{1/\sin t} = \dfrac{\cos^2 t}{\sin t} = \dfrac{1 - \sin^2 t}{\sin t}$

33. $\sec x - \cos x = \dfrac{1}{\cos x} - \cos x$

$= \dfrac{1 - \cos^2 x}{\cos x}$

$= \dfrac{\sin^2 x}{\cos x}$

$= \sin x \cdot \dfrac{\sin x}{\cos x}$

$= \sin x \tan x$

35. $\dfrac{\cot x}{\sec x} = \dfrac{\cos x/\sin x}{1/\cos x} = \dfrac{\cos^2 x}{\sin x} = \dfrac{1 - \sin^2 x}{\sin x} = \dfrac{1}{\sin x} - \dfrac{\sin^2 x}{\sin x} = \csc x - \sin x$

37. $\sin^{1/2} x \cos x - \sin^{5/2} x \cos x = \sin^{1/2} x \cos x\left(1 - \sin^2 x\right) = \sin^{1/2} x \cos x \cdot \cos^2 x = \cos^3 x\sqrt{\sin x}$

39. $\left(1 + \sin y\right)\left[1 + \sin(-y)\right] = \left(1 + \sin y\right)\left(1 - \sin y\right)$

$= 1 - \sin^2 y$

$= \cos^2 y$

43. $\cot(-x) \neq \cot x$

The correct substitution is $\cot(-x) = -\cot x$.

$\dfrac{1}{\tan x} + \cot(-x) = \cot x - \cot x = 0$

41. $\sqrt{\dfrac{1 + \sin \theta}{1 - \sin \theta}} = \sqrt{\dfrac{1 + \sin \theta}{1 - \sin \theta} \cdot \dfrac{1 + \sin \theta}{1 + \sin \theta}}$

$= \sqrt{\dfrac{(1 + \sin \theta)^2}{1 - \sin^2 \theta}}$

$= \sqrt{\dfrac{(1 + \sin \theta)^2}{\cos^2 \theta}}$

$= \dfrac{1 + \sin \theta}{|\cos \theta|}$

45. (a)

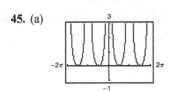

Identity

(b)

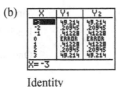

Identity

(c) $\left(1 + \cot^2 x\right)\left(\cos^2 x\right) = \csc^2 x \cos^2 x = \dfrac{1}{\sin^2 x} \cdot \cos^2 x = \cot^2 x$

47. (a)

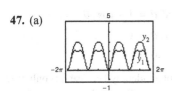

Not an identity

(b)

Not an identity

(c) $2 + \cos^2 x - 3 \cos^4 x = \left(1 - \cos^2 x\right)\left(2 + 3 \cos^2 x\right) = \sin^2 x\left(2 + 3 \cos^2 x\right) \neq \sin^2 x\left(3 + 2 \cos^2 x\right)$

49. (a)

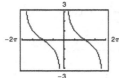

Identity

(b)

Identity

(c)
$$\frac{1 + \cos x}{\sin x} = \frac{(1 + \cos x)(1 - \cos x)}{\sin x(1 - \cos x)}$$

$$= \frac{1 - \cos^2 x}{\sin x(1 - \cos x)}$$

$$= \frac{\sin^2 x}{\sin x(1 - \cos x)}$$

$$= \frac{\sin x}{1 - \cos x}$$

51. $\tan^3 x \sec^2 x - \tan^3 x = \tan^3 x(\sec^2 x - 1)$

$$= \tan^3 x \tan^2 x$$

$$= \tan^5 x$$

53. $(\sin^2 x - \sin^4 x)\cos x = \sin^2 x(1 - \sin^2 x)\cos x$

$$= \sin^2 x \cos^2 x \cos x$$

$$= \sin^2 x \cos^3 x$$

55. $\sin^2 25° + \sin^2 65° = \sin^2 25° + \cos^2(90° - 65°)$

$$= \sin^2 25° + \cos^2 25°$$

$$= 1$$

57. Let $\theta = \sin^{-1} x \Rightarrow \sin \theta = x = \dfrac{x}{1}$.

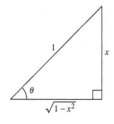

From the diagram,

$$\tan(\sin^{-1} x) = \tan \theta = \frac{x}{\sqrt{1 - x^2}}.$$

59. Let $\theta = \sin^{-1} \dfrac{x - 1}{4} \Rightarrow \sin \theta = \dfrac{x - 1}{4}$.

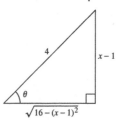

From the diagram,

$$\tan\left(\sin^{-1} \frac{x - 1}{4}\right) = \tan \theta = \frac{x - 1}{\sqrt{16 - (x - 1)^2}}.$$

61. $\cos x - \csc x \cot x = \cos x - \dfrac{1}{\sin x}\dfrac{\cos x}{\sin x}$

$$= \cos x\left(1 - \frac{1}{\sin^2 x}\right)$$

$$= \cos x(1 - \csc^2 x)$$

$$= -\cos x(\csc^2 x - 1)$$

$$= -\cos x \cot^2 x$$

63. False. $\tan x^2 = \tan(x \cdot x)$ and

$\tan^2 x = (\tan x)(\tan x),\ \tan x^2 \neq \tan^2 x.$

65. False. For the equation to be an identity, it must be true for all values of θ in the domain.

67. Because $\sin^2 \theta = 1 - \cos^2 \theta$, then

$\sin \theta = \pm\sqrt{1 - \cos^2 \theta};\ \sin \theta \neq \sqrt{1 - \cos^2 \theta}$ if θ lies in Quadrant III or IV.

One such angle is $\theta = \dfrac{7\pi}{4}$.

69.
$$1 - \cos \theta = \sin \theta$$
$$(1 - \cos \theta)^2 = (\sin \theta)^2$$
$$1 - 2\cos \theta + \cos^2 \theta = \sin^2 \theta$$
$$1 - 2\cos \theta + \cos^2 \theta = 1 - \cos^2 \theta$$
$$2\cos^2 \theta - 2\cos \theta = 0$$
$$2\cos \theta(\cos \theta - 1) = 0$$

The equation is not an identity because it is only true when $\cos \theta = 0$ or $\cos \theta = 1$. So, one angle for which the equation is not true is $-\dfrac{\pi}{2}$.

Section 2.3   Solving Trigonometric Equations

1. isolate

3. quadratic

5. $\tan x - \sqrt{3} = 0$

(a) $x = \dfrac{\pi}{3}$

$$\tan \dfrac{\pi}{3} - \sqrt{3} = \sqrt{3} - \sqrt{3} = 0$$

(b) $x = \dfrac{4\pi}{3}$

$$\tan \dfrac{4\pi}{3} - \sqrt{3} = \sqrt{3} - \sqrt{3} = 0$$

7. $3 \tan^2 2x - 1 = 0$

(a) $x = \dfrac{\pi}{12}$

$$3\left[\tan 2\left(\dfrac{\pi}{12}\right)\right]^2 - 1 = 3 \tan^2 \dfrac{\pi}{6} - 1$$
$$= 3\left(\dfrac{1}{\sqrt{3}}\right)^2 - 1$$
$$= 0$$

(b) $x = \dfrac{5\pi}{12}$

$$3\left[\tan 2\left(\dfrac{5\pi}{12}\right)\right]^2 - 1 = 3 \tan^2 \dfrac{5\pi}{6} - 1$$
$$= 3\left(-\dfrac{1}{\sqrt{3}}\right)^2 - 1$$
$$= 0$$

9. $2 \sin^2 x - \sin x - 1 = 0$

(a) $x = \dfrac{\pi}{2}$

$$2 \sin^2 \dfrac{\pi}{2} - \sin \dfrac{\pi}{2} - 1 = 2(1)^2 - 1 - 1$$
$$= 0$$

(b) $x = \dfrac{7\pi}{6}$

$$2 \sin^2 \dfrac{7\pi}{6} - \sin \dfrac{7\pi}{6} - 1 = 2\left(-\dfrac{1}{2}\right)^2 - \left(-\dfrac{1}{2}\right) - 1$$
$$= \dfrac{1}{2} + \dfrac{1}{2} - 1$$
$$= 0$$

11. $\sqrt{3} \csc x - 2 = 0$

$$\sqrt{3} \csc x = 2$$
$$\csc x = \dfrac{2}{\sqrt{3}}$$
$$x = \dfrac{\pi}{3} + 2n\pi$$
or $x = \dfrac{2\pi}{3} + 2n\pi$

13. $\cos x + 1 = -\cos x$

$$2 \cos x + 1 = 0$$
$$\cos x = -\dfrac{1}{2}$$
$$x = \dfrac{2\pi}{3} + 2n\pi \text{ or } x = \dfrac{4\pi}{3} + 2n\pi$$

15. $3 \sec^2 x - 4 = 0$

$$\sec^2 x = \dfrac{4}{3}$$
$$\sec x = \pm \dfrac{2}{\sqrt{3}}$$
$$x = \dfrac{\pi}{6} + n\pi$$
or $x = \dfrac{5\pi}{6} + n\pi$

17. $4 \cos^2 x - 1 = 0$

$$\cos^2 x = \dfrac{1}{4}$$
$$\cos x = \pm \dfrac{1}{2}$$
$$x = \dfrac{\pi}{3} + n\pi \quad \text{or} \quad x = \dfrac{2\pi}{3} + n\pi$$

19. $\sin x(\sin x + 1) = 0$

$\sin x = 0 \quad$ or $\quad \sin x = -1$

$x = n\pi \qquad\qquad x = \dfrac{3\pi}{2} + 2n\pi$

21. $\cos^3 x - \cos x = 0$

$$\cos x(\cos^2 x - 1) = 0$$

$\cos x = 0 \qquad$ or $\cos^2 x - 1 = 0$

$x = \dfrac{\pi}{2} + n\pi \qquad\qquad \cos x = \pm 1$

$\qquad\qquad\qquad\qquad x = n\pi$

Both of these answers can be represented as $x = \dfrac{n\pi}{2}$.

23. $3 \tan^3 x = \tan x$

$3 \tan^3 x - \tan x = 0$

$\tan x \left(3 \tan^2 x - 1 \right) = 0$

$\tan x = 0$ or $3 \tan^2 x - 1 = 0$

$x = n\pi$

$\tan x = \pm \dfrac{\sqrt{3}}{3}$

$x = \dfrac{\pi}{6} + n\pi, \dfrac{5\pi}{6} + n\pi$

25. $2 \cos^2 x + \cos x - 1 = 0$

$(2 \cos x - 1)(\cos x + 1) = 0$

$2 \cos x - 1 = 0$ or $\cos x + 1 = 0$

$\cos x = \dfrac{1}{2}$ $\cos x = -1$

$x = \pi + 2n\pi$

$x = \dfrac{\pi}{3} + 2n\pi, \dfrac{5\pi}{3} + 2n\pi$

27. $\sec^2 x - \sec x = 2$

$\sec^2 x - \sec x - 2 = 0$

$(\sec x - 2)(\sec x + 1) = 0$

$\sec x - 2 = 0$ or $\sec x + 1 = 0$

$\sec x = 2$ $\sec x = -1$

$x = \pi + 2n\pi$

$x = \dfrac{\pi}{3} + 2n\pi, \dfrac{5\pi}{3} + 2n\pi$

29. $\sin x - 2 = \cos x - 2$

$\sin x = \cos x$

$\dfrac{\sin x}{\cos x} = 1$

$\tan x = 1$

$x = \tan^{-1} 1$

$x = \dfrac{\pi}{4}, \dfrac{5\pi}{4}$

31. $2 \sin^2 x = 2 + \cos x$

$2 - 2 \cos^2 x = 2 + \cos x$

$2 \cos^2 x + \cos x = 0$

$\cos x (2 \cos x + 1) = 0$

$\cos x = 0$ or $2 \cos x + 1 = 0$

$x = \dfrac{\pi}{2}, \dfrac{3\pi}{2}$ $2 \cos x = -1$

$\cos x = -\dfrac{1}{2}$

$x = \dfrac{2\pi}{3}, \dfrac{4\pi}{3}$

33. $\sin^2 x = 3 \cos^2 x$

$\sin^2 x - 3 \cos^2 x = 0$

$\sin^2 x - 3 \left(1 - \sin^2 x \right) = 0$

$4 \sin^2 x = 3$

$\sin x = \pm \dfrac{\sqrt{3}}{2}$

$x = \dfrac{\pi}{3}, \dfrac{2\pi}{3}, \dfrac{4\pi}{3}, \dfrac{5\pi}{3}$

35. $2 \sin x + \csc x = 0$

$2 \sin x + \dfrac{1}{\sin x} = 0$

$2 \sin^2 x + 1 = 0$

$\sin^2 x = -\dfrac{1}{2} \Rightarrow$ No solution

37.
$$\csc x + \cot x = 1$$
$$(\csc x + \cot x)^2 = 1^2$$
$$\csc^2 x + 2\csc x \cot x + \cot^2 x = 1$$
$$\cot^2 x + 1 + 2\csc x \cot x + \cot^2 x = 1$$
$$2\cot^2 x + 2\csc x \cot x = 0$$
$$2\cot x(\cot x + \csc x) = 0$$

$2\cot x = 0$ or $\cot x + \csc x = 0$

$$x = \frac{\pi}{2}, \frac{3\pi}{2} \qquad\qquad \frac{\cos x}{\sin x} = -\frac{1}{\sin x}$$

$\left(\dfrac{3\pi}{2}\text{ is extraneous.}\right)$ $\qquad\qquad \cos x = -1$

$$x = \pi$$

$$(\pi \text{ is extraneous.})$$

$x = \pi/2$ is the only solution.

39. $2\cos 2x - 1 = 0$

$$\cos 2x = \frac{1}{2}$$

$$2x = \frac{\pi}{3} + 2n\pi \quad\text{or}\quad 2x = \frac{5\pi}{3} + 2n\pi$$

$$x = \frac{\pi}{6} + n\pi \qquad\qquad x = \frac{5\pi}{6} + n\pi$$

41. $\tan 3x - 1 = 0$

$$\tan 3x = 1$$

$$3x = \frac{\pi}{4} + n\pi$$

$$x = \frac{\pi}{12} + \frac{n\pi}{3}$$

43. $2\cos\dfrac{x}{2} = \sqrt{2} = 0$

$$\cos\frac{x}{2} = \frac{\sqrt{2}}{2}$$

$$\frac{x}{2} = \frac{\pi}{4} + 2n\pi \quad\text{or}\quad \frac{x}{2} = \frac{7\pi}{4} + 2n\pi$$

$$x = \frac{\pi}{2} + 4n\pi \qquad\qquad x = \frac{7\pi}{2} + 4n\pi$$

45. $3\tan\dfrac{x}{2} - \sqrt{3} = 0$

$$\tan\frac{x}{2} = \frac{\sqrt{3}}{3}$$

$$\frac{x}{2} = \frac{\pi}{6} + n\pi \Rightarrow x = \frac{\pi}{3} + 2n\pi$$

47. $y = \sin\dfrac{\pi x}{2} + 1$

$$\sin\left(\frac{\pi x}{2}\right) + 1 = 0$$

$$\sin\left(\frac{\pi x}{2}\right) = -1$$

$$\frac{\pi x}{2} = \frac{3\pi}{2} + 2n\pi$$

$$x = 3 + 4n$$

For $-2 < x < 4$, the intercepts are -1 and 3.

49. $5\sin x + 2 = 0$

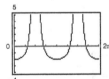

$x \approx 3.553$ and $x \approx 5.872$

51. $\sin x - 3\cos x = 0$

$x \approx 1.249$ and $x \approx 4.391$

53. $\cos x = x$

$x \approx 0.739$

55. $\sec^2 x - 3 = 0$

$x \approx 0.955,\ x \approx 2.186,\ x \approx 4.097$ and $x \approx 5.328$

57. $2\tan^2 x = 15$

$x \approx 1.221,\ x \approx 1.921,\ x \approx 4.362$ and $x \approx 5.062$

59. $\tan^2 x + \tan x - 12 = 0$

$(\tan x + 4)(\tan x - 3) = 0$

$\tan x + 4 = 0 \qquad\qquad$ or $\quad \tan x - 3 = 0$

$\qquad \tan x = -4 \qquad\qquad\qquad \tan x = 3$

$\qquad\qquad x = \arctan(-4) + n\pi \qquad\qquad x = \arctan 3 + n\pi$

61. $\qquad\quad \sec^2 x - 6 \tan x = -4$

$1 + \tan^2 x - 6 \tan x + 4 = 0$

$\qquad \tan^2 x - 6 \tan x + 5 = 0$

$\qquad (\tan x - 1)(\tan x - 5) = 0$

$\tan x - 1 = 0 \qquad\quad \tan x - 5 = 0$

$\quad \tan x = 1 \qquad\qquad \tan x = 5$

$\qquad\quad x = \dfrac{\pi}{4} + n\pi \qquad\quad x = \arctan 5 + n\pi$

63. $\qquad\quad 2\sin^2 x + 5\cos x = 4$

$2\left(1 - \cos^2 x\right) + 5\cos x - 4 = 0$

$\quad -2\cos^2 x + 5\cos x - 2 = 0$

$\quad -(2\cos x - 1)(\cos x - 2) = 0$

$2\cos x - 1 = 0 \qquad\qquad$ or $\quad \cos x - 2 = 0$

$\qquad \cos x = \dfrac{1}{2} \qquad\qquad\qquad\qquad \cos x = 2$

$\qquad\qquad x = \dfrac{\pi}{3} + 2n\pi, \dfrac{5\pi}{3} + 2n\pi \qquad$ No solution

65. $\cot^2 x - 9 = 0$

$\quad \cot^2 x = 9$

$\qquad \dfrac{1}{9} = \tan^2 x$

$\qquad \pm\dfrac{1}{3} = \tan x$

$\qquad\qquad x = \arctan \tfrac{1}{3} + n\pi, \arctan\left(-\tfrac{1}{3}\right) + n\pi$

67. $\sec^2 x - 4 \sec x = 0$

$\sec x(\sec x - 4) = 0$

$\sec x = 0 \quad\quad \sec x - 4 = 0$

No solution $\qquad \sec x = 4$

$\qquad\qquad\qquad \dfrac{1}{4} = \cos x$

$\qquad\qquad\qquad\quad x = \arccos\dfrac{1}{4} + 2n\pi, -\arccos\dfrac{1}{4} + 2n\pi$

69. $\csc^2 x + 3 \csc x - 4 = 0$

$(\csc x + 4)(\csc x - 1) = 0$

$\csc x + 4 = 0$ $\qquad\qquad$ or $\quad \csc x - 1 = 0$

$\qquad \csc x = -4$ $\qquad\qquad\qquad\qquad \csc x = 1$

$\qquad -\dfrac{1}{4} = \sin x$ $\qquad\qquad\qquad\qquad 1 = \sin x$

$\qquad x = \arcsin\!\left(\dfrac{1}{4}\right) + 2n\pi,\ \arcsin\!\left(-\dfrac{1}{4}\right) + 2n\pi$ $\qquad x = \dfrac{\pi}{2} + 2n\pi$

71. $12 \sin^2 x - 13 \sin x + 3 = 0$

$$\sin x = \frac{-(-13) \pm \sqrt{(-13)^2 - 4(12)(3)}}{2(12)} = \frac{13 \pm 5}{24}$$

$\sin x = \dfrac{1}{3}$ $\quad$ or $\qquad \sin x = \dfrac{3}{4}$

$x \approx 0.3398,\ 2.8018$ $\qquad x \approx 0.8481,\ 2.2935$

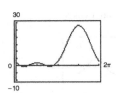

The x-intercepts occur at $x \approx 0.3398$,

$x \approx 0.8481,\ x \approx 2.2935,$ and $x \approx 2.8018.$

73. $\tan^2 x + 3 \tan x + 1 = 0$

$$\tan x = \frac{-3 \pm \sqrt{3^2 - 4(1)(1)}}{2(1)} = \frac{-3 \pm \sqrt{5}}{2}$$

$\tan x = \dfrac{-3 - \sqrt{5}}{2}$ $\quad$ or $\quad \tan x = \dfrac{-3 + \sqrt{5}}{2}$

$x \approx 1.9357,\ 5.0773$ $\qquad x \approx 2.7767,\ 5.9183$

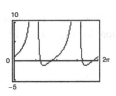

The x-intercepts occur at $x \approx 1.9357,\ x \approx 2.7767,$

$x \approx 5.0773,$ and $x \approx 5.9183.$

75. $3 \tan^2 x + 5 \tan x - 4 = 0,\ \left[-\dfrac{\pi}{2}, \dfrac{\pi}{2}\right]$

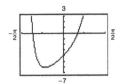

$x \approx -1.154,\ 0.534$

77. $4 \cos^2 x - 2 \sin x + 1 = 0,\ \left[-\dfrac{\pi}{2}, \dfrac{\pi}{2}\right]$

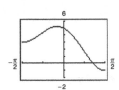

$x \approx 1.110$

79. (a) $f(x) = \sin^2 x + \cos x$

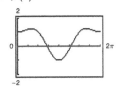

Maximum: $(1.0472, 1.25)$

Maximum: $(5.2360, 1.25)$

Minimum: $(0, 1)$

Minimum: $(3.1416, -1)$

(b) $2 \sin x \cos x - \sin x = 0$

$\sin x (2 \cos x - 1) = 0$

$\sin x = 0$ $\quad$ or $\qquad 2 \cos x - 1 = 0$

$x = 0,\ \pi$ $\qquad\qquad\qquad \cos x = \dfrac{1}{2}$

$\approx 0,\ 3.1416$

$\qquad\qquad\qquad\qquad x = \dfrac{\pi}{3},\ \dfrac{5\pi}{3}$

$\qquad\qquad\qquad\qquad \approx 1.0472,\ 5.2360$

81. (a) $f(x) = \sin x + \cos x$

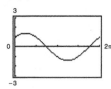

Maximum: $(0.7854, 1.4142)$

Minimum: $(3.9270, -1.4142)$

(b) $\cos x - \sin x = 0$

$$\cos x = \sin x$$

$$1 = \frac{\sin x}{\cos x}$$

$$\tan x = 1$$

$$x = \frac{\pi}{4}, \frac{5\pi}{4}$$

$$\approx 0.7854, 3.9270$$

83. (a) $f(x) = \sin x \cos x$

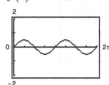

Maximum: $(0.7854, 0.5)$

Maximum: $(3.9270, 0.5)$

Minimum: $(2.3562, -0.5)$

Minimum: $(5.4978, -0.5)$

(b) $-\sin^2 x + \cos^2 x = 0$

$$-\sin^2 x + 1 - \sin^2 x = 0$$

$$-2\sin^2 x + 1 = 0$$

$$\sin^2 x = \frac{1}{2}$$

$$\sin x = \pm\sqrt{\frac{1}{2}} = \pm\frac{\sqrt{2}}{2}$$

$$x = \frac{\pi}{4}, \frac{3\pi}{4}, \frac{5\pi}{4}, \frac{7\pi}{4}$$

$$\approx 0.7854, 2.3562, 3.9270, 5.4978$$

85. The graphs of $y_1 = 2\sin x$ and $y_2 = 3x + 1$ appear to have one point of intersection. This implies there is one solution to the equation $2\sin x = 3x + 1$.

87. $f(x) = \dfrac{\sin x}{x}$

(a) Domain: all real numbers except $x = 0$.

(b) The graph has y-axis symmetry.

(c) As $x \to 0$, $f(x) \to 1$.

(d) $\dfrac{\sin x}{x} = 0$ has four solutions in the interval $[-8, 8]$.

$$\sin x\left(\frac{1}{x}\right) = 0$$

$$\sin x = 0$$

$$x = -2\pi, -\pi, \pi, 2\pi$$

89.
$$y = \frac{1}{12}(\cos 8t - 3\sin 8t)$$

$$\frac{1}{12}(\cos 8t - 3\sin 8t) = 0$$

$$\cos 8t = 3\sin 8t$$

$$\frac{1}{3} = \tan 8t$$

$$8t \approx 0.32175 + n\pi$$

$$t \approx 0.04 + \frac{n\pi}{8}$$

In the interval $0 \le t \le 1$, $t \approx 0.04, 0.43,$ and 0.83.

91. Graph $y_1 = 58.3 + 32\cos\left(\dfrac{\pi t}{6}\right)$

$y_2 = 75$.

Left point of intersection: $(1.95, 75)$

Right point of intersection: $(10.05, 75)$

So, sales exceed 7500 in January, November, and December.

93. (a) and (c)

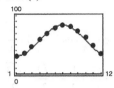

The model fits the data well.

(b) $C = a \cos(bt - c) + d$

$$a = \frac{1}{2}[\text{high} - \text{low}] = \frac{1}{2}[84.1 - 31.0] = 26.55$$

$$p = 2[\text{high time} - \text{low time}] = 2[7 - 1] = 12$$

$$b = \frac{2\pi}{p} = \frac{2\pi}{12} = \frac{\pi}{6}$$

The maximum occurs at 7, so the left end point is

$$\frac{c}{b} = 7 \Rightarrow c = 7\left(\frac{\pi}{6}\right) = \frac{7\pi}{6}$$

$$d = \frac{1}{2}[\text{high} + \text{low}] = \frac{1}{2}[93.6 + 62.3] = 57.55$$

$$C = 26.55 \cos\left(\frac{\pi}{6}t - \frac{7\pi}{6}\right) + 57.55$$

(d) The constant term, d, gives the average maximum temperature.

The average maximum temperature in Chicago is 57.55°F.

(e) The average maximum temperature is above 72°F from June through September. The average maximum temperature is below 70°F from October through May.

95. $A = 2x \cos x, 0 < x < \dfrac{\pi}{2}$

(a)

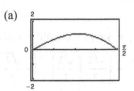

The maximum area of $A \approx 1.12$ occurs when $x \approx 0.86$.

(b) $A \geq 1$ for $0.6 < x < 1.1$

97. $f(x) = \tan \dfrac{\pi x}{4}$

Because $\tan \pi/4 = 1$, $x = 1$ is the smallest nonnegative fixed point.

99. True. The period of $2 \sin 4t - 1$ is $\dfrac{\pi}{2}$ and the period of $2 \sin t - 1$ is 2π.

In the interval $[0, 2\pi)$ the first equation has four cycles whereas the second equation has only one cycle, so the first equation has four times the x-intercepts (solutions) as the second equation.

101. $\cot x \cos^2 x = 2 \cot x$

$$\cos^2 x = 2$$

$$\cos x = \pm\sqrt{2}$$

No solution

Because you solved this problem by first dividing by $\cot x$, you do not get the same solution as Example 3.

When solving equations, you do not want to divide each side by a variable expression that will cancel out because you may accidentally remove one of the solutions.

103. (a)

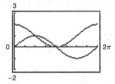

The graphs intersect when $x = \dfrac{\pi}{2}$ and $x = \pi$.

(b)

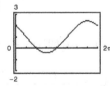

The x-intercepts are $\left(\dfrac{\pi}{2}, 0\right)$ and $(\pi, 0)$.

(c) Both methods produce the same x-values. Answers will vary on which method is preferred.

Section 2.4 Sum and Difference Formulas

1. $\sin u \cos v - \cos u \sin v$

3. $\dfrac{\tan u + \tan v}{1 - \tan u \tan v}$

5. $\cos u \cos v + \sin u \sin v$

7. (a) $\cos\!\left(\dfrac{\pi}{4} + \dfrac{\pi}{3}\right) = \cos\dfrac{\pi}{4}\cos\dfrac{\pi}{3} - \sin\dfrac{\pi}{4}\sin\dfrac{\pi}{3}$

$$= \dfrac{\sqrt{2}}{2}\cdot\dfrac{1}{2} - \dfrac{\sqrt{2}}{2}\cdot\dfrac{\sqrt{3}}{2}$$

$$= \dfrac{\sqrt{2} - \sqrt{6}}{4}$$

(b) $\cos\dfrac{\pi}{4} + \cos\dfrac{\pi}{3} = \dfrac{\sqrt{2}}{2} + \dfrac{1}{2} = \dfrac{\sqrt{2}+1}{2}$

9. (a) $\sin(135° - 30°) = \sin 135° \cos 30° - \cos 135° \sin 30°$

$$= \left(\dfrac{\sqrt{2}}{2}\right)\!\left(\dfrac{\sqrt{3}}{2}\right) - \left(-\dfrac{\sqrt{2}}{2}\right)\!\left(\dfrac{1}{2}\right) = \dfrac{\sqrt{6}+\sqrt{2}}{4}$$

(b) $\sin 135° - \cos 30° = \dfrac{\sqrt{2}}{2} - \dfrac{\sqrt{3}}{2} = \dfrac{\sqrt{2}-\sqrt{3}}{2}$

11. $\sin\dfrac{11\pi}{12} = \sin\!\left(\dfrac{3\pi}{4} + \dfrac{\pi}{6}\right) = \sin\dfrac{3\pi}{4}\cos\dfrac{\pi}{6} + \cos\dfrac{3\pi}{4}\sin\dfrac{\pi}{6} = \dfrac{\sqrt{2}}{2}\cdot\dfrac{\sqrt{3}}{2} + \left(-\dfrac{\sqrt{2}}{2}\right)\dfrac{1}{2} = \dfrac{\sqrt{2}}{4}(\sqrt{3}-1)$

$\cos\dfrac{11\pi}{12} = \cos\!\left(\dfrac{3\pi}{4} + \dfrac{\pi}{6}\right) = \cos\dfrac{3\pi}{4}\cos\dfrac{\pi}{6} - \sin\dfrac{3\pi}{4}\sin\dfrac{\pi}{6} = -\dfrac{\sqrt{2}}{2}\cdot\dfrac{\sqrt{3}}{2} - \dfrac{\sqrt{2}}{2}\cdot\dfrac{1}{2} = -\dfrac{\sqrt{2}}{4}(\sqrt{3}+1)$

$\tan\dfrac{11\pi}{4} = \tan\!\left(\dfrac{3\pi}{4} + \dfrac{\pi}{6}\right) = \dfrac{\tan\dfrac{3\pi}{4} + \tan\dfrac{\pi}{6}}{1 - \tan\dfrac{3\pi}{4}\tan\dfrac{\pi}{6}} = \dfrac{-1 + \dfrac{\sqrt{3}}{3}}{1 - (-1)\dfrac{\sqrt{3}}{3}} = \dfrac{-3 + \sqrt{3}}{3 + \sqrt{3}}\cdot\dfrac{3 - \sqrt{3}}{3 - \sqrt{3}} = \dfrac{-12 + 6\sqrt{3}}{6} = -2 + \sqrt{3}$

13. $\sin\dfrac{17\pi}{12} = \sin\!\left(\dfrac{9\pi}{4} - \dfrac{5\pi}{6}\right) = \sin\dfrac{9\pi}{4}\cos\dfrac{5\pi}{6} - \cos\dfrac{9\pi}{4}\sin\dfrac{5\pi}{6} = \dfrac{\sqrt{2}}{2}\left(-\dfrac{\sqrt{3}}{2}\right) - \left(\dfrac{\sqrt{2}}{2}\right)\!\left(\dfrac{1}{2}\right) = -\dfrac{\sqrt{2}}{4}(\sqrt{3}+1)$

$\cos\dfrac{17\pi}{12} = \cos\!\left(\dfrac{9\pi}{4} - \dfrac{5\pi}{6}\right) = \cos\dfrac{9\pi}{4}\cos\dfrac{5\pi}{6} + \sin\dfrac{9\pi}{4}\sin\dfrac{5\pi}{6} = \dfrac{\sqrt{2}}{2}\left(-\dfrac{\sqrt{3}}{2}\right) + \dfrac{\sqrt{2}}{2}\left(\dfrac{1}{2}\right) = \dfrac{\sqrt{2}}{4}(1 - \sqrt{3})$

$\tan\dfrac{17\pi}{12} = \tan\!\left(\dfrac{9\pi}{4} - \dfrac{5\pi}{6}\right) = \dfrac{\tan(9\pi/4) - \tan(5\pi/6)}{1 + \tan(9\pi/4)\tan(5\pi/6)} = \dfrac{1 - \left(-\sqrt{3}/3\right)}{1 + \left(-\sqrt{3}/3\right)} = \dfrac{3 + \sqrt{3}}{3 - \sqrt{3}}\cdot\dfrac{3 + \sqrt{3}}{3 + \sqrt{3}} = \dfrac{12 + 6\sqrt{3}}{6} = 2 + \sqrt{3}$

15. $\sin 105° = \sin(60° + 45°) = \sin 60°\cos 45° + \cos 60°\sin 45° = \dfrac{\sqrt{3}}{2}\cdot\dfrac{\sqrt{2}}{2} + \dfrac{1}{2}\cdot\dfrac{\sqrt{2}}{2} = \dfrac{\sqrt{2}}{4}(\sqrt{3}+1)$

$\cos 105° = \cos(60° + 45°) = \cos 60°\cos 45° - \sin 60°\sin 45° = \dfrac{1}{2}\cdot\dfrac{\sqrt{2}}{2} - \dfrac{\sqrt{3}}{2}\cdot\dfrac{\sqrt{2}}{2} = \dfrac{\sqrt{2}}{4}(1 - \sqrt{3})$

$\tan 105° = \tan(60° + 45°) = \dfrac{\tan 60° + \tan 45°}{1 - \tan 60°\tan 45°} = \dfrac{\sqrt{3}+1}{1 - \sqrt{3}} = \dfrac{\sqrt{3}+1}{1 - \sqrt{3}}\cdot\dfrac{1 + \sqrt{3}}{1 + \sqrt{3}} = \dfrac{4 + 2\sqrt{3}}{-2} = -2 - \sqrt{3}$

17. $\sin(-195°) = \sin(30° - 225°) = \sin 30° \cos 225° - \cos 30° \sin 225° = \sin 30°(-\cos 45°) - \cos 30°(-\sin 45°)$

$$= \frac{1}{2}\left(-\frac{\sqrt{2}}{2}\right) - \frac{\sqrt{3}}{2}\left(-\frac{\sqrt{2}}{2}\right) = -\frac{\sqrt{2}}{4}\left(1 - \sqrt{3}\right) = \frac{\sqrt{2}}{4}\left(\sqrt{3} - 1\right)$$

$\cos(-195°) = \cos(30° - 225°) = \cos 30° \cos 225° + \sin 30° \sin 225° = \cos 30°(-\cos 45°) + \sin 30°(-45°)$

$$= \frac{\sqrt{3}}{2}\left(-\frac{\sqrt{2}}{2}\right) + \frac{1}{2}\left(-\frac{\sqrt{2}}{2}\right) = -\frac{\sqrt{2}}{4}\left(\sqrt{3} + 1\right)$$

$\tan(-195°) = \tan(30° - 225°) = \dfrac{\tan 30° - \tan 225°}{1 + \tan 30° \tan 225°} = \dfrac{\tan 30° - \tan 45°}{1 + \tan 30° \tan 45°}$

$$= \frac{\left(\dfrac{\sqrt{3}}{3}\right) - 1}{1 + \left(\dfrac{\sqrt{3}}{3}\right)} = \frac{\sqrt{3} - 3}{3 + \sqrt{3}} \cdot \frac{3 - \sqrt{3}}{3 - \sqrt{3}} = \frac{-12 + 6\sqrt{3}}{6} = -2 + \sqrt{3}$$

19. $\dfrac{13\pi}{12} = \dfrac{3\pi}{4} + \dfrac{\pi}{3}$

$\sin \dfrac{13\pi}{12} = \sin\left(\dfrac{3\pi}{4} + \dfrac{\pi}{3}\right)$

$\qquad = \sin \dfrac{3\pi}{4} \cos \dfrac{\pi}{3} + \cos \dfrac{3\pi}{4} \sin \dfrac{\pi}{3}$

$\qquad = \dfrac{\sqrt{2}}{2} \cdot \dfrac{1}{2} + \left(-\dfrac{\sqrt{2}}{2}\right)\left(\dfrac{\sqrt{3}}{2}\right)$

$\qquad = \dfrac{\sqrt{2}}{4}\left(1 - \sqrt{3}\right)$

$\cos \dfrac{13\pi}{12} = \cos\left(\dfrac{3\pi}{4} + \dfrac{\pi}{3}\right)$

$\qquad = \cos \dfrac{3\pi}{4} \cos \dfrac{\pi}{3} - \sin \dfrac{3\pi}{4} \sin \dfrac{\pi}{3}$

$\qquad = -\dfrac{\sqrt{2}}{2} \cdot \dfrac{1}{2} - \dfrac{\sqrt{2}}{2} \cdot \dfrac{\sqrt{3}}{2} = -\dfrac{\sqrt{2}}{4}\left(1 + \sqrt{3}\right)$

$\tan \dfrac{13\pi}{12} = \tan\left(\dfrac{3\pi}{4} + \dfrac{\pi}{3}\right)$

$\qquad = \dfrac{\tan\left(\dfrac{3\pi}{4}\right) + \tan\left(\dfrac{\pi}{3}\right)}{1 - \tan\left(\dfrac{3\pi}{4}\right) \tan\left(\dfrac{\pi}{3}\right)}$

$\qquad = \dfrac{-1 + \sqrt{3}}{1 - (-1)\left(\sqrt{3}\right)}$

$\qquad = -\dfrac{1 - \sqrt{3}}{1 + \sqrt{3}} \cdot \dfrac{1 - \sqrt{3}}{1 - \sqrt{3}}$

$\qquad = -\dfrac{4 - 2\sqrt{3}}{-2}$

$\qquad = 2 - \sqrt{3}$

21. $-\dfrac{5\pi}{12} = -\dfrac{\pi}{4} - \dfrac{\pi}{6}$

$\sin\left(-\dfrac{\pi}{4} - \dfrac{\pi}{6}\right) = \sin\left(-\dfrac{\pi}{4}\right)\cos \dfrac{\pi}{6} - \cos\left(-\dfrac{\pi}{4}\right)\sin \dfrac{\pi}{6} = \left(-\dfrac{\sqrt{2}}{2}\right)\left(\dfrac{\sqrt{3}}{2}\right) - \left(\dfrac{\sqrt{2}}{2}\right)\left(\dfrac{1}{2}\right) = -\dfrac{\sqrt{2}}{4}\left(\sqrt{3} + 1\right)$

$\cos\left(-\dfrac{\pi}{4} - \dfrac{\pi}{6}\right) = \cos\left(-\dfrac{\pi}{4}\right)\cos \dfrac{\pi}{6} + \sin\left(-\dfrac{\pi}{4}\right)\sin \dfrac{\pi}{6} = \left(\dfrac{\sqrt{2}}{2}\right)\left(\dfrac{\sqrt{3}}{2}\right) + \left(-\dfrac{\sqrt{2}}{2}\right)\left(\dfrac{1}{2}\right) = \dfrac{\sqrt{2}}{4}\left(\sqrt{3} - 1\right)$

$\tan\left(-\dfrac{\pi}{4} - \dfrac{\pi}{6}\right) = \dfrac{\tan\left(-\dfrac{\pi}{4}\right) - \tan \dfrac{\pi}{6}}{1 + \tan\left(-\dfrac{\pi}{4}\right)\tan \dfrac{\pi}{6}} = \dfrac{-1 - \dfrac{\sqrt{3}}{3}}{1 + (-1)\left(\dfrac{\sqrt{3}}{3}\right)} = \dfrac{-3 - \sqrt{3}}{3 - \sqrt{3}} = \dfrac{-3 - \sqrt{3}}{3 - \sqrt{3}} \cdot \dfrac{3 + \sqrt{3}}{3 + \sqrt{3}} = \dfrac{-12 - 6\sqrt{3}}{6} = -2 - \sqrt{3}$

23. $285° = 225° + 60°$

$\sin 285° = \sin(225° + 60°) = \sin 225° \cos 60° + \cos 225° \sin 60°$

$$= -\frac{\sqrt{2}}{2}\left(\frac{1}{2}\right) - \frac{\sqrt{2}}{2}\left(\frac{\sqrt{3}}{2}\right) = -\frac{\sqrt{2}}{4}\left(\sqrt{3} + 1\right)$$

$\cos 285° = \cos(225° + 60°) = \cos 225° \cos 60° - \sin 225° \sin 60°$

$$= -\frac{\sqrt{2}}{2}\left(\frac{1}{2}\right) - \left(-\frac{\sqrt{2}}{2}\right)\left(\frac{\sqrt{3}}{2}\right) = \frac{\sqrt{2}}{4}\left(\sqrt{3} - 1\right)$$

$\tan 285° = \tan(225° + 60°) = \dfrac{\tan 225° + \tan 60°}{1 - \tan 225° \tan 60°}$

$$= \frac{1 + \sqrt{3}}{1 - \sqrt{3}} \cdot \frac{1 + \sqrt{3}}{1 + \sqrt{3}} = \frac{4 + 2\sqrt{3}}{-2} = -2 - \sqrt{3} = -\left(2 + \sqrt{3}\right)$$

25. $-165° = -(120° + 45°)$

$\sin(-165°) = \sin\left[-(120° + 45°)\right] = -\sin(120° + 45°) = -\left[\sin 120° \cos 45° + \cos 120° \sin 45°\right]$

$$= -\left[\frac{\sqrt{3}}{2} \cdot \frac{\sqrt{2}}{2} - \frac{1}{2} \cdot \frac{\sqrt{2}}{2}\right] = -\frac{\sqrt{2}}{4}\left(\sqrt{3} - 1\right)$$

$\cos(-165°) = \cos\left[-(120° + 45°)\right] = \cos(120° + 45°) = \cos 120° \cos 45° - \sin 120° \sin 45°$

$$= -\frac{1}{2} \cdot \frac{\sqrt{2}}{2} - \frac{\sqrt{3}}{2} \cdot \frac{\sqrt{2}}{2} = -\frac{\sqrt{2}}{4}\left(1 + \sqrt{3}\right)$$

$\tan(-165°) = \tan\left[-(120° + 45°)\right] = -\tan(120° + \tan 45°) = -\dfrac{\tan 120° + \tan 45°}{1 - \tan 120° \tan 45°}$

$$= -\frac{-\sqrt{3} + 1}{1 - \left(-\sqrt{3}\right)(1)} = -\frac{1 - \sqrt{3}}{1 + \sqrt{3}} \cdot \frac{1 - \sqrt{3}}{1 - \sqrt{3}} = -\frac{4 - 2\sqrt{3}}{-2} = 2 - \sqrt{3}$$

27. $\sin 3 \cos 1.2 - \cos 3 \sin 1.2 = \sin(3 - 1.2) = \sin 1.8$

29. $\sin 60° \cos 15° + \cos 60° \sin 15° = \sin(60° + 15°)$
$$= \sin 75°$$

31. $\dfrac{\tan\left(\pi/15\right) + \tan\left(2\pi/5\right)}{1 - \tan\left(\pi/15\right)\tan\left(2\pi/5\right)} = \tan(\pi/15 + 2\pi/5)$

$$= \tan(7\pi/15)$$

33. $\cos 3x \cos 2y + \sin 3x \sin 2y = \cos(3x - 2y)$

35. $\sin \dfrac{\pi}{12} \cos \dfrac{\pi}{4} + \cos \dfrac{\pi}{12} \sin \dfrac{\pi}{4} = \sin\left(\dfrac{\pi}{12} + \dfrac{\pi}{4}\right)$

$$= \sin \frac{\pi}{3}$$

$$= \frac{\sqrt{3}}{2}$$

37. $\cos 130° \cos 10° + \sin 130° \sin 10° = \cos(130° - 10°)$

$$= \cos 120°$$

$$= -\frac{1}{2}$$

39. $\dfrac{\tan(9\pi/8) - \tan(\pi/8)}{1 + \tan(9\pi/8)\tan(\pi/8)} = \tan\left(\dfrac{9\pi}{8} - \dfrac{\pi}{8}\right)$

$$= \tan \pi$$

$$= 0$$

For Exercises 41–45, you have:

$\sin u = -\frac{3}{5}$, u in Quadrant IV $\Rightarrow \cos u = \frac{4}{5}$, $\tan u = -\frac{4}{3}$

$\cos v = \frac{15}{17}$, v in Quadrant I $\Rightarrow \sin v = \frac{8}{17}$, $\tan v = \frac{8}{15}$

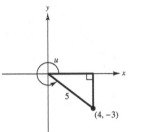 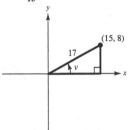

Figures for Exercises 41–45

41. $\sin(u + v) = \sin u \cos v + \cos u \sin v = \left(-\frac{3}{5}\right)\left(\frac{15}{17}\right) + \left(\frac{4}{5}\right)\left(\frac{8}{17}\right) = -\frac{13}{85}$

43. $\tan(u + v) = \dfrac{\tan u + \tan v}{1 - \tan u \tan v} = \dfrac{-\dfrac{3}{4} + \left(\dfrac{8}{15}\right)}{1 - \left(-\dfrac{3}{4}\right)\left(\dfrac{8}{15}\right)} = \dfrac{-\dfrac{13}{60}}{1 + \dfrac{32}{60}} = \left(-\dfrac{13}{60}\right)\left(\dfrac{5}{7}\right) = -\dfrac{13}{84}$

45. $\sec(v - u) = \dfrac{1}{\cos(v - u)} = \dfrac{1}{\cos v \cos u + \sin v \sin u} = \dfrac{1}{\left(\dfrac{15}{17}\right)\left(\dfrac{4}{5}\right) + \left(\dfrac{8}{17}\right)\left(-\dfrac{3}{5}\right)} = \dfrac{1}{\left(\dfrac{60}{85}\right) + \left(-\dfrac{24}{85}\right)} = \dfrac{1}{\dfrac{36}{85}} = \dfrac{85}{36}$

For Exercises 47–51, you have:

$\sin u = -\frac{7}{25}$, u in Quadrant III $\Rightarrow \cos u = -\frac{24}{25}$, $\tan u = \frac{7}{24}$

$\cos v = -\frac{4}{5}$, v in Quadrant III $\Rightarrow \sin v = -\frac{3}{5}$, $\tan v = \frac{3}{4}$

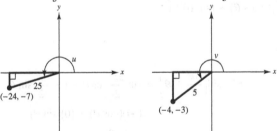

Figures for Exercises 47–51

47. $\cos(u + v) = \cos u \cos v - \sin u \sin v = \left(-\frac{24}{25}\right)\left(-\frac{4}{5}\right) - \left(-\frac{7}{25}\right)\left(-\frac{3}{5}\right) = \frac{3}{5}$

49. $\tan(u - v) = \dfrac{\tan u - \tan v}{1 + \tan u \tan v} = \dfrac{\dfrac{7}{24} - \dfrac{3}{4}}{1 + \left(\dfrac{7}{24}\right)\left(\dfrac{3}{4}\right)} = \dfrac{-\dfrac{11}{24}}{\dfrac{39}{32}} = -\dfrac{44}{117}$

51. $\csc(u - v) = \dfrac{1}{\sin(u - v)} = \dfrac{1}{\sin u \cos v - \cos u \sin v} = \dfrac{1}{\left(-\dfrac{7}{25}\right)\left(-\dfrac{4}{5}\right) - \left(-\dfrac{24}{25}\right)\left(-\dfrac{3}{5}\right)} = \dfrac{1}{-\dfrac{44}{125}} = -\dfrac{125}{44}$

53. $\sin(\arcsin x + \arccos x) = \sin(\arcsin x)\cos(\arccos x) + \sin(\arccos x)\cos(\arcsin x)$

$$= x \cdot x + \sqrt{1-x^2} \cdot \sqrt{1-x^2}$$

$$= x^2 + 1 - x^2$$

$$= 1$$

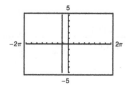

$\theta = \arcsin x$ $\theta = \arccos x$

55. $\cos(\arccos x + \arcsin x) = \cos(\arccos x)\cos(\arcsin x) - \sin(\arccos x)\sin(\arcsin x)$

$$= x \cdot \sqrt{1-x^2} - \sqrt{1-x^2} \cdot x$$

$$= 0$$

(Use the triangles in Exercise 53.)

57. $\sin\left(\dfrac{\pi}{2} - x\right) = \sin\dfrac{\pi}{2}\cos x - \cos\dfrac{\pi}{2}\sin x = (1)(\cos x) - (0)(\sin x) = \cos x$

59. $\sin\left(\dfrac{\pi}{6} + x\right) = \sin\dfrac{\pi}{6}\cos x + \cos\dfrac{\pi}{6}\sin x = \dfrac{1}{2}\left(\cos x + \sqrt{3}\sin x\right)$

61. $\tan(\theta + \pi) = \dfrac{\tan\theta + \tan\pi}{1 - \tan\theta\tan\pi} = \dfrac{\tan\theta + 0}{1 - (\tan\theta)(0)} = \dfrac{\tan\theta}{1} = \tan\theta$

63. $\cos(\pi - \theta) + \sin\left(\dfrac{\pi}{2} + \theta\right) = \cos\pi\cos\theta + \sin\pi\sin\theta + \sin\dfrac{\pi}{2}\cos\theta + \cos\dfrac{\pi}{2}\sin\theta$

$$= (-1)(\cos\theta) + (0)(\sin\theta) + (1)(\cos\theta) + (\sin\theta)(0)$$

$$= -\cos\theta + \cos\theta$$

$$= 0$$

65. $\cos\left(\dfrac{3\pi}{2} - \theta\right) = \cos\dfrac{3\pi}{2}\cos\theta + \sin\dfrac{3\pi}{2}\sin\theta$

$$= (0)(\cos\theta) + (-1)(\sin\theta)$$

$$= -\sin\theta$$

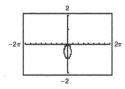

The graphs appear to coincide, so

$\cos\left(\dfrac{3\pi}{2} - \theta\right) = -\sin\theta.$

67. $\sin\left(\dfrac{3\pi}{2} + \theta\right) = \sin\dfrac{3\pi}{2}\cos\theta + \cos\dfrac{3\pi}{2}\sin\theta$

$$= (-1)(\cos\theta) + (0)(\sin\theta)$$

$$= -\cos\theta$$

$\csc\left(\dfrac{3\pi}{2} + \theta\right) = \dfrac{1}{\sin\left(\frac{3\pi}{2} + \theta\right)} = \dfrac{1}{-\cos\theta} = -\sec\theta$

The graphs appear to coincide, so

$\csc\left(\dfrac{3\pi}{2} + \theta\right) = -\sec\theta.$

69.
$$\sin(x + \pi) - \sin x + 1 = 0$$
$$\sin x \cos \pi + \cos x \sin \pi - \sin x + 1 = 0$$
$$(\sin x)(-1) + (\cos x)(0) - \sin x + 1 = 0$$
$$-2 \sin x + 1 = 0$$
$$\sin x = \frac{1}{2}$$
$$x = \frac{\pi}{6}, \frac{5\pi}{6}$$

71.
$$\cos\left(x + \frac{\pi}{4}\right) - \cos\left(x - \frac{\pi}{4}\right) = 1$$
$$\cos x \cos \frac{\pi}{4} - \sin x \sin \frac{\pi}{4} - \left(\cos x \cos \frac{\pi}{4} + \sin x \sin \frac{\pi}{4}\right) = 1$$
$$-2 \sin x \left(\frac{\sqrt{2}}{2}\right) = 1$$
$$-\sqrt{2} \sin x = 1$$
$$\sin x = -\frac{1}{\sqrt{2}}$$
$$\sin x = -\frac{\sqrt{2}}{2}$$
$$x = \frac{5\pi}{4}, \frac{7\pi}{4}$$

73.
$$\tan(x + \pi) + 2\sin(x + \pi) = 0$$
$$\frac{\tan x + \tan \pi}{1 - \tan x \tan \pi} + 2(\sin x \cos \pi + \cos x \sin \pi) = 0$$
$$\frac{\tan x + 0}{1 - \tan x(0)} + 2\left[\sin x(-1) + \cos x(0)\right] = 0$$
$$\frac{\tan x}{1} - 2 \sin x = 0$$
$$\frac{\sin x}{\cos x} = 2 \sin x$$
$$\sin x = 2 \sin x \cos x$$
$$\sin x(1 - 2 \cos x) = 0$$
$$\sin x = 0 \quad \text{or} \quad \cos x = \frac{1}{2}$$
$$x = 0, \pi \qquad x = \frac{\pi}{3}, \frac{5\pi}{3}$$

75. $\cos\left(x + \frac{\pi}{4}\right) + \cos\left(x - \frac{\pi}{4}\right) = 1$

Graph $y_1 = \cos\left(x + \frac{\pi}{4}\right) + \cos\left(x - \frac{\pi}{4}\right)$ and $y_2 = 1$.

$$x = \frac{\pi}{4}, \frac{7\pi}{4}$$

77. $\sin\left(x + \frac{\pi}{2}\right) + \cos^2 x = 0$

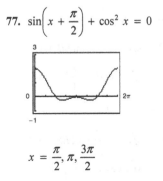

$$x = \frac{\pi}{2}, \pi, \frac{3\pi}{2}$$

79. $y = \dfrac{1}{3}\sin 2t + \dfrac{1}{4}\cos 2t$

   (a) $a = \dfrac{1}{3}, b = \dfrac{1}{4}, B = 2$

      $C = \arctan \dfrac{b}{a} = \arctan \dfrac{3}{4} \approx 0.6435$

      $y \approx \sqrt{\left(\dfrac{1}{3}\right)^2 + \left(\dfrac{1}{4}\right)^2}\; \sin(2t + 0.6435) = \dfrac{5}{12}\sin(2t + 0.6435)$

   (b) Amplitude: $\dfrac{5}{12}$ feet

   (c) Frequency: $\dfrac{1}{\text{period}} = \dfrac{B}{2\pi} = \dfrac{2}{2\pi} = \dfrac{1}{\pi}$ cycle per second

81. True.

    $\sin(u + v) = \sin u \cos v + \cos u \sin v$

    $\sin(u - v) = \sin u \cos v - \cos u \sin v$

    So, $\sin(u \pm v) = \sin u \cos v \pm \cos u \sin v.$

83. $\sin(\alpha + \beta) = \sin \alpha \cos \beta + \sin \beta \cos \alpha = 0$

    $\sin \alpha \cos \beta + \sin \beta \cos \alpha = 0$

           $\sin \alpha \cos \beta = -\sin \beta \cos \alpha$

    False. When α and β are supplementary, $\sin \alpha \cos \beta = -\cos \alpha \sin \beta.$

85. The denominator should be $1 + \tan x \tan(\pi/4)$.

    $\tan\left(x - \dfrac{\pi}{4}\right) = \dfrac{\tan x - \tan(\pi/4)}{1 + \tan x \tan(\pi/4)} = \dfrac{\tan x - 1}{1 + \tan x}$

87. $\cos(n\pi + \theta) = \cos n\pi \cos \theta - \sin n\pi \sin \theta = (-1)^n(\cos \theta) - (0)(\sin \theta) = (-1)^n(\cos \theta)$, where n is an integer.

89. $C = \arctan \dfrac{b}{a} \Rightarrow \sin C = \dfrac{b}{\sqrt{a^2 + b^2}}, \cos C = \dfrac{a}{\sqrt{a^2 + b^2}}$

    $\sqrt{a^2 + b^2}\,\sin(B\theta + C) = \sqrt{a^2 + b^2}\left(\sin B\theta \cdot \dfrac{a}{\sqrt{a^2 + b^2}} + \dfrac{b}{\sqrt{a^2 + b^2}} \cdot \cos B\theta\right) = a \sin B\theta + b \cos B\theta$

91. $\sin \theta + \cos \theta$

    $a = 1, b = 1, B = 1$

   (a) $C = \arctan \dfrac{b}{a} = \arctan 1 = \dfrac{\pi}{4}$

      $\sin \theta + \cos \theta = \sqrt{a^2 + b^2}\,\sin(B\theta + C)$

                 $= \sqrt{2}\,\sin\left(\theta + \dfrac{\pi}{4}\right)$

   (b) $C = \arctan \dfrac{a}{b} = \arctan 1 = \dfrac{\pi}{4}$

      $\sin \theta + \cos \theta = \sqrt{a^2 + b^2}\,\cos(B\theta - C)$

                 $= \sqrt{2}\,\cos\left(\theta - \dfrac{\pi}{4}\right)$

93. $12 \sin 3\theta + 5 \cos 3\theta$

$a = 12, b = 5, B = 3$

(a) $C = \arctan \dfrac{b}{a} = \arctan \dfrac{5}{12} \approx 0.3948$

$12 \sin 3\theta + 5 \cos 3\theta = \sqrt{a^2 + b^2} \sin(B\theta + C)$

$\approx 13 \sin(3\theta + 0.3948)$

(b) $C = \arctan \dfrac{a}{b} = \arctan \dfrac{12}{5} \approx 1.1760$

$12 \sin 3\theta + 5 \cos 3\theta = \sqrt{a^2 + b^2} \cos(B\theta - C)$

$\approx 13 \cos(3\theta - 1.1760)$

95. $C = \arctan \dfrac{b}{a} = \dfrac{\pi}{4} \Rightarrow a = b, a > 0, b > 0$

$\sqrt{a^2 + b^2} = 2 \Rightarrow a = b = \sqrt{2}$

$B = 1$

$2 \sin\left(\theta + \dfrac{\pi}{4}\right) = \sqrt{2} \sin \theta + \sqrt{2} \cos \theta$

97.

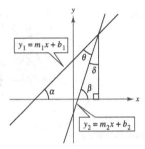

$m_1 = \tan \alpha$ and $m_2 = \tan \beta$

$\beta + \delta = 90° \Rightarrow \delta = 90° - \beta$

$\alpha + \theta + \delta = 90° \Rightarrow \alpha + \theta + (90° - \beta)$

$= 90° \Rightarrow \theta = \beta - \alpha$

So, $\theta = \arctan m_2 - \arctan m_1$. For $y = x$ and $y = \sqrt{3}x$ you have $m_1 = 1$ and $m_2 = \sqrt{3}$.

$\theta = \arctan\sqrt{3} - \arctan 1 = 60° - 45° = 15°$

99. $y_1 = \cos(x + 2), y_2 = \cos x + \cos 2$

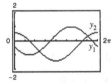

No, $y_1 \neq y_2$ because their graphs are different.

101. (a) To prove the identity for $\sin(u + v)$ you first need to prove the identity for $\cos(u - v)$.

Assume $0 < v < u < 2\pi$ and locate u, v, and $u - v$ on the unit circle.
The coordinates of the points on the circle are:

$A = (1, 0)$, $B = (\cos v, \sin v)$, $C = (\cos(u - v), \sin(u - v))$, and $D = (\cos u, \sin u)$.

Because $\angle DOB = \angle COA$, chords AC and BD are equal. By the Distance Formula:

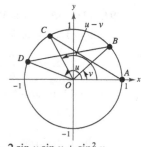

$$\sqrt{\left[\cos(u - v) - 1\right]^2 + \left[\sin(u - v) - 0\right]^2} = \sqrt{(\cos u - \cos v)^2 + (\sin u - \sin v)^2}$$

$$\cos^2(u - v) - 2\cos(u - v) + 1 + \sin^2(u - v) = \cos^2 u - 2\cos u \cos v + \cos^2 v + \sin^2 u - 2\sin u \sin v + \sin^2 v$$

$$\left[\cos^2(u - v) + \sin^2(u - v)\right] + 1 - 2\cos(u - v) = (\cos^2 u + \sin^2 u) + (\cos^2 v + \sin^2 v) - 2\cos u \cos v - 2\sin u \sin v$$

$$2 - 2\cos(u - v) = 2 - 2\cos u \cos v - 2\sin u \sin v$$

$$-2\cos(u - v) = -2(\cos u \cos v + \sin u \sin v)$$

$$\cos(u - v) = \cos u \cos v + \sin u \sin v$$

Now, to prove the identity for $\sin(u + v)$, use cofunction identities.

$$\sin(u + v) = \cos\left[\frac{\pi}{2} - (u + v)\right] = \cos\left[\left(\frac{\pi}{2} - u\right) - v\right]$$

$$= \cos\left(\frac{\pi}{2} - u\right)\cos v + \sin\left(\frac{\pi}{2} - u\right)\sin v$$

$$= \sin u \cos v + \cos u \sin v$$

(b) First, prove $\cos(u - v) = \cos u \cos v + \sin u \sin v$ using the figure containing points

$A(1, 0)$

$B(\cos(u - v), \sin(u - v))$

$C(\cos v, \sin v)$

$D(\cos u, \sin u)$

on the unit circle.

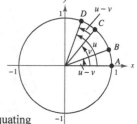

Because chords AB and CD are each subtended by angle $u - v$, their lengths are equal. Equating $[d(A, B)]^2 = [d(C, D)]^2$ you have $(\cos(u - v) - 1)^2 + \sin^2(u - v) = (\cos u - \cos v)^2 + (\sin u - \sin v)^2$.

Simplifying and solving for $\cos(u - v)$, you have $\cos(u - v) = \cos u \cos v + \sin u \sin v$.

Using $\sin \theta = \cos\left(\frac{\pi}{2} - \theta\right)$,

$$\sin(u - v) = \cos\left[\frac{\pi}{2} - (u - v)\right] = \cos\left[\left(\frac{\pi}{2} - u\right) - (-v)\right] = \cos\left(\frac{\pi}{2} - u\right)\cos(-v) + \sin\left(\frac{\pi}{2} - u\right)\sin(-v)$$

$$= \sin u \cos v - \cos u \sin v$$

Section 2.5 Multiple-Angle and Product-to-Sum Formulas

1. $2 \sin u \cos u$

3. $\dfrac{1}{2}\left[\sin(u + v) + \sin(u - v)\right]$

5. $\pm\sqrt{\dfrac{1 - \cos u}{2}}$

7.
$$\sin 2x - \sin x = 0$$
$$2 \sin x \cos x - \sin x = 0$$
$$\sin x(2 \cos x - 1) = 0$$
$$\sin x = 0 \quad \text{or} \quad 2 \cos x - 1 = 0$$
$$x = n\pi \qquad\qquad \cos x = \frac{1}{2}$$
$$x = \frac{\pi}{3} + 2n\pi, \frac{5\pi}{3} + 2n\pi$$

9.
$$\cos 2x - \cos x = 0$$
$$\cos 2x = \cos x$$
$$\cos^2 x - \sin^2 x = \cos x$$
$$\cos^2 x - (1 - \cos^2 x) - \cos x = 0$$
$$2\cos^2 x - \cos x - 1 = 0$$
$$(2\cos x + 1)(\cos x - 1) = 0$$

$$2\cos x + 1 = 0 \quad \text{or} \quad \cos x - 1 = 0$$

$$\cos x = -\frac{1}{2} \qquad \cos x = 1$$

$$x = \frac{2n\pi}{3} \qquad x = 0$$

11.
$$\sin 4x = -2\sin 2x$$
$$\sin 4x + 2\sin 2x = 0$$
$$2\sin 2x \cos 2x + 2\sin 2x = 0$$
$$2\sin 2x(\cos 2x + 1) = 0$$

$$2\sin 2x = 0 \quad \text{or} \quad \cos 2x + 1 = 0$$

$$\sin 2x = 0 \qquad \cos 2x = -1$$

$$2x = n\pi \qquad 2x = \pi + 2n\pi$$

$$x = \frac{n}{2}\pi \qquad x = \frac{\pi}{2} + n\pi$$

13.
$$\tan 2x - \cot x = 0$$

$$\frac{2\tan x}{1 - \tan^2 x} = \cot x$$

$$2\tan x = \cot x(1 - \tan^2 x)$$

$$2\tan x = \cot x - \cot x \tan^2 x$$

$$2\tan x = \cot x - \tan x$$

$$3\tan x = \cot x$$

$$3\tan x - \cot x = 0$$

$$3\tan x - \frac{1}{\tan x} = 0$$

$$\frac{3\tan^2 x - 1}{\tan x} = 0$$

$$\frac{1}{\tan x}(3\tan^2 x - 1) = 0$$

$$\cot x(3\tan^2 x - 1) = 0$$

$$\cot x = 0 \qquad \text{or} \quad 3\tan^2 x - 1 = 0$$

$$x = \frac{\pi}{2} + n\pi \qquad \tan^2 x = \frac{1}{3}$$

$$\tan x = \pm\frac{\sqrt{3}}{3}$$

$$x = \frac{\pi}{6} + n\pi, \frac{5\pi}{6} + n\pi$$

15. $6\sin x \cos x = 3(2\sin x \cos x)$
$$= 3\sin 2x$$

17. $6\cos^2 x - 3 = 3(2\cos^2 x - 1)$
$$= 3\cos 2x$$

19. $4 - 8\sin^2 x = 4(1 - 2\sin^2 x)$
$$= 4\cos 2x$$

21. $\sin u = -\dfrac{3}{5}, \dfrac{3\pi}{2} < u < 2\pi$

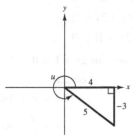

$\sin 2u = 2 \sin u \cos u = 2\left(-\dfrac{3}{5}\right)\left(\dfrac{4}{5}\right) = -\dfrac{24}{25}$

$\cos 2u = \cos^2 u - \sin^2 u = \dfrac{16}{25} - \dfrac{9}{25} = \dfrac{7}{25}$

$\tan 2u = \dfrac{2 \tan u}{1 - \tan^2 u} = \dfrac{2\left(-\dfrac{3}{4}\right)}{1 - \dfrac{9}{16}} = -\dfrac{3}{2}\left(\dfrac{16}{7}\right) = -\dfrac{24}{7}$

23. $\tan u = \dfrac{3}{5}, 0 < u < \dfrac{\pi}{2}$

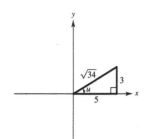

$\sin 2u = 2 \sin u \cos u = 2\left(\dfrac{3}{\sqrt{34}}\right)\left(\dfrac{5}{\sqrt{34}}\right) = \dfrac{15}{17}$

$\cos 2u = \cos^2 u - \sin^2 u = \dfrac{25}{34} - \dfrac{9}{34} = \dfrac{8}{17}$

$\tan 2u = \dfrac{2 \tan u}{1 - \tan^2 u} = \dfrac{2\left(\dfrac{3}{5}\right)}{1 - \dfrac{9}{25}} = \dfrac{6}{5}\left(\dfrac{25}{16}\right) = \dfrac{15}{8}$

25. $\cos 4x = \cos(2x + 2x)$

$= \cos 2x \cos 2x - \sin 2x \sin 2x$

$= \cos^2 2x - \sin^2 2x$

$= \cos^2 2x - \left(1 - \cos^2 2x\right)$

$= 2 \cos^2 2x - 1$

$= 2\left(\cos 2x\right)^2 - 1$

$= 2\left(2 \cos^2 x - 1\right)^2 - 1$

$= 2\left(4 \cos^4 x - 4 \cos x + 1\right) - 1$

$= 8 \cos^4 x - 8 \cos x + 1$

27. $\cos^4 x = \left(\cos^2 x\right)\left(\cos^2 x\right) = \left(\dfrac{1 + \cos 2x}{2}\right)\left(\dfrac{1 + \cos 2x}{2}\right) = \dfrac{1 + 2 \cos 2x + \cos^2 2x}{4}$

$= \dfrac{1 + 2 \cos 2x + \dfrac{1 + \cos 4x}{2}}{4}$

$= \dfrac{2 + 4 \cos 2x + 1 + \cos 4x}{8}$

$= \dfrac{3 + 4 \cos 2x + \cos 4x}{8}$

$= \dfrac{1}{8}\left(3 + 4 \cos 2x + \cos 4x\right)$

29. $\sin^4 2x = \left(\sin^2 2x\right)^2$

$$= \left(\frac{1 - \cos 4x}{2}\right)^2$$

$$= \frac{1}{4}\left(1 - 2\cos 4x + \cos^2 4x\right)$$

$$= \frac{1}{4}\left(1 - 2\cos 4x + \frac{1 + \cos 8x}{2}\right)$$

$$= \frac{1}{4} - \frac{1}{2}\cos 4x + \frac{1}{8} + \frac{1}{8}\cos 8x$$

$$= \frac{3}{8} - \frac{1}{2}\cos 4x + \frac{1}{8}\cos 8x$$

$$= \frac{1}{8}(3 - 4\cos 4x + \cos 8x)$$

31. $\tan^4 2x = \left(\tan^2 2x\right)^2$

$$= \left(\frac{1 - \cos 4x}{1 + \cos 4x}\right)^2$$

$$= \frac{1 - 2\cos 4x + \cos^2 4x}{1 + 2\cos 4x + \cos^2 4x}$$

$$= \frac{1 - 2\cos 4x + \dfrac{1 + \cos 8x}{2}}{1 + 2\cos 4x + \dfrac{1 + \cos 8x}{2}}$$

$$= \frac{\dfrac{1}{2}(2 - 4\cos 4x + 1 + \cos 8x)}{\dfrac{1}{2}(2 + 4\cos 4x + 1 + \cos 8x)}$$

$$= \frac{3 - 4\cos 4x + \cos 8x}{3 + 4\cos 4x + \cos 8x}$$

33. $\sin^2 2x \cos^2 2x = \left(\dfrac{1 - \cos 4x}{2}\right)\left(\dfrac{1 + \cos 4x}{2}\right)$

$$= \frac{1}{4}\left(1 - \cos^2 4x\right)$$

$$= \frac{1}{4}\left(1 - \frac{1 + \cos 8x}{2}\right)$$

$$= \frac{1}{4} - \frac{1}{8} - \frac{1}{8}\cos 8x$$

$$= \frac{1}{8} - \frac{1}{8}\cos 8x$$

$$= \frac{1}{8}(1 - \cos 8x)$$

35. $\sin 75° = \sin\left(\dfrac{1}{2} \cdot 150°\right) = \sqrt{\dfrac{1 - \cos 150°}{2}} = \sqrt{\dfrac{1 + \left(\sqrt{3}/2\right)}{2}}$

$$= \frac{1}{2}\sqrt{2 + \sqrt{3}}$$

$\cos 75° = \cos\left(\dfrac{1}{2} \cdot 150°\right) = \sqrt{\dfrac{1 + \cos 150°}{2}} = \sqrt{\dfrac{1 - \left(\sqrt{3}/2\right)}{2}}$

$$= \frac{1}{2}\sqrt{2 - \sqrt{3}}$$

$\tan 75° = \tan\left(\dfrac{1}{2} \cdot 150°\right) = \dfrac{\sin 150°}{1 + \cos 150°} = \dfrac{1/2}{1 - \left(\sqrt{3}/2\right)}$

$$= \frac{1}{2 - \sqrt{3}} \cdot \frac{2 + \sqrt{3}}{2 + \sqrt{3}} = \frac{2 + \sqrt{3}}{4 - 3} = 2 + \sqrt{3}$$

37. $\sin 112° \, 30' = \sin\left(\frac{1}{2} \cdot 225°\right) = \sqrt{\dfrac{1 - \cos 225°}{2}} = \sqrt{\dfrac{1 - \left(-\sqrt{2}/2\right)}{2}} = \dfrac{1}{2}\sqrt{2 + \sqrt{2}}$

$\cos 112° \, 30' = \cos\left(\frac{1}{2} \cdot 225°\right) = -\sqrt{\dfrac{1 + \cos 225°}{2}} = -\sqrt{\dfrac{1 + \left(-\sqrt{2}/2\right)}{2}} = \dfrac{1}{2} - \sqrt{2 - 2}$

$\tan 112° \, 30' = \tan\left(\frac{1}{2} \cdot 225°\right) = \dfrac{\sin 225°}{1 + \cos 225°} = \dfrac{-\sqrt{2}/2}{1 + \left(-\sqrt{2}/2\right)} = \dfrac{-\sqrt{2}}{2 - \sqrt{2}} \cdot \dfrac{2 + \sqrt{2}}{2 + \sqrt{2}} = \dfrac{-2\sqrt{2} - 2}{2} = -1 - \sqrt{2}$

39. $\sin \dfrac{\pi}{8} = \sin\left[\dfrac{1}{2}\left(\dfrac{\pi}{4}\right)\right] = \sqrt{\dfrac{1 - \cos \dfrac{\pi}{4}}{2}} = \dfrac{1}{2}\sqrt{2 - \sqrt{2}}$

$\cos \dfrac{\pi}{8} = \cos\left[\dfrac{1}{2}\left(\dfrac{\pi}{4}\right)\right] = \sqrt{\dfrac{1 + \cos \dfrac{\pi}{4}}{2}} = \dfrac{1}{2}\sqrt{2 + \sqrt{2}}$

$\tan \dfrac{\pi}{8} = \tan\left[\dfrac{1}{2}\left(\dfrac{\pi}{4}\right)\right] = \dfrac{\sin \dfrac{\pi}{4}}{1 + \cos \dfrac{\pi}{4}} = \dfrac{\dfrac{\sqrt{2}}{2}}{1 + \dfrac{\sqrt{2}}{2}} = \sqrt{2} - 1$

41. $\cos u = \dfrac{7}{25},\ 0 < u < \dfrac{\pi}{2}$

(a) Because u is in Quadrant I, $\dfrac{u}{2}$ is also in Quadrant I.

(b) $\sin \dfrac{u}{2} = \sqrt{\dfrac{1 - \cos u}{2}} = \sqrt{\dfrac{1 - \dfrac{7}{25}}{2}} = \sqrt{\dfrac{9}{25}} = \dfrac{3}{5}$

$\cos \dfrac{u}{2} = \sqrt{\dfrac{1 + \cos u}{2}} = \sqrt{\dfrac{1 + \dfrac{7}{25}}{2}} = \sqrt{\dfrac{16}{25}} = \dfrac{4}{5}$

$\tan \dfrac{u}{2} = \dfrac{1 - \cos u}{\sin u} = \dfrac{1 - \dfrac{7}{25}}{\dfrac{24}{25}} = \dfrac{3}{4}$

43. $\tan u = -\dfrac{5}{12},\ \dfrac{3\pi}{2} < u < 2\pi$

(a) Because u is in Quadrant IV, $\dfrac{u}{2}$ is in Quadrant II.

(b) $\sin \dfrac{u}{2} = \sqrt{\dfrac{1 - \cos u}{2}} = \sqrt{\dfrac{1 - \dfrac{12}{13}}{2}} = \sqrt{\dfrac{1}{26}} = \dfrac{\sqrt{26}}{26}$

$\cos \dfrac{u}{2} = -\sqrt{\dfrac{1 + \cos u}{2}} = -\sqrt{\dfrac{1 + \dfrac{12}{13}}{2}} = -\sqrt{\dfrac{25}{26}} = -\dfrac{5\sqrt{26}}{26}$

$\tan \dfrac{u}{2} = \dfrac{1 - \cos u}{\sin u} = \dfrac{1 - \dfrac{12}{13}}{\left(-\dfrac{5}{13}\right)} = -\dfrac{1}{5}$

45. $\sin\dfrac{x}{2} + \cos x = 0$

$$\pm\sqrt{\dfrac{1 - \cos x}{2}} = -\cos x$$

$$\dfrac{1 - \cos x}{2} = \cos^2 x$$

$$0 = 2\cos^2 x + \cos x - 1$$

$$= (2\cos x - 1)(\cos x + 1)$$

$$\cos x = \dfrac{1}{2} \quad\text{or}\quad \cos x = -1$$

$$x = \dfrac{\pi}{3}, \dfrac{5\pi}{3} \qquad x = \pi$$

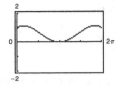

By checking these values in the original equation, $x = \pi/3$ and $x = 5\pi/3$ are extraneous, and $x = \pi$ is the only solution.

47. $\cos\dfrac{x}{2} - \sin x = 0$

$$\pm\sqrt{\dfrac{1 + \cos x}{2}} = \sin x$$

$$\dfrac{1 + \cos x}{2} = \sin^2 x$$

$$1 + \cos x = 2\sin^2 x$$

$$1 + \cos x = 2 - 2\cos^2 x$$

$$2\cos^2 x + \cos x - 1 = 0$$

$$(2\cos x - 1)(\cos x + 1) = 0$$

$$2\cos x - 1 = 0 \quad\text{or}\quad \cos x + 1 = 0$$

$$\cos x = \dfrac{1}{2} \qquad \cos x = -1$$

$$x = \dfrac{\pi}{3}, \dfrac{5\pi}{3} \qquad x = \pi$$

$$x = \dfrac{\pi}{3}, \pi, \dfrac{5\pi}{3}$$

$\pi/3, \pi,$ and $5\pi/3$ are all solutions to the equation.

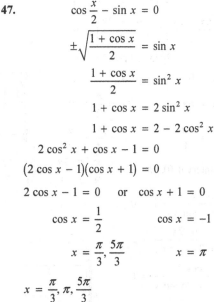

49. $\sin 5\theta \sin 3\theta = \tfrac{1}{2}\big[\cos(5\theta - 3\theta) - \cos(5\theta + 3\theta)\big] = \tfrac{1}{2}(\cos 2\theta - \cos 8\theta)$

51. $\cos 2\theta \cos 4\theta = \tfrac{1}{2}\big[\cos(2\theta - 4\theta) + \cos(2\theta + 4\theta)\big] = \tfrac{1}{2}\big[\cos(-2\theta) + \cos 6\theta\big]$

53. $\sin 5\theta - \sin 3\theta = 2\cos\left(\dfrac{5\theta + 3\theta}{2}\right)\sin\left(\dfrac{5\theta - 3\theta}{2}\right)$

$$= 2\cos 4\theta \sin\theta$$

55. $\cos 6x + \cos 2x = 2\cos\left(\dfrac{6x + 2x}{2}\right)\cos\left(\dfrac{6x - 2x}{2}\right)$

$$= 2\cos 4x \cos 2x$$

57. $\sin 75° + \sin 15° = 2\sin\left(\dfrac{75° + 15°}{2}\right)\cos\left(\dfrac{75° - 15°}{2}\right) = 2\sin 45° \cos 30° = 2\left(\dfrac{\sqrt{2}}{2}\right)\left(\dfrac{\sqrt{3}}{2}\right) = \dfrac{\sqrt{6}}{2}$

59. $\cos\dfrac{3\pi}{4} - \cos\dfrac{\pi}{4} = -2\sin\left(\dfrac{\dfrac{3\pi}{4} + \dfrac{\pi}{4}}{2}\right)\sin\left(\dfrac{\dfrac{3\pi}{4} - \dfrac{\pi}{4}}{2}\right) = -2\sin\dfrac{\pi}{2}\sin\dfrac{\pi}{4}$

$$\cos\dfrac{3\pi}{4} - \cos\dfrac{\pi}{4} = -\dfrac{\sqrt{2}}{2} - \dfrac{\sqrt{2}}{2} = -\sqrt{2}$$

61.
$$\sin 6x + \sin 2x = 0$$

$$2 \sin\left(\frac{6x + 2x}{2}\right) \cos\left(\frac{6x - 2x}{2}\right) = 0$$

$$2(\sin 4x) \cos 2x = 0$$

$$\sin 4x = 0 \quad \text{or} \quad \cos 2x = 0$$

$$4x = n\pi \qquad\qquad 2x = \frac{\pi}{2} + n\pi$$

$$x = \frac{n\pi}{4} \qquad\qquad x = \frac{\pi}{4} + \frac{n\pi}{2}$$

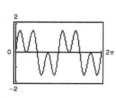

In the interval $[0, 2\pi)$ $x = 0, \dfrac{\pi}{4}, \dfrac{\pi}{2}, \dfrac{3\pi}{4}, \pi, \dfrac{5\pi}{4}, \dfrac{3\pi}{2}, \dfrac{7\pi}{4}.$

63.
$$\frac{\cos 2x}{\sin 3x - \sin x} - 1 = 0$$

$$\frac{\cos 2x}{\sin 3x - \sin x} = 1$$

$$\frac{\cos 2x}{2 \cos 2x \sin x} = 1$$

$$2 \sin x = 1$$

$$\sin x = \frac{1}{2}$$

$$x = \frac{\pi}{6}, \frac{5\pi}{6}$$

65.
$$\csc 2\theta = \frac{1}{\sin 2\theta} = \frac{1}{2 \sin \theta \cos \theta}$$

$$= \frac{1}{\sin \theta} \cdot \frac{1}{2 \cos \theta} = \frac{\csc \theta}{2 \cos \theta}$$

67.
$$(\sin x + \cos x)^2 = \sin^2 x + 2 \sin x \cos x + \cos^2 x$$

$$= (\sin^2 x + \cos^2 x) + 2 \sin x \cos x$$

$$= 1 + \sin 2x$$

69.
$$\frac{\sin x \pm \sin y}{\cos x + \cos y} = \frac{2 \sin\left(\dfrac{x \pm y}{2}\right) \cos\left(\dfrac{x \mp y}{2}\right)}{2 \cos\left(\dfrac{x + y}{2}\right) \cos\left(\dfrac{x - y}{2}\right)}$$

$$= \tan\left(\frac{x \pm y}{2}\right)$$

71. (a)
$$\sin\left(\frac{\theta}{2}\right) = \pm\sqrt{\frac{1 - \cos \theta}{2}} = \frac{1}{M}$$

$$\left(\pm\sqrt{\frac{1 - \cos \theta}{2}}\right)^2 = \left(\frac{1}{M}\right)^2$$

$$\frac{1 - \cos \theta}{2} = \frac{1}{M^2}$$

$$M^2(1 - \cos \theta) = 2$$

$$1 - \cos \theta = \frac{2}{M^2}$$

$$-\cos \theta = \frac{2}{M^2} - 1$$

$$\cos \theta = 1 - \frac{2}{M^2}$$

$$\cos \theta = \frac{M^2 - 2}{M^2}$$

(b) When $M = 2$, $\cos \theta = \dfrac{2^2 - 2}{2^2} = \dfrac{1}{2}$. So,

$$\theta = \frac{\pi}{3}.$$

(d) When $M = 2$, $\dfrac{\text{speed of object}}{\text{speed of sound}} = M$

$$\frac{\text{speed of object}}{760 \text{ mph}} = 2$$

$$\text{speed of object} = 1520 \text{ mph.}$$

When $M = 4.5$, $\dfrac{\text{speed of object}}{\text{speed of sound}} = M$

$$\frac{\text{speed of object}}{760 \text{ mph}} = 4.5$$

$$\text{speed of object} = 3420 \text{ mph.}$$

(c) When $M = 4.5$, $\cos \theta = \dfrac{(4.5)^2 - 2}{(4.5)^2}$ So, $\theta \approx 0.4482$ radian.

$$\cos \theta \approx 0.901235.$$

73. $\dfrac{x}{2} = 2r\sin^2\dfrac{\theta}{2} = 2r\left(\dfrac{1-\cos\theta}{2}\right)$

$$= r(1 - \cos\theta)$$

So, $x = 2r(1 - \cos\theta)$.

75. True. Using the double angle formula and that sine is an odd function and cosine is an even function,

$$\sin(-2x) = \sin\left[2(-x)\right]$$

$$= 2\sin(-x)\cos(-x)$$

$$= 2(-\sin x)\cos x$$

$$= -2\sin x \cos x.$$

77. Because ϕ and θ are complementary angles, $\sin\phi = \cos\theta$ and $\cos\phi = \sin\theta$.

(a) $\sin(\phi - \theta) = \sin\phi\cos\theta - \sin\theta\cos\phi$

$$= (\cos\theta)(\cos\theta) - (\sin\theta)(\sin\theta)$$

$$= \cos^2\theta - \sin^2\theta$$

$$= \cos 2\theta$$

(b) $\cos(\phi - \theta) = \cos\phi\cos\theta + \sin\phi\sin\theta$

$$= (\sin\theta)(\cos\theta) + (\cos\theta)(\sin\theta)$$

$$= 2\sin\theta\cos\theta$$

$$= \sin 2\theta$$

Review Exercises for Chapter 2

1. $\cot x$

3. $\cos x$

5. $\cos\theta = -\dfrac{2}{5}$, $\tan\theta > 0$, θ is in Quadrant III.

$\sec\theta = \dfrac{1}{\cos\theta} = -\dfrac{5}{2}$

$\sin\theta = -\sqrt{1 - \cos^2\theta} = -\sqrt{1 - \dfrac{4}{25}} = -\sqrt{\dfrac{21}{25}} = -\dfrac{\sqrt{21}}{5}$

$\csc\theta = \dfrac{1}{\sin\theta} = -\dfrac{5}{\sqrt{21}} = -\dfrac{5\sqrt{21}}{21}$

$\tan\theta = \dfrac{\sin\theta}{\cos\theta} = \dfrac{-\dfrac{\sqrt{21}}{5}}{-\dfrac{2}{5}} = \dfrac{\sqrt{21}}{2}$

$\cot\theta = \dfrac{1}{\tan\theta} = \dfrac{2}{\sqrt{21}} = \dfrac{2\sqrt{21}}{21}$

7. $\dfrac{1}{\cot^2 x + 1} = \dfrac{1}{\csc^2 x} = \sin^2 x$

9. $\tan^2 x\left(\csc^2 x - 1\right) = \tan^2 x\left(\cot^2 x\right)$

$$= \tan^2 x\left(\dfrac{1}{\tan^2 x}\right)$$

$$= 1$$

11. $\dfrac{\cot\left(\dfrac{\pi}{2} - u\right)}{\cos u} = \dfrac{\tan u}{\cos u} = \tan u \sec u$

13. $\cos^2 x + \cos^2 x \cot^2 x = \cos^2 x\left(1 + \cot^2 x\right)$

$$= \cos^2 x\left(\csc^2 x\right)$$

$$= \cos^2 x\left(\dfrac{1}{\sin^2 x}\right)$$

$$= \dfrac{\cos^2 x}{\sin^2 x}$$

$$= \cot^2 x$$

15. $\dfrac{1}{\csc\theta + 1} - \dfrac{1}{\csc\theta - 1} = \dfrac{(\csc\theta - 1) - (\csc\theta + 1)}{(\csc\theta + 1)(\csc\theta - 1)}$

$$= \dfrac{-2}{\csc^2\theta - 1}$$

$$= \dfrac{-2}{\cot^2\theta}$$

$$= -2\tan^2\theta$$

17. Let $x = 5 \sin \theta$, then

$$\sqrt{25 - x^2} = \sqrt{25 - (5 \sin \theta)^2} = \sqrt{25 - 25 \sin^2 \theta} = \sqrt{25(1 - \sin^2 \theta)} = \sqrt{25 \cos^2 \theta} = 5 \cos \theta.$$

19.
$$\cos x (\tan^2 x + 1) = \cos x \sec^2 x$$
$$= \frac{1}{\sec x} \sec^2 x$$
$$= \sec x$$

21.
$$\sin\left(\frac{\pi}{2} - \theta\right) \tan \theta = \cos \theta \tan \theta$$
$$= \cos \theta \left(\frac{\sin \theta}{\cos \theta}\right)$$
$$= \sin \theta$$

23.
$$\frac{1}{\tan \theta \csc \theta} = \frac{1}{\frac{\sin \theta}{\cos \theta} \cdot \frac{1}{\sin \theta}} = \cos \theta$$

25.
$$\sin^5 x \cos^2 x = \sin^4 x \cos^2 x \sin x$$
$$= (1 - \cos^2 x)^2 \cos^2 x \sin x$$
$$= (1 - 2\cos^2 x + \cos^4 x) \cos^2 x \sin x$$
$$= (\cos^2 x - 2\cos^4 x + \cos^6 x) \sin x$$

27.
$$\sin x = \sqrt{3} - \sin x$$
$$\sin x = \frac{\sqrt{3}}{2}$$
$$x = \frac{\pi}{3} + 2\pi n, \frac{2\pi}{3} + 2\pi n$$

29.
$$3\sqrt{3} \tan u = 3$$
$$\tan u = \frac{1}{\sqrt{3}}$$
$$u = \frac{\pi}{6} + n\pi$$

31.
$$3 \csc^2 x = 4$$
$$\csc^2 x = \frac{4}{3}$$
$$\sin x = \pm\frac{\sqrt{3}}{2}$$
$$x = \frac{\pi}{3} + 2\pi n, \frac{2\pi}{3} + 2\pi n, \frac{4\pi}{3} + 2\pi n, \frac{5\pi}{3} + 2\pi n$$

These can be combined as:

$$x = \frac{\pi}{3} + n\pi \quad \text{or} \quad x = \frac{2\pi}{3} + n\pi$$

33.
$$\sin^3 x = \sin x$$
$$\sin^3 x - \sin x = 0$$
$$\sin x (\sin^2 x - 1) = 0$$
$$\sin x = 0 \Rightarrow x = 0, \pi$$
$$\sin^2 x = 1$$
$$\sin x = \pm 1 \Rightarrow x = \frac{\pi}{2}, \frac{3\pi}{2}$$

35.
$$\cos^2 x + \sin x = 1$$
$$1 - \sin^2 x + \sin x - 1 = 0$$
$$-\sin x (\sin x - 1) = 0$$
$$\sin x = 0 \qquad \sin x - 1 = 0$$
$$x = 0, \pi \qquad \sin x = 1$$
$$x = \frac{\pi}{2}$$

37.
$$2 \sin 2x - \sqrt{2} = 0$$
$$\sin 2x = \frac{\sqrt{2}}{2}$$
$$2x = \frac{\pi}{4} + 2\pi n, \frac{3\pi}{4} + 2\pi n$$
$$x = \frac{\pi}{8} + \pi n, \frac{3\pi}{8} + \pi n$$
$$x = \frac{\pi}{8}, \frac{3\pi}{8}, \frac{9\pi}{8}, \frac{11\pi}{8}$$

39.
$$3 \tan^2\left(\frac{x}{3}\right) - 1 = 0$$
$$\tan^2\left(\frac{x}{3}\right) = \frac{1}{3}$$
$$\tan \frac{x}{3} = \pm\sqrt{\frac{1}{3}}$$
$$\tan \frac{x}{3} = \pm\frac{\sqrt{3}}{3}$$
$$\frac{x}{3} = \frac{\pi}{6}, \frac{5\pi}{6}, \frac{7\pi}{6}$$
$$x = \frac{\pi}{2}, \frac{5\pi}{2}, \frac{7\pi}{2}$$

$\frac{5\pi}{2}$ and $\frac{7\pi}{2}$ are greater than 2π, so they are not solutions. The solution is $x = \frac{\pi}{2}$.

41. $\cos 4x(\cos x - 1) = 0$

$\cos 4x = 0$ $\qquad\qquad\qquad\qquad$ $\cos x - 1 = 0$

$\qquad 4x = \dfrac{\pi}{2} + 2\pi n, \dfrac{3\pi}{2} + 2\pi n$ $\qquad$ $\cos x = 1$

$\qquad x = \dfrac{\pi}{8} + \dfrac{\pi}{2} n, \dfrac{3\pi}{8} + \dfrac{\pi}{2} n$ $\qquad$ $x = 0$

$x = 0, \dfrac{\pi}{8}, \dfrac{3\pi}{8}, \dfrac{5\pi}{8}, \dfrac{7\pi}{8}, \dfrac{9\pi}{8}, \dfrac{11\pi}{8}, \dfrac{13\pi}{8}, \dfrac{15\pi}{8}$

43. $\tan^2 x - 2\tan x = 0$

$\tan x(\tan x - 2) = 0$

$\tan x = 0 \quad$ or $\quad \tan x - 2 = 0$

$\qquad x = n\pi \qquad\qquad \tan x = 2$

$\qquad\qquad\qquad\qquad x = \arctan 2 + n\pi$

45. $\quad \tan^2 \theta + \tan \theta - 6 = 0$

$(\tan \theta + 3)(\tan \theta - 2) = 0$

$\tan \theta + 3 = 0 \qquad\qquad$ or $\quad \tan \theta - 2 = 0$

$\qquad \tan \theta = -3 \qquad\qquad\qquad \tan \theta = 2$

$\qquad \theta = \arctan(-3) + n\pi \qquad\qquad \theta = \arctan 2 + n\pi$

47. $\sin 75° = \sin(120° - 45°)$

$\qquad\quad = \sin 120° \cos 45° - \cos 120° \sin 45°$

$\qquad\quad = \left(\dfrac{\sqrt{3}}{2}\right)\left(\dfrac{\sqrt{2}}{2}\right) - \left(-\dfrac{1}{2}\right)\left(\dfrac{\sqrt{2}}{2}\right)$

$\qquad\quad = \dfrac{\sqrt{2}}{4}\left(\sqrt{3} + 1\right)$

$\cos 75° = \cos(120° - 45°)$

$\qquad\quad = \cos 120° \cos 45° + \sin 120° \sin 45°$

$\qquad\quad = \left(-\dfrac{1}{2}\right)\left(\dfrac{\sqrt{2}}{2}\right) + \left(\dfrac{\sqrt{3}}{2}\right)\left(\dfrac{\sqrt{2}}{2}\right)$

$\qquad\quad = \dfrac{\sqrt{2}}{4}\left(\sqrt{3} - 1\right)$

$\tan 75° = \tan(120° - 45°) = \dfrac{\tan 120° - \tan 45°}{1 + \tan 120° \tan 45°}$

$\qquad\quad = \dfrac{-\sqrt{3} - 1}{1 + \left(-\sqrt{3}\right)(1)} = \dfrac{-\sqrt{3} - 1}{1 - \sqrt{3}}$

$\qquad\quad = \dfrac{-\sqrt{3} - 1}{1 - \sqrt{3}} \cdot \dfrac{1 + \sqrt{3}}{1 + \sqrt{3}}$

$\qquad\quad = \dfrac{-4 - 2\sqrt{3}}{-2} = 2 + \sqrt{3}$

49. $\sin \dfrac{25\pi}{12} = \sin\left(\dfrac{11\pi}{6} + \dfrac{\pi}{4}\right) = \sin \dfrac{11\pi}{6} \cos \dfrac{\pi}{4} + \cos \dfrac{11\pi}{6} \sin \dfrac{\pi}{4}$

$\qquad = \left(-\dfrac{1}{2}\right)\!\left(\dfrac{\sqrt{2}}{2}\right) + \left(\dfrac{\sqrt{3}}{2}\right)\!\left(\dfrac{\sqrt{2}}{2}\right) = \dfrac{\sqrt{2}}{4}\left(\sqrt{3} - 1\right)$

$\quad \cos \dfrac{25\pi}{12} = \cos\left(\dfrac{11\pi}{6} + \dfrac{\pi}{4}\right) = \cos \dfrac{11\pi}{6} \cos \dfrac{\pi}{4} - \sin \dfrac{11\pi}{6} \sin \dfrac{\pi}{4}$

$\qquad = \left(\dfrac{\sqrt{3}}{2}\right)\!\left(\dfrac{\sqrt{2}}{2}\right) - \left(-\dfrac{1}{2}\right)\!\left(\dfrac{\sqrt{2}}{2}\right) = \dfrac{\sqrt{2}}{4}\left(\sqrt{3} + 1\right)$

$\quad \tan \dfrac{25\pi}{12} = \tan\left(\dfrac{11\pi}{6} + \dfrac{\pi}{4}\right) = \dfrac{\tan \dfrac{11\pi}{6} + \tan \dfrac{\pi}{4}}{1 - \tan \dfrac{11\pi}{6} \tan \dfrac{\pi}{4}}$

$\qquad = \dfrac{\left(-\dfrac{\sqrt{3}}{3}\right) + 1}{1 - \left(-\dfrac{\sqrt{3}}{3}\right)(1)} = 2 - \sqrt{3}$

51. $\sin 60° \cos 45° - \cos 60° \sin 45° = \sin(60° - 45°)$

$\qquad\qquad\qquad\qquad\qquad\qquad\quad = \sin 15°$

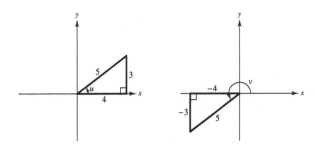

Figures for Exercises 53–55

53. $\sin(u + v) = \sin u \cos v + \cos u \sin v = \dfrac{3}{5}\left(-\dfrac{4}{5}\right) + \dfrac{4}{5}\left(-\dfrac{3}{5}\right) = -\dfrac{24}{25}$

55. $\cos(u - v) = \cos u \cos v + \sin u \sin v = \dfrac{4}{5}\left(-\dfrac{4}{5}\right) + \dfrac{3}{5}\left(-\dfrac{3}{5}\right) = -1$

57. $\cos\left(x + \dfrac{\pi}{2}\right) = \cos x \cos \dfrac{\pi}{2} - \sin x \sin \dfrac{\pi}{2} = \cos x(0) - \sin x(1) = -\sin x$

59. $\tan(\pi - x) = \dfrac{\tan \pi - \tan x}{1 - \tan \pi \tan x} = -\tan x$

61. $\sin\left(x + \dfrac{\pi}{4}\right) - \sin\left(x - \dfrac{\pi}{4}\right) = 1$

$\qquad\qquad 2 \cos x \sin \dfrac{\pi}{4} = 1$

$\qquad\qquad\qquad \cos x = \dfrac{\sqrt{2}}{2}$

$\qquad\qquad\qquad\quad x = \dfrac{\pi}{4}, \dfrac{7\pi}{4}$

63. $\sin u = -\dfrac{4}{5}, \pi < u < \dfrac{3\pi}{2}$

$\cos u = -\sqrt{1 - \sin^2 u} = \dfrac{-3}{5}$

$\tan u = \dfrac{\sin u}{\cos u} = \dfrac{4}{3}$

$\sin 2u = 2 \sin u \cos u = 2\left(-\dfrac{4}{5}\right)\left(-\dfrac{3}{5}\right) = \dfrac{24}{25}$

$\cos 2u = \cos^2 u - \sin^2 u = \left(-\dfrac{3}{5}\right)^2 - \left(-\dfrac{4}{5}\right)^2 = -\dfrac{7}{25}$

$\tan 2u = \dfrac{2 \tan u}{1 - \tan^2 u} = \dfrac{2\left(\dfrac{4}{3}\right)}{1 - \left(\dfrac{4}{3}\right)^2} = -\dfrac{24}{7}$

65. $\sin 4x = 2 \sin 2x \cos 2x$

$\qquad = 2\left[2 \sin x \cos x\left(\cos^2 x - \sin^2 x\right)\right]$

$\qquad = 4 \sin x \cos x\left(2 \cos^2 x - 1\right)$

$\qquad = 8 \cos^3 x \sin x - 4 \cos x \sin x$

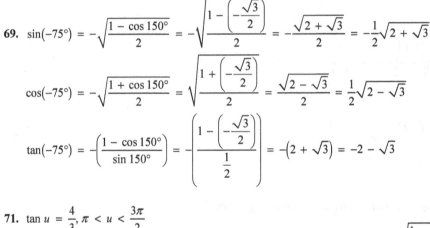

67. $\tan^2 3x = \dfrac{\sin^2 3x}{\cos^2 3x} = \dfrac{\dfrac{1 - \cos 6x}{2}}{\dfrac{1 + \cos 6x}{2}} = \dfrac{1 - \cos 6x}{1 + \cos 6x}$

69. $\sin(-75°) = -\sqrt{\dfrac{1 - \cos 150°}{2}} = -\sqrt{\dfrac{1 - \left(-\dfrac{\sqrt{3}}{2}\right)}{2}} = -\dfrac{\sqrt{2 + \sqrt{3}}}{2} = -\dfrac{1}{2}\sqrt{2 + \sqrt{3}}$

$\cos(-75°) = -\sqrt{\dfrac{1 + \cos 150°}{2}} = \sqrt{\dfrac{1 + \left(-\dfrac{\sqrt{3}}{2}\right)}{2}} = \dfrac{\sqrt{2 - \sqrt{3}}}{2} = \dfrac{1}{2}\sqrt{2 - \sqrt{3}}$

$\tan(-75°) = -\left(\dfrac{1 - \cos 150°}{\sin 150°}\right) = -\left(\dfrac{1 - \left(-\dfrac{\sqrt{3}}{2}\right)}{\dfrac{1}{2}}\right) = -\left(2 + \sqrt{3}\right) = -2 - \sqrt{3}$

71. $\tan u = \dfrac{4}{3}, \pi < u < \dfrac{3\pi}{2}$

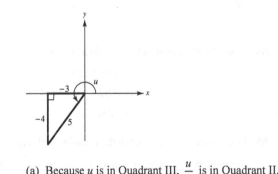

(a) Because u is in Quadrant III, $\dfrac{u}{2}$ is in Quadrant II.

(b) $\sin \dfrac{u}{2} = \sqrt{\dfrac{1 - \cos u}{2}} = \sqrt{\dfrac{1 - \left(-\dfrac{3}{5}\right)}{2}} = \sqrt{\dfrac{4}{5}}$

$\qquad = \dfrac{2\sqrt{5}}{5}$

$\cos \dfrac{u}{2} = -\sqrt{\dfrac{1 + \cos u}{2}} = -\sqrt{\dfrac{1 + \left(-\dfrac{3}{5}\right)}{2}} = -\sqrt{\dfrac{1}{5}}$

$\qquad = -\dfrac{\sqrt{5}}{5}$

$\tan \dfrac{u}{2} = \dfrac{1 - \cos u}{\sin u} = \dfrac{1 - \left(-\dfrac{3}{5}\right)}{\left(-\dfrac{4}{5}\right)} = -2$

73. $\cos u = -\dfrac{2}{7}, \dfrac{\pi}{2} < u < \pi$

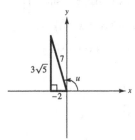

(a) Because u is in Quadrant II, $\dfrac{u}{2}$ is in Quadrant I.

(b) $\sin \dfrac{u}{2} = \sqrt{\dfrac{1 - \cos u}{2}} = \sqrt{\dfrac{1 - \left(-\dfrac{2}{7}\right)}{2}} = \sqrt{\dfrac{9}{14}}$

$\qquad = \dfrac{3\sqrt{14}}{14}$

$\cos \dfrac{u}{2} = \sqrt{\dfrac{1 + \cos u}{2}} = \sqrt{\dfrac{1 + \left(-\dfrac{2}{7}\right)}{2}} = \sqrt{\dfrac{5}{14}}$

$\qquad = \dfrac{\sqrt{70}}{14}$

$\tan \dfrac{u}{2} = \dfrac{1 - \cos u}{\sin u} = \dfrac{1 - \left(-\dfrac{2}{7}\right)}{\dfrac{3\sqrt{5}}{7}} = \dfrac{3\sqrt{5}}{5}$

75. $\cos 4\theta \sin 6\theta = \frac{1}{2}\left[\sin(4\theta + 6\theta) - \sin(4\theta - 6\theta)\right] = \frac{1}{2}\left[\sin 10\theta - \sin(-2\theta)\right]$

77. $\cos 6\theta + \cos 5\theta = 2 \cos\left(\dfrac{6\theta + 5\theta}{2}\right)\cos\left(\dfrac{6\theta - 5\theta}{2}\right) = 2 \cos \dfrac{11\theta}{2} \cos \dfrac{\theta}{2}$

79. $\qquad r = \dfrac{1}{32}{v_0}^2 \sin 2\theta$

$\qquad \text{range} = 100 \text{ feet}$

$\qquad v_0 = 80 \text{ feet per second}$

$\qquad r = \dfrac{1}{32}(80)^2 \sin 2\theta = 100$

$\qquad \sin 2\theta = 0.5$

$\qquad 2\theta = 30°$

$\qquad \theta = 15° \text{ or } \dfrac{\pi}{12}$

81. False. If $\dfrac{\pi}{2} < \theta < \pi$, then $\dfrac{\pi}{4} < \dfrac{\theta}{2} < \dfrac{\pi}{2}$, and $\dfrac{\theta}{2}$ is in

Quadrant I. $\cos\dfrac{\theta}{2} > 0$

83. True. $4 \sin(-x)\cos(-x) = 4(-\sin x) \cos x$

$\qquad\qquad\qquad\qquad\quad = -4 \sin x \cos x$

$\qquad\qquad\qquad\qquad\quad = -2(2 \sin x \cos x)$

$\qquad\qquad\qquad\qquad\quad = -2 \sin 2x$

85. Yes. *Sample Answer.* When the domain is all real

numbers, the solutions of $\sin x = \dfrac{1}{2}$ are $x = \dfrac{\pi}{6} + 2n\pi$

and $x = \dfrac{5\pi}{6} + 2n\pi$, so there are infinitely many

solutions.

Problem Solving for Chapter 2

1. $\sin\theta = \pm\sqrt{1 - \cos^2\theta}$

$\tan\theta = \dfrac{\sin\theta}{\cos\theta} = \pm\dfrac{\sqrt{1 - \cos^2\theta}}{\cos\theta}$

$\csc\theta = \dfrac{1}{\sin\theta} = \pm\dfrac{1}{\sqrt{1 - \cos^2\theta}}$

$\sec\theta = \dfrac{1}{\cos\theta}$

$\cot\theta = \dfrac{1}{\tan\theta} = \pm\dfrac{\cos\theta}{\sqrt{1 - \cos^2\theta}}$

You also have the following relationships:

$\sin\theta = \cos\left(\dfrac{\pi}{2} - \theta\right)$

$\tan\theta = \dfrac{\cos\left[(\pi/2) - \theta\right]}{\cos\theta}$

$\csc\theta = \dfrac{1}{\cos\left[(\pi/2) - \theta\right]}$

$\sec\theta = \dfrac{1}{\cos\theta}$

$\cot\theta = \dfrac{\cos\theta}{\cos\left[(\pi/2) - \theta\right]}$

3. $\sin\left[\dfrac{(12n + 1)\pi}{6}\right] = \sin\left[\dfrac{1}{6}(12n\pi + \pi)\right]$

$\qquad = \sin\left(2n\pi + \dfrac{\pi}{6}\right)$

$\qquad = \sin\dfrac{\pi}{6} = \dfrac{1}{2}$

So, $\sin\left[\dfrac{(12n + 1)\pi}{6}\right] = \dfrac{1}{2}$ for all integers n.

5. From the figure, it appears that $u + v = w$. Assume that u, v, and w are all in Quadrant I.

From the figure:

$\tan u = \dfrac{s}{3s} = \dfrac{1}{3}$

$\tan v = \dfrac{s}{2s} = \dfrac{1}{2}$

$\tan w = \dfrac{s}{s} = 1$

$\tan(u + v) = \dfrac{\tan u + \tan v}{1 - \tan u \tan v} = \dfrac{1/3 + 1/2}{1 - (1/3)(1/2)} = \dfrac{5/6}{1 - (1/6)} = 1 = \tan w.$

So, $\tan(u + v) = \tan w$. Because u, v, and w are all in Quadrant I, you have

$\arctan\left[\tan(u + v)\right] = \arctan\left[\tan w\right]u + v = w.$

7. (a)

$\sin\dfrac{\theta}{2} = \dfrac{\frac{1}{2}b}{10}$ and $\cos\dfrac{\theta}{2} = \dfrac{h}{10}$

$\qquad b = 20\sin\dfrac{\theta}{2}$ $\qquad h = 10\cos\dfrac{\theta}{2}$

$A = \dfrac{1}{2}bh = \dfrac{1}{2}\left(20\sin\dfrac{\theta}{2}\right)\left(10\cos\dfrac{\theta}{2}\right) = 100\sin\dfrac{\theta}{2}\cos\dfrac{\theta}{2}$

(b) $A = 50\left(2\sin\dfrac{\theta}{2}\cos\dfrac{\theta}{2}\right)$

$\qquad = 50\sin\left(2\left(\dfrac{\theta}{2}\right)\right)$

$\qquad = 50\sin\theta$

Because $\sin\dfrac{\pi}{2} = 1$ is a maximum, $\theta = \dfrac{\pi}{2}$.

So, the area is a maximum at $A = 50\sin\dfrac{\pi}{2} = 50$

square meters.

9. $F = \dfrac{0.6W \sin(\theta + 90°)}{\sin 12°}$

(a) $F = \dfrac{0.6W(\sin \theta \cos 90° + \cos \theta \sin 90°)}{\sin 12°}$

$= \dfrac{0.6W[(\sin \theta)(0) + (\cos \theta)(1)]}{\sin 12°}$

$= \dfrac{0.6W \cos \theta}{\sin 12°}$

(b) Let $y_1 = \dfrac{0.6(185) \cos x}{\sin 12°}$.

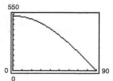

(c) The force is maximum (533.88 pounds) when $\theta = 0°$.
The force is minimum (0 pounds) when $\theta = 90°$.

11. $d = 35 - 28 \cos \dfrac{\pi}{6.2}t$ when $t = 0$ corresponds to 12:00 A.M.

(a) The high tides occur when $\cos \dfrac{\pi}{6.2}t = -1$. Solving

yields $t = 6.2$ or $t = 18.6$.

These t-values correspond to 6:12 A.M. and 6:36 P.M.

The low tide occurs when $\cos \dfrac{\pi}{6.2}t = 1$. Solving

yields $t = 0$ and $t = 12.4$ which corresponds to 12:00 A.M. and 12:24 P.M.

(b) The water depth is never 3.5 feet. At low tide, the depth is $d = 35 - 28 = 7$ feet.

(c)

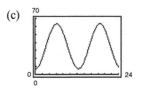

13. (a) $n = \dfrac{\sin\left(\dfrac{\theta}{2} + \dfrac{\alpha}{2}\right)}{\sin \dfrac{\theta}{2}}$

$= \dfrac{\sin\left(\dfrac{\theta}{2}\right)\cos\left(\dfrac{\alpha}{2}\right) + \cos\left(\dfrac{\theta}{2}\right)\sin\left(\dfrac{\alpha}{2}\right)}{\sin\left(\dfrac{\theta}{2}\right)}$

$= \cos\left(\dfrac{\alpha}{2}\right) + \cot\left(\dfrac{\theta}{2}\right)\sin\left(\dfrac{\alpha}{2}\right)$

For $\alpha = 60°$, $n = \cos 30° + \cot\left(\dfrac{\theta}{2}\right)\sin 30°$

$n = \dfrac{\sqrt{3}}{2} + \dfrac{1}{2}\cot\left(\dfrac{\theta}{2}\right)$.

(b) For glass, $n = 1.50$.

$1.50 = \dfrac{\sqrt{3}}{2} + \dfrac{1}{2}\cot\left(\dfrac{\theta}{2}\right)$

$2\left(1.50 - \dfrac{\sqrt{3}}{2}\right) = \cot\left(\dfrac{\theta}{2}\right)$

$\dfrac{1}{3 - \sqrt{3}} = \tan\left(\dfrac{\theta}{2}\right)$

$\theta = 2 \tan^{-1}\left(\dfrac{1}{3 - \sqrt{3}}\right)$

$\theta \approx 76.5°$

15. (a) Let $y_1 = \sin x$ and $y_2 = 0.5$.

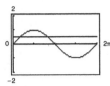

$\sin x \geq 0.5$ on the interval $\left[\dfrac{\pi}{6}, \dfrac{5\pi}{6}\right]$.

(b) Let $y_1 = \cos x$ and $y_2 = -0.5$.

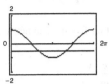

$\cos x \leq -0.5$ on the interval $\left[\dfrac{2\pi}{3}, \dfrac{4\pi}{3}\right]$.

(c) Let $y_1 = \tan x$ and $y_2 = \sin x$.

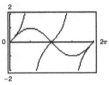

$\tan x < \sin x$ on the intervals $\left(\dfrac{\pi}{2}, \pi\right)$ and $\left(\dfrac{3\pi}{2}, 2\pi\right)$.

(d) Let $y_1 = \cos x$ and $y_2 = \sin x$.

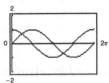

$\cos x \geq \sin x$ on the intervals $\left[0, \dfrac{\pi}{4}\right]$ and $\left[\dfrac{5\pi}{4}, 2\pi\right]$.

Practice Test for Chapter 2

1. Find the value of the other five trigonometric functions, given $\tan x = \frac{4}{11}$, $\sec x < 0$.

2. Simplify $\dfrac{\sec^2 x + \csc^2 x}{\csc^2 x(1 + \tan^2 x)}$.

3. Rewrite as a single logarithm and simplify $\ln|\tan \theta| - \ln|\cot \theta|$.

4. True or false:

$$\cos\left(\frac{\pi}{2} - x\right) = \frac{1}{\csc x}$$

5. Factor and simplify: $\sin^4 x + \left(\sin^2 x\right)\cos^2 x$

6. Multiply and simplify: $(\csc x + 1)(\csc x - 1)$

7. Rationalize the denominator and simplify:

$$\frac{\cos^2 x}{1 - \sin x}$$

8. Verify:

$$\frac{1 + \cos \theta}{\sin \theta} + \frac{\sin \theta}{1 + \cos \theta} = 2 \csc \theta$$

9. Verify:

$$\tan^4 x + 2 \tan^2 x + 1 = \sec^4 x$$

10. Use the sum or difference formulas to determine:

   (a) $\sin 105°$

   (b) $\tan 15°$

11. Simplify: $(\sin 42°) \cos 38° - (\cos 42°) \sin 38°$

12. Verify $\tan\left(\theta + \dfrac{\pi}{4}\right) = \dfrac{1 + \tan \theta}{1 - \tan \theta}$.

13. Write $\sin(\arcsin x - \arccos x)$ as an algebraic expression in x.

14. Use the double-angle formulas to determine:

   (a) $\cos 120°$

   (b) $\tan 300°$

15. Use the half-angle formulas to determine:

   (a) $\sin 22.5°$

   (b) $\tan \dfrac{\pi}{12}$

16. Given $\sin \theta = 4/5$, θ lies in Quadrant II, find $\cos(\theta/2)$.

17. Use the power-reducing identities to write $\left(\sin^2 x\right)\cos^2 x$ in terms of the first power of cosine.

18. Rewrite as a sum: $6(\sin 5\theta)\cos 2\theta$.

19. Rewrite as a product: $\sin(x + \pi) + \sin(x - \pi)$.

20. Verify $\dfrac{\sin 9x + \sin 5x}{\cos 9x - \cos 5x} = -\cot 2x$.

21. Verify:

$$(\cos u)\sin v = \tfrac{1}{2}\left[\sin(u + v) - \sin(u - v)\right].$$

22. Find all solutions in the interval $[0, 2\pi)$:

$$4\sin^2 x = 1$$

23. Find all solutions in the interval $[0, 2\pi)$:

$$\tan^2 \theta + \left(\sqrt{3} - 1\right)\tan \theta - \sqrt{3} = 0$$

24. Find all solutions in the interval $[0, 2\pi)$:

$$\sin 2x = \cos x$$

25. Use the quadratic formula to find all solutions in the interval $[0, 2\pi)$:

$$\tan^2 x - 6\tan x + 4 = 0$$

C H A P T E R 3
Additional Topics in Trigonometry

CHAPTER 3
Additional Topics in Trigonometry

Section 3.1   Law of Sines

1. oblique

3. angles; side

5.

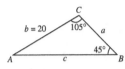

Given: $B = 45°, C = 105°, b = 20$

$A = 180° - B - C = 30°$

$a = \dfrac{b}{\sin B}(\sin A) = \dfrac{20 \sin 30°}{\sin 45°} = 10\sqrt{2} \approx 14.14$

$C = \dfrac{b}{\sin B}(\sin C) = \dfrac{20 \sin 105°}{\sin 45°} \approx 27.32$

7.

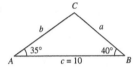

Given: $A = 35°, B = 40°, c = 10$

$C = 180° - A - B = 105°$

$a = \dfrac{c}{\sin C}(\sin A) = \dfrac{10 \sin 35°}{\sin 105°} \approx 5.94$

$b = \dfrac{c}{\sin C}(\sin B) = \dfrac{10 \sin 40°}{\sin 105°} \approx 6.65$

9. Given: $A = 102.4°, C = 16.7°, a = 21.6$

$B = 180° - A - C = 60.9°$

$b = \dfrac{a}{\sin A}(\sin B) = \dfrac{21.6}{\sin 102.4°}(\sin 60.9°) \approx 19.32$

$c = \dfrac{a}{\sin A}(\sin C) = \dfrac{21.6}{\sin 102.4°}(\sin 16.7°) \approx 6.36$

11. Given: $A = 83°20', C = 54.6°, c = 18.1$

$B = 180° - A - C = 180° - 83°20' - 54°36' = 42°4'$

$a = \dfrac{c}{\sin C}(\sin A) = \dfrac{18.1}{\sin 54.6°}(\sin 83°20') \approx 22.05$

$b = \dfrac{c}{\sin C}(\sin B) = \dfrac{18.1}{\sin 54.6°}(\sin 42°4') \approx 14.88$

13. Given: $A = 35°, B = 65°, c = 10$

$C = 180° - A - B = 80°$

$a = \dfrac{c}{\sin C}(\sin A) = \dfrac{10 \sin 35°}{\sin 80°} \approx 5.82$

$b = \dfrac{c}{\sin C}(\sin B) = \dfrac{10 \sin 65°}{\sin 80°} \approx 9.20$

15. Given: $A = 55°, B = 42°, c = \dfrac{3}{4}$

$C = 180° - A - B = 83°$

$a = \dfrac{c}{\sin C}(\sin A) = \dfrac{0.75}{\sin 83°}(\sin 55°) \approx 0.62$

$b = \dfrac{c}{\sin C}(\sin B) = \dfrac{0.75}{\sin 83°}(\sin 42°) \approx 0.51$

17. Given: $A = 36°, a = 8, b = 5$

$\sin B = \dfrac{b \sin A}{a} = \dfrac{5 \sin 36°}{8} \approx 0.36737 \Rightarrow B \approx 21.55°$

$C = 180° - A - B \approx 180° - 36° - 21.55 = 122.45°$

$c = \dfrac{a}{\sin A}(\sin C) = \dfrac{8}{\sin 36°}(\sin 122.45°) \approx 11.49$

19. Given: $A = 145°, a = 14, b = 4$

$\sin B = \dfrac{b \sin A}{a} = \dfrac{4 \sin 145°}{14} \approx 0.1639 \Rightarrow B \approx 9.43°$

$C = 180° - A - B \approx 25.57°$

$c = \dfrac{a}{\sin A}(\sin C) \approx \dfrac{14 \sin 25.57°}{\sin 145°} \approx 10.53$

21. Given: $B = 15°30'$, $a = 4.5$, $b = 6.8$

$$\sin A = \frac{a \sin B}{b} = \frac{4.5 \sin 15°30'}{6.8} \approx 0.17685 \Rightarrow A \approx 10°11'$$

$$C = 180° - A - B \approx 180° - 10°11' - 15°30' = 154°19'$$

$$c = \frac{b}{\sin B}(\sin C) = \frac{6.8}{\sin 15°30'}(\sin 154°19') \approx 11.03$$

23. Given: $A = 110°$, $a = 125$, $b = 100$

$$\sin B = \frac{b \sin A}{a} = \frac{100 \sin 110°}{125} \approx 0.75175 \Rightarrow B \approx 48.74°$$

$$C = 180° - A - B \approx 21.26°$$

$$c = \frac{a \sin C}{\sin A} \approx \frac{125 \sin 21.26°}{\sin 110°} \approx 48.23$$

25. Given: $a = 18$, $b = 20$, $A = 76°$

$$h = 20 \sin 76° \approx 19.41$$

Because $a < h$, no triangle is formed.

27. Given: $A = 58°$, $a = 11.4$, $c = 12.8$

$$\sin B = \frac{b \sin A}{a} = \frac{12.8 \sin 58°}{11.4} \approx 0.9522 \Rightarrow B \approx 72.21° \text{ or } B \approx 107.79°$$

Case 1	Case 2
$B \approx 72.21°$	$B \approx 107.79°$
$C = 180° - A - B \approx 49.79°$	$C = 180° - A - B \approx 14.21°$
$c = \frac{a}{\sin A}(\sin C) \approx \frac{11.4 \sin 49.79°}{\sin 58°} \approx 10.27$	$c = \frac{a}{\sin A}(\sin C) \approx \frac{11.4 \sin 14.21°}{\sin 58°} \approx 3.30$

29. Given: $A = 120°$, $a = b = 25$

No triangle is formed because A is obtuse and $a = b$.

31. Given: $A = 45°$, $a = b = 1$

Because $a = b = 1$, $B = 45°$.

$$C = 180° - A - B = 90°$$

$$c = \frac{a}{\sin A}(\sin C) = \frac{1 \sin 90°}{\sin 45°} = \sqrt{2} \approx 1.41$$

33. Given: $A = 36°$, $a = 5$

(a) One solution if $b \le 5$ or $b = \dfrac{5}{\sin 36°}$.

(b) Two solutions if $5 < b < \dfrac{5}{\sin 36°}$.

(c) No solution if $b > \dfrac{5}{\sin 36°}$.

35. Given: $A = 105°$, $a = 80$

(a) One solution if $b < 80$.

(b) Not possible for two solutions.

(c) No solution if $b \ge 80$.

37. $A = 125°$, $b = 9$, $c = 6$

$$\text{Area} = \frac{1}{2}bc \sin A$$

$$= \frac{1}{2}(9)(6) \sin 125° \approx 22.1$$

39. $B = 39°, a = 25, c = 12$

$\text{Area} = \dfrac{1}{2}ac \sin B$

$\qquad = \dfrac{1}{2}(25)(12) \sin 39° = 94.4$

41. $C = 103° \, 15', a = 16, b = 28$

$\text{Area} = \dfrac{1}{2}ab \sin C$

$\qquad = \dfrac{1}{2}(16)(28) \sin 103° \, 15' \approx 218.0$

43. $A = 67°, B = 43°, a = 8$

$b = \dfrac{a}{\sin A}(\sin B) = \dfrac{8 \sin 43°}{\sin 67°} \approx 5.927$

$C = 180° - A - B = 70°$

$\text{Area} = \dfrac{1}{2}ab \sin C$

$\qquad = \dfrac{1}{2}(8)(5.927) \sin 70° \approx 22.3$

45. (a) $\qquad C = 180° - 94° - 30° = 56°$

$\dfrac{h}{\sin 30°} = \dfrac{40}{\sin 56°}$

$h = \dfrac{40}{\sin 56°}(\sin 30°)$

(b) $h = \dfrac{40}{\sin 56°}(\sin 30°) \approx 24.1$ meters

47. Given: $A = 15°, B = 135°, c = 30$

$C = 180° - A - B = 30°$

From Pine Knob:

$b = \dfrac{c \sin B}{\sin C} = \dfrac{30 \sin 135°}{\sin 30°} \approx 42.4$ kilometers

From Colt Station:

$a = \dfrac{c \sin A}{\sin C} = \dfrac{30 \sin 15°}{\sin 30°} \approx 15.5$ kilometers

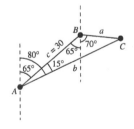

49. $\dfrac{\sin(42° - \theta)}{10} = \dfrac{\sin 48°}{17}$

$\sin(42° - \theta) \approx 0.43714$

$42° - \theta \approx 25.9°$

$\theta \approx 16.1°$

51. Given: $A = 55°, c = 2.2$

(a)

(b) $\qquad B = 180° - 72° = 108°$

$\qquad C = 180° - 55° - 108° = 17°$

$\dfrac{a}{\sin A} = \dfrac{c}{\sin c}$

$a = \dfrac{c}{\sin c}(\sin A) = \dfrac{2.2 \sin 55°}{\sin 17°} \approx 6.16$

(c) $h = a \sin 72° \approx 6.16 \sin 72° \approx 5.86$ miles

(d) The plane must travel a horizontal distance d to be directly above point A.

$\angle ACD = \angle ACB + \angle BCD$

$\qquad = 17° + (180° - 72° - 90°)$

$\qquad = 17° + 18° = 35°$

$\tan 35° = \dfrac{d}{5.86}$

$d = 5.86 \tan 35° \approx 4.10$ miles

53. $\qquad \alpha = 180 - \theta - (180 - \phi) = \phi - \theta$

$\dfrac{d}{\sin \theta} = \dfrac{2}{\sin \alpha}$

$d = \dfrac{2 \sin \theta}{\sin(\phi - \theta)}$

55. True. If one angle of a triangle is obtuse, then there is less than $90°$ left for the other two angles, so it cannot contain a right angle. It must be oblique.

57. False. To solve an oblique triangle using the Law of Sines, you need to know two angles and any side, or two sides and an angle opposite one of them.

59. To find the area using angle C, the formula should be $A = \frac{1}{2}ab \sin C$ and not $A = \frac{1}{2}bc \sin C$. So first find angle B, to find side a. Then the area can be calculated.

61. Yes.

$A = 180° - B - C = 40°$

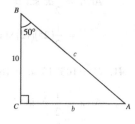

$$\frac{a}{\sin A} = \frac{c}{\sin C}$$

$$c = \frac{a}{\sin A} \sin C$$

$$\approx 15.6$$

$$\frac{b}{\sin B} = \frac{c}{\sin C}$$

$$b = \frac{c}{\sin C} \sin B$$

$$\approx 11.9$$

An alternative method is to use the trigonometric ratios of a right triangle.

That is $\sin A = \dfrac{opp}{hyp} = \dfrac{a}{c}$ and $\tan A = \dfrac{opp}{hyp} = \dfrac{a}{b}$.

Section 3.2 Law of Cosines

1. $b^2 = a^2 + c^2 - 2ac \cos B$

3. standard

5. Given: $a = 10, b = 12, c = 16$

$$\cos C = \frac{a^2 + b^2 - c^2}{2ab} = \frac{100 + 144 - 256}{2(10)(12)} = -0.05 \Rightarrow C \approx 92.87°$$

$$\sin B = \frac{b \sin C}{c} \approx \frac{12 \sin 92.87°}{16} \approx 0.749059 \Rightarrow B \approx 48.51°$$

$$A \approx 180° - 48.51° - 92.87° = 38.62°$$

7. Given: $a = 6, b = 8, c = 12$

$$\cos C = \frac{a^2 + b^2 + c^2}{2ab} = \frac{6^2 + 8^2 - 12^2}{2(6)(8)} \approx -0.458333 \Rightarrow C \approx 117.28°$$

$$\sin B = b\left(\frac{\sin C}{c}\right) = 8\left(\frac{\sin 117.28°}{12}\right) \approx 0.592518 \Rightarrow B \approx 36.34°$$

$$A = 180° - B - C \approx 180° - 36.34° - 117.28° = 26.38°$$

9. Given: $A = 30°, b = 15, c = 30$

$$a^2 = b^2 + c^2 - 2bc \cos A$$

$$= 225 + 900 - 2(15)(30) \cos 30° \approx 345.5771$$

$$a \approx 18.59$$

$$\cos C = \frac{a^2 + b^2 - c^2}{2ab} \approx \frac{(345.5771)^2 + 15^2 - 30^2}{2(18.59)(15)} \approx -0.590681 \Rightarrow C \approx 126.21°$$

$$B \approx 180° - 30° - 126.21° = 23.79°$$

11. Given: $A = 50°, b = 15, c = 30$

$$a^2 = b^2 + c^2 - 2bc \cos A = 15^2 + 30^2 - 2(15)(30) \cos 50°$$
$$\approx 546.4912 \Rightarrow a \approx 23.38$$

$$\sin C = c\left(\frac{\sin A}{a}\right) \approx 30\left(\frac{\sin 50°}{23.3772}\right) \approx 0.983066$$

There are two angles between $0°$ and $180°$ whose sine is 0.983066, $C_1 \approx 79.4408°$ and $C_2 \approx 180° - 79.4408° \approx 100.56°$.

Because side c is the longest side of the triangle, C must be the largest angle of the triangle. So, $C \approx 100.56°$ and $B = 180° - A - C \approx 180° - 50° - 100.56° = 29.44°$.

13. Given: $a = 11, b = 15, c = 21$

$$\cos C = \frac{a^2 + b^2 - c^2}{2ab} = \frac{121 + 225 - 441}{2(11)(15)} \approx -0.287879 \Rightarrow C \approx 106.73°$$

$$\sin B = \frac{b \sin C}{c} = \frac{15 \sin 106.73°}{21} \approx 0.684051 \Rightarrow B \approx 43.16°$$

$$A \approx 180° - 43.16° - 106.73° = 30.11°$$

15. Given: $a = 2.5, b = 1.8, c = 0.9$

$$\cos A = \frac{b^2 + c^2 - a^2}{2bc} = \frac{(1.8)^2 + (0.9)^2 - (2.5)^2}{2(1.8)(0.9)} = -0.679012 \Rightarrow A \approx 132.77°$$

$$\cos B = \frac{a^2 + c^2 - b^2}{2ac} = \frac{(2.5)^2 + (0.9)^2 - (1.8)^2}{2(2.5)(0.9)} \approx 0.848889 \Rightarrow B \approx 31.91°$$

$$C = 180° - 132.77° - 31.91° = 15.32°$$

17. Given: $A = 120°, b = 6, c = 7$

$$a^2 = b^2 + c^2 - 2bc \cos A = 36 + 49 - 2(6)(7) \cos 120° = 127 \Rightarrow a \approx 11.27$$

$$\sin B = \frac{b \sin A}{a} \approx \frac{6 \sin 120°}{11.27} \approx 0.461061 \Rightarrow B \approx 27.46°$$

$$C \approx 180° - 120° - 27.46° = 32.54°$$

19. Given: $B = 10° 35', a = 40, c = 30$

$$b^2 = a^2 + c^2 - 2ac \cos B = 1600 + 900 - 2(40)(30) \cos 10° 35' \approx 140.8268 \Rightarrow b \approx 11.87$$

$$\sin C = \frac{c \sin B}{b} = \frac{30 \sin 10° 35'}{11.87} \approx 0.464192 \Rightarrow C \approx 27.66° \approx 27° 40'$$

$$A \approx 180° - 10° 35' - 27° 40' = 141° 45'$$

21. Given: $B = 125° 40', a = 37, c = 37$

$$b^2 = a^2 + c^2 - 2ac \cos B = 1369 + 1369 - 2(37)(37) \cos 125° 40' \approx 4334.4420 \Rightarrow b \approx 65.84$$

$$A = C \Rightarrow 2A = 180 - 125° 40' = 54° 20' \Rightarrow A = C = 27° 10'$$

23. $C = 43°, a = \frac{4}{9}, b = \frac{7}{9}$

$$c^2 = a^2 + b^2 - 2ab \cos C = \left(\frac{4}{9}\right)^2 + \left(\frac{7}{9}\right)^2 - 2\left(\frac{4}{9}\right)\left(\frac{7}{9}\right) \cos 43° \approx 0.296842 \Rightarrow c \approx 0.54$$

$$\sin A = \frac{a \sin C}{c} = \frac{(4/9) \sin 43°}{0.544832} \approx 0.556337 \Rightarrow A \approx 33.80°$$

$$B \approx 180° - 43° - 33.8° = 103.20°$$

25. $d^2 = 5^2 + 8^2 - 2(5)(8) \cos 45° \approx 32.4315 \Rightarrow d \approx 5.69$

$2\phi = 360° - 2(45°) = 270° \Rightarrow \phi = 135°$

$c^2 = 5^2 + 8^2 - 2(5)(8) \cos 135° \approx 145.5685 \Rightarrow c \approx 12.07$

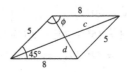

27.

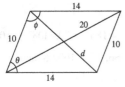

$$\cos \phi = \frac{10^2 + 14^2 - 20^2}{2(10)(14)}$$

$\phi \approx 111.8°$

$2\theta \approx 360° - 2(111.8°)$

$\theta = 68.2°$

$d^2 = 10^2 + 14^2 - 2(10)(14) \cos 68.2°$

$d \approx 13.86$

29. $\cos \alpha = \dfrac{(12.5)^2 + (15)^2 - 10^2}{2(12.5)(15)} = 0.75 \Rightarrow \alpha \approx 41.41°$

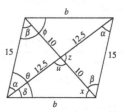

$\cos \beta = \dfrac{10^2 + 15^2 - (12.5)^2}{2(10)(15)} = 0.5625 \Rightarrow \beta \approx 55.77°$

$z = 180° - \alpha - \beta = 82.82°$

$u = 180° - z = 97.18°$

$b^2 = 12.5^2 + 10^2 - 2(12.5)(10) \cos 97.18° \approx 287.4967 \Rightarrow b \approx 16.96$

$\cos \delta = \dfrac{12.5^2 + 16.96^2 - 10^2}{2(12.5)(16.96)} \approx 0.8111 \Rightarrow \delta \approx 35.80°$

$\theta = \alpha + \delta = 41.41° + 35.80° = 77.2°$

$2\phi = 360° - 2\theta \Rightarrow \phi = \dfrac{360° - 2(77.21°)}{2} = 102.8°$

31. Given: $a = 8, c = 5, B = 40°$

Given two sides and included angle, use the Law of Cosines.

$$b^2 = a^2 + c^2 - 2ac \cos B = 64 + 25 - 2(8)(5) \cos 40° \approx 27.7164 \Rightarrow b \approx 5.26$$

$$\cos A = \frac{b^2 + c^2 - a^2}{2bc} \approx \frac{(5.26)^2 + 25 - 64}{2(5.26)(5)} \approx -0.2154 \Rightarrow A \approx 102.44°$$

$C \approx 180° - 102.44° - 40° = 37.56°$

33. Given: $A = 24°, a = 4, b = 18$

Given two sides and an angle opposite one of them, use the Law of Sines.

$h = b \sin A = 18 \sin 24° \approx 7.32$

Because $a < h$, no triangle is formed.

35. Given: $A = 42°, B = 35°, c = 1.2$

Given two angles and a side, use the Law of Sines.

$C = 180° - 42° - 35° = 103°$

$$a = \frac{c \sin A}{\sin C} = \frac{1.2 \sin 42°}{\sin 103°} \approx 0.82$$

$$b = \frac{c \sin B}{\sin C} = \frac{1.2 \sin 35°}{\sin 103°} \approx 0.71$$

37. $a = 6, b = 12, c = 17$

$$s = \frac{a + b + c}{2} = \frac{6 + 12 + 17}{2} = 17.5$$

$$\text{Area} = \sqrt{s(s - a)(s - b)(s - c)} = \sqrt{17.5(11.5)(5.5)(0.5)} \approx 23.53$$

39. $a = 2.5, b = 10.2, c = 8$

$$s = \frac{a + b + c}{2} = \frac{2.5 + 10.2 + 8}{2} = 10.35$$

$$\text{Area} = \sqrt{s(s - a)(s - b)(s - c)} = \sqrt{10.35(7.85)(0.15)(2.35)} \approx 5.35$$

41. Given: $a = 1, b = \frac{1}{2}, c = \frac{5}{4}$

$$s = \frac{a + b + c}{2} = \frac{1 + \frac{1}{2} + \frac{5}{4}}{2} = \frac{11}{8}$$

$$\text{Area} = \sqrt{s(s - a)(s - b)(s - c)} = \sqrt{\frac{11}{8}\left(\frac{3}{8}\right)\left(\frac{7}{8}\right)\left(\frac{1}{8}\right)} \approx 0.24$$

43. $\text{Area} = \frac{1}{2}bc \sin A$

$$= \frac{1}{2}(75)(41) \sin 80°$$

$$\approx 1514.14$$

45. $b^2 = 220^2 + 250^2 - 2(220)(250) \cos 105° \Rightarrow b \approx 373.3$ meters

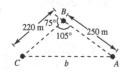

47. $d = \sqrt{330^2 + 420^2 - 2(330)(420) \cos 8°} \approx 103.9$ feet

49. The angles at the base of the tower are $96°$ and $84°$.

The longer guy wire g_1 is given by:

$$g_1{}^2 = 75^2 + 100^2 - 2(75)(100) \cos 96° \approx 17{,}192.9 \Rightarrow g_1 \approx 131.1 \text{ feet}$$

The shorter guy wire g_2 is given by:

$$g_2{}^2 = 75^2 + 100^2 - 2(75)(100) \cos 84 \approx 14{,}057.1 \Rightarrow g_2 \approx 118.6 \text{ feet}$$

51. $\cos B = \dfrac{1700^2 + 3700^2 - 3000^2}{2(1700)(3700)} \Rightarrow B \approx 52.9°$

Bearing: $90° - 52.9° = $ N $37.1°$ E

$\cos C = \dfrac{1700^2 + 3000^2 - 3700^2}{2(1700)(3000)} \Rightarrow C \approx 100.2°$

Bearing: $90° - 26.9° = $ S $63.1°$ E

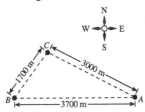

53.

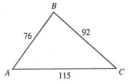

$\cos A = \dfrac{115^2 + 76^2 - 92^2}{2(115)(76)} \approx 0.6028 \Rightarrow A \approx 52.9°$

$\cos C = \dfrac{115^2 + 92^2 - 76^2}{2(115)(92)} \approx 0.75203 \Rightarrow c \approx 41.2°$

55. (a) $C = 180° - 53° - 67° = 60°$

$d^2 = a^2 + (3s)^2 - 2ab \cos C$

$\quad = 36^2 + 9s^2 - 2(36)(3s)(0.5)$

$d = \sqrt{9s^2 - 108s + 1296}$

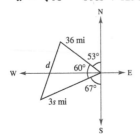

(b) $43 = \sqrt{9s^2 - 108s + 1296}$

$9s^2 - 108s - 553 = 0$

Using the quadratic formula, $s \approx 15.87$ mph.

57. $a = 200$

$b = 500$

$c = 600 \Rightarrow s = \dfrac{200 + 500 + 600}{2} = 650$

Area $= \sqrt{650(450)(150)(50)} \approx 46{,}837.5$ square feet

59. $s = \dfrac{510 + 840 + 1120}{2} = 1235$

Area $= \sqrt{1235(1235 - 510)(1235 - 840)(1235 - 1120)}$

$\quad \approx 201{,}674$ square yards

Cost $\approx \left(\dfrac{201{,}674.02}{4840}\right)(2000) \approx \$83{,}336.37$

61. False. The average of the three sides of a triangle is

$\dfrac{a + b + c}{3}$, not $\dfrac{a + b + c}{2} = s$.

63. $c^2 = a^2 + b^2 - 2ab \cos C$

$\quad = a^2 + b^2 - 2ab \cos 90°$

$\quad = a^2 + b^2 - 2ab(0)$

$\quad = a^2 + b^2$

When $C = 90°$, you obtain the Pythagorean Theorem. The Pythagorean Theorem is a special case of the Law of Cosines.

65. There is no method that can be used to solve the no-solution case of SSA.

The Law of Cosines can be used to solve the single-solution case of SSA. You can substitute values into $a^2 = b^2 + c^2 - 2bc \cos A$. The simplified quadratic equation in terms of c can be solved, with one positive solution and one negative solution. The negative solution can be discarded because length is positive. You can use the positive solution to solve the triangle.

67. (a) $\dfrac{1}{2}bc(1 + \cos A) = \dfrac{1}{2}bc\left[1 + \dfrac{b^2 + c^2 - a^2}{2bc}\right]$

$= \dfrac{1}{2}bc\left[\dfrac{2bc + b^2 + c^2 - a^2}{2bc}\right]$

$= \dfrac{1}{4}\left[(b + c)^2 - a^2\right]$

$= \dfrac{1}{4}\left[(b + c) + a\right]\left[(b + c) - a\right]$

$= \dfrac{b + c + a}{2} \cdot \dfrac{b + c - a}{2}$

$= \dfrac{a + b + c}{2} \cdot \dfrac{-a + b + c}{2}$

(b) $\dfrac{1}{2}bc(1 - \cos A) = \dfrac{1}{2}bc\left[1 + \dfrac{a^2 - (b^2 + c^2)}{2bc}\right]$

$\qquad\qquad\qquad = \dfrac{1}{2}bc\left[\dfrac{2bc + a^2 - b^2 - c^2}{2bc}\right]$

$\qquad\qquad\qquad = \dfrac{a^2 - (b^2 - 2bc + c^2)}{4}$

$\qquad\qquad\qquad = \dfrac{a^2 - (b - c)^2}{4}$

$\qquad\qquad\qquad = \left(\dfrac{a - (b - c)}{2}\right)\left(\dfrac{a + (b - c)}{2}\right)$

$\qquad\qquad\qquad = \dfrac{a - b + c}{2} \cdot \dfrac{a + b - c}{2}$

Section 3.3   Vectors in the Plane

1. directed line segment

3. vector

5. standard position

7. multiplication; addition

9. $\|\mathbf{u}\| = \sqrt{(6 - 2)^2 + (5 - 4)^2} = \sqrt{17}$

$\|\mathbf{v}\| = \sqrt{(4 - 0)^2 + (1 - 0)^2} = \sqrt{17}$

$\text{slope}_{\mathbf{u}} = \dfrac{5 - 4}{6 - 2} = \dfrac{1}{4}$

$\text{slope}_{\mathbf{v}} = \dfrac{1 - 0}{4 - 0} = \dfrac{1}{4}$

u and **v** have the same magnitude and direction so they are equivalent.

11. $\|\mathbf{u}\| = \sqrt{(-1 - 2)^2 + (4 - 2)^2} = \sqrt{13}$

$\|\mathbf{v}\| = \sqrt{(-5 - (-3))^2 + (2 - (-1))^2} = \sqrt{13}$

$\text{slope}_{\mathbf{u}} = \dfrac{4 - 2}{-1 - 2} = -\dfrac{2}{3}$

$\text{slope}_{\mathbf{v}} = \dfrac{2 - (-1)}{-5 - (-3)} = -\dfrac{3}{2}$

u and **v** have the same magnitude but not the same direction so they are not equivalent.

13. $\|\mathbf{u}\| = \sqrt{(5 - 2)^2 + (-10 - (-1))^2} = \sqrt{90} = 3\sqrt{10}$

$\|\mathbf{v}\| = \sqrt{(9 - 6)^2 + (-8 - 1)^2} = \sqrt{90} = 3\sqrt{10}$

$\text{slope}_{\mathbf{u}} = \dfrac{-10 - (-1)}{5 - 2} = -3$

$\text{slope}_{\mathbf{v}} = \dfrac{-8 - 1}{9 - 6} = -3$

u and **v** have the same magnitude and direction so they are equivalent.

15. Initial point: $(0, 0)$

Terminal point: $(1, 3)$

$\mathbf{v} = \langle 1 - 0, 3 - 0 \rangle = \langle 1, 3 \rangle$

$\|\mathbf{v}\| = \sqrt{1^2 + 3^2} = \sqrt{10}$

17. Initial point: $(3, -2)$

Terminal point: $(3, 3)$

$\mathbf{v} = \langle 3 - 3, 3 - (-2) \rangle = \langle 0, 5 \rangle$

$\|\mathbf{v}\| = \sqrt{0^2 + 5^2} = \sqrt{25} = 5$

19. Initial point: $(-3, -5)$

Terminal point: $(-11, 1)$

$\mathbf{v} = \langle -11 - (-3), 1 - (-5) \rangle = \langle -8, 6 \rangle$

$\|\mathbf{v}\| = \sqrt{(-8)^2 + 6^2} = \sqrt{100} = 10$

21. Initial point: $(1, 3)$

Terminal point: $(-8, -9)$

$\mathbf{v} = \langle -8 - 1, -9 - 3 \rangle = \langle -9, -12 \rangle$

$\|\mathbf{v}\| = \sqrt{(-9)^2 + (-12)^2} = \sqrt{225} = 15$

23. Initial point: $(-1, 5)$

Terminal point: $(15, -21)$

$$\mathbf{v} = \langle 15 - (-1), -21 - 5 \rangle = \langle 16, -26 \rangle$$

$$\|\mathbf{v}\| = \sqrt{(16)^2 + (-26)^2} = \sqrt{932} = 2\sqrt{233}$$

25. $-\mathbf{v}$

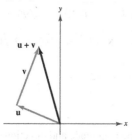

27. $\mathbf{u} + \mathbf{v}$

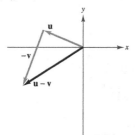

29. $\mathbf{u} - \mathbf{v}$

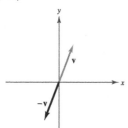

31. $\mathbf{u} = \langle 2, 1 \rangle$, $\mathbf{v} = \langle 1, 3 \rangle$

(a) $\mathbf{u} + \mathbf{v} = \langle 3, 4 \rangle$

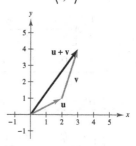

(b) $\mathbf{u} - \mathbf{v} = \langle 1, -2 \rangle$

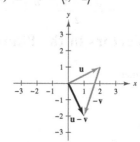

(c) $2\mathbf{u} - 3\mathbf{v} = \langle 4, 2 \rangle - \langle 3, 9 \rangle = \langle 1, -7 \rangle$

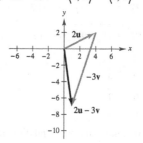

33. $\mathbf{u} = \langle -5, 3 \rangle$, $\mathbf{v} = \langle 0, 0 \rangle$

(a) $\mathbf{u} + \mathbf{v} = \langle -5, 3 \rangle = \mathbf{u}$

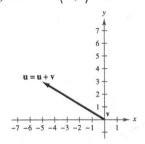

(b) $\mathbf{u} - \mathbf{v} = \langle -5, 3 \rangle = \mathbf{u}$

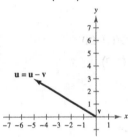

(c) $2\mathbf{u} - 3\mathbf{v} = \langle -10, 6 \rangle = 2\mathbf{u}$

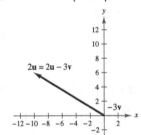

35. $\mathbf{u} = -7\mathbf{j}$, $\mathbf{v} = \mathbf{i} - 2\mathbf{j}$

(a) $\mathbf{u} + \mathbf{v} = \mathbf{i} - 9\mathbf{j}$

$\langle 1, -9 \rangle$

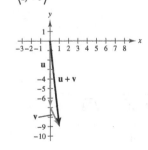

(b) $\mathbf{u} - \mathbf{v} = -\mathbf{i} - 5\mathbf{j}$

$\langle -1, -5 \rangle$

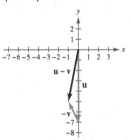

(c) $2\mathbf{u} - 3\mathbf{v} = (-14\mathbf{j}) - (3\mathbf{i} - 6\mathbf{j})$

$= -3\mathbf{i} - 8\mathbf{j}$

$\langle -3, -8 \rangle$

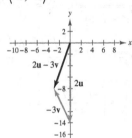

37. $\mathbf{u} = \langle 2, 0 \rangle$

$5\mathbf{u} = \langle 10, 0 \rangle$

$\| 5\mathbf{u} \| = \sqrt{(10)^2 + 0^2} = 10$

39. $\mathbf{v} = \langle -3, 6 \rangle$

$-3\mathbf{v} = \langle 9, -18 \rangle$

$\| 4\mathbf{v} \| = \sqrt{9^2 + (-18)^2} = \sqrt{405} = 9\sqrt{5}$

41. $\mathbf{v} = \langle 3, 0 \rangle$

$\mathbf{u} = \dfrac{1}{\| \mathbf{v} \|} \mathbf{v} = \dfrac{1}{\sqrt{3^2 + 0^2}} \langle 3, 0 \rangle = \dfrac{1}{3} \langle 3, 0 \rangle = \langle 1, 0 \rangle$

$\| \mathbf{u} \| = \sqrt{1^2 + 0^2} = 1$

43. $\mathbf{v} = \langle -2, 2 \rangle$

$$\mathbf{u} = \frac{1}{\|\mathbf{v}\|}\mathbf{v} = \frac{1}{\sqrt{(-2)^2 + 2^2}}\langle -2, 2\rangle = \frac{1}{2\sqrt{2}}\langle -2, 2\rangle$$

$$= \left\langle -\frac{1}{\sqrt{2}}, \frac{1}{\sqrt{2}}\right\rangle$$

$$= \left\langle -\frac{\sqrt{2}}{2}, \frac{\sqrt{2}}{2}\right\rangle$$

$$\|\mathbf{u}\| = \sqrt{\left(\frac{-\sqrt{2}}{2}\right)^2 + \left(\frac{\sqrt{2}}{2}\right)^2} = 1$$

45. $\mathbf{v} = \langle 1, -6 \rangle$

$$\mathbf{u} = \frac{1}{\|\mathbf{v}\|}\mathbf{v} = \frac{1}{\sqrt{1^2 + (-6)^2}}\langle 1, -6\rangle = \frac{1}{\sqrt{37}}\langle 1, -6\rangle$$

$$= \frac{1}{\sqrt{37}}\langle 1, -6\rangle = \left\langle \frac{\sqrt{37}}{37}, -\frac{6\sqrt{37}}{37}\right\rangle$$

$$\|\mathbf{u}\| = \sqrt{\left(\frac{\sqrt{37}}{37}\right)^2 + \left(\frac{-6\sqrt{37}}{37}\right)^2} = 1$$

47. $\mathbf{v} = 10\left(\frac{1}{\|\mathbf{u}\|}\mathbf{u}\right) = 10\left(\frac{1}{\sqrt{(-3)^2 + 4^2}}\langle -3, 4\rangle\right)$

$$= 2\langle -3, 4\rangle$$

$$= \langle -6, 8\rangle$$

49. $9\left(\frac{1}{\|\mathbf{u}\|}\mathbf{u}\right) = 9\left(\frac{1}{\sqrt{2^2 + 5^2}}\langle 2, 5\rangle\right) = \frac{9}{\sqrt{29}}\langle 2, 5\rangle$

$$= \left\langle \frac{18}{\sqrt{29}}, \frac{45}{\sqrt{29}}\right\rangle = \left\langle \frac{18\sqrt{29}}{29}, \frac{45\sqrt{29}}{29}\right\rangle$$

51. $\mathbf{u} = \langle 3 - (-2), -2 - 1 \rangle$

$$= \langle 5, -3\rangle$$

$$= 5\mathbf{i} - 3\mathbf{j}$$

53. $\mathbf{u} = \langle -6 - 0, 4 - 1 \rangle$

$$\mathbf{u} = \langle -6, 3\rangle$$

$$\mathbf{u} = -6\mathbf{i} + 3\mathbf{j}$$

55. $\mathbf{v} = \frac{3}{2}\mathbf{u}$

$$= \frac{3}{2}(2\mathbf{i} - \mathbf{j})$$

$$= 3\mathbf{i} - \frac{3}{2}\mathbf{j} = \left\langle 3, -\frac{3}{2}\right\rangle$$

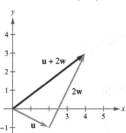

57. $\mathbf{v} = \mathbf{u} + 2\mathbf{w}$

$$= (2\mathbf{i} - \mathbf{j}) + 2(\mathbf{i} + 2\mathbf{j})$$

$$= 4\mathbf{i} + 3\mathbf{j} = \langle 4, 3\rangle$$

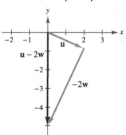

59. $\mathbf{v} = \mathbf{u} - 2\mathbf{w}$

$$= (2\mathbf{i} - \mathbf{j}) - 2(\mathbf{i} + 2\mathbf{j})$$

$$= -5\mathbf{j} = \langle 0, -5\rangle$$

61. $\mathbf{v} = 6\mathbf{i} - 6\mathbf{j}$

$$\|\mathbf{v}\| = \sqrt{6^2 + (-6)^2} = \sqrt{72} = 6\sqrt{2}$$

$$\tan \theta = \frac{-6}{6} = -1$$

Since $\mathbf{v}$ lies in Quadrant IV, $\theta = 315°$.

63. $\mathbf{v} = 3(\cos 60°\mathbf{i} + \sin 60°\mathbf{j})$

$$\|\mathbf{v}\| = 3, \theta = 60°$$

65. $\mathbf{v} = \langle 3\cos 0°, 3\sin 0° \rangle$

$\quad = \langle 3, 0 \rangle$

67. $\mathbf{v} = \left\langle \dfrac{7}{2}\cos 150°, \dfrac{7}{2}\sin 150° \right\rangle$

$\quad = \left\langle -\dfrac{7\sqrt{3}}{4}, \dfrac{7}{4} \right\rangle$

69. $\mathbf{v} = 3\left(\dfrac{1}{\sqrt{3^2 + 4^2}} \right)(3\mathbf{i} + 4\mathbf{j})$

$\quad = \dfrac{3}{5}(3\mathbf{i} + 4\mathbf{j})$

$\quad = \dfrac{9}{5}\mathbf{i} + \dfrac{12}{5}\mathbf{j} = \left\langle \dfrac{9}{5}, \dfrac{12}{5} \right\rangle$

71. $\mathbf{u} = \langle 4\cos 60°, 4\sin 60° \rangle = \langle 2, 2\sqrt{3} \rangle$

$\quad\quad \mathbf{v} = \langle 4\cos 90°, 4\sin 90° \rangle = \langle 0, 4 \rangle$

$\mathbf{u} + \mathbf{v} = \langle 2, 4 + 2\sqrt{3} \rangle$

73. $\mathbf{v} = \mathbf{i} + \mathbf{j}$

$\quad \mathbf{w} = 2\mathbf{i} - 2\mathbf{j}$

$\quad \mathbf{u} = \mathbf{v} - \mathbf{w} = -\mathbf{i} + 3\mathbf{j}$

$\quad \|\mathbf{v}\| = \sqrt{2}$

$\quad \|\mathbf{w}\| = 2\sqrt{2}$

$\quad \|\mathbf{v} - \mathbf{w}\| = \sqrt{10}$

$\cos\alpha = \dfrac{\|\mathbf{v}\|^2 + \|\mathbf{w}\|^2 - \|\mathbf{v} - \mathbf{w}\|^2}{2\|\mathbf{v}\|\,\|\mathbf{w}\|} = \dfrac{2 + 8 - 10}{2\sqrt{2}\cdot 2\sqrt{2}} = 0$

$\quad\quad \alpha = 90°$

75. Force One: $\mathbf{u} = 45\mathbf{i}$

Force Two: $\mathbf{v} = 60\cos\theta\mathbf{i} + 60\sin\theta\mathbf{j}$

Resultant Force: $\mathbf{u} + \mathbf{v} = (45 + 60\cos\theta)\mathbf{i} + 60\sin\theta\mathbf{j}$

$\|\mathbf{u} + \mathbf{v}\| = \sqrt{(45 + 60\cos\theta)^2 + (60\sin\theta)^2} = 90$

$2025 + 5400\cos\theta + 3600 = 8100$

$5400\cos\theta = 2475$

$\cos\theta = \frac{2475}{5400} \approx 0.4583$

$\theta \approx 62.7°$

77. Horizontal component of velocity: $1200 \cos 6° \approx 1193.4$ ft/sec

Vertical component of velocity: $1200 \sin 6° \approx 125.4$ ft/sec

79. $\mathbf{u} = 300\mathbf{i}$

$$\mathbf{v} = (125 \cos 45°)\mathbf{i} + (125 \sin 45°)\mathbf{j} = \frac{125}{\sqrt{2}}\mathbf{i} + \frac{125}{\sqrt{2}}\mathbf{j}$$

$$\mathbf{u} + \mathbf{v} = \left(300 + \frac{125}{\sqrt{2}}\right)\mathbf{i} + \frac{125}{\sqrt{2}}\mathbf{j}$$

$$\|\mathbf{u} + \mathbf{v}\| = \sqrt{\left(300 + \frac{125}{\sqrt{2}}\right)^2 + \left(\frac{125}{\sqrt{2}}\right)^2} \approx 398.32 \text{ newtons}$$

$$\tan \theta = \frac{\dfrac{125}{\sqrt{2}}}{300 + \left(\dfrac{125}{\sqrt{2}}\right)} \Rightarrow \theta \approx 12.8°$$

81.
$$\mathbf{u} = (75 \cos 30°)\mathbf{i} + (75 \sin 30°)\mathbf{j} \approx 64.95\mathbf{i} + 37.5\mathbf{j}$$
$$\mathbf{v} = (100 \cos 45°)\mathbf{i} + (100 \sin 45°)\mathbf{j} \approx 70.71\mathbf{i} + 70.71\mathbf{j}$$
$$\mathbf{w} = (125 \cos 120°)\mathbf{i} + (125 \sin 120°)\mathbf{j} \approx -62.5\mathbf{i} + 108.3\mathbf{j}$$
$$\mathbf{u} + \mathbf{v} + \mathbf{w} \approx 73.16\mathbf{i} + 216.5\mathbf{j}$$
$$\|\mathbf{u} + \mathbf{v} + \mathbf{w}\| \approx 228.5 \text{ pounds}$$
$$\tan \theta \approx \frac{216.5}{73.16} \approx 2.9593$$
$$\theta \approx 71.3°$$

83. Left crane: $\mathbf{u} = \|\mathbf{u}\|(\cos 155.7°\mathbf{i} + \sin 155.7°\mathbf{j})$

Right crane: $\mathbf{v} = \|\mathbf{v}\|(\cos 44.5°\mathbf{i} + \sin 44.5°\mathbf{j})$

Resultant: $\mathbf{u} + \mathbf{v} = -20{,}240\mathbf{j}$

System of equations:

$\|\mathbf{u}\| \cos 155.7° + \|\mathbf{v}\| \cos 44.5° = 0$

$\|\mathbf{u}\| \sin 155.7° + \|\mathbf{v}\| \sin 44.5° = 20{,}240$

Solving this system of equations yields the following:

Left crane $= \|\mathbf{u}\| \approx 15{,}484$ pounds

Right crane $= \|\mathbf{v}\| \approx 19{,}786$ pounds

85. Horizontal force: $\mathbf{u} = \|\mathbf{u}\|\mathbf{i}$

Weight: $\mathbf{w} = -\mathbf{j}$

Rope: $\mathbf{t} = \|\mathbf{t}\|(\cos 135°\mathbf{i} + \sin 135°\mathbf{j})$

$\mathbf{u} + \mathbf{w} + \mathbf{t} = 0 \Rightarrow \|\mathbf{u}\| + \|\mathbf{t}\| \cos 135° = 0$

$-1 + \|\mathbf{t}\| \sin 135° = 0$

$\|\mathbf{t}\| \approx \sqrt{2}$ pounds

$\|\mathbf{u}\| \approx 1$ pound

87. Towline 1: $\mathbf{u} = \|\mathbf{u}\|(\cos 18°\mathbf{i} + \sin 18°\mathbf{j})$

Towline 2: $\mathbf{v} = \|\mathbf{u}\|(\cos 18°\mathbf{i} - \sin 18°\mathbf{j})$

Resultant: $\mathbf{u} + \mathbf{v} = 6000\mathbf{i}$

$\|\mathbf{u}\| \cos 18° + \|\mathbf{u}\| \cos 18° = 6000$

$\|\mathbf{u}\| \approx 3154.4$

So, the tension on each towline is $\|\mathbf{u}\| \approx 3154.4$ pounds.

89. $W = 100, \theta = 12°$

$$\sin \theta = \frac{F}{W}$$
$$F = W \sin \theta = 100 \sin 12° \approx 20.8 \text{ pounds}$$

91. $F = 5000, W = 15{,}000$

$$\sin \theta = \frac{F}{W}$$
$$\sin \theta = \frac{5000}{15{,}000}$$
$$\theta = \sin^{-1} \frac{1}{3} \approx 19.5°$$

93. Airspeed: $\mathbf{u} = (875 \cos 58°)\mathbf{i} - (875 \sin 58°)\mathbf{j}$

Groundspeed: $\mathbf{v} = (800 \cos 50°)\mathbf{i} - (800 \sin 50°)\mathbf{j}$

Wind: $\mathbf{w} = \mathbf{v} - \mathbf{u} = (800 \cos 50° - 875 \cos 58°)\mathbf{i} + (-800 \sin 50° + 875 \sin 58°)\mathbf{j}$

$$\approx 50.5507\mathbf{i} + 129.2065\mathbf{j}$$

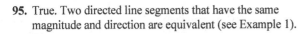

Wind speed: $\|\mathbf{w}\| \approx \sqrt{(50.5507)^2 + (129.2065)^2} \approx 138.7$ kilometers per hour

Wind direction: $\tan \theta \approx \dfrac{129.2065}{50.5507}$

$$\theta \approx 68.6°; 90° - \theta = 21.4°$$

Bearing: N 21.4° E

95. True. Two directed line segments that have the same magnitude and direction are equivalent (see Example 1).

97. True. If $\mathbf{v} = a\mathbf{i} + b\mathbf{j} = 0$ is the zero vector, then $a = b = 0$. So, $a = -b$.

99. The order of subtraction should be switched.

$\mathbf{u} = \langle 6 - (-3), -1 - 4 \rangle = \langle 9, -5 \rangle$

101. Let $\mathbf{v} = (\cos \theta)\mathbf{i} + (\sin \theta)\mathbf{j}$.

$$\|\mathbf{v}\| = \sqrt{\cos^2 \theta + \sin^2 \theta} = \sqrt{1} = 1$$

So, $\mathbf{v}$ is a unit vector for any value of θ.

103.
$$\mathbf{u} = \langle 5 - 1, 2 - 6 \rangle = \langle 4, -4 \rangle$$
$$\mathbf{v} = \langle 9 - 4, 4 - 5 \rangle = \langle 5, -1 \rangle$$
$$\mathbf{u} - \mathbf{v} = \langle -1, -3 \rangle \text{ or } \mathbf{v} - \mathbf{u} = \langle 1, 3 \rangle$$

105. $\mathbf{F}_1 = \langle 10, 0 \rangle$, $\mathbf{F}_2 = 5\langle \cos \theta, \sin \theta \rangle$

(a) $\mathbf{F}_1 + \mathbf{F}_2 = \langle 10 + 5 \cos \theta, 5 \sin \theta \rangle$

$\|\mathbf{F}_1 + \mathbf{F}_2\| = \sqrt{(10 + 5 \cos \theta)^2 + (5 \sin \theta)^2}$

$= \sqrt{100 + 100 \cos \theta + 25 \cos^2 \theta + 25 \sin^2 \theta}$

$= 5\sqrt{4 + 4 \cos \theta + \cos^2 \theta + \sin^2 \theta}$

$= 5\sqrt{4 + 4 \cos \theta + 1}$

$= 5\sqrt{5 + 4 \cos \theta}$

(b)
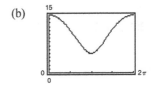

(c) Range: $[5, 15]$

Maximum is 15 when $\theta = 0$.

Minimum is 5 when $\theta = \pi$.

(d) The magnitude of the resultant is never 0 because the magnitudes of $\mathbf{F}_1$ and $\mathbf{F}_2$ are not the same.

107. (a) Answers will vary. *Sample answer:* To add two vectors $\mathbf{u}$ and $\mathbf{v}$ geometrically, first position them (without changing their lengths or directions) so that the initial point of the second vector $\mathbf{v}$ coincides with the terminal point of the first vector $\mathbf{u}$. The sum $\mathbf{u} + \mathbf{v}$ is the vector formed by joining the initial point of the first vector $\mathbf{u}$ with the terminal point of the second vector $\mathbf{v}$.

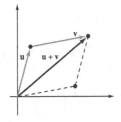

(b) Answers will vary. Sample Answer: Geometrically, the product of a vector $\mathbf{v}$ and a scalar k is the vector that is $|k|$ times as long as $\mathbf{v}$. When k is positive, $k\mathbf{v}$ has the same direction as $\mathbf{v}$, and when k is negative, $k\mathbf{v}$ has the direction opposite that of $\mathbf{v}$.

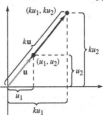

Section 3.4 Vectors and Dot Products

1. dot product

3. $\dfrac{\mathbf{u} \cdot \mathbf{v}}{\|\mathbf{u}\| \, \|\mathbf{v}\|}$

5. $\left(\dfrac{\mathbf{u} \cdot \mathbf{v}}{\|\mathbf{v}\|^2} \right) \mathbf{v}$

7. $\mathbf{u} = \langle 7, 1 \rangle, \mathbf{v} = \langle -3, 2 \rangle$

$\mathbf{u} \cdot \mathbf{v} = 7(-3) + 1(2) = -19$

9. $\mathbf{u} = \langle -6, 2 \rangle, \mathbf{v} = \langle 1, 3 \rangle$

$\mathbf{u} \cdot \mathbf{v} = -6(1) + 2(3) = 0$

11. $\mathbf{u} = 4\mathbf{i} - 2\mathbf{j}, \mathbf{v} = \mathbf{i} - \mathbf{j}$

$\mathbf{u} \cdot \mathbf{v} = 4(1) + (-2)(-1) = 6$

13. $\mathbf{u} = \langle 3, 3 \rangle$

$\mathbf{u} \cdot \mathbf{u} = 3(3) + 3(3) = 18$

The result is a scalar.

15. $\mathbf{u} = \langle 3, 3 \rangle, \mathbf{v} = \langle -4, 2 \rangle$

$\begin{aligned}
(\mathbf{u} \cdot \mathbf{v})\mathbf{v} &= \left[3(-4) + 3(2) \right] \langle -4, 2 \rangle \\
&= -6 \langle -4, 2 \rangle \\
&= \langle 24, -12 \rangle
\end{aligned}$

The result is a vector.

17. $\mathbf{u} = \langle 3, 3 \rangle, \mathbf{v} = \langle -4, 2 \rangle, \mathbf{w} = \langle 3, -1 \rangle$

$(\mathbf{v} \cdot \mathbf{0})\mathbf{w} = 0 \langle 3, -1 \rangle = \langle 0, 0 \rangle = \mathbf{0}$

The result is a vector.

19. $\mathbf{w} = \langle 3, -1 \rangle$

$\|\mathbf{w}\| - 1 = \sqrt{3^2 + (-1)^2} - 1 = \sqrt{10} - 1$

The result is a scalar.

21. $\mathbf{u} = \langle 3, 3 \rangle, \mathbf{v} = \langle -4, 2 \rangle, \mathbf{w} = \langle 3, -1 \rangle$

$\begin{aligned}
(\mathbf{u} \cdot \mathbf{v}) - (\mathbf{u} \cdot \mathbf{w}) &= \left[3(-4) + 3(2) \right] - \left[3(3) + 3(-1) \right] \\
&= -6 - 6 \\
&= -12
\end{aligned}$

The result is a scalar.

23. $\mathbf{u} = \langle -8, 15 \rangle$

$\|\mathbf{u}\| = \sqrt{\mathbf{u} \cdot \mathbf{u}} = \sqrt{(-8)(-8) + 15(15)} = \sqrt{289} = 17$

25. $\mathbf{u} = 20\mathbf{i} + 25\mathbf{j}$

$\|\mathbf{u}\| = \sqrt{\mathbf{u} \cdot \mathbf{u}} = \sqrt{(20)^2 + (25)^2} = \sqrt{1025} = 5\sqrt{41}$

27. $\mathbf{u} = 6\mathbf{j}$

$\|\mathbf{u}\| = \sqrt{\mathbf{u} \cdot \mathbf{u}} = \sqrt{(0)^2 + (6)^2} = \sqrt{36} = 6$

29. $\mathbf{u} = \langle 1, 0 \rangle, \mathbf{v} = \langle 0, -2 \rangle$

$\cos \theta = \dfrac{\mathbf{u} \cdot \mathbf{v}}{\|\mathbf{u}\| \, \|\mathbf{v}\|} = \dfrac{0}{(1)(2)} = 0$

$\theta = \dfrac{\pi}{2}$ radians

31. $\mathbf{u} = 3\mathbf{i} + 4\mathbf{j}, \mathbf{v} = -2\mathbf{j}$

$\cos \theta = \dfrac{\mathbf{u} \cdot \mathbf{v}}{\|\mathbf{u}\| \, \|\mathbf{v}\|} = -\dfrac{8}{(5)(2)}$

$\theta = \arccos \left(-\dfrac{4}{5} \right)$

$\theta \approx 2.50$ radians

33. $\mathbf{u} = 2\mathbf{i} - \mathbf{j}, \mathbf{v} = 6\mathbf{i} - 3\mathbf{j}$

$\begin{aligned}
\cos \theta &= \dfrac{\mathbf{u} \cdot \mathbf{v}}{\|\mathbf{u}\| \, \|\mathbf{v}\|} \\
&= \dfrac{2(6) + (-1)(-3)}{\sqrt{2^2 + (-1)^2} \sqrt{6^2 + (-3)^2}} \\
&= \dfrac{15}{\sqrt{225}} = 1
\end{aligned}$

$\theta = 0$

35. $\mathbf{u} = -6\mathbf{i} - 3\mathbf{j}, \mathbf{v} = -8\mathbf{i} + 4\mathbf{j}$

$\cos u = \dfrac{\mathbf{u} \cdot \mathbf{v}}{\|\mathbf{u}\| \, \|\mathbf{v}\|} = \dfrac{-6(-8) + (-3)(4)}{\sqrt{45}\sqrt{80}} = \dfrac{36}{60} = 0.6$

$\theta \approx 0.93$ radian

37. $\mathbf{u} = \left(\cos \dfrac{\pi}{3} \right)\mathbf{i} + \left(\sin \dfrac{\pi}{3} \right)\mathbf{j} = \dfrac{1}{2}\mathbf{i} + \dfrac{\sqrt{3}}{2}\mathbf{j}$

$\mathbf{v} = \left(\cos \dfrac{3\pi}{4} \right)\mathbf{i} + \left(\sin \dfrac{3\pi}{4} \right)\mathbf{j} = -\dfrac{\sqrt{2}}{2}\mathbf{i} + \dfrac{\sqrt{2}}{2}\mathbf{j}$

$\|\mathbf{u}\| = \|\mathbf{v}\| = 1$

$\cos \theta = \dfrac{\mathbf{u} \cdot \mathbf{v}}{\|\mathbf{u}\| \, \|\mathbf{v}\|} = \mathbf{u} \cdot \mathbf{v}$

$= \left(\dfrac{1}{2} \right)\left(-\dfrac{\sqrt{2}}{2} \right) + \left(\dfrac{\sqrt{3}}{2} \right)\left(\dfrac{\sqrt{2}}{2} \right) = \dfrac{-\sqrt{2} + \sqrt{6}}{4}$

$\theta = \arccos \left(\dfrac{-\sqrt{2} + \sqrt{6}}{4} \right) = \dfrac{5\pi}{12}$

39. $\mathbf{u} = 3\mathbf{i} + 4\mathbf{j}$

$\mathbf{v} = -7\mathbf{i} + 5\mathbf{j}$

$$\cos\theta = \frac{\mathbf{u}\cdot\mathbf{v}}{\|\mathbf{u}\|\,\|\mathbf{v}\|}$$

$$= \frac{3(-7)+4(5)}{3\sqrt{74}}$$

$$= \frac{-1}{5\sqrt{74}} \approx -0.0232$$

$\theta \approx 91.33°$

41. $\mathbf{u} = -5\mathbf{i} - 5\mathbf{j}$

$\mathbf{v} = -8\mathbf{i} + 8\mathbf{j}$

$$\cos\theta = \frac{\mathbf{u}\cdot\mathbf{v}}{\|\mathbf{u}\|\,\|\mathbf{v}\|}$$

$$= \frac{-5(-8)+(-5)(8)}{\sqrt{50}\sqrt{128}}$$

$$= 0$$

$\theta = 90°$

43. $P = (1, 2), Q = (3, 4), R = (2, 5)$

$\overline{PQ} = \langle 2, 2\rangle, \overline{PR} = \langle 1, 3\rangle, \overline{QR} = \langle -1, 1\rangle$

$$\cos\alpha = \frac{\overline{PQ}\cdot\overline{PR}}{\|\overline{PQ}\|\,\|\overline{PR}\|} = \frac{8}{(2\sqrt{2})(\sqrt{10})} \Rightarrow \alpha = \arccos\frac{2}{\sqrt{5}} \approx 26.57°$$

$$\cos\beta = \frac{\overline{PQ}\cdot\overline{QR}}{\|\overline{PQ}\|\,\|\overline{QR}\|} = 0 \Rightarrow \beta = 90°$$

$$\gamma = 180° - 26.57° - 90° = 63.43°$$

45. $P = (-3, 0), Q = (2, 2), R = (0, 6)$

$\overline{QP} = \langle -5, -2\rangle, \overline{PR} = \langle 3, 6\rangle, \overline{QR} = \langle -2, 4\rangle, \overline{PQ} = \langle 5, 2\rangle$

$$\cos\alpha = \frac{\overline{PQ}\cdot\overline{PR}}{\|\overline{PQ}\|\,\|\overline{PR}\|} = \frac{27}{\sqrt{29}\sqrt{45}} \Rightarrow \alpha \approx 41.63°$$

$$\cos\beta = \frac{\overline{QP}\cdot\overline{QR}}{\|\overline{QP}\|\,\|\overline{PR}\|} = \frac{2}{\sqrt{29}\sqrt{20}} \Rightarrow \beta \approx 85.24°$$

$$\delta = 180° - 41.63° - 85.24° = 53.13°$$

47. $\|\mathbf{u}\| = 4, \|\mathbf{v}\| = 10, \theta = \frac{2\pi}{3}$

$$\mathbf{u}\cdot\mathbf{v} = \|\mathbf{u}\|\,\|\mathbf{v}\|\cos\theta$$

$$= (4)(10)\cos\frac{2\pi}{3}$$

$$= 40\left(-\frac{1}{2}\right)$$

$$= -20$$

49. $\|\mathbf{u}\| = 100, \|\mathbf{v}\| = 250, \theta = \frac{\pi}{6}$

$$\mathbf{u}\cdot\mathbf{v} = \|\mathbf{u}\|\,\|\mathbf{v}\|\cos\theta$$

$$= (100)(250)\cos\frac{\pi}{6}$$

$$= 25{,}000\cdot\frac{\sqrt{3}}{2}$$

$$= 12{,}500\sqrt{3}$$

51. $\mathbf{u} = \langle 3, 15\rangle, \mathbf{v} = \langle -1, 5\rangle$

$\mathbf{u} \neq k\mathbf{v} \Rightarrow$ Not parallel

$\mathbf{u}\cdot\mathbf{v} \neq 0 \Rightarrow$ Not orthogonal

Neither

53. $\mathbf{u} = 2\mathbf{i} - 2\mathbf{j}, \mathbf{v} = -\mathbf{i} - \mathbf{j}$

$\mathbf{u}\cdot\mathbf{v} = 0 \Rightarrow \mathbf{u}$ and $\mathbf{v}$ are orthogonal.

55. $\mathbf{u} = 1, \mathbf{v} = -2\mathbf{i} + 2\mathbf{j}$

$\mathbf{u} \neq k\mathbf{v} \Rightarrow$ Not parallel

$\mathbf{u}\cdot\mathbf{v} \neq 0 \Rightarrow$ Not orthogonal

Neither

57. $\mathbf{u} = \langle 2, 2\rangle, \mathbf{v} = \langle 6, 1\rangle$

$$\mathbf{w}_1 = \text{proj}_{\mathbf{v}}\mathbf{u} = \left(\frac{\mathbf{u}\cdot\mathbf{v}}{\|\mathbf{v}\|^2}\right)\mathbf{v} = \frac{14}{37}\langle 6, 1\rangle = \frac{1}{37}\langle 84, 14\rangle$$

$$\mathbf{w}_2 = \mathbf{u} - \mathbf{w}_1 = \langle 2, 2\rangle - \frac{14}{37}\langle 6, 1\rangle = \left\langle -\frac{10}{37}, \frac{60}{37}\right\rangle = \frac{10}{37}\langle -1, 6\rangle = \frac{1}{37}\langle -10, 60\rangle$$

$$\mathbf{u} = \frac{1}{37}\langle 84, 14\rangle + \frac{1}{37}\langle -10, 60\rangle = \langle 2, 2\rangle$$

59. $\mathbf{u} = \langle 4, 2 \rangle$, $\mathbf{v} = \langle 1, -2 \rangle$

$\mathbf{w}_1 = \text{proj}_\mathbf{v}\mathbf{u} = \left(\dfrac{\mathbf{u} \cdot \mathbf{v}}{\|\mathbf{v}\|^2} \right)\mathbf{v} = 0\langle 1, -2 \rangle = \langle 0, 0 \rangle$

$\mathbf{w}_2 = \mathbf{u} - \mathbf{w}_1 = \langle 4, 2 \rangle - \langle 0, 0 \rangle = \langle 4, 2 \rangle$

$\mathbf{u} = \langle 4, 2 \rangle + \langle 0, 0 \rangle = \langle 4, 2 \rangle$

61. $\text{proj}_\mathbf{v}\mathbf{u} = \mathbf{u}$ because $\mathbf{u}$ and $\mathbf{v}$ are parallel.

$\text{proj}_\mathbf{v}\mathbf{u} = \dfrac{\mathbf{u} \cdot \mathbf{v}}{\|\mathbf{v}\|^2}\mathbf{v} = \dfrac{3(6) + 2(4)}{\left(\sqrt{6^2 + 4^2} \right)^2}\langle 6, 4 \rangle = \dfrac{1}{2}\langle 6, 4 \rangle = \langle 3, 2 \rangle = \mathbf{u}$

63. Because $\mathbf{u}$ and $\mathbf{v}$ are orthogonal, $\mathbf{u} \cdot \mathbf{v} = 0$ and $\text{proj}_\mathbf{v}\mathbf{u} = 0$.

$\text{proj}_\mathbf{v}\mathbf{u} = \dfrac{\mathbf{u} \cdot \mathbf{v}}{\|\mathbf{v}\|^2}\mathbf{v} = 0$, because $\mathbf{u} \cdot \mathbf{v} = 0$.

65. $\mathbf{u} = \langle 3, 5 \rangle$

For $\mathbf{v}$ to be orthogonal to $\mathbf{u}$, $\mathbf{u} \cdot \mathbf{v}$ must equal 0.

Two possibilities: $\langle -5, 3 \rangle$ and $\langle 5, -3 \rangle$

67. $\mathbf{u} = \frac{1}{2}\mathbf{i} - \frac{2}{3}\mathbf{j}$

For $\mathbf{u}$ and $\mathbf{v}$ to be orthogonal, $\mathbf{u} \cdot \mathbf{v}$ must equal 0.

Two possibilities: $\mathbf{v} = \frac{2}{3}\mathbf{i} + \frac{1}{2}\mathbf{j}$ and $\mathbf{v} = -\frac{2}{3}\mathbf{i} - \frac{1}{2}\mathbf{j}$

69. Work $= \left\| \text{proj}_{\overrightarrow{PQ}}\mathbf{v} \right\| \left\| \overrightarrow{PQ} \right\|$ where $\overrightarrow{PQ} = \langle 4, 7 \rangle$ and $\mathbf{v} = \langle 1, 4 \rangle$.

$\text{proj}_{\overrightarrow{PQ}}\mathbf{v} = \left(\dfrac{\mathbf{v} \cdot \overrightarrow{PQ}}{\left\| \overrightarrow{PQ} \right\|^2} \right)\overrightarrow{PQ} = \left(\dfrac{32}{65} \right)\langle 4, 7 \rangle$

Work $= \left\| \text{proj}_{\overrightarrow{PQ}}\mathbf{v} \right\| \left\| \overrightarrow{PQ} \right\| = \left(\dfrac{32\sqrt{65}}{65} \right)\left(\sqrt{65} \right) = 32$

71. (a) $\mathbf{u} \cdot \mathbf{v} = 1225(12.20) + 2445(8.50)$

$= 35{,}727.5$

The total amount paid to the employees is $35,727.50.

(b) To increase wages by 2%, use scalar multiplication to multiply 1.02 by $\mathbf{v}$.

73. (a) Force due to gravity:

$\mathbf{F} = -30{,}000\mathbf{j}$

Unit vector along hill:

$\mathbf{v} = (\cos d)\mathbf{i} + (\sin d)\mathbf{j}$

Projection of $\mathbf{F}$ onto $\mathbf{v}$:

$\mathbf{w}_1 = \text{proj}_\mathbf{v}\mathbf{F} = \left(\dfrac{\mathbf{F} \cdot \mathbf{v}}{\|\mathbf{v}\|^2} \right)\mathbf{v} = (\mathbf{F} \cdot \mathbf{v})\mathbf{v} = -30{,}000 \sin d\mathbf{v}$

The magnitude of the force is $30{,}000 \sin d$.

(b)

d	0°	1°	2°	3°	4°	5°	6°	7°	8°	9°	10°
Force	0	523.6	1047.0	1570.1	2092.7	2614.7	3135.9	3656.1	4175.2	4693.0	5209.4

(c) Force perpendicular to the hill when $d = 5°$:

Force $= \sqrt{(30{,}000)^2 - (2614.7)^2} \approx 29{,}885.8$ pounds

75. Work $= (245)(3) = 735$ newton-meters

77. Work $= (\cos 30°)(45)(20) \approx 779.4$ foot-pounds

79. Work $= (\cos 35°)(15{,}691)(800)$

$\approx 10{,}282{,}651.78$ newton-meters

81. Work $= (\cos\theta)\|\mathbf{F}\|\,\|\overrightarrow{PQ}\|$

$= (\cos 20°)(25 \text{ pounds})(50 \text{ feet})$

$\approx 1174.62 \text{ foot-pounds}$

83. False. Work is represented by a scalar.

85. A dot product is a scalar, not a vector.

$\mathbf{v} \cdot \mathbf{0} = \langle -3, 5 \rangle \cdot \langle 0, 0 \rangle = (-3)(0) + (5)(0) = 0$

87. $\quad \mathbf{u} \cdot \mathbf{v} = \langle 8, 4 \rangle \cdot \langle 2, -k \rangle = 16 - 4k = 0$

$16 - 4k = 0$

$-4k = -16$

$k = 4$

89. $\mathbf{u} \cdot \mathbf{u} = \|\mathbf{u}\|^2$

$\phantom{\mathbf{u} \cdot \mathbf{u}} = 1^2 = 1$

91. (a) $\text{proj}_{\mathbf{v}}\mathbf{u} = \mathbf{u} \Rightarrow \mathbf{u}$ and $\mathbf{v}$ are parallel.

(b) $\text{proj}_{\mathbf{v}}\mathbf{u} = 0 \Rightarrow \mathbf{u}$ and $\mathbf{v}$ are orthogonal.

93. Let $\mathbf{u} = \langle u_1, u_2 \rangle$ and $\mathbf{v} = \langle v_1, v_2 \rangle$.

$\mathbf{u} - \mathbf{v} = \langle u_1 - v_1, u_2 - v_2 \rangle$

$\|\mathbf{u} - \mathbf{v}\|^2 = (u_1 - v_1)^2 + (u_2 - v_2)^2$

$\phantom{\|\mathbf{u} - \mathbf{v}\|^2} = u_1^2 - 2u_1v_1 + v_1^2 + u_2^2 - 2u_2v_2 + v_2^2$

$\phantom{\|\mathbf{u} - \mathbf{v}\|^2} = u_1^2 + u_2^2 + v_1^2 + v_2^2 - 2u_1v_1 - 2u_2v_2$

$\phantom{\|\mathbf{u} - \mathbf{v}\|^2} = \|\mathbf{u}\|^2 + \|\mathbf{v}\|^2 - 2(u_1v_1 + u_2v_2)$

$\phantom{\|\mathbf{u} - \mathbf{v}\|^2} = \|\mathbf{u}\|^2 + \|\mathbf{v}\|^2 - 2\mathbf{u} \cdot \mathbf{v}$

Review Exercises for Chapter 3

1. Given: $A = 38°$, $B = 70°$, $a = 8$

$C = 180° - 38° - 70° = 72°$

$b = \dfrac{a \sin B}{\sin A} = \dfrac{8 \sin 70°}{\sin 38°} \approx 12.21$

$c = \dfrac{a \sin C}{\sin A} = \dfrac{8 \sin 72°}{\sin 38°} \approx 12.36$

3. Given: $B = 72°$, $C = 82°$, $b = 54$

$A = 180° - 72° - 82° = 26°$

$a = \dfrac{b \sin A}{\sin B} = \dfrac{54 \sin 26°}{\sin 72°} \approx 24.89$

$c = \dfrac{b \sin C}{\sin B} = \dfrac{54 \sin 82°}{\sin 72°} \approx 56.23$

9. Given: $B = 150°$, $b = 30$, $c = 10$

$\sin C = \dfrac{c \sin B}{b} = \dfrac{10 \sin 150°}{30} \approx 0.1667 \Rightarrow C \approx 9.59°$

$A \approx 180° - 150° - 9.59° = 20.41°$

$a = \dfrac{b \sin A}{\sin B} = \dfrac{30 \sin 20.41°}{\sin 150°} \approx 20.92$

11. $A = 75°$, $a = 51.2$, $b = 33.7$

$\sin B = \dfrac{b \sin A}{a} = \dfrac{33.7 \sin 75°}{51.2} \approx 0.6358 \Rightarrow B \approx 39.48°$

$C \approx 180° - 75° - 39.48° = 65.52°$

$c = \dfrac{a \sin C}{\sin A} = \dfrac{51.2 \sin 65.52°}{\sin 75°} \approx 48.24$

5. Given: $A = 16°$, $B = 98°$, $c = 8.4$

$C = 180° - 16° - 98° = 66°$

$a = \dfrac{c \sin A}{\sin C} = \dfrac{8.4 \sin 16°}{\sin 66°} \approx 2.53$

$b = \dfrac{c \sin B}{\sin C} = \dfrac{8.4 \sin 98°}{\sin 66°} \approx 9.11$

7. Given: $A = 24°$, $C = 48°$, $b = 27.5$

$B = 180° - 24° - 48° = 108°$

$a = \dfrac{b \sin A}{\sin B} = \dfrac{27.5 \sin 24°}{\sin 108°} \approx 11.76$

$c = \dfrac{b \sin C}{\sin B} = \dfrac{27.5 \sin 48°}{\sin 108°} \approx 21.49$

13. $A = 33°, b = 7, c = 10$

Area $= \frac{1}{2}bc \sin A = \frac{1}{2}(7)(10) \sin 33° \approx 19.06$

15. $C = 119°, a = 18, b = 6$

Area $= \frac{1}{2}ab \sin C = \frac{1}{2}(18)(6) \sin 119° \approx 47.23$

17. Area $= \frac{1}{2}ac \sin B = \frac{1}{2}(105)(64) \sin(72° 30') \approx 3204.5$

19. $\dfrac{h}{\sin 17°} = \dfrac{75}{\sin 45°}$

$h = \dfrac{75 \sin 17°}{\sin 45°}$

$h \approx 31.01$ feet

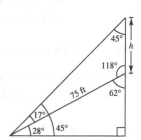

21. Given: $a = 8, b = 14, c = 17$

$\cos C = \dfrac{a^2 + b^2 - c^2}{2ab} = \dfrac{64 + 196 - 289}{2(8)(14)} \approx -0.1295 \Rightarrow C \approx 97.44°$

$\sin B = \dfrac{b \sin C}{c} \approx \dfrac{14 \sin 97.44}{17} \approx 0.8166 \Rightarrow B \approx 54.75°$

$A \approx 180° - 54.75° - 97.44° = 27.81°$

23. Given: $a = 6, b = 9, c = 14$

$\cos C = \dfrac{a^2 + b^2 - c^2}{2ab} = \dfrac{36 + 81 - 196}{2(6)(9)} \approx -0.7315 \Rightarrow C \approx 137.01°$

$\sin B = \dfrac{b \sin C}{c} \approx \dfrac{9 \sin 137.01°}{14} \approx 0.4383 \Rightarrow B \approx 26.00°$

$A \approx 180° - 26.00° - 137.01° = 16.99°$

25. Given: $a = 2.5, b = 5.0, c = 4.5$

$\cos B = \dfrac{a^2 + c^2 - b^2}{2ac} = 0.0667 \Rightarrow B \approx 86.18°$

$\cos C = \dfrac{a^2 + b^2 - c^2}{2ab} = 0.44 \Rightarrow C \approx 63.90°$

$A = 180° - B - C \approx 29.92°$

27. Given: $B = 108°, a = 11, c = 11$

$b^2 = a^2 + c^2 - 2ac \cos B = 11^2 + 11^2 - 2(11)(11) \cos 108° \Rightarrow b \approx 17.80$

$A = C = \frac{1}{2}(180° - 108°) = 36°$

29. Given: $C = 43°, a = 22.5, b = 31.4$

$c = \sqrt{a^2 + b^2 - 2ab \cos C} \approx 21.42$

$\cos B = \dfrac{a^2 + c^2 - b^2}{2ac} \approx -0.02169 \Rightarrow B \approx 91.24°$

$A = 180° - B - C \approx 45.76°$

31. Given: $C = 64°, b = 9, c = 13$.

Given two sides and an angle opposite one of them, the Law of Cosines cannot be used, so use the Law of Sines.

$$\sin B = \frac{b \sin C}{c} = \frac{9 \sin 64°}{13} \approx 0.62224 \implies B \approx 38.48°$$

$$A \approx 180° - 38.48° - 64° = 77.52°$$

$$a = \frac{c \sin A}{\sin C} \approx \frac{13 \sin 77.52°}{\sin 64°} \approx 14.12$$

33. Given: $a = 13, b = 15, c = 24$

Given three sides, the Law of Cosines can be used.

$$\cos C = \frac{a^2 + b^2 - c^2}{2ab} = \frac{169 + 225 - 576}{2(13)(15)} \approx -0.46667 \implies C \approx 117.82°$$

$$\sin A = \frac{a \sin C}{c} \approx \frac{13 \sin 117.82°}{24} \approx 0.47906 \implies A \approx 28.62°$$

$$B \approx 180° - 28.62° - 117.82° = 33.56°$$

35. Given: $a = 160, B = 12°, C = 7°$

Given two angles and a side, use the Law of Sines.

$$A = 180° - 12° - 7° = 161°$$

$$b = \frac{a \sin B}{\sin A} = \frac{160 \sin 12°}{\sin 161°} \approx 102.18$$

$$c = \frac{a \sin C}{\sin A} = \frac{160 \sin 7°}{\sin 161°} \approx 59.89$$

37.

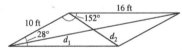

Let d_1 be the longer diagonal and d_2 be the shorter diagonal.

Using the Law of Cosines, you can find each of the diagonals.

$$d_1^2 = 16^2 + 10^2 - 2(16)(10) \cos 152° \approx 638.543$$

$$d_1 \approx 25.27 \text{ feet}$$

$$d_2^2 = 16^2 + 10^2 - 2(16)(10) \cos 28° \approx 73.457$$

$$d_2 \approx 8.57 \text{ feet}$$

39. Length of $AC = \sqrt{300^2 + 425^2 - 2(300)(425) \cos 115°}$

$$\approx 615.1 \text{ meters}$$

41. $a = 3, b = 6, c = 8$

$$s = \frac{a + b + c}{2} = \frac{3 + 6 + 8}{2} = 8.5$$

$$\text{Area} = \sqrt{s(s - a)(s - b)(s - c)}$$

$$= \sqrt{8.5(5.5)(2.5)(0.5)}$$

$$\approx 7.64$$

43. $a = 12.3, b = 15.8, c = 3.7$

$$s = \frac{a + b + c}{2} = \frac{12.3 + 15.8 + 3.7}{2} = 15.9$$

$$\text{Area} = \sqrt{s(s - a)(s - b)(s - c)}$$

$$= \sqrt{15.9(3.6)(0.1)(12.2)} = 8.36$$

45. $\|\mathbf{u}\| = \sqrt{(4 - (-2))^2 + (6 - 1)^2} = \sqrt{61}$

$$\|\mathbf{v}\| = \sqrt{(6 - 0)^2 + (3 - (-2))^2} = \sqrt{61}$$

$\mathbf{u}$ is directed along a line with a slope of $\dfrac{6 - 1}{4 - (-2)} = \dfrac{5}{6}$.

$\mathbf{v}$ is directed along a line with a slope of $\dfrac{3 - (-2)}{6 - 0} = \dfrac{5}{6}$.

Because $\mathbf{u}$ and $\mathbf{v}$ have identical magnitudes and directions, $\mathbf{u} = \mathbf{v}$.

47. Initial point: $(-5, 4)$

Terminal point: $(2, -1)$

$$\mathbf{v} = \langle 2 - (-5), -1 - 4 \rangle = \langle 7, -5 \rangle$$

$$\|\mathbf{v}\| = \sqrt{7^2 + (-5)^2} = \sqrt{74}$$

49. Initial point: $(0, 10)$

Terminal point: $(7, 3)$

$$\mathbf{v} = \langle 7 - 0, 3 - 10 \rangle = \langle 7, -7 \rangle$$

$$\|\mathbf{v}\| = \sqrt{7^2 + (-7)^2} = \sqrt{98}$$

$$= 7\sqrt{2}$$

51. u $= \langle -1, -3 \rangle$, **v** $= \langle -3, 6 \rangle$

    (a)   **u** + **v** $= \langle -1, -3 \rangle + \langle -3, 6 \rangle = \langle -4, 3 \rangle$

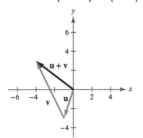

    (b)   **u** − **v** $= \langle -1, -3 \rangle - \langle -3, 6 \rangle = \langle 2, -9 \rangle$

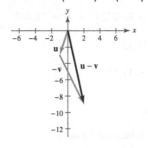

    (c) $4\mathbf{u} = 4\langle -1, -3 \rangle = \langle -4, -12 \rangle$

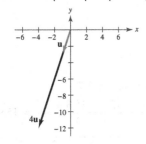

    (d) $3\mathbf{v} + 5\mathbf{u} = 3\langle -3, 6 \rangle + 5\langle -1, -3 \rangle = \langle -9, 18 \rangle + \langle -5, -15 \rangle = \langle -14, 3 \rangle$

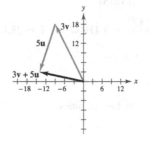

53. u $= \langle -5, 2 \rangle$, **v** $= \langle 4, 4 \rangle$

    (a)   **u** + **v** $= \langle -5, 2 \rangle + \langle 4, 4 \rangle = \langle -1, 6 \rangle$

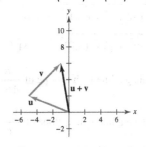

    (b)   **u** − **v** $= \langle -5, 2 \rangle - \langle 4, 4 \rangle = \langle -9, -2 \rangle$

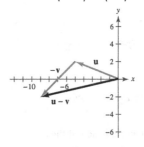

    (c) $4\mathbf{u} = 4\langle -5, 2 \rangle = \langle -20, 8 \rangle$

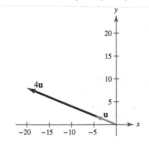

    (d) $3\mathbf{v} + 5\mathbf{u} = 3\langle 4, 4 \rangle + 5\langle -5, 2 \rangle = \langle 12, 12 \rangle + \langle -25, 10 \rangle = \langle -13, 22 \rangle$

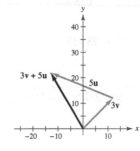

55. $u = 2i - j$, $v = 5i + 3j$

(a) $u + v = (2i - j) + (5i + 3j) = 7i + 2j$

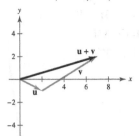

(b) $u - v = (2i - j) - (5i + 3j) = -3i - 4j$

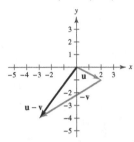

(c) $4u = 4(2i - j) = 8i - 4j$

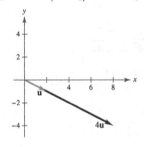

(d) $3v + 5u = 3(5i + 3j) + 5(2i - j) = 15i + 9j + 10i - 5j = 25i + 4j$

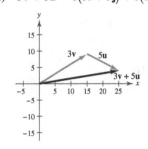

57. $u = 4i$, $v = -i + 6j$

(a) $u + v = 4i + (-i + 6j) = 3i + 6j$

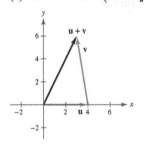

(b) $u - v = 4i - (-i + 6j) = 5i - 6j$

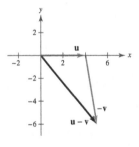

(c) $4u = 4(4i) = 16i$

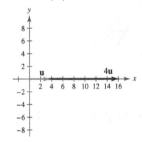

(d) $3v + 5u = 3(-i + 6j) + 5(4i) = -3i + 18j + 20i = 17i + 18j$

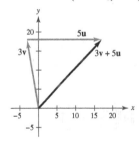

59. $P = (2, 3)$, $Q = (1, 8)$

$$\overrightarrow{PQ} = v = \langle 1 - 2, 8 - 3 \rangle$$
$$v = \langle -1, 5 \rangle$$
$$v = -i + 5j$$

61. $P = (3, 4)$, $Q = (9, 8)$

$$\overrightarrow{PQ} = v = \langle 9 - 3, 8 - 4 \rangle$$
$$v = \langle 6, 4 \rangle$$
$$v = 6i + 4j$$

63. $\mathbf{v} = 10\mathbf{i} + 3\mathbf{j}$

$3\mathbf{v} = 3(10\mathbf{i} + 3\mathbf{j})$

$= 30\mathbf{i} + 9\mathbf{j}$

$= \langle 30, 9 \rangle$

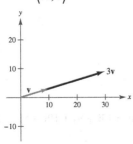

65. $\mathbf{u} = 6\mathbf{i} - 5\mathbf{j}, \mathbf{v} = 10\mathbf{i} + 3\mathbf{j}$

$2\mathbf{u} + \mathbf{v} = 2(6\mathbf{i} - 5\mathbf{j}) + (10\mathbf{i} + 3\mathbf{j})$

$= 22\mathbf{i} - 7\mathbf{j}$

$= \langle 22, -7 \rangle$

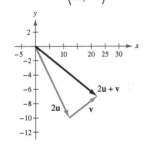

67. $\mathbf{u} = 6\mathbf{i} - 5\mathbf{j}, \mathbf{v} = 10\mathbf{i} + 3\mathbf{j}$

$5\mathbf{u} - 4\mathbf{v} = 5(6\mathbf{i} - 5\mathbf{j}) - 4(10\mathbf{i} + 3\mathbf{j})$

$= 30\mathbf{i} - 25\mathbf{j} - 40\mathbf{i} - 12\mathbf{j}$

$= -10\mathbf{i} - 37\mathbf{j}$

$= \langle -10, -37 \rangle$

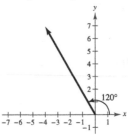

69. $\mathbf{v} = 5\mathbf{i} + 4\mathbf{j}$

$\|\mathbf{v}\| = \sqrt{5^2 + 4^2} = \sqrt{41}$

$\tan \theta = \frac{4}{5} \Rightarrow \theta \approx 38.7°$

71. $\mathbf{v} = -3\mathbf{i} - 3\mathbf{j}$

$\|\mathbf{v}\| = \sqrt{(-3)^2 + (-3)^2} = 3\sqrt{2}$

$\tan \theta = \frac{-3}{-3} = 1 \Rightarrow \theta = 225°$

73. $\mathbf{v} = 8(\cos 120° + i \sin 120°)$

$= 8\left(-\frac{1}{2} + \frac{\sqrt{3}}{2}i\right)$

$= -4 + 4\sqrt{3}i$

$= \langle -4, 4\sqrt{3} \rangle$

75. Rope One:

$\mathbf{u} = \|\mathbf{u}\|(\cos 30°\mathbf{i} - \sin 30°\mathbf{j}) = \|\mathbf{u}\|\left(\frac{\sqrt{3}}{2}\mathbf{i} - \frac{1}{2}\mathbf{j}\right)$

Rope Two:

$\mathbf{v} = \|\mathbf{u}\|(-\cos 30°\mathbf{i} - \sin 30°\mathbf{j}) = \|\mathbf{u}\|\left(-\frac{\sqrt{3}}{2}\mathbf{i} - \frac{1}{2}\mathbf{j}\right)$

Resultant: $\mathbf{u} + \mathbf{v} = -\|\mathbf{u}\|\mathbf{j} = -180\mathbf{j}$

$\|\mathbf{u}\| = 180$

So, the tension on each rope is $\|\mathbf{u}\| = 180$ lb.

77. $\mathbf{u} = \langle 6, 7 \rangle, \mathbf{v} = \langle -3, 9 \rangle$

$\mathbf{u} \cdot \mathbf{v} = 6(-3) + 7(9) = 45$

79. $\mathbf{u} = 3\mathbf{i} + 7\mathbf{j}, \mathbf{v} = 11\mathbf{i} - 5\mathbf{j}$

$\mathbf{u} \cdot \mathbf{v} = 3(11) + 7(-5) = -2$

81. $\mathbf{u} = \langle -4, 2 \rangle$

$2\mathbf{u} = \langle -8, 4 \rangle$

$2\mathbf{u} \cdot \mathbf{u} = -8(-4) + 4(2) = 40$

The result is a scalar.

83. $\mathbf{u} = \langle -4, 2 \rangle$

$4 - \|\mathbf{u}\| = 4 - \sqrt{(-4)^2 + 2^2} = 4 - \sqrt{20} = 4 - 2\sqrt{5}$

The result is a scalar.

85. $\mathbf{u} = \langle -4, 2 \rangle, \mathbf{v} = \langle 5, 1 \rangle$

$\mathbf{u}(\mathbf{u} \cdot \mathbf{v}) = \langle -4, 2 \rangle \left[-4(5) + 2(1) \right]$

$= -18\langle -4, 2 \rangle$

$= \langle 72, -36 \rangle$

The result is a vector.

87. $\mathbf{u} = \langle -4, 2 \rangle, \mathbf{v} = \langle 5, 1 \rangle$

$(\mathbf{u} \cdot \mathbf{u}) - (\mathbf{u} \cdot \mathbf{v}) = \left[-4(-4) + 2(2) \right] - \left[-4(5) + 2(1) \right]$

$= 20 - (-18)$

$= 38$

The result is a scalar.

89. $\mathbf{u} = \cos \dfrac{7\pi}{4} \mathbf{i} + \sin \dfrac{7\pi}{4} \mathbf{j} = \left\langle \dfrac{1}{\sqrt{2}}, -\dfrac{1}{\sqrt{2}} \right\rangle$

$\mathbf{v} = \cos \dfrac{5\pi}{6} \mathbf{i} + \sin \dfrac{5\pi}{6} \mathbf{j} = \left\langle -\dfrac{\sqrt{3}}{2}, \dfrac{1}{2} \right\rangle$

$\cos \theta = \dfrac{\mathbf{u} \cdot \mathbf{v}}{\|\mathbf{u}\|\|\mathbf{v}\|} = \dfrac{-\sqrt{3} - 1}{2\sqrt{2}} \Rightarrow \theta = \dfrac{11\pi}{12}$

91. $\mathbf{u} = \langle 2\sqrt{2}, -4 \rangle, \mathbf{v} = \langle -\sqrt{2}, 1 \rangle$

$\cos \theta = \dfrac{\mathbf{u} \cdot \mathbf{v}}{\|\mathbf{u}\|\|\mathbf{v}\|} = \dfrac{-8}{(\sqrt{24})(\sqrt{3})} \Rightarrow \theta \approx 160.5°$

93. $\mathbf{u} = \langle -3, 8 \rangle$

$\mathbf{v} = \langle 8, 3 \rangle$

$\mathbf{u} \cdot \mathbf{v} = -3(8) + 8(3) = 0$

$\mathbf{u}$ and $\mathbf{v}$ are orthogonal.

95. $\mathbf{u} = -\mathbf{i}$

$\mathbf{v} = \mathbf{i} + 2\mathbf{j}$

$\mathbf{u} \cdot \mathbf{v} \neq 0$

$\mathbf{u}$ and $\mathbf{v}$ are *not* orthogonal.

97. $\mathbf{u} = \langle -4, 3 \rangle, \mathbf{v} = \langle -8, -2 \rangle$

$\mathbf{w}_1 = \text{proj}_{\mathbf{v}} \mathbf{u} = \left(\dfrac{\mathbf{u} \cdot \mathbf{v}}{\|\mathbf{v}\|^2} \right) \mathbf{v} = \left(\dfrac{26}{68} \right) \langle -8, -2 \rangle = -\dfrac{13}{17} \langle 4, 1 \rangle$

$\mathbf{w}_2 = \mathbf{u} - \mathbf{w}_1 = \langle -4, 3 \rangle - \left(-\dfrac{13}{17} \right) \langle 4, 1 \rangle = \dfrac{16}{17} \langle -1, 4 \rangle$

$\mathbf{u} = \mathbf{w}_1 + \mathbf{w}_2 = -\dfrac{13}{17} \langle 4, 1 \rangle + \dfrac{16}{17} \langle -1, 4 \rangle$

99. $\mathbf{u} = \langle 2, 7 \rangle, \mathbf{v} = \langle 1, -1 \rangle$

$\mathbf{w}_1 = \text{proj}_{\mathbf{v}} \mathbf{u} = \left(\dfrac{\mathbf{u} \cdot \mathbf{v}}{\|\mathbf{v}\|^2} \right) \mathbf{v} = -\dfrac{5}{2} \langle 1, -1 \rangle = \dfrac{5}{2} \langle -1, 1 \rangle$

$\mathbf{w}_2 = \mathbf{u} - \mathbf{w}_1 = \langle 2, 7 \rangle - \left(\dfrac{5}{2} \right) \langle -1, 1 \rangle = \dfrac{9}{2} \langle 1, 1 \rangle$

$\mathbf{u} = \mathbf{w}_1 + \mathbf{w}_2 = \dfrac{5}{2} \langle -1, 1 \rangle + \dfrac{9}{2} \langle 1, 1 \rangle$

101. $P = (5, 3), Q = (8, 9) \Rightarrow \overrightarrow{PQ} = \langle 3, 6 \rangle$

Work $= \mathbf{v} \cdot \overrightarrow{PQ} = \langle 2, 7 \rangle \cdot \langle 3, 6 \rangle = 48$

103. Work $= (18,000)\left(\dfrac{48}{12} \right) = 72,000$ foot-pounds

105. True. Using the Law of Cosines, there is exactly one solution.

107. True. $\sin 90°$ is defined in the Law of Sines.

109. **A** and **C** appear to have the same magnitude and direction.

111. If $k > 0$, the direction of $k\mathbf{u}$ is the same, and the magnitude is $k\|\mathbf{u}\|$.

If $k < 0$, the direction of $k\mathbf{u}$ is the opposite direction of $\mathbf{u}$, and the magnitude is $|k|\|\mathbf{u}\|$.

113. A vector in the plane has both a magnitude and a direction.

Problem Solving for Chapter 3

1. $\left(\overline{PQ}\right)^2 = 4.7^2 + 6^2 - 2(4.7)(6)\cos 25° A$

$\overline{PQ} \approx 2.6409$ feet

$\dfrac{\sin \alpha}{4.7} = \dfrac{\sin 25°}{2.6409} \Rightarrow \alpha \approx 48.78°$

$\theta + \beta = 180° - 25° - 48.78° = 106.22°$

$(\theta + \beta) + \theta = 180° \Rightarrow \theta = 180° - 106.22° = 73.78°$

$\beta = 106.22° - 73.78° = 32.44°$

$\gamma = 180° - \alpha - \beta = 180° - 48.78° - 32.44° = 98.78°$

$\phi = 180° - \gamma = 180° - 98.78° = 81.22°$

$\dfrac{\overline{PT}}{\sin 25°} = \dfrac{4.7}{\sin 81.22°}$

$\overline{PT} \approx 2.01$ feet

3. (a)

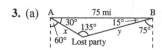

(b) $\dfrac{x}{\sin 15°} = \dfrac{75}{\sin 135°}$

$x \approx 27.45$ miles

and

$\dfrac{y}{\sin 30°} = \dfrac{75}{\sin 135°}$

$y \approx 53.03$ miles

(c)

$z^2 = (27.45)^2 + (20)^2 - 2(27.45)(20)\cos 20°$

$z \approx 11.03$ miles

$\dfrac{\sin \theta}{27.45} = \dfrac{\sin 20°}{11.03}$

$\sin \theta \approx 0.8511$

$\theta = 180° - \sin^{-1}(0.8511)$

$\theta \approx 121.7°$

To find the bearing, we have $\theta - 10° - 90° \approx 21.7°$. Bearing: S 21.7° E

5. If $\mathbf{u} \neq 0$, $\mathbf{v} \neq 0$, and $\mathbf{u} + \mathbf{v} \neq 0$, then $\left\|\dfrac{\mathbf{u}}{\|\mathbf{u}\|}\right\| = \left\|\dfrac{\mathbf{v}}{\|\mathbf{v}\|}\right\| = \left\|\dfrac{\mathbf{u} + \mathbf{v}}{\|\mathbf{u} + \mathbf{v}\|}\right\| = 1$ because all of these are magnitudes of unit vectors.

(a) $\mathbf{u} = \langle 1, -1 \rangle$, $\mathbf{v} = \langle -1, 2 \rangle$, $\mathbf{u} + \mathbf{v} = \langle 0, 1 \rangle$

    (i) $\|\mathbf{u}\| = \sqrt{2}$ (ii) $\|\mathbf{v}\| = \sqrt{5}$ (iii) $\|\mathbf{u} + \mathbf{v}\| = 1$ (iv) $\left\|\dfrac{\mathbf{u}}{\|\mathbf{u}\|}\right\| = 1$ (v) $\left\|\dfrac{\mathbf{v}}{\|\mathbf{v}\|}\right\| = 1$ (vi) $\left\|\dfrac{\mathbf{u} + \mathbf{v}}{\|\mathbf{u} + \mathbf{v}\|}\right\| = 1$

(b) $\mathbf{u} = \langle 0, 1 \rangle$, $\mathbf{v} = \langle 3, -3 \rangle$, $\mathbf{u} + \mathbf{v} = \langle 3, -2 \rangle$

    (i) $\|\mathbf{u}\| = 1$ (ii) $\|\mathbf{v}\| = \sqrt{18} = 3\sqrt{2}$ (iii) $\|\mathbf{u} + \mathbf{v}\| = \sqrt{13}$ (iv) $\left\|\dfrac{\mathbf{u}}{\|\mathbf{u}\|}\right\| = 1$ (v) $\left\|\dfrac{\mathbf{v}}{\|\mathbf{v}\|}\right\| = 1$ (vi) $\left\|\dfrac{\mathbf{u} + \mathbf{v}}{\|\mathbf{u} + \mathbf{v}\|}\right\| = 1$

(c) $\mathbf{u} = \left\langle 1, \dfrac{1}{2} \right\rangle$, $\mathbf{v} = \langle 2, 3 \rangle$, $\mathbf{u} + \mathbf{v} = \left\langle 3, \dfrac{7}{2} \right\rangle$

    (i) $\|\mathbf{u}\| = \dfrac{\sqrt{5}}{2}$ (ii) $\|\mathbf{v}\| = \sqrt{13}$ (iii) $\|\mathbf{u} + \mathbf{v}\| = \sqrt{9 + \dfrac{49}{4}} = \dfrac{\sqrt{85}}{2}$ (iv) $\left\|\dfrac{\mathbf{u}}{\|\mathbf{u}\|}\right\| = 1$

    (v) $\left\|\dfrac{\mathbf{v}}{\|\mathbf{v}\|}\right\| = 1$ (vi) $\left\|\dfrac{\mathbf{u} + \mathbf{v}}{\|\mathbf{u} + \mathbf{v}\|}\right\| = 1$

(d) $\mathbf{u} = \langle 2, -4 \rangle$, $\quad \mathbf{v} = \langle 5, 5 \rangle$, $\quad \mathbf{u} + \mathbf{v} = \langle 7, 1 \rangle$

   (i) $\|\mathbf{u}\| = \sqrt{20} = 2\sqrt{5}$ $\quad$ (ii) $\|\mathbf{v}\| = \sqrt{50} = 5\sqrt{2}$ $\quad$ (iii) $\|\mathbf{u} + \mathbf{v}\| = \sqrt{50} = 5\sqrt{2}$ $\quad$ (iv) $\left\| \dfrac{\mathbf{u}}{\|\mathbf{u}\|} \right\| = 1$

   (v) $\left\| \dfrac{\mathbf{v}}{\|\mathbf{v}\|} \right\| = 1$ $\quad$ (vi) $\left\| \dfrac{\mathbf{u} + \mathbf{v}}{\|\mathbf{u} + \mathbf{v}\|} \right\| = 1$

7. (a) The angle between them is $0°$.

   (b) The angle between them is $180°$.

   (c) No. At most it can be equal to the sum when the angle between them is $0°$.

9. (a) Since $0 < C < 180°$, $\cos\left(\dfrac{C}{2}\right) = \sqrt{\dfrac{1 + \cos C}{2}}$.

   Hence, $\cos\left(\dfrac{C}{2}\right) = \sqrt{\dfrac{1 + \left(a^2 + b^2 - c^2\right)/2ab}{2}} = \sqrt{\dfrac{2ab + a^2 + b^2 - c^2}{4ab}}$. On the other hand,

   $$s(s-c) = \frac{1}{2}(a+b+c)\left(\frac{1}{2}(a+b+c) - c\right)$$

   $$= \frac{1}{2}(a+b+c)\frac{1}{2}(a+b-c)$$

   $$= \frac{1}{4}\left((a+b)^2 - c^2\right)$$

   $$= \frac{1}{4}(a^2 + b^2 + 2ab - c^2).$$

   Thus, $\sqrt{\dfrac{s(s-c)}{ab}} = \sqrt{\dfrac{a^2 + b^2 + 2ab - c^2}{4ab}}$ and we have verified that $\cos\left(\dfrac{C}{2}\right) = \sqrt{\dfrac{s(s-c)}{ab}}$.

   (b) Since $0 < C < 180°$, $\sin\left(\dfrac{C}{2}\right) = \sqrt{\dfrac{1 - \cos C}{2}}$.

   Hence, $\sin\left(\dfrac{C}{2}\right) = \sqrt{\dfrac{1 - \left(a^2 + b^2 - c^2\right)/(2ab)}{2}} = \sqrt{\dfrac{2ab - a^2 - b^2 + c^2}{4ab}}$.

   On the other hand,

   $$(s-a)(s-b) = \left[\frac{1}{2}(a+b+c) - a\right]\left[\frac{1}{2}(a+b+c) - b\right]$$

   $$= \frac{1}{2}(b+c-a)\frac{1}{2}(a+c-b)$$

   $$= \frac{1}{4}[c - (a-b)][c + (a-b)]$$

   $$= \frac{1}{4}[c^2 - (a-b)^2]$$

   $$= \frac{1}{4}(c^2 - a^2 - b^2 + 2ab).$$

   Thus, $\sqrt{\dfrac{(s-a)(s-b)}{ab}} = \sqrt{\dfrac{c^2 - a^2 - b^2 + 2ab}{4ab}} = \sin\left(\dfrac{C}{2}\right)$.

11. Answers will vary.

13. (a) $\mathbf{u} = -120\mathbf{j}$

   $\mathbf{v} = 40\mathbf{i}$

(b) $\mathbf{s} = \mathbf{u} + \mathbf{v} = 40\mathbf{i} - 120\mathbf{j}$

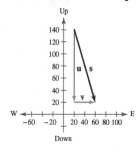

(c) $\|\mathbf{s}\| = \sqrt{40^2 + (-120)^2} = \sqrt{16{,}000} = 40\sqrt{10}$

   ≈ 126.5 miles per hour

This represents the actual rate of the skydiver's fall.

(d) $\tan\theta = \dfrac{120}{40} \Rightarrow \theta = \tan^{-1} 3 \Rightarrow \theta \approx 71.565°$

(e)

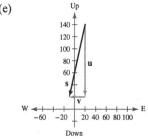

   $\mathbf{s} = 30\mathbf{i} - 120\mathbf{j}$

   $\|\mathbf{s}\| = \sqrt{30^2 + (-120)^2}$

   $\qquad = \sqrt{15{,}300}$

   $\qquad \approx 123.7$ miles per hour

Practice Test for Chapter 3

For Exercises 1 and 2, use the Law of Sines to find the remaining sides and angles of the triangle.

1. $A = 40°, B = 12°, b = 100$

2. $C = 150°, a = 5, c = 20$

3. Find the area of the triangle: $a = 3, b = 6, C = 130°$.

4. Determine the number of solutions to the triangle: $a = 10, b = 35, A = 22.5°$.

For Exercises 5 and 6, use the Law of Cosines to find the remaining sides and angles of the triangle.

5. $a = 49, b = 53, c = 38$

6. $C = 29°, a = 100, c = 300$

7. Use Heron's Formula to find the area of the triangle: $a = 4.1, b = 6.8, c = 5.5$.

8. A ship travels 40 miles due east, then adjusts its course $12°$ southward. After traveling 70 miles in that direction, how far is the ship from its point of departure?

9. $\mathbf{w} = 4\mathbf{u} - 7\mathbf{v}$ where $\mathbf{u} = 3\mathbf{i} + \mathbf{j}$ and $\mathbf{v} = -\mathbf{i} + 2\mathbf{j}$. Find $\mathbf{w}$.

10. Find a unit vector in the direction of $\mathbf{v} = 5\mathbf{i} - 3\mathbf{j}$.

11. Find the dot product and the angle between $\mathbf{u} = 6\mathbf{i} + 5\mathbf{j}$ and $\mathbf{v} = 2\mathbf{i} - 3\mathbf{j}$.

12. $\mathbf{v}$ is a vector of magnitude 4 making an angle of $30°$ with the positive x-axis. Find $\mathbf{v}$ in component form.

13. Find the projection of $\mathbf{u}$ onto $\mathbf{v}$ given $\mathbf{u} = \langle 3, -1 \rangle$ and $\mathbf{v} = \langle -2, 4 \rangle$.

14. Give the trigonometric form of $z = 5 - 5i$.

15. Give the standard form of $z = 6(\cos 225° + i \sin 225°)$.

16. Multiply $\left[7(\cos 23° + i \sin 23°) \right]\left[4(\cos 7° + i \sin 7°) \right]$.

17. Divide $\dfrac{9\left(\cos \dfrac{5\pi}{4} + i \sin \dfrac{5\pi}{4} \right)}{3(\cos \pi + i \sin \pi)}$.

18. Find $(2 + 2i)^8$.

19. Find the cube roots of $8\left(\cos \dfrac{\pi}{3} + i \sin \dfrac{\pi}{3} \right)$.

20. Find all the solutions to $x^4 + i = 0$.

C H A P T E R 4
Complex Numbers

CHAPTER 4
Complex Numbers

Section 4.1   Complex Numbers

1. real

3. pure imaginary

5. principal square

7. $a + bi = 9 + 8i$
$\quad a = 9$
$\quad b = 8$

9. $(a - 2) + (b + 1)i = 6 + 5i$
$\quad a - 2 = 6 \Rightarrow a = 8$
$\quad b + 1 = 5 \Rightarrow b = 4$

11. $2 + \sqrt{-25} = 2 + 5i$

13. $1 - \sqrt{-12} = 1 - 2\sqrt{3}\,i$

15. $\sqrt{-40} = 2\sqrt{10}\,i$

17. 23

19. $-6i + i^2 = -6i + (-1)$
$\qquad\quad = -1 - 6\,i$

21. $\sqrt{-0.04} = \sqrt{0.04}\,i$
$\qquad\quad = 0.2i$

23. $(5 + i) + (2 + 3i) = 5 + i + 2 + 3i$
$\qquad\qquad\qquad\quad = 7 + 4i$

25. $(9 - i) - (8 - i) = 1$

27. $\left(-2 + \sqrt{-8}\right) + \left(5 - \sqrt{-50}\right) = -2 + 2\sqrt{2}i + 5 - 5\sqrt{2}i = 3 - 3\sqrt{2}i$

29. $13i - (14 - 7i) = 13i - 14 + 7i$
$\qquad\qquad\qquad = -14 + 20i$

31. $(1 + i)(3 - 2i) = 3 - 2i + 3i - 2i^2$
$\qquad\qquad\qquad = 3 + i + 2 = 5 + i$

33. $12i(1 - 9i) = 12i - 108i^2$
$\qquad\qquad = 12i + 108$
$\qquad\qquad = 108 + 12i$

35. $\left(\sqrt{2} + 3i\right)\left(\sqrt{2} - 3i\right) = 2 - 9t^2$
$\qquad\qquad\qquad\qquad = 2 + 9 = 11$

37. $(6 + 7i)^2 = 36 + 84i + 49i^2$
$\qquad\qquad = 36 + 84i - 49$
$\qquad\qquad = -13 + 84i$

39. The complex conjugate of $9 + 2i$ is $9 - 2i$.
$\quad (9 + 2i)(9 - 2i) = 81 - 4i^2$
$\qquad\qquad\qquad = 81 + 4$
$\qquad\qquad\qquad = 85$

41. The complex conjugate of $-1 - \sqrt{5}i$ is $-1 + \sqrt{5}i$.
$\quad \left(-1 - \sqrt{5}i\right)\left(-1 + \sqrt{5}i\right) = 1 - 5i^2$
$\qquad\qquad\qquad\qquad\qquad = 1 + 5 = 6$

43. The complex conjugate of $\sqrt{-20} = 2\sqrt{5}i$ is $-2\sqrt{5}i$.
$\quad \left(2\sqrt{5}i\right)\left(-2\sqrt{5}i\right) = -20i^2 = 20$

45. The complex conjugate of $\sqrt{6}$ is $\sqrt{6}$.
$\quad \left(\sqrt{6}\right)\left(\sqrt{6}\right) = 6$

47. $\dfrac{2}{4 - 5i} = \dfrac{2}{4 - 5i} \cdot \dfrac{4 + 5i}{4 + 5i}$
$\qquad = \dfrac{2(4 + 5i)}{16 + 25} = \dfrac{8 + 10i}{41} = \dfrac{8}{41} + \dfrac{10}{41}i$

49. $\dfrac{5 + i}{5 - i} \cdot \dfrac{(5 + i)}{(5 + i)} = \dfrac{25 + 10i + i^2}{25 - i^2}$
$\qquad\qquad\qquad = \dfrac{24 + 10i}{26} = \dfrac{12}{13} + \dfrac{5}{13}i$

51. $\dfrac{9 - 4i}{i} \cdot \dfrac{-i}{-i} = \dfrac{-9i + 4i^2}{-i^2} = -4 - 9i$

53. $\dfrac{3i}{(4-5i)^2} = \dfrac{3i}{16-40i+25i^2} = \dfrac{3i}{-9-40i} \cdot \dfrac{-9+40i}{-9+40i}$

$= \dfrac{-27i+120i^2}{81+1600} = \dfrac{-120-27i}{1681}$

$= -\dfrac{120}{1681} - \dfrac{27}{1681}i$

55. $\dfrac{2}{1+i} - \dfrac{3}{1-i} = \dfrac{2(1-i)-3(1+i)}{(1+i)(1-i)}$

$= \dfrac{2-2i-3-3i}{1+1}$

$= \dfrac{-1-5i}{2}$

$= -\dfrac{1}{2} - \dfrac{5}{2}i$

57. $\dfrac{i}{3-2i} + \dfrac{2i}{3+8i} = \dfrac{i(3+8i)+2i(3-2i)}{(3-2i)(3+8i)}$

$= \dfrac{3i+8i^2+6i-4i^2}{9+24i-6i-16i^2}$

$= \dfrac{4i^2+9i}{9+18i+16}$

$= \dfrac{-4+9i}{25+18i} \cdot \dfrac{25-18i}{25-18i}$

$= \dfrac{-100+72i+225i-162i^2}{625+324}$

$= \dfrac{62+297i}{949} = \dfrac{62}{949} + \dfrac{297}{949}i$

59. $\sqrt{-6} \cdot \sqrt{-2} = \left(\sqrt{6}i\right)\left(\sqrt{2}i\right) = \sqrt{12}i^2 = \left(2\sqrt{3}\right)(-1)$

$= -2\sqrt{3}$

61. $\left(\sqrt{-15}\right)^2 = \left(\sqrt{15}i\right)^2 = 15i^2 = -15$

63. $\sqrt{-8} + \sqrt{-50} = \sqrt{8}i + \sqrt{50}i$

$= 2\sqrt{2}i + 5\sqrt{2}i$

$= 7\sqrt{2}i$

65. $\left(3+\sqrt{-5}\right)\left(7-\sqrt{-10}\right) = \left(3+\sqrt{5}i\right)\left(7-\sqrt{10}i\right)$

$= 21 - 3\sqrt{10}i + 7\sqrt{5}i - \sqrt{50}i^2$

$= \left(21+\sqrt{50}\right) + \left(7\sqrt{5}-3\sqrt{10}\right)i$

$= \left(21+5\sqrt{2}\right) + \left(7\sqrt{5}-3\sqrt{10}\right)i$

67. $x^2 - 2x + 2 = 0; a = 1, b = -2, c = 2$

$x = \dfrac{-(-2) \pm \sqrt{(-2)^2 - 4(1)(2)}}{2(1)}$

$= \dfrac{2 \pm \sqrt{-4}}{2}$

$= \dfrac{2 \pm 2i}{2}$

$= 1 \pm i$

69. $4x^2 + 16x + 17 = 0; a = 4, b = 16, c = 17$

$x = \dfrac{-16 \pm \sqrt{(16)^2 - 4(4)(17)}}{2(4)}$

$= \dfrac{-16 \pm \sqrt{-16}}{8}$

$= \dfrac{-16 \pm 4i}{8}$

$= -2 \pm \dfrac{1}{2}i$

71. $4x^2 + 16x + 21 = 0; a = 4, b = 16, c = 21$

$x = \dfrac{-16 \pm \sqrt{(16)^2 - 4(4)(21)}}{2(4)}$

$= \dfrac{-16 \pm \sqrt{-80}}{8}$

$= \dfrac{-16 \pm \sqrt{80}\,i}{8}$

$= \dfrac{-16 \pm 4\sqrt{5}\,i}{8}$

$= -2 \pm \dfrac{\sqrt{5}}{2}i$

73. $\frac{3}{2}x^2 - 6x + 9 = 0$ Multiply both sides by 2.

$3x^2 - 12x + 18 = 0; a = 3, b = -12, c = 18$

$x = \dfrac{-(-12) \pm \sqrt{(-12)^2 - 4(3)(18)}}{2(3)}$

$= \dfrac{12 \pm \sqrt{-72}}{6}$

$= \dfrac{12 \pm 6\sqrt{2}i}{6}$

$= 2 \pm \sqrt{2}i$

75. $1.4x^2 - 2x + 10 = 0 \Rightarrow 14x^2 - 20x + 100 = 0;$

$a = 14, b = -20, c = 100$

$x = \dfrac{-(-20) \pm \sqrt{(-20)^2 - 4(14)(100)}}{2(14)}$

$= \dfrac{20 \pm \sqrt{-5200}}{28}$

$= \dfrac{20 \pm 20\sqrt{13}\, i}{28}$

$= \dfrac{20}{28} \pm \dfrac{20\sqrt{13}\, i}{28}$

$= \dfrac{5}{7} \pm \dfrac{5\sqrt{13}}{7}i$

77. $-6i^3 + i^2 = -6i^2 i + i^2$

$= -6(-1)i + (-1)$

$= 6i - 1$

$= -1 + 6i$

79. $-14i^5 = -14i^2 i^2 i = -14(-1)(-1)(i) = -14i$

81. $\left(\sqrt{-72}\right)^3 = \left(6\sqrt{2}i\right)^3$

$= 6^3\left(\sqrt{2}\right)^3 i^3$

$= 216\left(2\sqrt{2}\right)i^2 i$

$= 432\sqrt{2}(-1)i$

$= -432\sqrt{2}i$

83. $\dfrac{1}{i^3} = \dfrac{1}{i^2 i} = \dfrac{1}{-i} = \dfrac{1}{-i} \cdot \dfrac{i}{i} = \dfrac{i}{-i^2} = i$

85. $(3i)^4 = 81i^4 = 81i^2 i^2 = 81(-1)(-1) = 81$

87. (a) $z_1 = 9 + 16i, z_2 = 20 - 10i$

(b) $\dfrac{1}{z} = \dfrac{1}{z_1} + \dfrac{1}{z_2} = \dfrac{1}{9 + 16i} + \dfrac{1}{20 - 10i} = \dfrac{20 - 10i + 9 + 16i}{(9 + 16i)(20 - 10i)} = \dfrac{29 + 6i}{340 + 230i}$

$z = \left(\dfrac{340 + 230i}{29 + 6i}\right)\left(\dfrac{29 - 6i}{29 - 6i}\right) = \dfrac{11,240 + 4630i}{877} = \dfrac{11,240}{877} + \dfrac{4630}{877}i$

89. False.

Sample answer: $(1 + i) + (3 + i) = 4 + 2i$ which is not a real number.

91. True.

$x^4 - x^2 + 14 = 56$

$\left(-i\sqrt{6}\right)^4 - \left(-i\sqrt{6}\right)^2 + 14 \overset{?}{=} 56$

$36 + 6 + 14 \overset{?}{=} 56$

$56 = 56$

93. $i = i$

$i^2 = -1$

$i^3 = -i$

$i^4 = 1$

$i^5 = i^4 i = i$

$i^6 = i^4 i^2 = -1$

$i^7 = i^4 i^3 = -i$

$i^8 = i^4 i^4 = 1$

$i^9 = i^4 i^4 i = i$

$i^{10} = i^4 i^4 i^2 = -1$

$i^{11} = i^4 i^4 i^3 = -i$

$i^{12} = i^4 i^4 i^4 = 1$

The pattern $i, -1, -i, 1$ repeats. Divide the exponent by 4.

If the remainder is 1, the result is i.

If the remainder is 2, the result is -1.

If the remainder is 3, the result is $-i$.

If the remainder is 0, the result is 1.

95. $\sqrt{-6}\sqrt{-6} = \sqrt{6}i\sqrt{6}i = 6i^2 = -6$

97. $(a_1 + b_1 i) + (a_2 + b_2 i) = (a_1 + a_2) + (b_1 + b_2)i$

The complex conjugate of this sum is
$(a_1 + a_2) - (b_1 + b_2)i.$

The sum of the complex conjugates is
$(a_1 - b_1 i) + (a_2 - b_2 i) = (a_1 + a_2) - (b_1 + b_2)i.$

So, the complex conjugate of the sum of two complex numbers is the sum of their complex conjugates.

Section 4.2 Complex Solutions of Equations

1. Fundamental Theorem; Algebra

3. complex conjugates

5. $2x^3 + 3x + 1 = 0$ has degree 3 so there are three solutions in the complex number system.

7. $50 - 2x^4 = 0$ has degree 4, so there are four solutions in the complex number system.

9. $2x^2 - 5x + 5 = 0$

$a = 2, b = -5, c = 5$

$b^2 - 4ac = (-5)^2 - 4(2)(5) = -15 < 0$

Both solutions are imaginary.

11. $4x^2 + 12x + 9 = 0$

$a = 4, b = 12, c = 9$

$b^2 - 4ac = (12)^2 - 4(4)(9) = 0$

There is one repeated real solution.

13. $\frac{1}{5}x^2 + \frac{6}{5}x - 8 = 0$

$a = \frac{1}{5}, b = \frac{6}{5}, c = -8$

$b^2 - 4ac = \left(\frac{6}{5}\right)^2 - 4\left(\frac{1}{5}\right)(-8) = \frac{196}{25} > 0$

There are two real solutions.

15. $x^2 - 5 = 0$

$x^2 = 5$

$x = \pm\sqrt{5}$

17. $2 - 2x - x^2 = 0$

$-x^2 - 2x + 2 = 0$

$x^2 + 2x = 2$

$x^2 + 2x + 1 = 2 + 1$

$(x + 1)^2 = 3$

$x + 1 = \pm\sqrt{3}$

$x = -1 \pm \sqrt{3}$

19. $x^2 - 8x + 16 = 0$

$(x - 4)^2 = 0$

$x = 4$

21. $x^2 + 2x + 5 = 0$

$x = \dfrac{-2 \pm \sqrt{2^2 - 4(1)(5)}}{2(1)}$

$= \dfrac{-2 \pm \sqrt{-16}}{2}$

$= \dfrac{-2 \pm 4i}{2}$

$= -1 \pm 2i$

23. $4x^2 - 4x + 5 = 0$

$$x = \frac{-(-4) \pm \sqrt{(-4)^2 - 4(4)(5)}}{2(4)}$$

$$= \frac{4 \pm \sqrt{-64}}{8}$$

$$= \frac{4 \pm 8i}{8}$$

$$= \frac{1}{2} \pm i$$

25.
$$x^4 - 6x^2 - 7 = 0$$
$$(x^2 - 7)(x^2 + 1) = 0$$
$$(x + \sqrt{7})(x - \sqrt{7})(x + i)(x - i) = 0$$

Setting each factor to zero yields the solutions $x = -\sqrt{7}$, $x = \sqrt{7}$, $x = i$, and $x = -i$.

27.
$$x^4 - 5x^2 - 6 = 0$$
$$(x^2 - 6)(x^2 + 1) = 0$$
$$(x + \sqrt{6})(x - \sqrt{6})(x + i)(x - i) = 0$$

Setting each factor to zero yields the solutions $x = -\sqrt{6}$, $x = \sqrt{6}$, $x = i$, and $x = -i$.

29. (a)
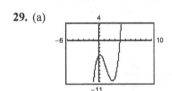

(b)
$$f(x) = x^3 - 4x^2 + x - 4$$
$$x^3 - 4x^2 + x - 4 = 0$$
$$x^2(x - 4) + 1(x - 4) = 0$$
$$(x^2 + 1)(x - 4) = 0$$
$$x^2 + 1 = 0 \Rightarrow x = \pm i$$
$$x - 4 = 0 \Rightarrow x = 4$$

Zeros: $x = \pm i, 4$

37. $h(x) = x^2 - 2x + 17$

By the Quadratic Formula, the zeros of $h(x)$ are:

$$x = \frac{-(-2) \pm \sqrt{(-2)^2 - 4(1)(17)}}{2(1)} = \frac{2 \pm \sqrt{-64}}{2} = \frac{2 \pm 8i}{2} = 1 \pm 4i$$

$$h(x) = \left[x - (1 + 4i)\right]\left[x - (1 - 4i)\right] = (x - 1 - 4i)(x - 1 + 4i)$$

(c) The graph has one *x*-intercept and the function has one real zero. The number of real zeros equals the number of *x*-intercepts. Each *x*-intercept represents a real solution of the equation $f(x) = 0$.

31. (a)
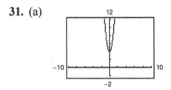

(b)
$$f(x) = x^4 + 4x^2 + 4$$
$$x^4 + 4x^2 + 4 = 0$$
$$(x^2 + 2)^2 = 0$$
$$x^2 + 2 = 0 \Rightarrow x = \pm\sqrt{2}i$$

(c) The graph has no *x*-intercepts and the function has no real zeros. The number of real zeros equals the number of *x*-intercepts. Each *x*-intercept represents a real solution of the equation $f(x) = 0$.

33. $f(x) = x^2 + 36$
$$= (x + 6i)(x - 6i)$$

The zeros of $f(x)$ are $x = \pm 6i$.

35. $f(x) = x^4 - 81$
$$= (x^2 - 9)(x^2 + 9)$$
$$= (x + 3)(x - 3)(x + 3i)(x - 3i)$$

The zeros of $f(x)$ are $x = \pm 3$ and $x = \pm 3i$.

39. $h(x) = x^2 - 6x - 10$

$$x = \frac{-(-6) \pm \sqrt{(-6)^2 - 4(1)(-10)}}{2(1)}$$

$$x = \frac{6 \pm \sqrt{76}}{2}$$

$$x = \frac{6 \pm 2\sqrt{19}}{2}$$

$$x = 3 \pm \sqrt{19}$$

The zeros of $h(x)$ are $x = 3 \pm \sqrt{19}$.

$$h(x) = \left(x - 3 - \sqrt{19}\right)\left(x - 3 + \sqrt{19}\right)$$

41. $g(x) = x^3 + 3x^2 - 3x - 9$

$$x^3 + 3x^2 - 3x - 9 = 0$$

$$x^2(x + 3) - 3(x + 3) = 0$$

$$\left(x^2 - 3\right)(x + 3) = 0$$

$$x^2 - 3 = 0 \Rightarrow x = \pm\sqrt{3}$$

$$x + 3 = 0 \Rightarrow x = -3$$

$$g(x) = (x + 3)\left(x + \sqrt{3}\right)\left(x - \sqrt{3}\right)$$

43. $f(x) = 2x^3 - x^2 + 36x - 18$

$$2x^3 - x^2 + 36x - 18 = 0$$

$$x^2(2x - 1) + 18(2x - 1) = 0$$

$$\left(x^2 + 18\right)(2x - 1) = 0$$

$$x^2 + 18 = 0 \Rightarrow x = \pm 3\sqrt{2}i$$

$$2x - 1 = 0 \Rightarrow x = \tfrac{1}{2}$$

$$f(x) = 2\left(x - \tfrac{1}{2}\right)\left(x + 3\sqrt{2}i\right)\left(x - 3\sqrt{2}i\right)$$

$$= (2x - 1)\left(x + 3\sqrt{2}i\right)\left(x - 3\sqrt{2}i\right)$$

45. $g(x) = x^4 - 6x^3 + 16x^2 - 96x$

$$x^4 - 6x^3 + 16x^2 - 96x = 0$$

$$x^3(x - 6) + 16x(x - 6) = 0$$

$$\left(x^3 + 16x\right)(x - 6) = 0$$

$$x\left(x^2 + 16\right)(x - 6) = 0$$

$$x = 0$$

$$x^2 + 16 = 0 \Rightarrow x = \pm\sqrt{-16} = \pm 4i$$

$$x - 6 = 0 \Rightarrow x = 6$$

Zeros: $0, 6, \pm 4i$

$$g(x) = x(x - 6)(x + 4i)(x - 4i)$$

47. $f(x) = x^4 + 10x^2 + 9$

$$= \left(x^2 + 1\right)\left(x^2 + 9\right)$$

$$= (x + i)(x - i)(x + 3i)(x - 3i)$$

The zeros of $f(x)$ are $x = \pm i$ and $x = \pm 3i$.

49. $f(x) = x^3 - x^2 + 4x - 4$

Because $2i$ is a zero, so is $-2i$.

$$
\begin{array}{r|rrrr}
2i & 1 & -1 & 4 & -4 \\
   &   & 2i & -4-2i & 4 \\
\hline
   & 1 & 2i-1 & -2i & 0
\end{array}
$$

$$
\begin{array}{r|rrr}
-2i & 1 & 2i-1 & -2i \\
    &   & -2i & 2i \\
\hline
    & 1 & -1   & 0
\end{array}
$$

$$f(x) = (x - 2i)(x + 2i)(x - 1)$$

The zeros of $f(x)$ are $x = 1, \pm 2i$.

51. $f(x) = 2x^4 - x^3 + 7x^2 - 4x - 4$

Because $2i$ is a zero, so is $-2i$.

$$
\begin{array}{r|rrrrr}
2i & 2 & -1 & 7 & -4 & -4 \\
   &   & 4i & -8 - 2i & 4 - 2i & 4 \\
\hline
   & 2 & -1 + 4i & -1 - 2i & -2i & 0
\end{array}
$$

$$
\begin{array}{r|rrrr}
-2i & 2 & -1 + 4i & -1 - 2i & -2i \\
    &   & -4i & 2i & 2i \\
\hline
    & 2 & -1 & -1 & 0
\end{array}
$$

The zeros of $2x^2 - x - 1 = (2x + 1)(x - 1)$ are

$x = -\frac{1}{2}$ and $x = 1$.

The zeros of $f(x)$ are $x = \pm 2i$, $x = -\frac{1}{2}$, and $x = 1$.

Alternate Solution:

Because $x = \pm 2i$ are zeros of $f(x)$,

$(x + 2i)(x - 2i) = x^2 + 4$ is a factor of $f(x)$.

By long division we have:

$$
\begin{array}{r}
2x^2 - x - 1 \\
x^2 + 0x + 4\overline{)2x^4 - x^3 + 7x^2 - 4x - 4} \\
\underline{2x^4 + 0x^3 + 8x^2} \\
-x^3 - x^2 - 4x \\
\underline{-x^3 + 0x^2 - 4x} \\
-x^2 + 0x - 4 \\
\underline{-x^2 + 0x - 4} \\
0
\end{array}
$$

Thus, $f(x) = (x^2 + 4)(2x^2 - x - 1)$

$= (x + 2i)(x - 2i)(2x + 1)(x - 1)$

and the zeros of $f(x)$ are $x = \pm 2i$, $x = -\frac{1}{2}$, and

$x = 1$.

53. $f(x) = x^3 - 2x^2 - 14x + 40$

Because $3 - i$ is a zero, so is $3 + i$.

$$
\begin{array}{r|rrrr}
3 - i & 1 & -2 & -14 & 40 \\
      &   & 3 - i & 2 - 4i & -40 \\
\hline
      & 1 & 1 - i & -12 - 4i & 0
\end{array}
$$

$$
\begin{array}{r|rrr}
3 + i & 1 & 1 - i & -12 - 4i \\
      &   & 3 + i & 12 + 4i \\
\hline
      & 1 & 4 & 0
\end{array}
$$

The zero of $x + 4$ is $x = -4$.

The zeros of f are $x = -4, 3 \pm i$.

55. $f(x) = x^3 - 8x^2 + 25x - 26$

Because $3 + 2i$ is a zero, so is $3 - 2i$.

$$
\begin{array}{r|rrrr}
3 + 2i & 1 & -8 & 25 & -26 \\
       &   & 3 + 2i & -19 - 4i & 26 \\
\hline
       & 1 & -5 + 2i & 6 - 4i & 0
\end{array}
$$

$$
\begin{array}{r|rrr}
3 - 2i & 1 & -5 + 2i & 6 - 4i \\
       &   & 3 - 2i & -6 + 4i \\
\hline
       & 1 & -2 & 0
\end{array}
$$

$f(x) = (x - (3 + 2i))(x - (3 - 2i))(x - 2)$

The zeros of $f(x)$ are $x = 3 \pm 2i, 2$.

Alternate Solution:

Because $x = 3 \pm 2i$ are zeros of

$f(x)$, $(x - (3 + 2i))(x - (3 - 2i)) = x^2 - 6x + 13$ is

a factor of $f(x)$.

By long division, you have:

$$
\begin{array}{r}
x - 2 \\
x^2 - 6x + 13\overline{)x^3 - 8x^2 + 25x - 26} \\
\underline{x^3 - 6x^2 + 13x} \\
-2x^2 + 12x - 26 \\
\underline{-2x^2 + 12x^2 - 26} \\
0
\end{array}
$$

$f(x) = (x^2 - 6x + 13)(x - 2)$

The zeros of $f(x)$ are $x = 3 \pm 2i, 2$.

57. $h(x) = x^4 + 2x^3 + 8x^2 - 8x + 16$

Because $1 + \sqrt{3}i$ is a zero, so is $1 - \sqrt{3}i$, and

$$\left[x - \left(1 + \sqrt{3}i\right)\right]\left[x - \left(1 - \sqrt{3}i\right)\right]$$

$$= \left[(x - 1) - \sqrt{3}i\right]\left[(x - 1) + \sqrt{3}i\right]$$

$$= (x - 1)^2 - \left(\sqrt{3}i\right)^2$$

$$= x^2 - 2x + 4$$

is a factor of $h(x)$. By long division, we have:

$$
\begin{array}{r}
x^2 \qquad\quad + 4 \\
x^2 - 2x + 4 \overline{\smash{\big)}\, x^4 - 2x^3 + 8x^2 - 8x + 16} \\
\underline{x^4 - 2x^3 + 4x^2} \\
4x^2 - 8x + 16 \\
\underline{4x^2 - 8x + 16} \\
0
\end{array}
$$

The zeros of $h(x) = \left(x^2 - 2x + 4\right)\left(x^2 + 4\right)$

$x^2 + 4 = 0$ are $x = \pm 2i$.

So, the zeros of $h(x)$ are $x = 1 \pm \sqrt{3}i$, $x = \pm 2$.

59. Because $5i$ is a zero, so is $-5i$.

$$f(x) = (x - 1)(x - 5i)(x + 5i)$$

$$= (x - 1)(x^2 + 25)$$

$$= x^3 - x^2 + 25x - 25$$

Note: $f(x) = a(x^3 - x^2 + 25x - 25)$, where a is any nonzero real number, has the zeros 1 and $\pm 5i$.

61. If $1 + i$ is a zero, so is its conjugate, $1 - i$.

$$f(x) = (x - 2)(x - 2)(x - (1 + i))(x - (1 - i))$$

$$= (x^2 - 4x + 4)(x^2 - 2x + 2)$$

$$= x^4 - 6x^3 + 14x^2 - 16x + 8$$

Note: $f(x) = a(x^4 - 6x^3 + 14x^2 - 16x + 8)$, where a is any nonzero real number, has the zeros 2, 2 and $1 \pm i$.

63. If $3 + \sqrt{2}i$ is a zero, so is $3 - \sqrt{2}i$.

$$f(x) = (3x - 2)(x + 1)\left[x - \left(3 + \sqrt{2}i\right)\right]\left[x - \left(3 - \sqrt{2}i\right)\right]$$

$$= (3x - 2)(x + 1)\left[(x - 3) - \sqrt{2}i\right]\left[(x - 3) + \sqrt{2}i\right]$$

$$= (3x^2 + x - 2)\left[(x - 3)^2 - \left(\sqrt{2}i\right)^2\right]$$

$$= (3x^2 + x - 2)(x^2 - 6x + 9 + 2)$$

$$= (3x^2 + x - 2)(x^2 - 6x + 11)$$

$$= 3x^4 - 17x^3 + 25x^2 + 23x - 22$$

Note: $f(x) = a(3x^4 - 17x^3 + 25x^2 + 23x - 22)$, where a is any nonzero real number, has the zeros $\frac{2}{3}, -1$, and $3 \pm \sqrt{2}i$.

65. Zeros: $1, 2i, -2i$ (Because $2i$ is a zero, so is $-2i$.)

$$f(x) = a(x - 1)(x - 2i)(x + 2i)$$

$$= a(x - 1)(x^2 + 4)$$

$$= a(x^3 - x^2 + 4x - 4)$$

Function value: $f(-1) = 10$

$f(-1) = a(-10) = 10 \Rightarrow a = -1$

$f(x) = -1(x^3 - x^2 + 4x - 4)$

$$= -x^3 + x^2 - 4x + 4$$

67. $f(x) = a(x + 2)(x - 1)(x - i)(x + i)$

$$= a(x^2 + x - 2)(x^2 + 1)$$

$$= a(x^4 + x^3 - x^2 + x - 2)$$

Since $f(0) = -4$

$-4 = a\left((0)^4 + (0)^3 - (0)^2 + (0) - 2\right)$

$-4 = -2a$

$a = 2$

So, $f(x) = 2(x^4 + x^3 - x^2 + x - 2)$

$$= 2x^4 + 2x^3 - 2x^2 + 2x - 4.$$

69. $f(x) = a(x + 3)\left(x - \left(1 + \sqrt{3}\,i\right)\right)\left(x - \left(1 - \sqrt{3}\,i\right)\right)$

$\qquad = a(x + 3)(x^2 - 2x + 4)$

$\qquad = a(x^3 + x^2 - 2x + 12)$

Since $f(-2) = 12$

$12 = a\left((-2)^3 + (-2)^2 - 2(-2) + 12\right)$

$12 = 12a$

$\quad a = 1$

So, $f(x) = (1)(x^4 - x^3 - 2x - 4)$

$\qquad = x^3 + x^2 - 2x + 12.$

73. *x*-intercept: $(2, 0) \Rightarrow x = 2$ is a zero.

Zeros: $2 - \sqrt{5}i, 2 + \sqrt{5}i, 2$

$f(x) = a\left[\left(x - \left(2 - \sqrt{5}i\right)\right)\left(x - \left(2 + \sqrt{5}i\right)\right)(x - 2)\right]$

$\qquad = a\left[\left(x^2 - x\left(2 - \sqrt{5}i\right) - x\left(2 + \sqrt{5}i\right) + \left(2 - \sqrt{5}i\right)\left(2 + \sqrt{5}i\right)\right)(x - 2)\right]$

$\qquad = a\left[\left(x^2 - 4x + 9\right)(x - 2)\right]$

$\qquad = a\left[x^3 - 6x^2 + 17x - 18\right], a \neq 0$

If $a = 1$, we have $f(x) = x^3 - 6x^2 + 17x - 18.$

75. *x*-intercepts: $(-3, 0)$ and $(2, 0) \Rightarrow x = -3$ and $x = 2$ are zeros.

Zeros: $-3, 2, \sqrt{2}i, -\sqrt{2}i$

$f(x) = a(x + 3)(x - 2)\left(x - \sqrt{2}i\right)\left(x + \sqrt{2}i\right)$

$\qquad = a(x^2 + x - 6)(x^2 + 2)$

$\qquad = a(x^4 + x^3 - 4x^2 + 2x - 12)$

Function value: $f(-2) = -12$

$f(-2) = a(-24) = -12 \Rightarrow a = \frac{1}{2}$

$f(x) = \frac{1}{2}(x^4 + x^3 - 4x^2 + 2x - 12)$

$\qquad = \frac{1}{2}x^4 + \frac{1}{2}x^3 - 2x^2 + x - 6$

71. *x*-intercept: $(-2, 0) \Rightarrow x = -2$ is a zero.

Zeros: $-2, 4 + 2i, 4 - 2i$

$f(x) = a(x + 2)\left[x - (4 + 2i)\right]\left[x - (4 - 2i)\right]$

$\qquad = a(x + 2)\left[(x - 4) - 2i\right]\left[(x - 4) + 2i\right]$

$\qquad = a(x + 2)\left[(x - 4)^2 + 4\right]$

$\qquad = a(x + 2)(x^2 - 8x + 20)$

$\qquad = a(x^3 - 6x^2 + 4x + 40), a \neq 0$

If $a = 1$, we have $f(x) = x^3 - 6x^2 + 4x + 40.$

77. $h(t) = -16t^2 + 48t, 0 \leq t \leq 3$

(a)

t	0	0.5	1	1.5	2	2.5	3
h	0	20	32	36	32	20	0

(b) No. The projectile reaches a maximum height of 36 feet.

(c) $\qquad -16t^2 + 48t = 64$

$\qquad -16t^2 + 48t - 64 = 0$

$\qquad -16(t^2 - 3t + 4) = 0$

$\qquad t^2 - 3t + 4 = 0 \Rightarrow t = \dfrac{3 \pm \sqrt{-7}}{2}$

$\qquad\qquad\qquad\qquad\qquad = \dfrac{3 \pm \sqrt{7}i}{2}$

This equation yields imaginary solutions. The projectile will not reach a height of 64 feet.

(d)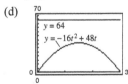

The graphs do not intersect, so the projectile does not reach 64 feet.

(e) All the results (numerical, algebraic, and graphical) show that it is not possible for the projectile to reach a height of 64 feet.

79. (a) Profit: $P = xp - C$

$$= x(140 - 0.0001x) - (80x + 150,000)$$
$$= 140x - 0.0001x^2 - 80x - 150,000$$
$$= -0.0001x^2 + 60x - 150,000$$

(b) $P(250,000) = -0.0001(250,000)^2 + 60(250,000) - 150,000 = \$8,600,000$

(c) $p(250,000) = 140 - 0.0001(250,000) = \$115$

(d) $-0.0001x^2 + 60x - 150,000 = 10,000,000$

$$0 = 0.0001x^2 - 60x + 10,150,000$$

By the Quadratic Formula, we obtain the complex roots $300,000 \pm 10,000\sqrt{115}i$. So, it is not possible to have a profit of 10 million dollars.

81. False. The most nonreal complex zeros it can have is two and the Linear Factorization Theorem guarantees that there are three linear factors, so one zero must be real.

83. Answers will vary.

85. (a) $f(x) = \left(x - \sqrt{b}i\right)\left(x + \sqrt{b}i\right) = x^2 + b$

(b) $f(x) = \left[x - (a + bi)\right]\left[x - (a - bi)\right]$

$$= \left[(x - a) - bi\right]\left[(x - a) + bi\right]$$
$$= (x - a)^2 - (bi)^2$$
$$= x^2 - 2ax + a^2 + b^2$$

87. $g(x) = -f(x)$. This function would have the same zeros as $f(x)$ so $r_1, r_2,$ and r_3 are also zeros of $g(x)$.

89. $g(x) = f(x - 5)$. The graph of $g(x)$ is a horizontal shift of the graph of $f(x)$ five units to the right so the zeros of $g(x)$ are $5 + r_1, 5 + r_2,$ and $5 + r_3$.

91. $g(x) = 3 + f(x)$. Since $g(x)$ is a vertical shift of the graph of $f(x)$, the zeros of $g(x)$ cannot be determined.

Section 4.3   The Complex Plane

1. real

3. absolute value

5. reflections

7. $2 = 2 + 0i$ matches (c)

8. $3i = 0 + 3i$ matches (f)

9. $1 + 2i$ matches (h)

10. $2 + i$ matches (a)

11. $3 - i$ matches (b)

12. $-3 + i$ matches (g)

13. $-2 - i$ matches (e)

14. $-1 - 3i$ matches (d)

15. $|-7i| = \sqrt{0^2 + (-7)^2}$

$$= \sqrt{49} = 7$$

17. $|-6 + 8i| = \sqrt{(-6)^2 + 8^2}$

$$= \sqrt{100} = 10$$

19. $|4 - 6i| = \sqrt{4^2 + (-6)^2}$

$\qquad = \sqrt{52} = 2\sqrt{13}$

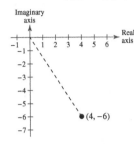

21. $(3 + i) + (2 + 5i) = 5 + 6i$

23. $(8 - 2i) + (2 + 6i) = 10 + 4i$

25. $(5 + 6i) + (1 - i) = 6 + 5i$

27. $(-3 + 4i) + (-2 + 3i) = -5 + 7i$

29. $(4 + 2i) - (6 + 4i) = -2 - 2i$

31. $(5 - i) - (-5 + 2i) = 10 - 3i$

33. $2 - (2 + 6i) = -6i$

35. $-2i - (3 - 5i) = -3 + 3i$

37.

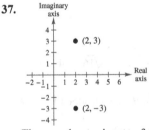

The complex conjugate of $2 + 3i$ is $2 - 3i$.

39.

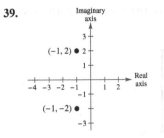

The complex conjugate of $-1 - 2i$ is $-1 + 2i$.

41. $d = \sqrt{(-1 - 1)^2 + (4 - 2)^2}$

$\qquad = \sqrt{8} = 2\sqrt{2} \approx 2.83$

43. $d = \sqrt{(3 - 0)^2 + (-4 - 6)^2}$

$\qquad = \sqrt{109} \approx 10.44$

45. Midpoint $= \left(\dfrac{2 + 6}{2}, \dfrac{1 + 5}{2} \right)$

$\qquad = 4 + 3i = (4, 3)$

47. Midpoint $= \left(\dfrac{0 + 9}{2}, \dfrac{7 - 10}{2} \right)$

$\qquad = \dfrac{9}{2} - \dfrac{3}{2}i$

$\qquad = \left(\dfrac{9}{2}, -\dfrac{3}{2} \right)$

49. (a) Ship A: $3 + 4i$

$\qquad$ Ship B: $-5 + 2i$

$\quad$ (b) To find the distance between the two ships using complex numbers, you can find the modulus of the difference of the two complex numbers.

$$d = \sqrt{(-5 - 3)^2 + (2 - 4)^2}$$

$$= \sqrt{68}$$

$$\approx 8.25 \text{ miles.}$$

51. False. The modulus of a complex number is always real.

53. False. The modulus of the sum of two complex numbers is not equal to the sum of their moduli.

$\left| 1 + i \right| + \left| 1 - i \right| = \sqrt{2} + \sqrt{2} = 2\sqrt{2} \neq \left| (1 + i) + (1 - i) \right| = \left| 2 \right| = 2$

55. The set of all points with the same modulus represent a circle in the complex plane. The modulus represents the distance from the origin, that is the radius of the circle.

57. If two complex conjugates are plotted in the complex plane, they will form an isosceles triangle because their moduli are equal.

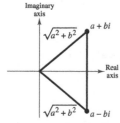

Section 4.4   Trigonometric Form of a Complex Number

1. trigonometric form; modulus; argument

3. $z = 3i$

$r = \sqrt{0^2 + 3^2} = \sqrt{9} = 3$

$\tan \theta = \dfrac{3}{0}$, undefined $\Rightarrow \theta = \dfrac{\pi}{2}$

$z = 3\left(\cos \dfrac{\pi}{2} + i \sin \dfrac{\pi}{2}\right)$

5. $z = -3 - 3i$

$r = \sqrt{(-3)^2 + (-3)^2} = \sqrt{18} = 3\sqrt{2}$

$\tan \theta = \dfrac{-3}{-3} = 1, \theta$ is in Quadrant III $\Rightarrow \theta = \dfrac{5\pi}{4}$.

$z = 3\sqrt{2}\left(\cos \dfrac{5\pi}{4} + i \sin \dfrac{5\pi}{4}\right)$

7. $z = 1 + i$

$r = \sqrt{1^2 + 1^2} = \sqrt{2}$

$\tan \theta = 1, \theta$ is in Quadrant I $\Rightarrow \theta = \dfrac{\pi}{4}$.

$z = \sqrt{2}\left(\cos \dfrac{\pi}{4} + i \sin \dfrac{\pi}{4}\right)$

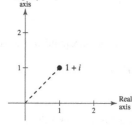

9. $z = 1 - \sqrt{3}i$

$r = \sqrt{1^2 + \left(-\sqrt{3}\right)^2} = \sqrt{4} = 2$

$\tan \theta = -\sqrt{3}, \theta$ is in Quadrant IV $\Rightarrow \theta = \dfrac{5\pi}{3}$.

$z = 2\left(\cos \dfrac{5\pi}{3} + i \sin \dfrac{5\pi}{3}\right)$

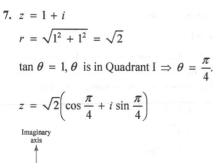

11. $z = -2\left(1 + \sqrt{3}i\right)$

$r = \sqrt{(-2)^2 + \left(-2\sqrt{3}\right)^2} = \sqrt{16} = 4$

$\tan \theta = \dfrac{\sqrt{3}}{1} = \sqrt{3}, \theta$ is in Quadrant III $\Rightarrow \theta = \dfrac{4\pi}{3}$.

$z = 4\left(\cos \dfrac{4\pi}{3} + i \sin \dfrac{4\pi}{3}\right)$

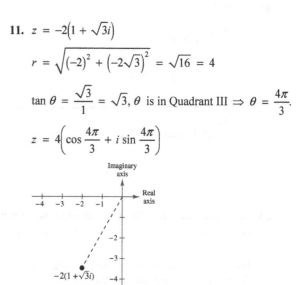

13. $z = -5i$

$r = \sqrt{0^2 + (-5)^2} = \sqrt{25} = 5$

$\tan \theta = \dfrac{-5}{0}$, undefined $\Rightarrow \theta = \dfrac{3\pi}{2}$

$z = 5\left(\cos \dfrac{3\pi}{2} + i \sin \dfrac{3\pi}{2}\right)$

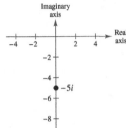

15. $z = 2$

$r = \sqrt{2^2 + 0^2} = \sqrt{4} = 2$

$\tan \theta = 0 \Rightarrow \theta = 0$

$z = 2(\cos 0 + i \sin 0)$

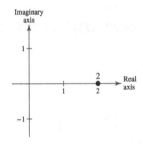

17. $z = -7 + 4i$

$r = \sqrt{(-7)^2 + (4)^2} = \sqrt{65}$

$\tan \theta = \dfrac{4}{-7}, \theta$ is in Quadrant II $\Rightarrow \theta \approx 2.62.$

$z \approx \sqrt{65}(\cos 2.62 + i \sin 2.62)$

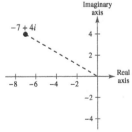

19. $z = 3 + \sqrt{3}i$

$r = \sqrt{(3)^2 + \left(\sqrt{3}\right)^2} = \sqrt{12} = 2\sqrt{3}$

$\tan \theta = \dfrac{\sqrt{3}}{3} \Rightarrow \theta = \dfrac{\pi}{6}$

$z = 2\sqrt{3}\left(\cos \dfrac{\pi}{6} + i \sin \dfrac{\pi}{6}\right)$

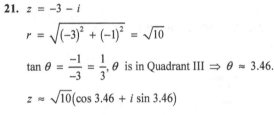

21. $z = -3 - i$

$r = \sqrt{(-3)^2 + (-1)^2} = \sqrt{10}$

$\tan \theta = \dfrac{-1}{-3} = \dfrac{1}{3}, \theta$ is in Quadrant III $\Rightarrow \theta \approx 3.46.$

$z \approx \sqrt{10}(\cos 3.46 + i \sin 3.46)$

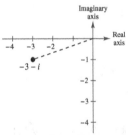

23. $z = 5 + 2i$

$r = \sqrt{5^2 + 2^2} = \sqrt{29}$

$\tan \theta = \dfrac{2}{5}$

$\theta \approx 0.38$

$z \approx \sqrt{29}(\cos 0.38 + i \sin 0.38)$

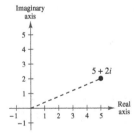

25. $z = -8 - 5\sqrt{3}i$

$r = \sqrt{(-8)^2 + (-5\sqrt{3})^2} = \sqrt{139}$

$\tan\theta = \dfrac{5\sqrt{3}}{8}$

$\theta \approx 3.97$

$z \approx \sqrt{139}(\cos 3.97 + i \sin 3.97)$

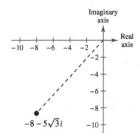

27. $2(\cos 60° + i \sin 60°) = 2\left(\dfrac{1}{2} + \dfrac{\sqrt{3}}{2}i\right)$

$= 1 + \sqrt{3}i$

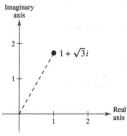

29. $\dfrac{9}{4}\left(\cos\dfrac{3\pi}{4} + i \sin\dfrac{3\pi}{4}\right) = \dfrac{9}{4}\left(-\dfrac{\sqrt{2}}{2} + \dfrac{\sqrt{2}}{2}i\right)$

$= -\dfrac{9\sqrt{2}}{8} + \dfrac{9\sqrt{2}}{8}i$

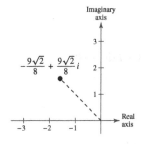

31. $\sqrt{48}\left[\cos(-30°) + i \sin(-30°)\right] = 4\sqrt{3}\left(\dfrac{\sqrt{3}}{2} - \dfrac{1}{2}i\right)$

$= 6 - 2\sqrt{3}i$

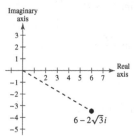

33. $7(\cos 0° + i \sin 0°) = 7$

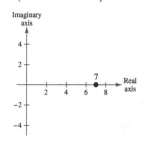

35. $5\left[\cos(198°\,45') + i \sin(198°\,45')\right] \approx -4.7347 - 1.6072i$

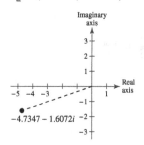

37. $5\left(\cos\dfrac{\pi}{9} + i \sin\dfrac{\pi}{9}\right) \approx 4.6985 + 1.7101i$

39. $2(\cos 155° + i \sin 155°) \approx -1.8126 + 0.8452i$

41. $\left[2\left(\cos\dfrac{\pi}{4} + i \sin\dfrac{\pi}{4}\right)\right]\left[6\left(\cos\dfrac{\pi}{12} + i \sin\dfrac{\pi}{12}\right)\right] = (2)(6)\left[\cos\left(\dfrac{\pi}{4} + \dfrac{\pi}{12}\right) + i \sin\left(\dfrac{\pi}{4} + \dfrac{\pi}{12}\right)\right]$

$= 12\left(\cos\dfrac{\pi}{3} + i \sin\dfrac{\pi}{3}\right)$

43. $\left[\frac{5}{3}(\cos 120° + i \sin 120°)\right]\left[\frac{2}{3}(\cos 30° + i \sin 30°)\right] = \frac{5}{3}\left(\frac{2}{3}\right)\left[\cos(120° + 30°) + i \sin(120° + 30°)\right]$

$$= \frac{10}{9}(\cos 150° + i \sin 150°)$$

45. $(\cos 80° + i \sin 80°)(\cos 330° + i \sin 330°) = \cos(80° + 330°) + i \sin(80° + 330°)$

$$= \cos 410° + i \sin 410°$$

$$= \cos 50° + i \sin 50°$$

47. $\dfrac{3(\cos 50° + i \sin 50°)}{9(\cos 20° + i \sin 20°)} = \dfrac{1}{3}\left[\cos(50° - 20°) + i \sin(50° - 20°)\right] = \dfrac{1}{3}(\cos 30° + i \sin 30°)$

49. $\dfrac{\cos \pi + i \sin \pi}{\cos(\pi/3) + i \sin(\pi/3)} = \cos\left(\pi - \dfrac{\pi}{3}\right) + i \sin\left(\pi - \dfrac{\pi}{3}\right) = \cos \dfrac{2\pi}{3} + i \sin \dfrac{2\pi}{3}$

51. $\dfrac{12(\cos 92° + i \sin 92°)}{2(\cos 122° + i \sin 122°)} = 6\left[\cos(92° - 122°) + i \sin(92° - 122°)\right]$

$$= 6\left[\cos(-30°) + i \sin(-30°)\right]$$

$$= 6(\cos 330° + i \sin 330°)$$

53. (a) $2 + 2i = 2\sqrt{2}\left(\cos \dfrac{\pi}{4} + i \sin \dfrac{\pi}{4}\right)$

$\qquad 1 - i = \sqrt{2}\left[\cos\left(-\dfrac{\pi}{4}\right) + i \sin\left(-\dfrac{\pi}{4}\right)\right] = \sqrt{2}\left(\cos \dfrac{7\pi}{4} + i \sin \dfrac{7\pi}{4}\right)$

(b) $(2 + 2i)(1 - i) = \left[2\sqrt{2}\left(\cos \dfrac{\pi}{4} + i \sin \dfrac{\pi}{4}\right)\right]\left[\sqrt{2}\left(\cos\left(\dfrac{7\pi}{4}\right) + i \sin\left(\dfrac{7\pi}{4}\right)\right)\right] = 4(\cos 2\pi + i \sin 2\pi)$

$$= 4(\cos 0 + i \sin 0) = 4$$

(c) $(2 + 2i)(1 - i) = 2 - 2i + 2i - 2i^2 = 2 + 2 = 4$

55. (a) $-2i = 2\left[\cos\left(-\dfrac{\pi}{2}\right) + i \sin\left(-\dfrac{\pi}{2}\right)\right] = 2\left(\cos \dfrac{3\pi}{2} + i \sin \dfrac{3\pi}{2}\right)$

$\qquad 1 + i = \sqrt{2}\left(\cos \dfrac{\pi}{4} + i \sin \dfrac{\pi}{4}\right)$

(b) $-2i(1 + i) = 2\left[\cos\left(\dfrac{3\pi}{2}\right) + i \sin\left(\dfrac{3\pi}{2}\right)\right]\left[\sqrt{2}\left(\cos \dfrac{\pi}{4} + i \sin \dfrac{\pi}{4}\right)\right]$

$$= 2\sqrt{2}\left[\cos\left(\dfrac{7\pi}{4}\right) + i \sin\left(\dfrac{7\pi}{4}\right)\right]$$

$$= 2\sqrt{2}\left[\dfrac{1}{\sqrt{2}} - \dfrac{1}{\sqrt{2}}i\right] = 2 - 2i$$

(c) $-2i(1 + i) = -2i - 2i^2 = -2i + 2 = 2 - 2i$

57. (a) $\quad 3 + 4i \approx 5(\cos 0.93 + i \sin 0.93)$

$\qquad 1 - \sqrt{3}i = 2\left(\cos \dfrac{5\pi}{3} + i \sin \dfrac{5\pi}{3}\right)$

(b) $\dfrac{3 + 4i}{1 - \sqrt{3}i} \approx \dfrac{5(\cos 0.93 + i \sin 0.93)}{2\left(\cos \dfrac{5\pi}{3} + i \sin \dfrac{5\pi}{3}\right)} \approx 2.5\big[\cos(-4.31) + i \sin(-4.31)\big] = \dfrac{5}{2}(\cos 1.97 + i \sin 1.97) \approx -0.982 + 2.299i$

(c) $\dfrac{3 + 4i}{1 - \sqrt{3}i} = \dfrac{3 + 4i}{1 - \sqrt{3}i} \cdot \dfrac{1 + \sqrt{3}i}{1 + \sqrt{3}i} = \dfrac{3 + \left(4 + 3\sqrt{3}\right)i + 4\sqrt{3}i^2}{1 + 3} = \dfrac{3 - 4\sqrt{3}}{4} + \dfrac{4 + 3\sqrt{3}}{4}i \approx -0.982 + 2.299i$

59. (a) $\qquad 5 = 5(\cos 0 + i \sin 0)$

$\qquad 2 + 3i \approx \sqrt{13}(\cos 0.98 + i \sin 0.98)$

(b) $\dfrac{5}{2 + 3i} \approx \dfrac{5(\cos 0 + i \sin 0)}{\sqrt{13}(\cos 0.98 + i \sin 0.98)} = \dfrac{5}{\sqrt{13}}\big[\cos(-0.98) + i \sin(-0.98)\big]$

$\qquad = \dfrac{5}{\sqrt{13}}(\cos 5.30 + i \sin 5.30) \approx 0.769 - 1.154i$

(c) $\dfrac{5}{2 + 3i} = \dfrac{5}{2 + 3i} \cdot \dfrac{2 - 3i}{2 - 3i} = \dfrac{10 - 15i}{13} = \dfrac{10}{13} - \dfrac{15}{13}i \approx 0.769 - 1.154i$

61. $\left[2\left(\cos \dfrac{2\pi}{3} + i \sin \dfrac{2\pi}{3}\right)\right]\left[\dfrac{1}{2}\left(\cos \dfrac{\pi}{3} + i \sin \dfrac{\pi}{3}\right)\right] = (2)\left(\dfrac{1}{2}\right)\left[\cos\left(\dfrac{2\pi}{3} + \dfrac{\pi}{3}\right) + i \sin\left(\dfrac{2\pi}{3} + \dfrac{\pi}{3}\right)\right]$

$\qquad = (\cos \pi + i \sin \pi)$

$\qquad = -1 + 0i$

$\qquad = -1$

63. $\left[4\left(\cos \dfrac{\pi}{4} + i \sin \dfrac{\pi}{4}\right)\right]\left[5\left(\cos \dfrac{\pi}{2} + i \sin \dfrac{\pi}{2}\right)\right] = (4)(5)\left[\cos\left(\dfrac{\pi}{4} + \dfrac{\pi}{2}\right) + i \sin\left(\dfrac{\pi}{4} + \dfrac{\pi}{2}\right)\right]$

$\qquad = 20\left(\cos \dfrac{3\pi}{4} + i \sin \dfrac{3\pi}{4}\right)$

$\qquad = 20\left(-\dfrac{\sqrt{2}}{2} + i\dfrac{\sqrt{2}}{2}\right)$

$\qquad = -10\sqrt{2} + 10\sqrt{2}\, i$

65. Let $z = x + iy$ such that:

$\qquad |z| = 2 \Rightarrow 2 = \sqrt{x^2 + y^2}$

$\qquad\qquad \Rightarrow 4 = x^2 + y^2$

Circle with radius of 2

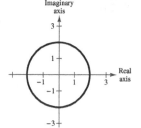

67. Let $\theta = \dfrac{\pi}{6}$.

Since $r \geq 0$, we have the portion of the line $\theta = \pi/6$ in Quadrant I.

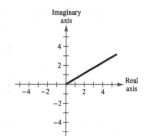

69. (a) $E = I \cdot Z$

$$= \left[6(\cos 41° + i \sin 41°)\right]\left(4\left[\cos(-11°) + i \sin(-11°)\right]\right)$$

$$= 24(\cos 30° + i \sin 30°) \text{ volts}$$

(b) $E = 24\left(\dfrac{\sqrt{3}}{2} + \dfrac{1}{2}i\right) = 12\sqrt{3} + 12i \text{ volts}$

(c) $|E| = \sqrt{\left(12\sqrt{3}\right)^2 + (12)^2} = \sqrt{576} = 24 \text{ volts}$

71. $z_1 = r_1(\cos \theta_1 + i \sin \theta_1),\ z_2 = r_2(\cos \theta_2 + i \sin \theta_2)$

$z_1 z_2 = r_1 r_2\left[\cos(\theta_1 + \theta_2) + i \sin(\theta_1 + \theta_2)\right]$ and $z_1 z_2 = 0$ if and only if $r_1 = 0$ and/or $r_2 = 0$.

True.

73. $\dfrac{z_1}{z_2} = \dfrac{r_1(\cos \theta_1 + i \sin \theta_1)}{r_2(\cos \theta_2 + i \sin \theta_2)} \cdot \dfrac{\cos \theta_2 - i \sin \theta_2}{\cos \theta_2 - i \sin \theta_2}$

$$= \dfrac{r_1}{r_2(\cos^2 \theta_2 + \sin^2 \theta_2)}\left[\cos \theta_1 \cos \theta_2 + \sin \theta_1 \sin \theta_2 + i(\sin \theta_1 \cos \theta_2 - \sin \theta_2 \cos \theta_1)\right]$$

$$= \dfrac{r_1}{r_2}\left[\cos(\theta_1 - \theta_2) + i \sin(\theta_1 - \theta_2)\right]$$

Section 4.5   DeMoivre's Theorem

1. DeMoivre's

3. $\dfrac{2\pi}{n}$

5. $\left[5(\cos 20° + i \sin 20°)\right]^3 = 5^3(\cos 60° + i \sin 60°)$

$$= \dfrac{125}{2} + \dfrac{125\sqrt{3}}{2}i$$

7. $\left(\cos \dfrac{\pi}{4} + i \sin \dfrac{\pi}{4}\right)^{12} = \cos \dfrac{12\pi}{4} + i \sin \dfrac{12\pi}{4}$

$$= \cos 3\pi + i \sin 3\pi$$

$$= -1$$

9. $\left[5(\cos 3.2 + i \sin 3.2)\right]^4 = 5^4(\cos 12.8 + i \sin 12.8)$

$$\approx 608.0 + 144.7i$$

11. $\left[3(\cos 15° + i \sin 15°)\right]^4 = 81(\cos 60° + i \sin 60°)$

$$= \dfrac{81}{2} + \dfrac{81\sqrt{3}}{2}i$$

13. $\left[5(\cos 95° + i \sin 95°)\right]^3 = 125(\cos 285° + i \sin 285°)$

$$\approx 32.3524 - 120.7407i$$

15. $\left[2\left(\cos \dfrac{\pi}{10} + i \sin \dfrac{\pi}{10}\right)\right]^5 = 2^5\left(\cos \dfrac{\pi}{2} + i \sin \dfrac{\pi}{2}\right)$

$$\approx 32i$$

17. $\left[3\left(\cos \dfrac{2\pi}{3} + i \sin \dfrac{2\pi}{3}\right)\right]^3 = 27(\cos \pi + i \sin 2\pi)$

$$= 27$$

19. $(1 + i)^5 = \left[\sqrt{2}\left(\cos \dfrac{\pi}{4} + i \sin \dfrac{\pi}{4}\right)\right]^5$

$$= \left(\sqrt{2}\right)^5\left(\cos \dfrac{5\pi}{4} + i \sin \dfrac{5\pi}{4}\right)$$

$$= 4\sqrt{2}\left(-\dfrac{\sqrt{2}}{2} - \dfrac{\sqrt{2}}{2}i\right)$$

$$= -4 - 4i$$

21. $(-1 + i)^6 = \left[\sqrt{2}\left(\cos \dfrac{3\pi}{4} + i \sin \dfrac{3\pi}{4}\right)\right]^6$

$$= \left(\sqrt{2}\right)^6\left(\cos \dfrac{18\pi}{4} + i \sin \dfrac{18\pi}{4}\right)$$

$$= 8\left(\cos \dfrac{9\pi}{2} + i \sin \dfrac{9\pi}{2}\right)$$

$$= 8(0 + i)$$

$$= 8i$$

23. $2(\sqrt{3} + i)^{10} = 2\left[2\left(\cos\dfrac{\pi}{6} + i\sin\dfrac{\pi}{6}\right)\right]^{16}$

$= 2\left[2^{10}\left(\cos\dfrac{10\pi}{6} + i\sin\dfrac{10\pi}{6}\right)\right]$

$= 2048\left(\cos\dfrac{5\pi}{3} + i\sin\dfrac{5\pi}{3}\right)$

$= 2048\left(\dfrac{1}{2} - \dfrac{\sqrt{3}}{2}i\right)$

$= 1024 - 1024\sqrt{3}i$

25. $(3 - 2i)^5 \approx \left[3.6056\left[\cos(-0.588) + i\sin(-0.588)\right]\right]^5$

$\approx (3.6056)^5\left[\cos(-2.94) + i\sin(-2.94)\right]$

$\approx -597 - 122i$

27. $(\sqrt{5} - 4i)^3 \approx \left[\sqrt{21}\left(\cos(-1.06106) + i\sin(-1.06106)\right)\right]^3$

$\approx (\sqrt{21})^3\left(\cos\left[(3)(-1.06106)\right] + i\sin\left[(3)(-1.06106)\right]\right)$

$\approx -43\sqrt{5} + 4i$

29. $z = \dfrac{\sqrt{2}}{2}(1 + i) = \cos 45° + i\sin 45°$

$z^2 = \cos 90° + i\sin 90° = i$

$z^3 = \cos 135° + i\sin 135° = \dfrac{\sqrt{2}}{2}(-1 + i)$

$z^4 = \cos 180° + i\sin 180° = -1$

The absolute value of each is 1, and consecutive powers of z are each 45° apart.

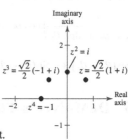

31. $2i = 2\left(\cos\dfrac{\pi}{2} + i\sin\dfrac{\pi}{2}\right)$

Square roots:

$\sqrt{2}\left(\cos\dfrac{\pi}{4} + i\sin\dfrac{\pi}{4}\right) = 1 + i$

$\sqrt{2}\left(\cos\dfrac{5\pi}{4} + i\sin\dfrac{5\pi}{4}\right) = -1 - i$

33. $-3i = 3\left(\cos\dfrac{3\pi}{2} + i\sin\dfrac{3\pi}{2}\right)$

Square roots:

$\sqrt{3}\left(\cos\dfrac{3\pi}{4} + i\sin\dfrac{3\pi}{4}\right) = -\dfrac{\sqrt{6}}{2} + \dfrac{\sqrt{6}}{2}i$

$\sqrt{3}\left(\cos\dfrac{7\pi}{4} + i\sin\dfrac{7\pi}{4}\right) = \dfrac{\sqrt{6}}{2} - \dfrac{\sqrt{6}}{2}i$

35. $2 - 2i = 2\sqrt{2}\left(\cos\dfrac{7\pi}{4} + i\sin\dfrac{7\pi}{4}\right)$

Square roots:

$8^{1/4}\left(\cos\dfrac{7\pi}{8} + i\sin\dfrac{7\pi}{8}\right) \approx -1.554 + 0.644i$

$8^{1/4}\left(\cos\dfrac{15\pi}{8} + i\sin\dfrac{15\pi}{8}\right) \approx 1.554 - 0.644i$

37. $1 + \sqrt{3}i = 2\left(\cos\dfrac{\pi}{3} + i\sin\dfrac{\pi}{3}\right)$

Square roots:

$\sqrt{2}\left(\cos\dfrac{\pi}{6} + i\sin\dfrac{\pi}{6}\right) = \dfrac{\sqrt{6}}{2} + \dfrac{\sqrt{2}}{2}i$

$\sqrt{2}\left(\cos\dfrac{7\pi}{6} + i\sin\dfrac{7\pi}{6}\right) = -\dfrac{\sqrt{6}}{2} - \dfrac{\sqrt{2}}{2}i$

39. (a) Square roots of $5(\cos 120° + i\sin 120°)$:

$\sqrt{5}\left[\cos\left(\dfrac{120° + 360°k}{2}\right) + i\sin\left(\dfrac{120° + 360°k}{2}\right)\right]$, $k = 0, 1$

$k = 0$: $\sqrt{5}(\cos 60° + i\sin 60°)$

$k = 1$: $\sqrt{5}(\cos 240° + i\sin 240°)$

(b) $\dfrac{\sqrt{5}}{2} + \dfrac{\sqrt{15}}{2}i, -\dfrac{\sqrt{5}}{2} - \dfrac{\sqrt{15}}{2}i$

(c)

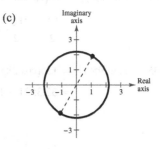

41. (a) Cube roots of $8\left(\cos \dfrac{2\pi}{3} + i \sin \dfrac{2\pi}{3}\right)$:

(c)

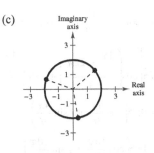

$$\sqrt[3]{8}\left[\cos\left(\frac{(2\pi/3) + 2\pi k}{3}\right) + i \sin\left(\frac{(2\pi/3) + 2\pi k}{3}\right)\right], \, k = 0, 1, 2$$

$k = 0: \, 2\left(\cos \dfrac{2\pi}{9} + i \sin \dfrac{2\pi}{9}\right)$

$k = 1: \, 2\left(\cos \dfrac{8\pi}{9} + i \sin \dfrac{8\pi}{9}\right)$

$k = 2: \, 2\left(\cos \dfrac{14\pi}{9} + i \sin \dfrac{14\pi}{9}\right)$

(b) $1.5321 + 1.2856i, -1.8794 + 0.6840i, 0.3473 - 1.9696i$

43. (a) Fifth roots of $243\left(\cos \dfrac{\pi}{6} + i \sin \dfrac{\pi}{6}\right)$:

(c)

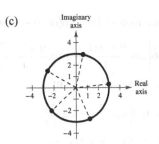

$$\sqrt[5]{243}\left[\cos\left(\frac{\frac{\pi}{6} + 2\pi k}{5}\right) + i \sin\left(\frac{\frac{\pi}{6} + 2\pi k}{5}\right)\right], \, k = 0, 1, 2, 3, 4$$

$k = 0: \, 3\left(\cos \dfrac{\pi}{30} + i \sin \dfrac{\pi}{30}\right)$

$k = 1: \, 3\left(\cos \dfrac{13\pi}{30} + i \sin \dfrac{13\pi}{30}\right)$

$k = 2: \, 3\left(\cos \dfrac{5\pi}{6} + i \sin \dfrac{5\pi}{6}\right)$

$k = 3: \, 3\left(\cos \dfrac{37\pi}{30} + i \sin \dfrac{37\pi}{30}\right)$

$k = 4: \, 3\left(\cos \dfrac{49\pi}{30} + i \sin \dfrac{49\pi}{30}\right)$

(b) $2.9836 + 0.3136i, 0.6237 + 2.9344i, -2.5981 + 1.5i, -2.2294 - 2.0074i, 1.2202 - 2.7406i$

45. (a) Fourth roots of $81i = 81\left(\cos \dfrac{\pi}{2} + i \sin \dfrac{\pi}{2}\right)$:

(c)

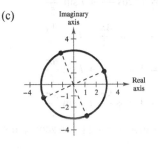

$$\sqrt[4]{81}\left[\cos\left(\frac{\frac{\pi}{2} + 2\pi k}{4}\right) + i \sin\left(\frac{\frac{\pi}{2} + 2\pi k}{4}\right)\right], \, k = 0, 1, 2, 3$$

$k = 0: \, 3\left(\cos \dfrac{\pi}{8} + i \sin \dfrac{\pi}{8}\right)$

$k = 1: \, 3\left(\cos \dfrac{5\pi}{8} + i \sin \dfrac{5\pi}{8}\right)$

$k = 2: \, 3\left(\cos \dfrac{9\pi}{8} + i \sin \dfrac{9\pi}{8}\right)$

$k = 3: \, 3\left(\cos \dfrac{13\pi}{8} + i \sin \dfrac{13\pi}{8}\right)$

(b) $2.7716 + 1.1481i, -1.1481 + 2.7716i, -2.7716 - 1.1481i, 1.1481 - 2.7716i$

47. (a) Cube roots of $-\dfrac{125}{2}\left(1 + \sqrt{3}i\right) = 125\left(\cos\dfrac{4\pi}{3} + i\sin\dfrac{4\pi}{3}\right)$:

$$\sqrt[3]{125}\left[\cos\left(\dfrac{\dfrac{4\pi}{3} + 2k\pi}{3}\right) + i\sin\left(\dfrac{\dfrac{4\pi}{3} + 2k\pi}{3}\right)\right], \quad k = 0, 1, 2$$

$k = 0$: $5\left(\cos\dfrac{4\pi}{9} + i\sin\dfrac{4\pi}{9}\right)$

$k = 1$: $5\left(\cos\dfrac{10\pi}{9} + i\sin\dfrac{10\pi}{9}\right)$

$k = 2$: $5\left(\cos\dfrac{16\pi}{9} + i\sin\dfrac{16\pi}{9}\right)$

(b) $0.8682 + 4.9240i, -4.6985 - 1.7101i, 3.8302 - 3.2140i$

(c)

49. (a) Fourth roots of $16 = 16(\cos 0 + i\sin 0)$:

$$\sqrt[4]{16}\left[\cos\dfrac{0 + 2\pi k}{4} + i\sin\dfrac{0 + 2\pi k}{4}\right], \quad k = 0, 1, 2, 3$$

$k = 0$: $2(\cos 0 + i\sin 0)$

$k = 1$: $2\left(\cos\dfrac{\pi}{2} + i\sin\dfrac{\pi}{2}\right)$

$k = 2$: $2(\cos \pi + i\sin \pi)$

$k = 3$: $2\left(\cos\dfrac{3\pi}{2} + i\sin\dfrac{3\pi}{2}\right)$

(b) $2, 2i, -2, -2i$

(c)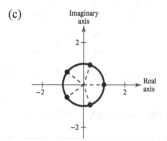

51. (a) Fifth roots of $1 = \cos 0 + i\sin 0$:

$$\cos\left(\dfrac{2k\pi}{5}\right) + i\sin\left(\dfrac{2k\pi}{5}\right), \quad k = 0, 1, 2, 3, 4$$

$k = 0$: $\cos 0 + i\sin 0$

$k = 1$: $\cos\dfrac{2\pi}{5} + i\sin\dfrac{2\pi}{5}$

$k = 2$: $\cos\dfrac{4\pi}{5} + i\sin\dfrac{4\pi}{5}$

$k = 3$: $\cos\dfrac{6\pi}{5} + i\sin\dfrac{6\pi}{5}$

$k = 4$: $\cos\dfrac{8\pi}{5} + i\sin\dfrac{8\pi}{5}$

(b) $1, 0.3090 + 0.9511i, -0.8090 + 0.5878i, -0.8090 - 0.5878i, 0.3090 - 0.9511i$

(c)

53. (a) Cube roots of $-125 = 125(\cos \pi + i \sin \pi)$:

(c)
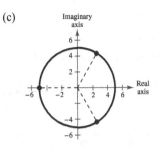

$$\sqrt[3]{125}\left[\cos\left(\frac{\pi + 2\pi k}{3}\right) + i \sin\left(\frac{\pi + 2\pi k}{3}\right)\right], \, k = 0, 1, 2$$

$k = 0$: $5\left(\cos\dfrac{\pi}{3} + i \sin\dfrac{\pi}{3}\right)$

$k = 1$: $5(\cos \pi + i \sin \pi)$

$k = 2$: $5\left(\cos\dfrac{5\pi}{3} + i \sin\dfrac{5\pi}{3}\right)$

(b) $\dfrac{5}{2} + \dfrac{5\sqrt{3}}{2}i, \, -5, \, \dfrac{5}{2} - \dfrac{5\sqrt{3}}{2}i$

55. (a) Fifth roots of $4(1 - i) = 4\sqrt{2}\left(\cos\dfrac{7\pi}{4} + i \sin\dfrac{7\pi}{4}\right)$:

(c)
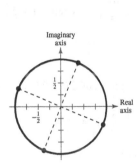

$$\sqrt[5]{4\sqrt{2}}\left[\cos\left(\frac{\frac{7\pi}{4} + 2\pi k}{5}\right) + i \sin\left(\frac{\frac{7\pi}{4} + 2\pi k}{5}\right)\right], \, k = 0, 1, 2, 3, 4$$

$k = 0$: $\sqrt{2}\left(\cos\dfrac{7\pi}{20} + i \sin\dfrac{7\pi}{20}\right)$

$k = 1$: $\sqrt{2}\left(\cos\dfrac{3\pi}{4} + i \sin\dfrac{3\pi}{4}\right)$

$k = 2$: $\sqrt{2}\left(\cos\dfrac{23\pi}{20} + i \sin\dfrac{23\pi}{20}\right)$

$k = 3$: $\sqrt{2}\left(\cos\dfrac{31\pi}{20} + i \sin\dfrac{31\pi}{20}\right)$

$k = 4$: $\sqrt{2}\left(\cos\dfrac{39\pi}{20} + i \sin\dfrac{39\pi}{20}\right)$

(b) $0.6420 + 1.2601i, \, -1 + 1i, \, -1.2601 - 0.6420i, \, 0.2212 - 1.3968i, \, 1.3968 - 0.2212i$

57. $x^4 + i = 0$

$x^4 = -i$

The solutions are the fourth roots of $i = \cos\dfrac{3\pi}{2} + i \sin\dfrac{3\pi}{2}$:

$$\sqrt[4]{1}\left[\cos\left(\frac{\frac{3\pi}{2} + 2k\pi}{4}\right) + i \sin\left(\frac{\frac{3\pi}{2} + 2k\pi}{4}\right)\right], \, k = 0, 1, 2, 3$$

$k = 0$: $\cos\dfrac{3\pi}{8} + i \sin\dfrac{3\pi}{8} \approx 0.3827 + 0.9239i$

$k = 1$: $\cos\dfrac{7\pi}{8} + i \sin\dfrac{7\pi}{8} \approx -0.9239 + 0.3827i$

$k = 2$: $\cos\dfrac{11\pi}{8} + i \sin\dfrac{11\pi}{8} \approx -0.3827 - 0.9239i$

$k = 3$: $\cos\dfrac{15\pi}{8} + i \sin\dfrac{15\pi}{8} \approx 0.9239 - 0.3827i$

59. $x^6 + 1 = 0$

$\qquad x^6 = -1$

The solutions are the sixth roots of $-1 = \cos \pi + i \sin \pi$:

$$\sqrt[6]{1}\left[\cos\left(\frac{\pi + 2\pi k}{6}\right) + i \sin\left(\frac{\pi + 2\pi k}{6}\right)\right], \, k = 0, 1, 2, 3, 4, 5$$

$k = 0: \cos\dfrac{\pi}{6} + i \sin\dfrac{\pi}{6} = \dfrac{\sqrt{3}}{2} + \dfrac{1}{2}i$

$k = 1: \cos\dfrac{\pi}{2} + i \sin\dfrac{\pi}{2} = i$

$k = 2: \cos\dfrac{5\pi}{6} + i \sin\dfrac{5\pi}{6} = -\dfrac{\sqrt{3}}{2} + \dfrac{1}{2}i$

$k = 3: \cos\dfrac{7\pi}{6} + i \sin\dfrac{7\pi}{6} = -\dfrac{\sqrt{3}}{2} - \dfrac{1}{2}i$

$k = 4: \cos\dfrac{3\pi}{2} + i \sin\dfrac{3\pi}{2} = -i$

$k = 5: \cos\dfrac{11\pi}{6} + i \sin\dfrac{11\pi}{6} = \dfrac{\sqrt{3}}{2} - \dfrac{1}{2}i$

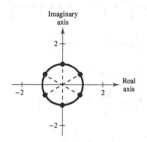

61. $x^5 + 32 = 0$

$\qquad x^5 = -32$

The solutions are the fifth roots of $-32 = 32(\cos \pi + i \sin \pi)$:

$$\sqrt[5]{32}\left[\cos\left(\frac{\pi + 2k\pi}{5}\right) + i \sin\frac{\pi + 2k\pi}{5}\right], \, k = 0, 1, 2, 3, 4$$

$k = 0: 2\left(\cos\dfrac{\pi}{5} + i \sin\dfrac{\pi}{5}\right) \approx 1.6180 + 1.1756i$

$k = 1: 2\left(\cos\dfrac{3\pi}{5} + i \sin\dfrac{3\pi}{5}\right) \approx -0.6180 + 1.902i$

$k = 2: 2(\cos \pi + i \sin \pi) = -2$

$k = 3: 2\left(\cos\dfrac{7\pi}{5} + i \sin\dfrac{7\pi}{5}\right) \approx -0.6180 - 1.9021i$

$k = 4: 2\left(\cos\dfrac{9\pi}{5} + i \sin\dfrac{9\pi}{5}\right) \approx 1.6180 - 1.1756i$

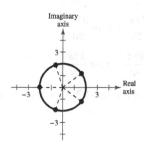

63. $x^3 - 27 = 0$

$\qquad x^3 = 27$

The solutions are the cube roots of $27 = 27(\cos 0 + i \sin 0)$:

$$\sqrt[3]{27}\left[\cos\left(\frac{2k\pi}{3}\right) + i \sin\left(\frac{2k\pi}{3}\right)\right], \, k = 0, 1, 2$$

$k = 0: 3(\cos 0 + i \sin 0) = 3$

$k = 1: 3\left(\cos\dfrac{2\pi}{3} + i \sin\dfrac{2\pi}{3}\right) = -\dfrac{3}{2} + \dfrac{3\sqrt{3}}{2}i$

$k = 2: 3\left(\cos\dfrac{4\pi}{3} + i \sin\dfrac{4\pi}{3}\right) = -\dfrac{3}{2} - \dfrac{3\sqrt{3}}{2}i$

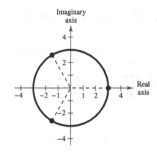

65. $x^4 + 16i = 0$

$$x^4 = -16i$$

The solutions are the fourth roots of $-16i = 16\left(\cos\dfrac{3\pi}{2} + i \sin\dfrac{3\pi}{2}\right)$:

$$\sqrt[4]{16}\left[\cos\dfrac{\dfrac{3\pi}{2} + 2\pi k}{4} + i \sin\dfrac{\dfrac{3\pi}{2} + 2\pi k}{4}\right],\ k = 0, 1, 2, 3$$

$k = 0:\ 2\left(\cos\dfrac{3\pi}{8} + i \sin\dfrac{3\pi}{8}\right) \approx 0.7654 + 1.8478i$

$k = 1:\ 2\left(\cos\dfrac{7\pi}{8} + i \sin\dfrac{7\pi}{8}\right) \approx -1.8478 + 0.7654i$

$k = 2:\ 2\left(\cos\dfrac{11\pi}{8} + i \sin\dfrac{11\pi}{8}\right) \approx -0.7654 - 1.8478i$

$k = 3:\ 2\left(\cos\dfrac{15\pi}{8} + i \sin\dfrac{15\pi}{8}\right) \approx 1.8478 - 0.7654i$

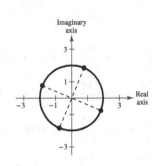

67. $x^4 - 16i = 0$

$$x^4 = 16i$$

The solutions are the fourth roots of $16i = 16\left(\cos\dfrac{\pi}{2} + i \sin\dfrac{\pi}{2}\right)$:

$$\sqrt[4]{16}\left[\cos\left(\dfrac{\dfrac{\pi}{2} + 2\pi k}{4}\right) + i \sin\left(\dfrac{\dfrac{\pi}{2} + 2\pi k}{4}\right)\right],\ k = 0, 1, 2, 3$$

$k = 0:\ 2\left(\cos\dfrac{\pi}{8} + i \sin\dfrac{\pi}{8}\right) \approx 1.8478 + 0.7654i$

$k = 1:\ 2\left(\cos\dfrac{5\pi}{8} + i \sin\dfrac{5\pi}{8}\right) \approx -0.7654 + 1.8478i$

$k = 2:\ 2\left(\cos\dfrac{9\pi}{8} + i \sin\dfrac{9\pi}{8}\right) \approx -1.8478 - 0.7654i$

$k = 3:\ 2\left(\cos\dfrac{13\pi}{8} + i \sin\dfrac{13\pi}{8}\right) \approx 0.7654 - 1.8478i$

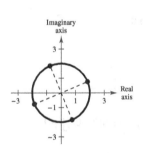

69. $x^3 - (1 - i) = 0$

$$x^3 = 1 - i = \sqrt{2}\left(\cos\dfrac{7\pi}{4} + i \sin\dfrac{7\pi}{4}\right)$$

The solutions are the cube roots of $1 - i$:

$$\sqrt[3]{\sqrt{2}}\left[\cos\left(\dfrac{(7\pi/4) + 2\pi k}{3}\right) + i \sin\left(\dfrac{(7\pi/4) + 2\pi k}{3}\right)\right],\ k = 0, 1, 2$$

$k = 0:\ \sqrt[6]{2}\left(\cos\dfrac{7\pi}{12} + i \sin\dfrac{7\pi}{12}\right) \approx -0.2905 + 1.0842i$

$k = 1:\ \sqrt[6]{2}\left(\cos\dfrac{5\pi}{4} + i \sin\dfrac{5\pi}{4}\right) \approx -0.7937 - 0.7937i$

$k = 2:\ \sqrt[6]{2}\left(\cos\dfrac{23\pi}{12} + i \sin\dfrac{23\pi}{12}\right) \approx 1.0842 - 0.2905i$

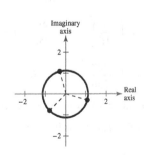

71. $x^6 + (1 + i) = 0$

$$x^6 = -(1 + i) = -1 - i$$

The solutions are the sixth roots of $-1 - i = \sqrt{2}\left(\cos\dfrac{5\pi}{4} + i \sin\dfrac{5\pi}{4}\right)$:

$$\sqrt[6]{\sqrt{2}}\left[\cos\left(\dfrac{\dfrac{5\pi}{4} + 2\pi k}{6}\right) + i \sin\left(\dfrac{\dfrac{5\pi}{4} + 2\pi k}{6}\right)\right], \ k = 0, 1, 2, 3, 4, 5$$

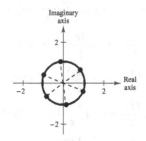

$k = 0$: $\sqrt[12]{2}\left(\cos\dfrac{5\pi}{24} + i \sin\dfrac{5\pi}{24}\right) \approx 0.8405 + 0.6450i$

$k = 1$: $\sqrt[12]{2}\left(\cos\dfrac{13\pi}{24} + i \sin\dfrac{13\pi}{24}\right) \approx -0.1383 + 1.0504i$

$k = 2$: $\sqrt[12]{2}\left(\cos\dfrac{7\pi}{8} + i \sin\dfrac{7\pi}{8}\right) \approx -0.9788 + 0.4054i$

$k = 3$: $\sqrt[12]{2}\left(\cos\dfrac{29\pi}{24} + i \sin\dfrac{29\pi}{24}\right) \approx -0.8405 - 0.6450i$

$k = 4$: $\sqrt[12]{2}\left(\cos\dfrac{37\pi}{24} + i \sin\dfrac{37\pi}{24}\right) \approx 0.1383 - 1.0504i$

$k = 5$: $\sqrt[12]{2}\left(\cos\dfrac{15\pi}{8} + i \sin\dfrac{15\pi}{8}\right) \approx 0.9788 - 0.4054i$

73. (a) $z_0 = \dfrac{1}{2}(\cos 0° + i \sin 0°)$

$z_1 = f(z_0)$

$\quad = \left[\dfrac{1}{2}(\cos 0° + i \sin 0°)\right]^2 - 1$

$\quad = \left(\dfrac{1}{2}\right)^2\left[\cos(2 \cdot 0°) + i \sin(2 \cdot 0°)\right] - 1$

$\quad = \dfrac{1}{4}(\cos 0° + i \sin 0°) - 1$

$\quad = \dfrac{1}{4} - 1$

$\quad = -\dfrac{3}{4}$

$z_2 = f(z_1) = \left(-\dfrac{3}{4}\right)^2 - 1 = \dfrac{9}{16} - 1 = -\dfrac{7}{16}$

$z_3 = f(z_2) = \left(-\dfrac{7}{16}\right)^2 - 1 = \dfrac{49}{256} - 1 = -\dfrac{207}{256}$

$z_4 = f(z_3) = \left(-\dfrac{207}{256}\right)^2 - 1 \approx -0.35$

The absolute values of the terms of the sequence are all less than 1. So, the sequence is bounded and $z_0 = \dfrac{1}{2}(\cos 0° + i \sin 0°)$ is in the prisoner set of $f(z) = z^2 - 1$.

(b) $z_0 = \sqrt{2}(\cos 30° + i \sin 30°)$

$z_1 = f(z_0)$

$= \left[\sqrt{2}(\cos 30° + i \sin 30°)\right]^2 - 1$

$= \left(\sqrt{2}\right)^2\left[\cos(2 \cdot 30°) + i \sin(2 \cdot 30°)\right] - 1$

$= 2(\cos 60° + i \sin 60°) - 1$

$= 2\left(\frac{1}{2} + \frac{\sqrt{3}}{2}i\right) - 1$

$= 1 + \frac{\sqrt{3}}{2}i - 1$

$= \frac{\sqrt{3}}{2}i$

$z_2 = f(z_1) = \left(\frac{\sqrt{3}}{2}i\right)^2 - 1 = -\frac{3}{4} - 1 = -\frac{7}{4}$

$z_3 = f(z_2) = \left(-\frac{7}{4}\right)^2 - 1 = \frac{49}{16} - 1 = \frac{33}{16}$

$z_4 = f(z_3) = \left(\frac{33}{16}\right)^2 - 1 \approx 3.25$

The absolute values of the terms are increasing. So, the sequence is unbounded and $z_0 = \sqrt{2}(\cos 30° + i \sin 30°)$ is in the escape set of $f(z) = z^2 - 1$.

(c) $z_0 = \sqrt[4]{2}\left(\cos \frac{\pi}{8} + i \sin \frac{\pi}{8}\right)$

$z_1 = f(z_0)$

$= \left[\sqrt[4]{2}\left(\cos \frac{\pi}{8} + i \sin \frac{\pi}{8}\right)\right]^2 - 1$

$= \left(\sqrt[4]{2}\right)^2\left[\cos\left(2 \cdot \frac{\pi}{8}\right) + i \sin\left(2 \cdot \frac{\pi}{8}\right)\right] - 1$

$= \sqrt{2}\left(\cos \frac{\pi}{4} + i \sin \frac{\pi}{4}\right) - 1$

$= \sqrt{2}\left(\frac{\sqrt{2}}{2} + \frac{\sqrt{2}}{2}i\right) - 1$

$= 1 + i - 1$

$= i$

$z_2 = f(z_1) = (i)^2 - 1 = -1 - 1 = -2$

$z_3 = f(z_2) = (-2)^2 - 1 = 4 - 1 = 3$

$z_4 = f(z_3) = (3)^2 - 1 = 9 - 1 = 8$

The absolute values of the terms are increasing. So, the sequence is unbounded and $z_0 = \sqrt[4]{2}\left(\cos \frac{\pi}{8} + i \sin \frac{\pi}{8}\right)$ is in the escape set of $f(z) = z^2 - 1$.

(d) $z_0 = \sqrt{2}(\cos \pi + i \sin \pi)$

$z_1 = f(z_0)$

$\quad = \left[\sqrt{2}(\cos \pi + i \sin \pi)\right]^2 - 1$

$\quad = \left(\sqrt{2}\right)^2 \left[\cos(2 \cdot \pi) + i \sin(2 \cdot \pi)\right] - 1$

$\quad = 2(\cos 2\pi + i \sin 2\pi) - 1$

$\quad = 2 - 1$

$\quad = 1$

$z_2 = f(z_1) = (1)^2 - 1 = 0$

$z_3 = f(z_2) = (0)^2 - 1 = -1$

$z_4 = f(z_3) = (-1)^2 - 1 = 0$

The absolute values of the terms alternate between 0 and 1. So, the sequence is bounded and $z_0 = \sqrt{2}(\cos \pi + i \sin \pi)$ is in the prisoner set of $f(z) = z^2 - 1$.

75. False. They are equally spaced along the circle centered at the origin with radius $\sqrt[n]{r}$.

77. True. $(i)^3 = -i$ and $(-i)^3 = i$

79. $\dfrac{1}{2}(1 - \sqrt{3}i) = \cos \dfrac{5\pi}{3} + i \sin \dfrac{5\pi}{3}$

$\left[\dfrac{1}{2}(1 - \sqrt{3}i)\right]^9 = \left(\cos \dfrac{5\pi}{3} + i \sin \dfrac{5\pi}{3}\right)^9$

$\quad = \cos 15\pi + i \sin 15\pi$

$\quad = -1$

81. (a) $x^2 - 4x + 8 = 0;\ x = 2\sqrt{2}(\cos 45° + i \sin 45°)$

$\quad = 2\sqrt{2}\left(\dfrac{\sqrt{2}}{2} + \dfrac{\sqrt{2}}{2}i\right)$

$\quad = 2 + 2i$

$(2 + 2i)^2 - 4(2 + 2i) + 8 \overset{?}{=} 0$

$(4 + 8i - 4) - 8 - 8i + 8 \overset{?}{=} 0$

$\quad\quad\quad\quad 0 = 0\ \checkmark$

(b)

$$2 + 2i \ \Big|\ \begin{array}{ccc} 1 & -4 & 8 \\ & 2 + 2i & -8 \\ \hline 1 & -2 + 2i & 0 \end{array}$$

$x - 2 + 2i = 0$

$x = 2 - 2i$

The other solution is

$x = 2 - 2i = 2\sqrt{2}(\cos 315° + i \sin 315°).$

(c) Used synthetic division with the given solution to find the other solution.

(d) $x = 2 - 2i$

$(2 - 2i)^2 - 4(2 - 2i) + 8 \overset{?}{=} 0$

$(4 - 8i - 4) - 8 + 8i + 8 \overset{?}{=} 0$

$\quad\quad\quad\quad 0 = 0$

83. (a) $x^2 + ix + 2 = 0$

$$x = \frac{-i \pm \sqrt{i^2 - 4(1)(2)}}{2(1)}$$

$$= \frac{-i \pm \sqrt{-9}}{2}$$

$$= \frac{-i \pm 3i}{2}$$

$$= -2i, i$$

(b) $x^2 + 2ix + 1 = 0$

$$x = \frac{-2i \pm \sqrt{(2i)^2 - 4(1)(1)}}{2(1)}$$

$$= \frac{-2i \pm \sqrt{-8}}{2}$$

$$= \frac{-2i \pm 2\sqrt{2}i}{2}$$

$$= -i \pm \sqrt{2}i$$

$$= \left(-1 \pm \sqrt{2}\right)i$$

(c) $x^2 + 2ix + \sqrt{3}i = 0$

$$x = \frac{-2i \pm \sqrt{(2i)^2 - 4(1)\left(\sqrt{3}i\right)}}{2(1)}$$

$$= \frac{-2i \pm \sqrt{-4 - 4\sqrt{3}i}}{2}$$

$$= \frac{-2i \pm \left(\sqrt{2} - \sqrt{6}i\right)}{2}$$

$$= -\frac{\sqrt{2}}{2} + \left(-1 + \frac{\sqrt{6}}{2}\right)i, \frac{\sqrt{2}}{2} - \left(1 + \frac{\sqrt{6}}{2}\right)i$$

Note: The square roots of $-4 - 4\sqrt{3}i$ are $\sqrt{2} - \sqrt{6}i$ and $-\sqrt{2} + \sqrt{6}i$, by the formula on page 362. Both of these roots yield the same two solutions to the equation.

85. $\left[\sqrt{2}(\cos 7.5° + i \sin 7.5°)\right]^4 = 4(\cos 30° + i \sin 30°) = 4\left(\frac{\sqrt{3}}{2} + \frac{1}{2}i\right) = 2\sqrt{3} + 2i$

Fourth roots of $2\sqrt{3} + 2i = 4(\cos 30° + i \sin 30°)$:

$$\sqrt[4]{4}\left[\cos\left(\frac{30° + 360°k}{4}\right) + i \sin\left(\frac{30° + 360°k}{4}\right)\right], k = 0, 1, 2, 3$$

$k = 0$: $\sqrt{2}(\cos 7.5° + i \sin 7.5°)$

$k = 1$: $\sqrt{2}(\cos 97.5° + i \sin 97.5°)$

$k = 2$: $\sqrt{2}(\cos 187.5° + i \sin 187.5°)$

$k = 3$: $\sqrt{2}(\cos 277.5° + i \sin 277.5°)$

Review Exercises for Chapter 4

1. $6 + \sqrt{-4} = 6 + 2i$

3. $i^2 + 3i = -1 + 3i$

5. $(6 - 4i) + (-9 + i) = \left(6 + (-9)\right) + (-4i + i) = -3 - 3i$

7. $-3i(-2 + 5i) = 6i - 15i^2$

$$= 6i - 15(-1)$$

$$= 15 + 6i$$

9. $(1 + 7i)(1 - 7i) = 1 - 49i^2$

$$= 1 - 49(-1)$$

$$= 1 + 49$$

$$= 50$$

11. $\dfrac{4}{1 - 2i} = \dfrac{4}{1 - 2i} \cdot \dfrac{1 + 2i}{1 + 2i}$

$$= \frac{4 + 8i}{1 - 4i^2}$$

$$= \frac{4 + 8i}{5}$$

$$= \frac{4}{5} + \frac{8}{5}i$$

13. $\dfrac{3 + 2i}{5 + i} = \dfrac{3 + 2i}{5 + i} \cdot \dfrac{5 - i}{5 - i}$

$\qquad = \dfrac{15 - 3i + 10i - 2i^2}{25 - i^2}$

$\qquad = \dfrac{17 + 7i}{26}$

$\qquad = \dfrac{17}{26} + \dfrac{7i}{26}$

15. $\dfrac{4}{2 - 3i} + \dfrac{2}{1 + i} = \dfrac{4}{2 - 3i} \cdot \dfrac{2 + 3i}{2 + 3i} + \dfrac{2}{1 + i} \cdot \dfrac{1 - i}{1 - i}$

$\qquad = \dfrac{8 + 12i}{4 + 9} + \dfrac{2 - 2i}{1 + 1}$

$\qquad = \dfrac{8}{13} + \dfrac{12}{13}i + 1 - i$

$\qquad = \left(\dfrac{8}{13} + 1\right) + \left(\dfrac{12}{13}i - i\right)$

$\qquad = \dfrac{21}{13} - \dfrac{1}{13}i$

21. $10i^2 - i^3 = 10i^2 - i^2i = 10(-1) - (-1)i = -10 + i$

23. $\dfrac{1}{i^7} = \dfrac{1}{i^2i^2i^2i} = \dfrac{1}{(-1)(-1)(-1)i} = -\dfrac{1}{i} \cdot \dfrac{i}{i} = \dfrac{-i}{i^2} = \dfrac{-i}{-i} = i$

25. $-2x^6 + 7x^3 + x^2 + 4x - 19 = 0$

Six solutions

27. $6x^2 + x - 2 = 0$

$b^2 - 4ac = 1^2 - 4(6)(-2) = 49 > 0$

Two real solutions

29. $0.13x^2 - 0.45x + 0.65 = 0$

$b^2 - 4ac = (-0.45)^2 - 4(0.13)(0.65) = -0.1355 < 0$

Two imaginary solutions

31. $\quad x^2 - 2x = 0$

$\quad x(x - 2) = 0$

$\quad x = 0, x = 2$

33. $x^2 - 3x + 5 = 0$

$\quad x = \dfrac{-(-3) \pm \sqrt{(-3)^2 - 4(1)(5)}}{2(1)}$

$\qquad = \dfrac{3 \pm \sqrt{-11}}{2}$

$\qquad = \dfrac{3}{2} \pm \dfrac{\sqrt{11}}{2}i$

17. $x^2 - 2x + 10 = 0$

$\quad x^2 - 2x + 1 = -10 + 1$

$\quad (x - 1)^2 = -9$

$\quad x - 1 = \pm\sqrt{-9}$

$\quad x = 1 \pm 3i$

19. $4x^2 + 4x + 7 = 0$

$\quad x = \dfrac{-b \pm \sqrt{b^2 - 4ac}}{2a}$

$\qquad = \dfrac{-4 \pm \sqrt{(4)^2 - 4(4)(7)}}{2(4)}$

$\qquad = \dfrac{-4 \pm \sqrt{-96}}{8}$

$\qquad = \dfrac{-4 \pm 4\sqrt{6}i}{8}$

$\qquad = -\dfrac{1}{2} \pm \dfrac{\sqrt{6}}{2}i$

35. $2x^2 + 3x + 6 = 0$

$\quad x = \dfrac{-3 \pm \sqrt{(3)^2 - 4(2)(6)}}{2(2)}$

$\qquad = \dfrac{-3 \pm \sqrt{-39}}{4} = -\dfrac{3}{4} \pm \dfrac{\sqrt{39}}{4}i$

37. $\quad x^4 + 8x^2 + 7 = 0$

$\quad (x^2 + 7)(x^2 + 1) = 0$

$\quad x^2 + 7 = 0 \Rightarrow x = \pm\sqrt{7}\,i$

$\quad x^2 + 1 = 0 \Rightarrow x = \pm i$

Zeros: $x = \pm i, \pm \sqrt{7}\,i$

39. $150 = 0.45x^2 - 1.65x + 50.75$

$\quad 0 = 0.45x^2 - 1.65x - 99.25$

$\quad x = \dfrac{1.65 \pm \sqrt{(-1.65)^2 - 4(0.45)(-99.25)}}{2(0.45)}$

$\qquad \approx -13.1, 16.8$

Because $10 \le x \le 25$, choose $16.8°C$.

$x = 10$: $0.45(10)^2 - 1.65(10) + 50.75 = 79.25$

41. $2x^2 + 2x + 3 = 0$

$$x = \frac{-2 \pm \sqrt{2^2 - 4(2)(3)}}{2(2)}$$

$$= \frac{-2 \pm \sqrt{-20}}{4}$$

$$= -\frac{2}{4} \pm \frac{2\sqrt{5}}{4}i$$

$$= -\frac{1}{2} \pm \frac{\sqrt{5}}{2}i$$

Zeros: $-\frac{1}{2} \pm \frac{\sqrt{5}}{2}i$

$$r(x) = 2\left[x - \left(-\frac{1}{2} + \frac{\sqrt{5}}{2}i\right)\right]\left[x - \left(-\frac{1}{2} - \frac{\sqrt{5}}{2}i\right)\right]$$

$$= 2\left(x + \frac{1}{2} - \frac{\sqrt{5}}{2}i\right)\left(x + \frac{1}{2} + \frac{\sqrt{5}}{2}i\right)$$

43. $2x^3 - 3x^2 + 50x - 75 = 0$

$$x^2(2x - 3) + 25(2x - 3) = 0$$

$$(x^2 + 25)(2x - 3) = 0$$

$$x^2 + 25 = 0 \Rightarrow x = \pm 5i$$

$$2x - 3 = 0 \Rightarrow x = \frac{3}{2}$$

Zeros: $\pm 5i, \frac{3}{2}$

$$f(x) = (2x - 3)(x - 5i)(x + 5i)$$

45. $4x^4 + 3x^2 - 10 = 0$

$$(4x^2 - 5)(x^2 + 2) = 0$$

$$4x^2 - 5 = 0 \Rightarrow x = \pm\frac{\sqrt{5}}{2}$$

$$x^2 + 2 = 0 \Rightarrow x = \pm\sqrt{2}i$$

Zeros: $\pm\frac{\sqrt{5}}{2}, \pm\sqrt{2}i$

$$f(x) = \left(2x - \sqrt{5}\right)\left(2x + \sqrt{5}\right)\left(x - \sqrt{2}i\right)\left(x + \sqrt{2}i\right)$$

47. $f(x) = x^3 + 3x^2 - 24x + 28$

Zeros: $2 \Rightarrow x - 2$ is a factor of $f(x)$.

$$
\begin{array}{r}
x^2 + 5x - 14 \\
x - 2 \overline{)x^3 + 3x^2 - 24x + 28} \\
\underline{x^3 - 2x^2} \\
5x^2 - 24x \\
\underline{5x^2 - 10x} \\
-14x + 28 \\
\underline{-14x + 28} \\
0
\end{array}
$$

Thus, $f(x) = (x - 2)(x^2 + 5x - 14)$

$$= (x - 2)(x - 2)(x + 7)$$

$$= (x - 2)^2(x + 7).$$

The zeros of $f(x)$ are $x = 2$ and $x = -7$.

49. $h(x) = -x^3 + 2x^2 - 16x + 32$

Because $-4i$ is a zero, so is $4i$.

$$
\begin{array}{r|rrrr}
-4i & -1 & 2 & -16 & 32 \\
& & 4i & 16 - 8i & -32 \\
\hline
& -1 & 2 + 4i & -8i & 0
\end{array}
$$

$$
\begin{array}{r|rrr}
4i & -1 & 2 + 4i & -8i \\
& & -4i & 8i \\
\hline
& -1 & 2 & 0
\end{array}
$$

$$h(x) = (x + 4i)(x - 4i)(-x + 2)$$

Zeros: $x = \pm 4i, 2$

51. $g(x) = 2x^4 - 3x^3 - 13x^2 + 37x - 15$, Zero: $2 + i$

Because $2 + i$ is a zero, so is $2 - i$.

$$
\begin{array}{r|rrrrr}
2 + i & 2 & -3 & -13 & 37 & -15 \\
& & 4 + 2i & 5i & -31 - 3i & 15 \\
\hline
& 2 & 1 + 2i & -13 + 5i & 6 - 3i & 0
\end{array}
$$

$$
\begin{array}{r|rrrr}
2 - i & 2 & 1 + 2i & -13 + 5i & 6 - 3i \\
& & 4 - 2i & 10 - 5i & -6 + 3i \\
\hline
& 2 & 5 & -3 & 0
\end{array}
$$

$$g(x) = \left[x - (2 + i)\right]\left[x - (2 - i)\right](2x^2 + 5x - 3)$$

$$= (x - 2 - i)(x - 2 + i)(2x - 1)(x + 3)$$

Zeros: $x = 2 \pm i, \frac{1}{2}, -3$

53. Zeros: $\frac{2}{3}, 4, \sqrt{3}i$

Because $\sqrt{3}\,i$ is a zero, so is $-\sqrt{3}\,i$.

$$f(x) = (3x - 2)(x - 4)\left(x - \sqrt{3}i\right)\left(x + \sqrt{3}i\right)$$
$$= \left(3x^2 - 14x + 8\right)\left(x^2 + 3\right)$$
$$= 3x^4 - 14x^3 + 17x^2 - 42x + 24$$

Note: Any nonzero multiple of $f(x)$ has the given zeros.

55. Zeros: $\sqrt{2}i, -5i$

Because $\sqrt{2}\,i$ and $-5i$ are zeros, so are $-\sqrt{2}\,i$ and $5i$.

$$f(x) = \left(x + \sqrt{2}i\right)\left(x - \sqrt{2}i\right)(x + 5i)(x - 5i)$$
$$= \left(x^2 + 2\right)\left(x^2 + 25\right) = x^4 + 27x^2 + 50$$

Note: Any nonzero multiple of $f(x)$ has the given zeros..

57. Zeros: $5, 1 - i$

Because $1 - i$ is a zero, so is $1 + i$.

$$f(x) = a(x - 5)\big[x - (1 - i)\big]\big[x - (1 + i)\big] = a(x - 5)\big[(x - 1) + i\big]\big[(x - 1) - i\big]$$
$$= a(x - 5)\big[(x - 1)^2 + 1\big] = a(x - 5)\left(x^2 - 2x + 2\right) = a\left(x^3 - 7x^2 + 12x - 10\right)$$

$$f(1) = -8 \Rightarrow -4a = -8 \Rightarrow a = 2$$

$$f(x) = 2\left(x^3 - 7x^2 + 12x - 10\right) = 2x^3 - 14x^2 + 24x - 20$$

59. $|7i| = \sqrt{0^2 + 7^2} = 7$

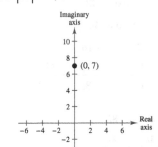

61. $|5 + 3i| = \sqrt{5^2 + 3^2}$
$$= \sqrt{34}$$

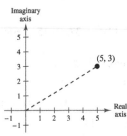

63. $(2 + 3i) + (1 - 2i) = 3 + i$

65. $(1 + 2i) - (3 + i) = -2 + i$

67. The complex conjugate of $3 + i$ is $3 - i$

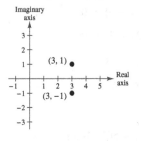

69. $d = \sqrt{(2 - 3)^2 + (-1 - 2)^2}$
$$= \sqrt{10}$$

71. Midpoint $= \left(\dfrac{1 + 4}{2}, \dfrac{1 + 3}{2}i\right)$
$$= \frac{5}{2} + 2i$$
$$= \left(\frac{5}{2}, 2\right)$$

73. $z = 4i$

$$r = \sqrt{0^2 + 4^2} = \sqrt{16} = 4$$

$$\tan \theta = \frac{4}{0}, \text{ undefined} \Rightarrow \theta = \frac{\pi}{2}$$

$$z = 4\left(\cos \frac{\pi}{2} + i \sin \frac{\pi}{2}\right)$$

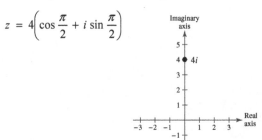

75. $z = 7 - 7i$

$r = \sqrt{(7)^2 + (-7)^2} = \sqrt{98} = 7\sqrt{2}$

$\tan\theta = \dfrac{-7}{7} = -1 \Rightarrow \theta = \dfrac{7\pi}{4}$ because the complex

number lies in Quadrant IV.

$7 - 7i = 7\sqrt{2}\left(\cos\dfrac{7\pi}{4} + i\sin\dfrac{7\pi}{4}\right)$

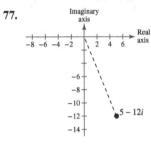

77.

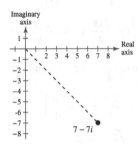

$z = 5 - 12i$

$r = \sqrt{5^2 + (-12)^2}$

$\quad = \sqrt{169} = 13$

$\tan\theta = \dfrac{-12}{5} \Rightarrow \theta \approx 5.107$

$z \approx 13(\cos 5.107 + i\sin 5.107)$

79. $2(\cos 30° + i\sin 30°) = \sqrt{3} + i$

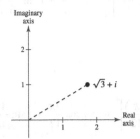

81. $\sqrt{2}\left[\cos(-45°) + i\sin(-45°)\right] = 1 - i$

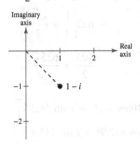

83. $2\left(\cos\dfrac{5\pi}{6} + i\sin\dfrac{5\pi}{6}\right) = -\sqrt{3} + i$

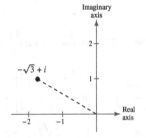

85. $\left[2\left(\cos\dfrac{\pi}{4} + i\sin\dfrac{\pi}{4}\right)\right]\left[2\left(\cos\dfrac{\pi}{3} + i\sin\dfrac{\pi}{3}\right)\right] = (2)(2)\left[\cos\left(\dfrac{\pi}{4} + \dfrac{\pi}{3}\right) + i\sin\left(\dfrac{\pi}{4} + \dfrac{\pi}{3}\right)\right]$

$= 4\left(\cos\dfrac{7\pi}{12} + i\sin\dfrac{7\pi}{12}\right)$

87. $\dfrac{2[\cos 60° + i\sin 60°]}{3[\cos 15° + i\sin 15°]} = \dfrac{2}{3}\left(\cos(60° - 15°) + i\sin(60° - 15°)\right)$

$= \dfrac{2}{3}\left(\cos 45° + i\sin 45°\right)$

89. $\left[4\left(\cos\dfrac{2\pi}{3} + i\sin\dfrac{2\pi}{3}\right)\right]\left[2\left(\cos\dfrac{5\pi}{6} + i\sin\dfrac{5\pi}{6}\right)\right] = (4)(2)\left[\cos\left(\dfrac{2\pi}{3} + \dfrac{5\pi}{6}\right) + i\sin\left(\dfrac{2\pi}{3} + \dfrac{5\pi}{6}\right)\right]$

$$= 8\left(\cos\dfrac{3\pi}{2} + i\sin\dfrac{3\pi}{2}\right)$$

$$= 8\big[0 + i(-1)\big] = -8i$$

91. $\left[5\left(\cos\dfrac{\pi}{12} + i\sin\dfrac{\pi}{12}\right)\right]^4 = 5^4\left(\cos\dfrac{4\pi}{12} + i\sin\dfrac{4\pi}{12}\right)$

$$= 625\left(\cos\dfrac{\pi}{3} + i\sin\dfrac{\pi}{3}\right)$$

$$= 625\left(\dfrac{1}{2} + \dfrac{\sqrt{3}}{2}i\right)$$

$$= \dfrac{625}{2} + \dfrac{625\sqrt{3}}{2}i$$

95. $(-1 + i)^7 = \left[\sqrt{2}\left(\cos\dfrac{3\pi}{4} + i\sin\dfrac{3\pi}{4}\right)\right]^7$

$$= \left(\sqrt{2}\right)^7\left(\cos\dfrac{21\pi}{4} + i\sin\dfrac{21\pi}{4}\right)$$

$$= 8\sqrt{2}\left[\cos\left(\dfrac{5\pi}{4} + 4\pi\right) + i\sin\left(\dfrac{5\pi}{4} + 4\pi\right)\right]$$

$$= 8\sqrt{2}\left(\cos\dfrac{5\pi}{4} + i\sin\dfrac{5\pi}{4}\right)$$

$$= 8\sqrt{2}\left(-\dfrac{\sqrt{2}}{2} - \dfrac{\sqrt{2}}{2}i\right)$$

$$= -8 - 8i$$

93. $(2 + 3i)^6 \approx \left[\sqrt{13}\left(\cos 56.3° + i\sin 56.3°\right)\right]^6$

$$= 13^3\left(\cos 337.9° + i\sin 337.9°\right)$$

$$\approx 13^3\left(0.9263 - 0.3769i\right)$$

$$\approx 2035 - 828i$$

97. (a) and (b)

Sixth roots of $-729i = 729\left(\cos\dfrac{3\pi}{2} + i\sin\dfrac{3\pi}{2}\right)$:

$$\sqrt[6]{729}\left[\cos\left(\dfrac{\dfrac{3\pi}{2} + 2k\pi}{6}\right) + i\sin\left(\dfrac{\dfrac{3\pi}{2} + 2k\pi}{6}\right)\right], k = 0, 1, 2, 3, 4, 5$$

$k = 0:\ 3\left(\cos\dfrac{\pi}{4} + i\sin\dfrac{\pi}{4}\right) = \dfrac{3\sqrt{2}}{2} + \dfrac{3\sqrt{2}}{2}i$

$k = 1:\ 3\left(\cos\dfrac{7\pi}{12} + i\sin\dfrac{7\pi}{12}\right) = -0.776 + 2.898i$

$k = 2:\ 3\left(\cos\dfrac{11\pi}{12} + i\sin\dfrac{11\pi}{12}\right) = -2.898 + 0.776i$

$k = 3:\ 3\left(\cos\dfrac{5\pi}{4} + i\sin\dfrac{5\pi}{4}\right) = \dfrac{-3\sqrt{2}}{2} - \dfrac{3\sqrt{2}}{2}i$

$k = 4:\ 3\left(\cos\dfrac{19\pi}{12} + i\sin\dfrac{19\pi}{12}\right) = 0.776 - 2.898i$

$k = 5:\ 3\left(\cos\dfrac{23\pi}{12} + i\sin\dfrac{23\pi}{12}\right) = 2.898 - 0.776i$

(c)

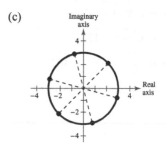

99. Cube roots of $8 = 8(\cos 0 + i \sin 0)$, $k = 0, 1, 2$

(a) $\sqrt[3]{8}\left[\cos\left(\dfrac{0 + 2\pi k}{3}\right) + i \sin\left(\dfrac{0 + 2\pi k}{3}\right)\right]$

$k = 0$: $2(\cos 0 + i \sin 0)$

$k = 1$: $2\left(\cos \dfrac{2\pi}{3} + i \sin \dfrac{2\pi}{3}\right)$

$k = 2$: $2\left(\cos \dfrac{4\pi}{3} + i \sin \dfrac{4\pi}{3}\right)$

(b) 2

$-1 + \sqrt{3}i$

$-1 - \sqrt{3}i$

(c)

101. $x^4 + 81 = 0$

$x^4 = -81$ \qquad Solve by finding the fourth roots of -81.

$-81 = 81(\cos \pi + i \sin \pi)$

$\sqrt[4]{-81} = \sqrt[4]{81}\left[\cos\left(\dfrac{\pi + 2\pi k}{4}\right) + i \sin\left(\dfrac{\pi + 2\pi k}{4}\right)\right]$, $k = 0, 1, 2, 3$

$k = 0$: $3\left(\cos \dfrac{\pi}{4} + i \sin \dfrac{\pi}{4}\right) = \dfrac{3\sqrt{2}}{2} + \dfrac{3\sqrt{2}}{2}i$

$k = 1$: $3\left(\cos \dfrac{3\pi}{4} + i \sin \dfrac{3\pi}{4}\right) = -\dfrac{3\sqrt{2}}{2} + \dfrac{3\sqrt{2}}{2}i$

$k = 2$: $3\left(\cos \dfrac{5\pi}{4} + i \sin \dfrac{5\pi}{4}\right) = -\dfrac{3\sqrt{2}}{2} - \dfrac{3\sqrt{2}}{2}i$

$k = 3$: $3\left(\cos \dfrac{7\pi}{4} + i \sin \dfrac{7\pi}{4}\right) = \dfrac{3\sqrt{2}}{2} - \dfrac{3\sqrt{2}}{2}i$

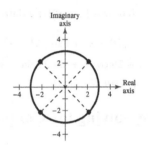

103. $x^3 + 8i = 0$

$x^3 = -8i$ \qquad Solve by finding the cube roots of $-8i$.

$-8i = 8\left(\cos \dfrac{3\pi}{2} + i \sin \dfrac{3\pi}{2}\right)$

$\sqrt[3]{-8i} = \sqrt[3]{8}\left[\cos\left(\dfrac{\dfrac{3\pi}{2} + 2\pi k}{3}\right) + i \sin\left(\dfrac{\dfrac{3\pi}{2} + 2\pi k}{3}\right)\right]$, $k = 0, 1, 2$

$k = 0$: $2\left(\cos \dfrac{\pi}{2} + i \sin \dfrac{\pi}{2}\right) = 2i$

$k = 1$: $2\left(\cos \dfrac{7\pi}{6} + i \sin \dfrac{7\pi}{6}\right) = -\sqrt{3} - i$

$k = 2$: $2\left(\cos \dfrac{11\pi}{6} + i \sin \dfrac{11\pi}{6}\right) = \sqrt{3} - i$

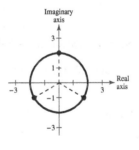

105. False.

$$\sqrt{-18}\sqrt{-2} = \left(3\sqrt{2}i\right)\left(\sqrt{2}i\right) = 6i^2 = -6, \text{ whereas}$$

$$\sqrt{(-18)(-2)} = \sqrt{36} = 6.$$

107. False. A fourth-degree polynomial with real coefficients has four zeros, and complex zeros occur in conjugate pairs.

109. (a) From the graph, the three roots are:

$$4(\cos 60° + i \sin 60°)$$

$$4(\cos 180° + i \sin 180°)$$

$$4(\cos 300° + i \sin 300°)$$

111. From the graph,

$$z_1 = 2(\cos \theta + i \sin \theta) \text{ and } z_2 = 2(\cos(\pi - \theta) + i \sin(\pi - \theta)).$$

$$z_1 z_2 = 2 \cdot 2\big(\cos(\theta + \pi - \theta) + i \sin(\theta + \pi - \theta)\big)$$

$$= 4(\cos \pi + i \sin \pi) = -4$$

$$\frac{z_1}{z_2} = \frac{2}{2}\Big(\cos\big(\theta - (\pi - \theta)\big) + i \sin\big(\theta - (\pi - \theta)\big)\Big)$$

$$= \cos(2\theta - \pi) + i \sin(2\theta - \pi)$$

$$= \cos 2\theta \cos \pi + \sin 2\theta \sin \pi + i(\sin 2\theta \cos \pi - \cos 2\theta \sin \pi)$$

$$= -\cos 2\theta - i \sin 2\theta$$

Problem Solving for Chapter 4

1. (a) $\left(\dfrac{-2 + 2\sqrt{3}i}{2}\right)^3 = 8$ (b) $\left(\dfrac{-3 + 3\sqrt{3}i}{2}\right)^3 = 27$

$$\left(\dfrac{-2 - 2\sqrt{3}i}{2}\right)^3 = 8 \qquad \left(\dfrac{-3 - 3\sqrt{3}i}{2}\right)^3 = 27$$

$$(2)^3 = 8 \qquad\qquad (3)^3 = 27$$

3. $(a + bi)(a - bi) = a^2 - abi + abi - b^2i^2 = a^2 + b^2$

Since a and b are real numbers, $a^2 + b^2$ is also a real number.

5. $x^2 - 2kx + k = 0$

$$x^2 - 2kx = -k$$

$$x^2 - 2kx + k^2 = k^2 - k$$

$$(x - k)^2 = k(k - 1)$$

$$x = k \pm \sqrt{k(k - 1)}$$

(a) If the equation has two real solutions, then $k(k - 1) > 0$. This means that $k < 0$ or $k > 1$.

(b) Since there are three evenly spaced roots on the circle of a radius 4, they are cube roots of a complex number of modulus $4^3 = 64$. Cubing them yields -64.

$$\big[4(\cos 60° + i \sin 60°)\big]^3 = -64$$

$$\big[4(\cos 180° + i \sin 180°)\big]^3 = -64$$

$$\big[4(\cos 300° + i \sin 300°)\big]^3 = -64$$

By using a calculator, show that each of the roots raised to the fourth power yields -64.

(c) The cube roots of a positive real number "a" are:

(i) $\sqrt[3]{a}$

(ii) $\dfrac{-\sqrt[3]{a} + \sqrt[3]{a}\sqrt{3}i}{2}$

(iii) $\dfrac{-\sqrt[3]{a} - \sqrt[3]{a}\sqrt{3}i}{2}$

(b) If the equation has two complex solutions, then $k(k - 1) < 0$. This means that $0 < k < 1$.

7. (a) $g(x) = f(x - 2)$

No. This function is a horizontal shift of $f(x)$. Note that x is a zero of g if and only if $x - 2$ is a zero of f; the number of real and complex zeros is not affected by a horizontal shift.

(b) $g(x) = f(2x)$

No. Since x is a zero of g if and only if $2x$ is a zero of f, the number of real and complex zeros of g is the same as the number of real and complex zeros of f.

9. (a) No. The graph is not of $f(x) = x^2(x + 2)(x - 3.5)$ because this function has $(0, 0)$ as an intercept and the given graph does not go through $(0, 0)$.

(b) No. The graph is not of $g(x) = (x + 2)(x - 3.5)$ because this function's graph is a parabola and the graph given is not. The function shown in the graph must be at least a fourth-degree polynomial.

(c) Yes. The function shown in the graph is $h(x) = (x + 2)(x - 3.5)(x^2 + 1)$.

(d) The graph is not of $k(x) = (x + 1)(x + 2)(x - 3.5)$ because this function has $(-1, 0)$ as an intercept and the given graph does not.

11. *Interval:* $(-\infty, -2)$, $(-2, 1)$, $(1, 4)$, $(4, \infty)$

Value of $f(x)$: Positive   Negative   Negative   Positive

(a) Zeros of $f(x)$: $x = -2, x = 1, x = 4$

(b) The graph touches the *x*-axis at $x = 1$.

(c) The least possible degree of the function is 4 because there are at least four real zeros (1 is repeated) and a function can have at most the number of real zeros equal to the degree of the function. The degree cannot be odd by the definition of multiplicity.

(d) The leading coefficient of f is positive. From the information in the table, you can conclude that the graph will eventually rise to the left and to the right.

(e) $f(x) = (x + 2)(x - 1)^2(x - 4) = x^4 - 4x^3 - 3x^2 + 14x - 8$

(This answer is not unique.)

(f)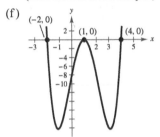

13. Let $z = a + bi$ and $\bar{z} = a - bi$.

$$z^2 - \bar{z}^2 = 0$$
$$(a + bi)^2 - (a - bi)^2 = 0$$
$$4abi = 0 \Rightarrow b = 0$$

$z = a$, a pure real number

$$z^2 + \bar{z}^2 = 0$$
$$(a + bi)^2 + (a - bi)^2 = 0$$
$$2a^2 - 2b^2 = 0 \Rightarrow b = \pm a$$

$z = a + ai$ or $z = a - ai$, both are complex solutions.

15. Let $z = x + yi$ and $\bar{z} = x - yi$.

$$|z - 1| \cdot |\bar{z} - 1| = 1$$
$$\sqrt{(x - 1)^2 + y^2}\sqrt{(x - 1)^2 + (-y)^2} = 1$$
$$(x - 1)^2 + y^2 = 1$$

The graph of the solution set is a circle centered at $(1, 0)$ of radius 1.

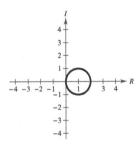

Practice Test for Chapter 4

1. Write $4 + \sqrt{-81} - 3i^2$ in standard form.

2. Write the result in standard form: $\dfrac{3 + i}{5 - 4i}$

3. Use the Quadratic Formula to solve $x^2 - 4x + 7 = 0$.

4. True or false: $\sqrt{-6}\sqrt{-6} = \sqrt{36} = 6$

5. Use the discriminant to determine the type of solutions of $3x^2 - 8x + 7 = 0$.

6. Find all the zeros of $f(x) = x^4 + 13x^2 + 36$.

7. Find a polynomial function that has the following zeros: $3, -1 \pm 4i$

8. Use the zero $x = 4 + i$ to find all the zeros of $f(x) = x^3 - 10x^2 + 33x - 34$.

9. Give the trigonometric form of $z = 5 - 5i$.

10. Give the standard form of $z = 6(\cos 225° + i \sin 225°)$.

11. Multiply $\left[7(\cos 23° + i \sin 23°)\right]\left[4(\cos 7° + i \sin 7°)\right]$.

12. Divide $\dfrac{9\left(\cos \dfrac{5\pi}{4} + i \sin \dfrac{5\pi}{4}\right)}{3(\cos \pi + i \sin \pi)}$.

13. Find $(2 + 2i)^8$.

14. Find the cube roots of $8\left(\cos \dfrac{\pi}{3} + i \sin \dfrac{\pi}{3}\right)$.

15. Find all the solutions to $x^4 + i = 0$.

C H A P T E R 5
Exponential and Logarithmic Functions

CHAPTER 5
Exponential and Logarithmic Functions

Section 5.1   Exponential Functions and Their Graphs

1. algebraic

3. One-to-One

5. $A = P\left(1 + \dfrac{r}{n}\right)^{nt}$

7. $f(1.4) = (0.9)^{1.4} \approx 0.863$

9. $f\left(\frac{2}{5}\right) = 3^{2/5} \approx 1.552$

11. $f(-1.5) = 5000\left(2^{-1.5}\right)$
    ≈ 1767.767

13. $f(x) = 2^x$

   Increasing

   Asymptote: $y = 0$

   Intercept: $(0, 1)$

   Matches graph (d).

14. $f(x) = 2^x + 1$

   Increasing

   Asymptote: $y = 1$

   Intercept: $(0, 2)$

   Matches graph (c).

15. $f(x) = 2^{-x}$

   Decreasing

   Asymptote: $y = 0$

   Intercept: $(0, 1)$

   Matches graph (a).

16. $f(x) = 2^{x-2}$

   Increasing

   Asymptote: $y = 0$

   Intercept: $\left(0, \frac{1}{4}\right)$

   Matches graph (b).

17. $f(x) = 7^x$

x	-2	-1	0	1	2
$f(x)$	0.020	0.143	1	7	49

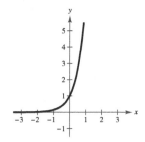

19. $f(x) = \left(\frac{1}{4}\right)^{-x}$

x	-2	-1	0	1	2
$f(x)$	0.063	0.25	1	4	16

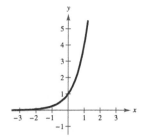

21. $f(x) = 4^{x-1}$

x	-2	-1	0	1	2
$f(x)$	0.016	0.063	0.25	1	4

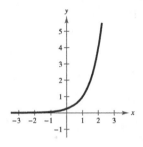

23. $f(x) = 2^{x+1} + 3$

x	-3	-2	-1	0	1
$f(x)$	3.25	3.5	4	5	7

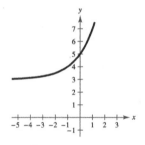

25. $3^{x+1} = 27$

  $3^{x+1} = 3^3$

  $x + 1 = 3$

  $x = 2$

27. $\left(\frac{1}{2}\right)^x = 32$

  $\left(\frac{1}{2}\right)^x = \left(\frac{1}{2}\right)^{-5}$

  $x = -5$

29. $f(x) = 3^x, g(x) = 3^x + 1$

Because $g(x) = f(x) + 1$, the graph of g can be obtained by shifting the graph of f one unit upward.

31. $f(x) = 10^x, g(x) = 10^{-x+3}$

Because $g(x) = f(-x + 3)$, the graph of g can be obtained by reflecting the graph of f in the y-axis and shifting f three units to the right. (**Note:** This is equivalent to shifting f three units to the left and then reflecting the graph in the y-axis.)

33. $f(x) = e^x$

  $f(1.9) = e^{1.9} \approx 6.686$

35. $f(6) = 5000e^{0.06(6)} \approx 7166.647$

37. $f(x) = 3e^{x+4}$

x	-8	-7	-6	-5	-4
$f(x)$	0.055	0.149	0.406	1.104	3

Asymptote: $y = 0$

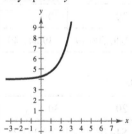

39. $f(x) = 2e^{x-2} + 4$

x	-2	-1	0	1	2
$f(x)$	4.037	4.100	4.271	4.736	6

Asymptote: $y = 4$

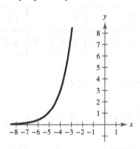

41. $s(t) = 2e^{0.5t}$

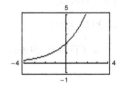

43. $g(x) = 1 + e^{-x}$

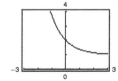

45. $e^{3x+2} = e^3$

$3x + 2 = 3$

$3x = 1$

$x = \frac{1}{3}$

47. $e^{x^2-3} = e^{2x}$

$x^2 - 3 = 2x$

$x^2 - 2x - 3 = 0$

$(x - 3)(x + 1) = 0$

$x = 3$ or $x = -1$

49. $P = \$1500, r = 2\%, t = 10$ years

Compounded n times per year: $A = P\left(1 + \dfrac{r}{n}\right)^{nt} = 1500\left(1 + \dfrac{0.02}{n}\right)^{10n}$

Compounded continuously: $A = Pe^{rt} = 1500e^{0.02(10)}$

n	1	2	4	12	365	Continuous
A	\$1828.49	\$1830.29	\$1831.19	\$1831.80	\$1832.09	\$1832.10

51. $P = \$2500, r = 4\%, t = 20$ years

Compounded n times per year: $A = P\left(1 + \dfrac{r}{n}\right)^{nt} = 2500\left(1 + \dfrac{0.04}{n}\right)^{20n}$

Compounded continuously: $A = Pe^{rt} = 2500e^{0.04(20)}$

n	1	2	4	12	365	Continuous
A	\$5477.81	\$5520.10	\$5541.79	\$5556.46	\$5563.61	\$5563.85

53. $A = Pe^{rt} = 12{,}000e^{0.04t}$

t	10	20	30	40	50
A	\$17,901.90	\$26,706.49	\$39,841.40	\$59,436.39	\$88,668.67

55. $A = Pe^{rt} = 12{,}000e^{0.065t}$

t	10	20	30	40	50
A	\$22,986.49	\$44,031.56	\$84,344.25	\$161,564.86	\$309,484.08

57. $A = 30{,}000e^{(0.05)(25)} \approx \$104{,}710.29$

59. $C(t) = 29.88(1.04)^t$

Ten years from today, $t = 10$: $C(10) = 29.88(1.04)^{10} \approx \$44.23$

61. (a)

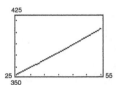

(b)

t	25	26	27	28
P (in millions)	350.281	352.107	353.943	355.788

t	29	30	31	32
P (in millions)	357.643	359.508	361.382	363.266

t	33	34	35	36
P (in millions)	365.160	367.064	368.977	370.901

t	37	38	39	40
P (in millions)	372.835	374.779	376.732	378.697

t	41	42	43	44
P (in millions)	380.671	382.656	384.651	386.656

t	45	46	47	48
P (in millions)	388.672	390.698	392.735	394.783

t	49	50	51	52
P (in millions)	396.841	398.910	400.989	403.080

t	53	54	55
P (in millions)	405.182	407.294	409.417

(c) Using the model and extending the table beyond the year 2055, the population will exceed 430 million in 2064.

t	55	56	57	58	59	60	61	62	63	64	65
P (in millions)	409.417	411.552	413.698	415.854	418.022	420.202	422.393	424.595	426.808	429.034	431.270

63. $Q = 16\left(\frac{1}{2}\right)^{t/24,100}$

(a) $Q(0) = 16$ grams

(b) $Q(75{,}000) \approx 1.85$ grams

(c)

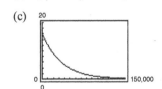

65. (a) $V(t) = 49{,}810\left(\frac{7}{8}\right)^{t}$ where t is the number of years since it was purchased.

(b) $V(4) = 49{,}810\left(\frac{7}{8}\right)^{4} \approx 29{,}197.71$

After 4 years, the value of the van is about \$29,198.

67. True. The line $y = -2$ is a horizontal asymptote for the graph of $f(x) = 10^{x} - 2$. As $x \to -\infty$, $f(x) \to -2$ but never reaches -2.

69. $f(x) = 3^{x-2}$

$\qquad = 3^x 3^{-2}$

$\qquad = 3^x \left(\dfrac{1}{3^2}\right)$

$\qquad = \dfrac{1}{9}(3^x)$

$\qquad = h(x)$

So, $f(x) \neq g(x)$, but $f(x) = h(x)$.

71. $f(x) = 16(4^{-x})$ and $f(x) = 16(4^{-x})$

$\qquad = 4^2(4^{-x}) \qquad\qquad = 16(2^2)^{-x}$

$\qquad = 4^{2-x} \qquad\qquad\quad = 16(2^{-2x})$

$\qquad = \left(\dfrac{1}{4}\right)^{-(2-x)} \qquad\quad = h(x)$

$\qquad = \left(\dfrac{1}{4}\right)^{x-2}$

$\qquad = g(x)$

So, $f(x) = g(x) = h(x)$.

73. $y = 3^x$ and $y = 4^x$

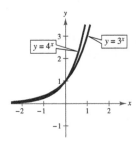

x	-2	-1	0	1	2
3^x	$\frac{1}{9}$	$\frac{1}{3}$	1	3	9
4^x	$\frac{1}{16}$	$\frac{1}{4}$	1	4	16

(a) $4^x < 3^x$ when $x < 0$.

(b) $4^x > 3^x$ when $x > 0$.

79. The functions (c) $h(x) = 3^x$ and (d) $k(x) = 2^{-x}$ are exponential.

75.

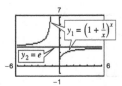

As x increases, the graph of y_1 approaches e, which is y_2.

77. (a)

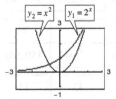

At $x = 2$, both functions have a value of 4. The function y_1 increases for all values of x. The function y_2 is symmetric with respect to the y-axis.

(b)

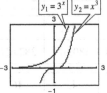

Both functions are increasing for all values of x. For $x > 0$, both functions have a similar shape. The function y_2 is symmetric with respect to the origin.

In both viewing windows, the constant raised to a variable power increases more rapidly than the variable raised to a constant power.

Section 5.2 Logarithmic Functions and Their Graphs

1. logarithmic

3. natural; e

5. $x = y$

7. $\log_4 16 = 2 \Rightarrow 4^2 = 16$

9. $\log_{12} 12 = 1 \Rightarrow 12^1 = 12$

11. $5^3 = 125 \Rightarrow \log_5 125 = 3$

13. $4^{-3} = \frac{1}{64} \Rightarrow \log_4 \frac{1}{64} = -3$

15. $f(x) = \log_2 x$

$\qquad f(64) = \log_2 64 = 6$ because $2^6 = 64$

17. $f(x) = \log_8 x$

$f(1) = \log_8 1 = 0$ because $8^0 = 1$

19. $g(x) = \log_a x$

$g(a^2) = \log_a a^{-2}$

$= -2$

21. $f(x) = \log x$

$f\left(\frac{7}{8}\right) = \log\left(\frac{7}{8}\right) \approx -0.058$

23. $f(x) = \log x$

$f(12.5) = \log 12.5 \approx 1.097$

25. $\log_8 8 = 1$ because $8^1 = 8$

27. $\log_{7.5} 1 = 0$ because $7.5^0 = 1$

29. $\log_5(x + 1) = \log_5 6$

$x + 1 = 6$

$x = 5$

31. $\log 11 = \log(x^2 + 7)$

$11 = x^2 + 7$

$x^2 = 4$

$x = \pm 2$

33.

x	-2	-1	0	1	2
$f(x) = 7^x$	$\frac{1}{49}$	$\frac{1}{7}$	1	7	49

x		$\frac{1}{49}$	$\frac{1}{7}$	1	7	49
$g(x) = \log_7 x$		-2	-1	0	1	2

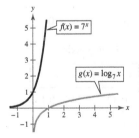

35.

x	-2	-1	0	1	2
$f(x) = 6^x$	$\frac{1}{36}$	$\frac{1}{6}$	1	6	36

x		$\frac{1}{36}$	$\frac{1}{6}$	1	6	36
$g(x) = \log_6 x$		-2	-1	0	1	2

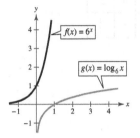

37. $f(x) = \log_3 x + 2$

Asymptote: $x = 0$

Point on graph: $(1, 2)$

Matches graph (a).

The graph of $f(x)$ is obtained from $g(x)$ by shifting the graph two units upward.

38. $f(x) = \log_3(x - 1)$

Asymptote: $x = 1$

Point on graph: $(2, 0)$

Matches graph (d).

$f(x)$ shifts $g(x)$ one unit to the right.

39. $f(x) = \log_3(1 - x) = \log_3\left[-(x - 1)\right]$

Asymptote: $x = 1$

Point on graph: $(0, 0)$

Matches graph (b).

The graph of $f(x)$ is obtained by reflecting the graph of $g(x)$ in the y-axis and shifting the graph one unit to the right.

40. $f(x) = -\log_3 x$

Asymptote: $x = 0$

Point on graph: $(1, 0)$

Matches graph (c).

$f(x)$ reflects $g(x)$ in the x-axis.

41. $f(x) = \log_4 x$

Domain: $(0, \infty)$

x-intercept: $(1, 0)$

Vertical asymptote: $x = 0$

$y = \log_4 x \Rightarrow 4^y = x$

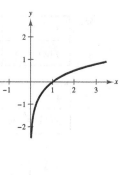

x	$\frac{1}{4}$	1	4	2
$f(x)$	-1	0	1	$\frac{1}{2}$

43. $y = \log_3 x + 1$

Domain: $(0, \infty)$

x-intercept:

$\log_3 x + 1 = 0$

$\log_3 x = -1$

$3^{-1} = x$

$\frac{1}{3} = x$

The x-intercept is $\left(\frac{1}{3}, 0\right)$.

Vertical asymptote: $x = 0$

$y = \log_3 x + 1$

$\log_3 x = y - 1 \Rightarrow 3^{y-1} = x$

x	$\frac{1}{9}$	$\frac{1}{3}$	0	3	9
y	-1	0	1	2	3

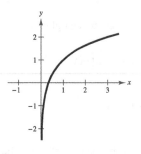

45. $f(x) = -\log_6(x + 2)$

Domain: $x + 2 > 0 \Rightarrow x > -2$

The domain is $(-2, \infty)$.

x-intercept:

$0 = -\log_6(x + 2)$

$0 = \log_6(x + 2)$

$6^0 = x + 2$

$1 = x + 2$

$-1 = x$

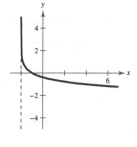

The x-intercept is $(-1, 0)$.

Vertical asymptote: $x + 2 = 0 \Rightarrow x = -2$

$y = -\log_6(x + 2)$

$-y = \log_6(x + 2)$

$6^{-y} - 2 = x$

x	4	-1	$-1\frac{5}{6}$	$-1\frac{35}{36}$
$f(x)$	-1	0	1	2

47. $y = \log\left(\frac{x}{7}\right)$

Domain: $\frac{x}{7} > 0 \Rightarrow x > 0$

The domain is $(0, \infty)$.

x-intercept: $\log\left(\frac{x}{7}\right) = 0$

$\frac{x}{7} = 10^0$

$\frac{x}{7} = 1$

$x = 7$

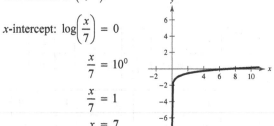

The x-intercept is $(7, 0)$.

Vertical asymptote: $\frac{x}{7} = 0 \Rightarrow x = 0$

The vertical asymptote is the y-axis.

x	1	2	3	4	5
y	-0.85	-0.54	-0.37	-0.24	-0.15

x	6	7	8
y	-0.069	0	0.06

49. $\ln \frac{1}{2} = -0.693... \Rightarrow e^{-0.693...} = \frac{1}{2}$

51. $\ln 250 = 5.521... \Rightarrow e^{5.521...} = 250$

53. $e^2 = 7.3890... \Rightarrow \ln 7.3890... = 2$

55. $e^{-4x} = \frac{1}{2} \Rightarrow \ln \frac{1}{2} = -4x$

57. $f(x) = \ln x$

$f(18.42) = \ln 18.42 \approx 2.913$

59. $g(x) = 8 \ln x$

$g(\sqrt{5}) = 8 \ln \sqrt{5} \approx 6.438$

61. $e^{\ln 4} = 4$

63. $2.5 \ln 1 = 2.5(0) = 0$

65. $\ln e^{\ln e} = \ln e^1 = 1$

67. $f(x) = \ln(x - 4)$

Domain: $x - 4 > 0 \Rightarrow x > 4$

The domain is $(4, \infty)$.

x-intercept: $0 = \ln(x - 4)$

$e^0 = x - 4$

$5 = x$

The x-intercept is $(5, 0)$.

Vertical asymptote: $x - 4 = 0 \Rightarrow x = 4$

x	4.5	5	6	7
$f(x)$	-0.69	0	0.69	1.10

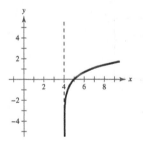

69. $g(x) = \ln(-x)$

Domain: $-x > 0 \Rightarrow x < 0$

The domain is $(-\infty, 0)$.

x-intercept:

$0 = \ln(-x)$

$e^0 = -x$

$-1 = x$

The x-intercept is $(-1, 0)$.

Vertical asymptote: $-x = 0 \Rightarrow x = 0$

x	-0.5	-1	-2	-3
$g(x)$	-0.69	0	0.69	1.10

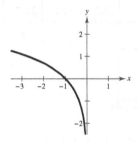

71. $f(x) = \ln(x - 1)$

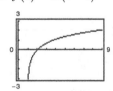

73. $f(x) = -\ln x + 8$

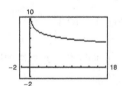

75. $\ln(x + 4) = \ln 12$

$x + 4 = 12$

$x = 8$

77. $\ln(x^2 - x) = \ln 6$

$x^2 - x = 6$

$x^2 - x - 6 = 0$

$(x - 3)(x + 2) = 0$

$x = -2 \text{ or } x = 3$

79. $t = 16.625 \ln\left(\dfrac{x}{x - 750}\right), x > 750$

   (a) When $x = \$897.72$:

$$t = 16.625 \ln\left(\frac{897.72}{897.72 - 750}\right) \approx 30 \text{ years}$$

    When $x = \$1659.24$:

$$t = 16.625 \ln\left(\frac{1659.24}{1659.24 - 750}\right) \approx 10 \text{ years}$$

   (b) Total amounts:

     $(897.72)(12)(30) = \$323,179.20 \approx \$323,179$

     $(1659.24)(12)(10) = \$199,108.80 \approx \$199,109$

    Interest charges:

     $323,179.20 - 150,000 = \$173,179.20 \approx \$173,179$

     $199,108.80 - 150,000 = \$49,108.80 \approx \$49,109$

   (c) The vertical asymptote is $x = 750$. The closer the payment is to $750 per month, the longer the length of the mortgage will be. Also, the monthly payment must be greater than $750.

81. $t = \dfrac{\ln 2}{r}$

   (a)

r	0.005	0.010	0.015	0.020	0.025	0.030
t	138.6	69.3	46.2	34.7	27.7	23.1

   (b)

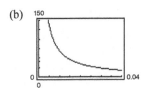

83. $f(t) = 80 - 17 \log(t + 1), 0 \le t \le 12$

   (a)

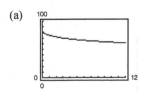

   (b) $f(0) = 80 - 17 \log 1 = 80.0$

   (c) $f(4) = 80 - 17 \log 5 \approx 68.1$

   (d) $f(10) = 80 - 17 \log 11 \approx 62.3$

85. False. Reflecting $g(x)$ about the line $y = x$ will determine the graph of $f(x)$.

87. (a) $f(x) = \ln x, g(x) = \sqrt{x}$

    The natural log function grows at a slower rate than the square root function.

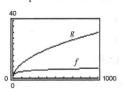

   (b) $f(x) = \ln x, g(x) = \sqrt[4]{x}$

    The natural log function grows at a slower rate than the fourth root function.

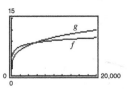

89.

x	1	2	8
y	0	1	3

y is not an exponential function of x, but it is a logarithmic function of x, $y = \log_2 x$.

91. $f(x) = \dfrac{\ln x}{x}$

(a)

x	1	5	10	10^2	10^4	10^6
$f(x)$	0	0.322	0.230	0.046	0.00092	0.0000138

(b) As $x \to \infty$, $f(x) \to 0$.

(c)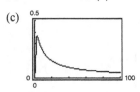

Section 5.3   Properties of Logarithms

1. change-of-base

3. $\dfrac{1}{\log_b a}$

5. (a) $\log_5 16 = \dfrac{\log 16}{\log 5}$

   (b) $\log_5 16 = \dfrac{\ln 16}{\ln 5}$

7. (a) $\log_x \dfrac{3}{10} = \dfrac{\log(3/10)}{\log x}$

   (b) $\log_x \dfrac{3}{10} = \dfrac{\ln(3/10)}{\ln x}$

9. $\log_3 17 = \dfrac{\log 17}{\log 3} = \dfrac{\ln 17}{\ln 3} \approx 2.579$

11. $\log_\pi 0.5 = \dfrac{\log 0.5}{\log \pi} = \dfrac{\ln 0.5}{\ln \pi} \approx -0.606$

13. $\log_3 35 = \log_3 (5 \cdot 7)$
   $= \log_3 5 + \log_3 7$

15. $\log_3 \left(\frac{7}{25}\right) = \log_3 7 - \log_3 25$
   $= \log_3 7 - \log_3 5^2$
   $= \log_3 7 - 2 \log_3 5$

17. $\log_3 \left(\frac{21}{5}\right) = \log_3 21 - \log_3 5$
   $= \log_3 (3 \cdot 7) - \log_3 5$
   $= \log_3 3 + \log_3 7 - \log_3 5$
   $= 1 + \log_3 7 - \log_3 5$

19. $\log_3 9 = 2 \log_3 3 = 2$

21. $\log_6 \sqrt[3]{\frac{1}{6}} = \log_6 \left(\frac{1}{6}\right)^{1/3} = \frac{1}{3} \log_6 \left(\frac{1}{6}\right) = \frac{1}{3} \log_6 6^{-1} = \frac{1}{3}(-1) = -\frac{1}{3}$

23. $\log_2(-2)$ is undefined. -2 is not in the domain of $\log_2 x$.

25. $\ln \sqrt[4]{e^3} = \ln e^{3/4}$
   $= \frac{3}{4} \ln e$
   $= \frac{3}{4}(1)$
   $= \frac{3}{4}$

27. $\ln e^2 + \ln e^5 = 2 + 5 = 7$

29. $\log_5 75 - \log_5 3 = \log_5 \frac{75}{3}$
   $= \log_5 25$
   $= \log_5 5^2$
   $= 2 \log_5 5$
   $= 2$

31. $\log_4 8 = \log_4 (4 \cdot 2) = \log_4 4 + \log_4 2 = \log_4 4 + \log_4 4^{1/2} = 1 + \frac{1}{2} = \frac{3}{2}$

33. $\log_b 10 = \log_b 2.5$
$= \log_b 2 + \log_b 5$
$\approx 0.3562 + 0.8271$
$= 1.1833$

35. $\log_b 0.04 = \log_b \frac{4}{100} = \log_b \frac{1}{25}$
$= \log_b 1 - \log_b 25$
$= \log_b 1 - \log_b 5^2$
$= 0 - 2\log_b 5$
$\approx -2(0.8271)$
$= -1.6542$

37. $\log_b 45 = \log_b 9.5$
$= \log_b 9 + \log_b 5$
$= \log_b 3^2 + \log_b 5$
$= 2\log_b 3 + \log_b 5$
$\approx 2(0.5646) + 0.8271$
$= 1.9563$

39. $\log_b (2b)^{-2} = -2\log_b 2b$
$= -2(\log_b 2 + \log_b b)$
$\approx -2(0.3562 + 1)$
$= -2.7124$

41. $\ln 7x = \ln 7 + \ln x$

43. $\log_8 x^4 = 4\log_8 x$

45. $\log_5 \frac{5}{x} = \log_5 5 - \log_5 x$
$= 1 - \log_5 x$

47. $\ln \sqrt{z} = \ln z^{1/2} = \frac{1}{2}\ln z$

49. $\ln xyz^2 = \ln x + \ln y + \ln z^2$
$= \ln x + \ln y + 2\ln z$

51. $\ln z(z-1)^2 = \ln z + \ln(z-1)^2$
$= \ln z + 2\ln(z-1), \, z > 1$

53. $\log_2 \left(\frac{\sqrt{a^2-4}}{7} \right) = \log_2 \sqrt{a^2-4} - \log_2 7$

$= \log_2 (a^2-4)^{1/2} - \log_2 7$
$= \frac{1}{2}\log_2 (a^2-4) - \log_2 7$
$= \frac{1}{2}\log_2 [(a-2)(a+2)] - \log_2 7$
$= \frac{1}{2}\big[\log_2 (a-2) + \log_2(a+2)\big] - \log_2 7$
$= \frac{1}{2}\log_2 (a-2) + \frac{1}{2}\log_2 (a+2) - \log_2 7$

55. $\log_5 \left(\frac{x^2}{y^2 z^3} \right) = \log_5 x^2 - \log_5 y^2 z^3$

$= \log_5 x^2 - (\log_5 y^2 + \log_5 z^3)$
$= 2\log_5 x - 2\log_5 y - 3\log_5 z$

57. $\ln \sqrt[3]{\frac{yz}{x^2}} = \ln \left(\frac{yz}{x^2} \right)^{1/3}$

$= \frac{1}{3}\ln \left(\frac{yz}{x^2} \right)$
$= \frac{1}{3}\big[\ln(yz) - \ln x^2\big]$
$= \frac{1}{3}\big[\ln(yz) - 2\ln x\big]$
$= \frac{1}{3}[\ln y + \ln z - 2\ln x]$
$= \frac{1}{3}\ln y + \frac{1}{3}\ln z - \frac{2}{3}\ln x$

59. $\ln \sqrt[4]{x^3(x^2+3)} = \frac{1}{4}\ln x^3(x^2+3)$
$= \frac{1}{4}\big[\ln x^3 + \ln(x^2+3)\big]$
$= \frac{1}{4}\big[3\ln x + \ln(x^2+3)\big]$
$= \frac{3}{4}\ln x + \frac{1}{4}\ln(x^2+3)$

61. $\ln 3 + \ln x = \ln(3x)$

63. $\frac{2}{3}\log_7(z-2) = \log_7(z-2)^{2/3}$

65. $\log_3 5x - 4\log_3 x = \log_3 5x - \log_3 x^4$
$= \log_3 \left(\frac{5x}{x^4} \right)$
$= \log_3 \left(\frac{5}{x^3} \right)$

67. $\log x + 2 \log(x + 1) = \log x + \log(x + 1)^2$

$$= \log\left[x(x + 1)^2\right]$$

69. $\log x - 2 \log y + 3 \log z = \log x - \log y^2 + \log z^3$

$$= \log \frac{x}{y^2} + \log z^3$$

$$= \log \frac{xz^3}{y^2}$$

71. $\ln x - \left[\ln(x + 1) + \ln(x - 1)\right] = \ln x - \ln(x + 1)(x - 1) = \ln \frac{x}{(x + 1)(x - 1)}$

73. $\dfrac{1}{2}\left[2 \ln(x + 3) + \ln x - \ln\left(x^2 - 1\right)\right] = \dfrac{1}{2}\left[\ln(x + 3)^2 + \ln x - \ln\left(x^2 - 1\right)\right]$

$$= \frac{1}{2}\left[\ln\left[x(x + 3)^2\right] - \ln\left(x^2 - 1\right)\right]$$

$$= \frac{1}{2}\left[\ln\left(\frac{x(x + 3)^2}{x^2 - 1}\right)\right]$$

$$= \frac{1}{2}\ln\left[\frac{x(x + 3)^2}{x^2 - 1}\right]$$

$$= \ln \sqrt{\frac{x(x + 3)^2}{x^2 - 1}}$$

75. $\dfrac{1}{3}\left[\log_8 y + 2 \log_8(y + 4)\right] - \log_8(y - 1) = \dfrac{1}{3}\left[\log_8 y + \log_8(y + 4)^2\right] - \log_8(y - 1)$

$$= \frac{1}{3} \log_8 y(y + 4)^2 - \log_8(y - 1)$$

$$= \log_8 \sqrt[3]{y(y + 4)^2} - \log_8(y - 1)$$

$$= \log_8 \left(\frac{\sqrt[3]{y(y + 4)^2}}{y - 1}\right)$$

77. $\log_2 \dfrac{32}{4} = \log_2 32 - \log_2 4 \neq \dfrac{\log_2 32}{\log_2 4}$

The second and third expressions are equal by Property 2.

79. $\beta = 10 \log\left(\dfrac{I}{10^{-12}}\right) = 10\left[\log I - \log 10^{-12}\right] = 10\left[\log I + 12\right] = 120 + 10 \log I$

When $I = 10^{-6}$:

$\beta = 120 + 10 \log 10^{-6} = 120 + 10(-6) = 60$ decibels

81. $\beta = 10 \log\left(\dfrac{I}{10^{-12}}\right)$

$\text{Difference} = 10 \log\left(\dfrac{10^{-4}}{10^{-12}}\right) - 10 \log\left(\dfrac{10^{-11}}{10^{-12}}\right)$

$$= 10\left[\log 10^8 - \log 10\right]$$

$$= 10(8 - 1)$$

$$= 10(7)$$

$$= 70 \text{ dB}$$

83.

x	1	2	3	4	5	6
y	1.000	1.189	1.316	1.414	1.495	1.565
$\ln x$	0	0.693	1.099	1.386	1.609	1.792
$\ln y$	0	0.173	0.275	0.346	0.402	0.448

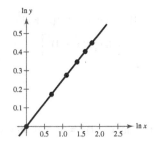

The slope of the line is $\frac{1}{4}$. So, $\ln y = \frac{1}{4} \ln x$

85.

x	1	2	3	4	5	6
y	2.500	2.102	1.900	1.768	1.672	1.597
$\ln x$	0	0.693	1.099	1.386	1.609	1.792
$\ln y$	0.916	0.743	0.642	0.570	0.514	0.468

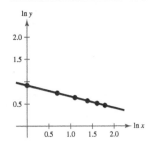

The slope of the line is $-\frac{1}{4}$. So, $\ln y = -\frac{1}{4} \ln x + \ln \frac{5}{2}$.

87.

Weight, x	25	35	50	75	500	1000
Galloping Speed, y	191.5	182.7	173.8	164.2	125.9	114.2
$\ln x$		3.219	3.555	3.912	4.317	6.215	6.908
$\ln y$		5.255	5.208	5.158	5.101	4.835	4.738

$\ln y = -0.14 \ln x + 5.7$

89. (a)

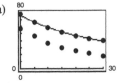

(b) $T - 21 = 54.4(0.964)^t$

$$T = 54.4(0.964)^t + 21$$

See graph in (a).

(c)

t (in minutes)	T (°C)	$T - 21$ (°C)	$\ln(T - 21)$	$1/(T - 21)$
0	78	57	4.043	0.0175
5	66	45	3.807	0.0222
10	57.5	36.5	3.597	0.0274
15	51.2	30.2	3.408	0.0331
20	46.3	25.3	3.231	0.0395
25	42.5	21.5	3.068	0.0465
30	39.6	18.6	2.923	0.0538

$\ln(T - 21) = -0.037t + 4$

$$T = e^{-0.037t + 3.997} + 21$$

This graph is identical to T in (b).

(d) $\dfrac{1}{T - 21} = 0.0012t + 0.016$

$$T = \dfrac{1}{0.001t + 0.016} + 21$$

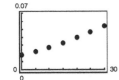 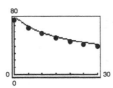

91. $f(x) = \ln x$

False, $f(0) \neq 0$ because 0 is not in the domain of $f(x)$.

$f(1) = \ln 1 = 0$

93. False.

$$f(x) - f(2) = \ln x - \ln 2 = \ln \dfrac{x}{2} \neq \ln(x - 2)$$

95. False.

$$f(u) = 2f(v) \Rightarrow \ln u = 2 \ln v \Rightarrow \ln u = \ln v^2 \Rightarrow u = v^2$$

97. $f(x) = \log_2 x = \dfrac{\log x}{\log 2} = \dfrac{\ln x}{\ln 2}$

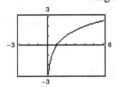

99. $f(x) = \log_{1/4} x$

$= \dfrac{\log x}{\log(1/4)} = \dfrac{\ln x}{\ln(1/4)}$

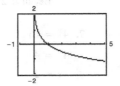

103.

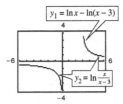

The graphing utility does not show the functions with the same domain. The domain of $y_1 = \ln x - \ln(x - 3)$ is $(3, \infty)$ and the domain of $y_2 = \ln \dfrac{x}{x - 3}$ is $(-\infty, 0) \cup (3, \infty)$.

101. The power property cannot be used because $\ln e$ is raised to the second power, not just e.

A correct statement is $(\ln e)^2 = (1)^2 = 1$.

105. $\ln 2 \approx 0.6931$, $\ln 3 \approx 1.0986$, $\ln 5 \approx 1.6094$

$\ln 1 = 0$

$\ln 2 \approx 0.6931$

$\ln 3 \approx 1.0986$

$\ln 4 = \ln(2 \cdot 2) = \ln 2 + \ln 2 \approx 0.6931 + 0.6931 = 1.3862$

$\ln 5 \approx 1.6094$

$\ln 6 = \ln(2 \cdot 3) = \ln 2 + \ln 3 \approx 0.6931 + 1.0986 = 1.7917$

$\ln 8 = \ln 2^3 = 3 \ln 2 \approx 3(0.6931) = 2.0793$

$\ln 9 = \ln 3^2 = 2 \ln 3 \approx 2(1.0986) = 2.1972$

$\ln 10 = \ln(5 \cdot 2) = \ln 5 + \ln 2 \approx 1.6094 + 0.6931 = 2.3025$

$\ln 12 = \ln(2^2 \cdot 3) = \ln 2^2 + \ln 3 = 2 \ln 2 + \ln 3 \approx 2(0.6931) + 1.0986 = 2.4848$

$\ln 15 = \ln(5 \cdot 3) = \ln 5 + \ln 3 \approx 1.6094 + 1.0986 = 2.7080$

$\ln 16 = \ln 2^4 = 4 \ln 2 \approx 4(0.6931) = 2.7724$

$\ln 18 = \ln(3^2 \cdot 2) = \ln 3^2 + \ln 2 = 2 \ln 3 + \ln 2 \approx 2(1.0986) + 0.6931 = 2.8903$

$\ln 20 = \ln(5 \cdot 2^2) = \ln 5 + \ln 2^2 = \ln 5 + 2 \ln 2 \approx 1.6094 + 2(0.6931) = 2.9956$

Section 5.4   Exponential and Logarithmic Equations

1. (a) $x = y$

(b) $x = y$

(c) x

(d) x

3. $4^{2x-7} = 64$

(a) $x = 5$

$4^{2(5)-7} = 4^3 = 64$

Yes, $x = 5$ *is* a solution.

(b) $x = 2$

$4^{2(2)-7} = 4^{-3} = \frac{1}{64} \neq 64$

No, $x = 2$ *is not* a solution.

(c) $\qquad x = \frac{1}{2}(\log_4 64 + 7)$

$4^{2(1/2(\log_4 64+7))-7} = 64$

$4^{(\log_4 64+7)-7} = 64$

$4^{(3+7)-7} = 64$

$4^3 = 64$

Yes, $x = \frac{1}{2}(\log_4 64 + 7)$ *is* a solution.

5. $\log_2(x + 3) = 10$

(a) $x = 1021$

$\log_2(1021 + 3) = \log_2(1024)$

Because $2^{10} = 1024$, $x = 1021$ *is* a solution.

(b) $x = 17$

$\log_2(17 + 3) = \log_2(20)$

Because $2^{10} \neq 20$, $x = 17$ *is not* a solution.

(c) $x = 10^2 - 3 = 97$

$\log_2(97 + 3) = \log_2(100)$

Because $2^{10} \neq 100$, $10^2 - 3$ *is not* a solution.

7. $4^x = 16$

$4^x = 4^2$

$x = 2$

9. $\ln x - \ln 2 = 0$

$\ln x = \ln 2$

$x = 2$

11. $e^x = 2$

$\ln e^x = \ln 2$

$x = \ln 2$

$x \approx 0.693$

13. $\ln x = -1$

$e^{\ln x} = e^{-1}$

$x = e^{-1}$

$x \approx 0.368$

15. $\log_4 x = 3$

$\quad 4^{\log_4 x} = 4^3$

$\quad\quad x = 4^3$

$\quad\quad x = 64$

17. $f(x) = g(x)$

$\quad 2^x = 8$

$\quad 2^x = 2^3$

$\quad\quad x = 3$

Point of intersection:

$(3, 8)$

19. $e^x = e^{x^2-2}$

$\quad x = x^2 - 2$

$\quad 0 = x^2 - x - 2$

$\quad 0 = (x + 1)(x - 2)$

$\quad x = -1, x = 2$

21. $\quad 4(3^x) = 20$

$\quad\quad 3^x = 5$

$\quad \log_3 3^x = \log_3 5$

$\quad\quad\quad x = \log_3 5 = \dfrac{\log 5}{\log 3} \text{ or } \dfrac{\ln 5}{\ln 3}$

$\quad\quad\quad x \approx 1.465$

23. $e^x - 8 = 31$

$\quad\quad e^x = 39$

$\quad \ln e^x = \ln 39$

$\quad\quad\quad x = \ln 39 \approx 3.664$

25. $\quad 3^{2x} = 80$

$\quad \ln 3^{2x} = \ln 80$

$\quad 2x \ln 3 = \ln 80$

$\quad\quad\quad x = \dfrac{\ln 80}{2 \ln 3} \approx 1.994$

27. $\quad\quad 3^{2-x} = 400$

$\quad\quad \ln 3^{2-x} = \ln 400$

$\quad (2 - x) \ln 3 = \ln 400$

$\quad 2 \ln 3 - x \ln 3 = \ln 400$

$\quad\quad -x \ln 3 = \ln 400 - 2 \ln 3$

$\quad\quad x \ln 3 = 2 \ln 3 - \ln 400$

$\quad\quad\quad x = \dfrac{2 \ln 3 - \ln 400}{\ln 3}$

$\quad\quad\quad x = 2 - \dfrac{\ln 400}{\ln 3} \approx -3.454$

29. $\quad 8(10^{3x}) = 12$

$\quad\quad 10^{3x} = \dfrac{12}{8}$

$\quad \log 10^{3x} = \log\left(\dfrac{3}{2}\right)$

$\quad\quad 3x = \log\left(\dfrac{3}{2}\right)$

$\quad\quad x = \tfrac{1}{3} \log\left(\dfrac{3}{2}\right)$

$\quad\quad x \approx 0.059$

31. $e^{3x} = 12$

$\quad 3x = \ln 12$

$\quad\quad x = \dfrac{\ln 12}{3} \approx 0.828$

33. $7 - 2e^x = 5$

$\quad -2e^x = -2$

$\quad\quad e^x = 1$

$\quad\quad x = \ln 1 = 0$

35. $6(2^{3x-1}) - 7 = 9$

$\quad 6(2^{3x-1}) = 16$

$\quad\quad 2^{3x-1} = \dfrac{8}{3}$

$\quad \log_2 2^{3x-1} = \log_2\left(\dfrac{8}{3}\right)$

$\quad 3x - 1 = \log_2\left(\dfrac{8}{3}\right) = \dfrac{\log(8/3)}{\log 2} \text{ or } \dfrac{\ln(8/3)}{\ln 2}$

$\quad\quad x = \dfrac{1}{3}\left[\dfrac{\log(8/3)}{\log 2} + 1\right] \approx 0.805$

37. $\quad\quad 3^x = 2^{x-1}$

$\quad\quad \ln 3^x = \ln 2^{x-1}$

$\quad x \ln 3 = (x - 1) \ln 2$

$\quad x \ln 3 = x \ln 2 - \ln 2$

$\quad x \ln 3 - x \ln 2 = -\ln 2$

$\quad x(\ln 3 - \ln 2) = -\ln 2$

$\quad\quad x = \dfrac{\ln 2}{\ln 2 - \ln 3} \approx -1.710$

39.
$$4^x = 5^{x^2}$$
$$\ln 4^x = \ln 5^{x^2}$$
$$x \ln 4 = x^2 \ln 5$$
$$x^2 \ln 5 - x \ln 4 = 0$$
$$x\left(x \ln 5 - \ln 4\right) = 0$$
$$x = 0$$
$$x \ln 5 - \ln 4 = 0 \Rightarrow x = \frac{\ln 4}{\ln 5} \approx 0.861$$

41.
$$e^{2x} - 4e^x - 5 = 0$$
$$\left(e^x + 1\right)\left(e^x - 5\right) = 0$$
$$e^x = -1 \quad \text{or} \quad e^x = 5$$
$$\text{(No solution)} \quad x = \ln 5 \approx 1.609$$

43.
$$\frac{1}{1 - e^x} = 5$$
$$1 = 5\left(1 - e^x\right)$$
$$\frac{1}{5} = 1 - e^x$$
$$\frac{1}{5} - 1 = -e^x$$
$$-\frac{4}{5} = -e^x$$
$$\frac{4}{5} = e^x$$
$$\ln \frac{4}{5} = \ln e^x$$
$$\ln\frac{4}{5} = x$$
$$x \approx -0.223$$

45.
$$\left(1 + \frac{0.065}{365}\right)^{365t} = 4$$
$$\ln\left(1 + \frac{0.065}{365}\right)^{365t} = \ln 4$$
$$365t \ln\left(1 + \frac{0.065}{365}\right) = \ln 4$$
$$t = \frac{\ln 4}{365 \ln\left(1 + \dfrac{0.065}{365}\right)} \approx 21.330$$

47.
$$\ln x = -3$$
$$x = e^{-3} \approx 0.050$$

49.
$$2.1 = \ln 6x$$
$$e^{2.1} = 6x$$
$$\frac{e^{2.1}}{6} = x$$
$$1.361 \approx x$$

51.
$$3 - 4 \ln x = 11$$
$$-4 \ln x = 8$$
$$\ln x = -2$$
$$x = e^{-2} = \frac{1}{e^2} \approx 0.135$$

53.
$$6 \log_3\left(0.5x\right) = 11$$
$$\log_3\left(0.5x\right) = \frac{11}{6}$$
$$3^{\log_3(0.5x)} = 3^{11/6}$$
$$0.5x = 3^{11/6}$$
$$x = 2\left(3^{11/6}\right) \approx 14.988$$

55.
$$\ln x - \ln(x + 1) = 2$$
$$\ln\left(\frac{x}{x + 1}\right) = 2$$
$$\frac{x}{x + 1} = e^2$$
$$x = e^2(x + 1)$$
$$x = e^2 x + e^2$$
$$x - e^2 x = e^2$$
$$x\left(1 - e^2\right) = e^2$$
$$x = \frac{e^2}{1 - e^2} \approx -1.157$$

This negative value is extraneous. The equation has no solution.

57.
$$\ln\left(x + 5\right) = \ln(x - 1) - \ln(x + 1)$$
$$\ln(x + 5) = \ln\left(\frac{x - 1}{x + 1}\right)$$
$$x + 5 = \frac{x - 1}{x + 1}$$
$$(x + 5)(x + 1) = x - 1$$
$$x^2 + 6x + 5 = x - 1$$
$$x^2 + 5x + 6 = 0$$
$$(x + 2)(x + 3) = 0$$
$$x = -2 \quad \text{or} \quad x = -3$$

Both of these solutions are extraneous, so the equation has no solution.

59. $\log(3x + 4) = \log(x - 10)$

$3x + 4 = x - 10$

$2x = -14$

$x = -7$

The negative value is extraneous.

The equation has no solution.

61. $\log_4 x - \log_4(x - 1) = \dfrac{1}{2}$

$\log_4\!\left(\dfrac{x}{x-1}\right) = \dfrac{1}{2}$

$4^{\log_4[x/(x-1)]} = 4^{1/2}$

$\dfrac{x}{x-1} = 4^{1/2}$

$x = 2(x - 1)$

$x = 2x - 2$

$-x = -2$

$x = 2$

63. $f(x) = 5^x - 212$

Algebraically:

$5^x = 212$

$\ln 5^x = \ln 212$

$x \ln 5 = \ln 212$

$x = \dfrac{\ln 212}{\ln 5}$

$x \approx 3.328$

The zero is $x \approx 3.328$.

65. $g(x) = 8e^{-2x/3} - 11$

Algebraically:

$8e^{-2x/3} = 11$

$e^{-2x/3} = 1.375$

$-\dfrac{2x}{3} = \ln 1.375$

$x = -1.5 \ln 1.375$

$x \approx -0.478$

The zero is $x \approx -0.478$.

67. $y_1 = 3$

$y_2 = \ln x$

From the graph,

$x \approx 20.086$ when $y = 3$.

Algebraically:

$3 - \ln x = 0$

$\ln x = 3$

$x = e^3 \approx 20.086$

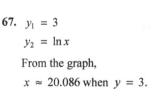

69. $y_1 = 2\ln(x + 3)$

$y_2 = 3$

From the graph, $x \approx 1.482$ when $y = 3$.

Algebraically:

$2\ln(x + 3) = 3$

$\ln(x + 3) = \dfrac{3}{2}$

$x + 3 = e^{3/2}$

$x = e^{3/2} - 3 \approx 1.482$

71. (a) $r = 0.025$

$A = Pe^{rt}$

$5000 = 2500e^{0.025t}$

$2 = e^{0.025t}$

$\ln 2 = 0.025t$

$\dfrac{\ln 2}{0.025} = t$

$t \approx 27.73$ years

(b) $r = 0.025$

$A = Pe^{rt}$

$7500 = 2500e^{0.025t}$

$3 = e^{0.025t}$

$\ln 3 = 0.025t$

$\dfrac{\ln 3}{0.025} = t$

$t \approx 43.94$ years

73. $2x^2 e^{2x} + 2xe^{2x} = 0$

$\left(2x^2 + 2x\right)e^{2x} = 0$

$2x^2 + 2x = 0 \qquad \left(\text{because } e^{2x} \neq 0\right)$

$2x(x + 1) = 0$

$x = 0, -1$

75. $-xe^{-x} + e^{-x} = 0$

$(-x + 1)e^{-x} = 0$

$-x + 1 = 0 \qquad \left(\text{because } e^{-x} \neq 0\right)$

$x = 1$

77. $\dfrac{1 + \ln x}{2} = 0$

$1 + \ln x = 0$

$\ln x = -1$

$x = e^{-1} = \dfrac{1}{e} \approx 0.368$

79. $2x \ln x + x = 0$

$x(2 \ln x + 1) = 0$

$2 \ln x + 1 = 0 \quad$ (because $x > 0$)

$\ln x = -\frac{1}{2}$

$x = e^{-1/2} \approx 0.607$

81. (a)

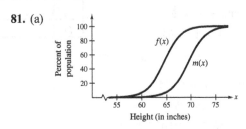

From the graph you see horizontal asymptotes at $y = 0$ and $y = 100$.

These represent the lower and upper percent bounds; the range falls between 0% and 100%.

(b) Males:

$$50 = \frac{100}{1 + e^{-0.5536(x-69.51)}}$$

$1 + e^{-0.5536(x-69.51)} = 2$

$e^{-0.5536(x-69.51)} = 1$

$-0.5536(x - 69.51) = \ln 1$

$-0.5536(x - 69.51) = 0$

$x = 69.51$

The average height of an American male is 69.51 inches.

Females:

$$50 = \frac{100}{1 + e^{-0.5834(x-64.49)}}$$

$1 + e^{-0.5834(x-64.49)} = 2$

$e^{-0.5834(x-64.49)} = 1$

$-0.5834(x - 64.49) = \ln 1$

$-0.5834(x - 64.49) = 0$

$x = 64.49$

The average height of an American female is 64.49 inches.

83. $N = 5.5 \cdot 10^{0.23x}$

When $N = 78$:

$78 = 5.5 \cdot 10^{0.23x}$

$\dfrac{78}{5.5} = 10^{0.23x}$

$\log_{10} \dfrac{78}{5.5} = 0.23x$

$x = \dfrac{\log_{10}(78/5.5)}{0.23} \approx 5.008$ years

The beaver population will reach 78 in about 5 years.

85. $P = 75 \ln t + 540$

Let $P = 720$

$720 = 75 \ln t + 540$

$180 = 75 \ln t$

$\dfrac{180}{75} = \ln t$

$\ln t = 2.4$

$t = e^{2.4} \approx 11.02$ or 2011

87. $T = 20 + 60e^{-0.06m}$

Let $T = 70$

$70 = 20 + 60e^{-0.06m}$

$50 = 60e^{-0.06m}$

$\frac{5}{6} = e^{-0.06m}$

$\ln \frac{5}{6} = -0.06m$

$m = -\frac{1}{0.06} \ln \frac{5}{6}$

$m \approx 3.039$ minutes

89. $\log_a(uv) = \log_a u + \log_a v$

True by Property 1 in Section 5.3.

91. $\log_a(u - v) = \log_a u - \log_a v$

False.

$$1.95 = \log(100 - 10)$$

$$\neq \log 100 - \log 10 = 1$$

93. Yes, a logarithmic equation can have more than one extraneous solution. See Exercise 57.

95. $A = Pe^{rt}$

(a) $A = (2P)e^{rt} = 2(Pe^{rt})$ This doubles your money.

(b) $A = Pe^{(2r)t} = Pe^{rt}e^{rt} = e^{rt}(Pe^{rt})$

(c) $A = Pe^{r(2t)} = Pe^{rt}e^{rt} = e^{rt}(Pe^{rt})$

Doubling the interest rate yields the same result as doubling the number of years.

If $2 > e^{rt}$ (i.e., $rt < \ln 2$), then doubling your investment would yield the most money. If $rt > \ln 2$, then doubling either the interest rate or the number of years would yield more money.

97. (a) $P = 1000, r = 0.07$, compounded annually, $n = 1$

Effective yield: $A = P\left(1 + \dfrac{r}{n}\right)^{nt} = 1000\left(1 + \dfrac{0.07}{1}\right)^{1} = \$1070$

$\dfrac{1070 - 1000}{1000} = 7\%$

The effective yield is 7%.

Balance after 5 years: $A = P\left(1 + \dfrac{r}{n}\right)^{nt} = 1000\left(1 + \dfrac{0.07}{1}\right)^{1(5)} \approx \$1402.55$

(b) $P = 1000, r = 0.07$, compounded continuously

Effective yield: $A = Pe^{rt} = 1000e^{0.07(1)} \approx \$1072.51$

$\dfrac{1072.51 - 1000}{1000} = 7.25\%$

The effective yield is about 7.25%.

Balance after 5 years: $A = Pe^{rt} = 1000e^{0.07(5)} \approx \$1419.07$

(c) $P = 1000, r = 0.07$, compounded quarterly, $n = 4$

Effective yield: $A = P\left(1 + \dfrac{r}{n}\right)^{nt} = 1000\left(1 + \dfrac{0.07}{4}\right)^{4(1)} \approx \$1071.86$

$\dfrac{1071.86 - 1000}{1000} = 7.19\%$

The effective yield is about 7.19%.

Balance after 5 years: $A = P\left(1 + \dfrac{r}{n}\right)^{nt} = 1000\left(1 + \dfrac{0.07}{4}\right)^{4(5)} \approx \$1414.78$

(d) $P = 1000, r = 0.0725$, compounded quarterly, $n = 4$

Effective yield: $A = P\left(1 + \dfrac{r}{n}\right)^{nt} = 1000\left(1 + \dfrac{0.0725}{4}\right)^{4(1)} \approx \$1074.50$

$\dfrac{1074.50 - 1000}{1000} \approx 7.45\%$

The effective yield is about 7.45%.

Balance after 5 years: $A = P\left(1 + \dfrac{r}{n}\right)^{nt} = 1000\left(1 + \dfrac{0.0725}{4}\right)^{4(5)} \approx \$1432.26$

Savings plan (d) has the greatest effective yield and the highest balance after 5 years.

Section 5.5 Exponential and Logarithmic Models

1. $y = ae^{bx}$; $y = ae^{-bx}$

3. normally distributed

5. (a) $A = Pe^{rt}$

$\dfrac{A}{e^{rt}} = P$

(b) $A = Pe^{rt}$

$\dfrac{A}{P} = e^{rt}$

$\ln \dfrac{A}{P} = \ln e^{rt}$

$\ln \dfrac{A}{P} = rt$

$\dfrac{\ln(A/P)}{r} = t$

7. Because $A = 1000e^{0.035t}$, the time to double is given by $2000 = 1000e^{0.035t}$ and you have

$2 = e^{0.035t}$

$\ln 2 = \ln e^{0.035t}$

$\ln 2 = 0.035t$

$t = \dfrac{\ln 2}{0.035} \approx 19.8$ years.

Amount after 10 years: $A = 1000e^{0.35} \approx \$1419.07$

9. Because $A = 750e^{rt}$ and $A = 1500$ when $t = 7.75$, you have

$1500 = 750e^{7.75r}$

$2 = e^{7.75r}$

$\ln 2 = \ln e^{7.75r}$

$\ln 2 = 7.75r$

$r = \dfrac{\ln 2}{7.75} \approx 0.089438 = 8.9438\%.$

Amount after 10 years: $A = 750e^{0.089438(10)} \approx \$1834.37$

11. Because $A = Pe^{0.045t}$ and $A = 10,000.00$ when $t = 10,$ you have

$10,000.00 = Pe^{0.045(10)}$

$\dfrac{10,000.00}{e^{0.045(10)}} = P \approx \$6376.28.$

The time to double is given by

$t = \dfrac{\ln 2}{0.045} \approx 15.40$ years.

13. $A = 500,000, r = 0.05, n = 12, t = 10$

$A = P\left(1 + \dfrac{r}{n}\right)^{nt}$

$500,000 = P\left(1 + \dfrac{0.05}{12}\right)^{12(10)}$

$P = \dfrac{500,000}{\left(1 + \dfrac{0.05}{12}\right)^{12(10)}}$

$\approx \$303,580.52$

15. $P = 1000, r = 0.1, A = 2000$

$A = P\left(1 + \dfrac{r}{n}\right)^{nt}$

$2000 = 1000\left(1 + \dfrac{0.1}{n}\right)^{nt}$

$2 = \left(1 + \dfrac{0.1}{n}\right)^{nt}$

(a) $n = 1$

$(1 + 0.1)^{t} = 2$

$(1.1)^{t} = 2$

$\ln(1.1)^{t} = \ln 2$

$t \ln 1.1 = \ln 2$

$t = \dfrac{\ln 2}{\ln 1.1} \approx 7.27$ years

(b) $n = 12$

$\left(1 + \dfrac{0.1}{12}\right)^{12t} = 2$

$\ln\left(\dfrac{12.1}{12}\right)^{12t} = \ln 2$

$12t \ln\left(\dfrac{12.1}{12}\right) = \ln 2$

$12t = \dfrac{\ln 2}{\ln(12.1/12)}$

$t = \dfrac{\ln 2}{12 \ln(12.1/12)} \approx 6.96$ years

(c) $n = 365$

$$\left(1 + \frac{0.1}{365}\right)^{365t} = 2$$

$$\ln\left(\frac{365.1}{365}\right)^{365t} = \ln 2$$

$$365t \, \ln\left(\frac{365.1}{365}\right) = \ln 2$$

$$365t = \frac{\ln 2}{\ln(365.1/365)}$$

$$t = \frac{\ln 2}{365 \, \ln(365.1/365)} \approx 6.93 \text{ years}$$

(d) Compounded continuously

$$A = Pe^{rt}$$

$$2000 = 1000e^{0.1t}$$

$$2 = e^{0.1t}$$

$$\ln 2 = \ln e^{0.1t}$$

$$0.1t = \ln 2$$

$$t = \frac{\ln 2}{0.1} \approx 6.93 \text{ years}$$

17. (a) $3P = Pe^{rt}$

$$3 = e^{rt}$$

$$\ln 3 = rt$$

$$\frac{\ln 3}{r} = t$$

r	2%	4%	6%	8%	10%	12%
$t = \dfrac{\ln 3}{r}$ (years)	54.93	27.47	18.31	13.73	10.99	9.16

(b) $3P = P(1 + r)^t$

$$3 = (1 + r)^t$$

$$\ln 3 = \ln (1 + r)^t$$

$$\frac{\ln 3}{\ln (1 + r)} = t$$

r	2%	4%	6%	8%	10%	12%
$t = \dfrac{\ln 3}{\ln (1 + r)}$ (years)	55.48	28.01	18.85	14.27	11.53	9.69

19. Continuous compounding results in faster growth.

$$A = 1 + 0.075[\![t]\!] \quad \text{and} \quad A = e^{0.07t}$$

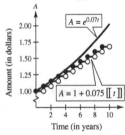

Time (in years)

21. $a = 10, \ y = \dfrac{1}{2}(10) = 5, \ t = 1599$

$$y = ae^{-bt}$$

$$5 = 10e^{-b(1599)}$$

$$0.5 = e^{-1599b}$$

$$\ln 0.5 = \ln e^{-1599b}$$

$$\ln 0.5 = -1599b$$

$$b = -\frac{\ln 0.5}{1599}$$

Given an initial quantity of 10 grams, after 1000 years, you have

$$y = 10e^{-\left[-(\ln 0.5)/1599\right](1000)} \approx 6.48 \text{ grams.}$$

23. $y = 2, \ a = 2(2) = 4, \ t = 5715$

$$y = ae^{-bt}$$

$$2 = 4e^{-b(5715)}$$

$$0.5 = e^{-5715b}$$

$$\ln 0.5 = \ln e^{-5715b}$$

$$\ln 0.5 = -5715b$$

$$b = -\frac{\ln 0.5}{5715}$$

Given 2 grams after 1000 years, the initial amount is

$$2 = ae^{-\left[-(\ln 0.5)/5715\right](1000)}$$

$$a \approx 2.26 \text{ grams.}$$

25.

$$y = ae^{bx}$$

$$1 = ae^{b(0)} \Rightarrow 1 = a$$

$$10 = e^{b(3)}$$

$$\ln 10 = 3b$$

$$\frac{\ln 10}{3} = b \Rightarrow b \approx 0.7675$$

So, $y = e^{0.7675x}$.

27.
$$y = ae^{bx}$$

$$5 = ae^{b(0)} \Rightarrow 5 = a$$

$$1 = 5e^{b(4)}$$

$$\frac{1}{5} = e^{4b}$$

$$\ln\left(\frac{1}{5}\right) = 4b$$

$$\frac{\ln(1/5)}{4} = b \Rightarrow b \approx -0.4024$$

So, $y = 5e^{-0.4024x}$.

29. (a) $P = 76.6e^{0.0313t}$

Year	1980	1990	2000	2010
P	104.752	143.251	195.899	267.896
Population	104,752	143,251	195,899	267,896

(b) Let $P = 360$, and solve for t.

$$360 = 76.6e^{0.0313t}$$

$$\frac{360}{76.6} = e^{0.0313t}$$

$$\ln\left(\frac{360}{76.6}\right) = 0.0313t$$

$$\frac{1}{0.0313}\ln\left(\frac{360}{76.6}\right) = t$$

$$49.4 \approx t$$

According to the model, the population will reach 360,000 in 2019.

(c) No; As t increases, the population increases rapidly.

31. $y = 4080e^{kt}$

When $t = 3$, $y = 10,000$:

$$10,000 = 4080e^{k(3)}$$

$$\frac{10,000}{4080} = e^{3k}$$

$$\ln\left(\frac{10,000}{4080}\right) = 3k$$

$$k = \frac{\ln(10,000/4080)}{3} \approx 0.2988$$

When $t = 24$: $y = 4080e^{0.2988(24)} \approx 5,309,734$ hits

33. $y = ae^{bt}$

When $t = 3$, $y = 100$:   When $t = 5$, $y = 400$:

$$100 = ae^{3b} \qquad\qquad 400 = ae^{5b}$$

$$\frac{100}{e^{3b}} = a$$

Substitute $\dfrac{100}{e^{3b}}$ for a in the equation on the right.

$$400 = \frac{100}{e^{3b}}e^{5b}$$

$$400 = 100e^{2b}$$

$$4 = e^{2b}$$

$$\ln 4 = 2b$$

$$\ln 2^2 = 2b$$

$$2\ln 2 = 2b$$

$$\ln 2 = b$$

$$a = \frac{100}{e^{3b}} = \frac{100}{e^{3\ln 2}} = \frac{100}{e^{\ln 2^3}} = \frac{100}{2^3} = \frac{100}{8} = 12.5$$

$$y = 12.5e^{(\ln 2)t}$$

After 6 hours, there are $y = 12.5e^{(\ln 2)(6)} = 800$ bacteria.

35. $(0, 575), (2, 275)$

(a) $m = \dfrac{275 - 575}{2 - 0} = -150$

$V = -150t + 575$

(b) Since $V = 575$, when

$t = 0, 575 = ae^{(b)(0)} \rightarrow a = 575$

Then $275 = 575e^{k(2)}$

$\ln\left(\dfrac{275}{575}\right) = 2k \Rightarrow k \approx -0.3688$

$V = 575e^{-0.3688t}$

(c)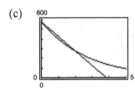

The exponential model depreciates faster in the first two years.

(d)

t	1	3
$V = -150t + 575$	\$425	\$125
$V = 575e^{-0.3688t}$	\$397.65	\$190.18

(e) Answers will vary. Sample Answer: The slope of the linear model means that the laptop depreciates \$150 per year, then loses all value late in the third year. The exponential model depreciates faster in the first three years but maintains value longer.

37. $R = \dfrac{1}{10^{12}}e^{-t/8223}$

$R = \dfrac{1}{8^{14}}$

$\dfrac{1}{10^{12}}e^{-t/8223} = \dfrac{1}{8^{14}}$

$e^{-t/8223} = \dfrac{10^{12}}{8^{14}}$

$-\dfrac{t}{8223} = \ln\left(\dfrac{10^{12}}{8^{14}}\right)$

$t = -8223 \ln\left(\dfrac{10^{12}}{8^{14}}\right) \approx 12{,}180$ years old

39. $y = 0.0266e^{-(x-100)^2/450}, 70 \le x \le 116$

(a)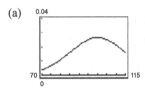

(b) The average IQ score of an adult student is 100.

41. (a) 1998: $t = 18, y = \dfrac{320{,}110}{1 + 374e^{-0.252(18)}}$

$\approx 63{,}992$ sites

2003: $t = 23, y = \dfrac{320{,}110}{1 + 374e^{-0.252(23)}}$

$\approx 149{,}805$ sites

2006: $t = 26, y = \dfrac{320{,}110}{1 + 374e^{-0.252(26)}}$

$\approx 208{,}705$ sites

(b)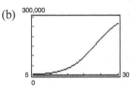

(c) When $y = 270{,}000, t \approx 30.2$. So, the number of cell sites will reach 270,000 in the year 2010.

(d) Let $y = 270{,}000$ and solve for t.

$270{,}000 = \dfrac{320{,}110}{1 + 374e^{-0.252t}}$

$1 + 374e^{-0.252t} = \dfrac{320{,}110}{270{,}000}$

$374e^{-0.252t} = 0.1855926$

$e^{-0.252t} \approx 0.000496237$

$-0.252t \approx \ln(0.000496237)$

$t \approx 30.2$

The number of cell sites will reach 270,000 during the year 2010.

43. $p(t) = \dfrac{1000}{1 + 9e^{-0.1656t}}$

   (a) $p(5) = \dfrac{1000}{1 + 9e^{-0.1656(5)}} \approx 203$ animals

   (b) $500 = \dfrac{1000}{1 + 9e^{-0.1656t}}$

   $1 + 9e^{-0.1656t} = 2$

   $9e^{-0.1656t} = 1$

   $e^{-0.1656t} = \dfrac{1}{9}$

   $t = -\dfrac{\ln(1/9)}{0.1656} \approx 13$ months

   (c)

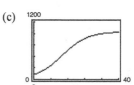

   The horizontal asymptotes are $p = 0$ and $p = 1000$.

   The asymptote with the larger p-value, $p = 1000$, indicates that the population size will approach 1000 as time increases.

45. $R = \log \dfrac{I}{I_0} = \log I$ because $I_0 = 1$.

   (a) $R = 7.6$

   $7.6 = \log I$

   $10^{7.6} = 10^{\log I}$

   $39{,}810{,}717 \approx I$

   (b) $R = 5.6$

   $5.6 = \log I$

   $10^{5.6} = 10^{\log I}$

   $10^{5.6} = I$

   $398{,}107 \approx I$

   (c) $R = 6.6$

   $6.6 = \log I$

   $10^{6.6} = 10^{\log I}$

   $3{,}981{,}072 \approx I$

47. $\beta = 10 \log \dfrac{I}{I_0}$ where $I_0 = 10^{-12}$ watt/m^2.

   (a) $\beta = 10 \log \dfrac{10^{-10}}{10^{-12}} = 10 \log 10^2 = 20$ decibels

   (b) $\beta = 10 \log \dfrac{10^{-5}}{10^{-12}} = 10 \log 10^7 = 70$ decibels

   (c) $\beta = 10 \log \dfrac{10^{-8}}{10^{-12}} = 10 \log 10^4 = 40$ decibels

   (d) $\beta = 10 \log \dfrac{10^{-3}}{10^{-12}} = 10 \log 10^9 = 90$ decibels

49. $\beta = 10 \log \dfrac{I}{I_0}$

   $\dfrac{\beta}{10} = \log \dfrac{I}{I_0}$

   $10^{\beta/10} = 10^{\log I/I_0}$

   $10^{\beta/10} = \dfrac{I}{I_0}$

   $I = I_0 10^{\beta/10}$

   % decrease $= \dfrac{I_0 10^{9.3} - I_0 10^{8.0}}{I_0 10^{9.3}} \times 100 \approx 95\%$

51. $\text{pH} = -\log\left[H^+\right]$

   $-\log\left(2.3 \times 10^{-5}\right) \approx 4.64$

53. $5.8 = -\log\left[H^+\right]$

   $-5.8 = \log\left[H^+\right]$

   $10^{-5.8} = 10^{\log\left[H^+\right]}$

   $10^{-5.8} = \left[H^+\right]$

   $\left[H^+\right] \approx 1.58 \times 10^{-6}$ moles per liter

55. $2.9 = -\log\left[H^+\right]$

   $-2.9 = \log\left[H^+\right]$

   $\left[H^+\right] = 10^{-2.9}$ for the apple juice

   $8.0 = -\log\left[H^+\right]$

   $-8.0 = \log\left[H^+\right]$

   $\left[H^+\right] = 10^{-8}$ for the drinking water

   $\dfrac{10^{-2.9}}{10^{-8}} = 10^{5.1}$ times the hydrogen ion concentration of drinking water

57. $t = -10 \ln \dfrac{T - 70}{98.6 - 70}$

At 9:00 A.M. you have: $t = -10 \ln \dfrac{85.7 - 70}{98.6 - 70} \approx 6$ hours

From this you can conclude that the person died at 3:00 A.M.

59. $u = 120,000 \left[\dfrac{0.075t}{1 - \left(\dfrac{1}{1 + 0.075/12} \right)^{12t}} - 1 \right]$

(a)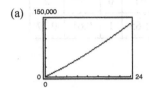

150,000

0   24

(b) From the graph, $u = \$120,000$ when $t \approx 21$ years. It would take approximately 37.6 years to pay $240,000 in interest. Yes, it is possible to pay twice as much in interest charges as the size of the mortgage. It is especially likely when the interest rates are higher.

61. False. The domain can be the set of real numbers for a logistic growth function.

63. False. The graph of $f(x)$ is the graph of $g(x)$ shifted upward five units.

65. Answers will vary.

Review Exercises for Chapter 5

1. $f(x) = 0.3^x$

$f(1.5) = 0.3^{1.5} \approx 0.164$

3. $f(x) = 2^x$

$f\left(\tfrac{2}{3}\right) = 2^{2/3} \approx 1.587$

5. $f(x) = 7\left(0.2^x\right)$

$f\left(-\sqrt{11}\right) = 7\left(0.2^{-\sqrt{11}}\right)$

≈ 1456.529

7. $f(x) = 4^{-x} + 4$

Horizontal asymptote: $y = 4$

x	-1	0	1	2	3
$f(x)$	8	5	4.25	4.063	4.016

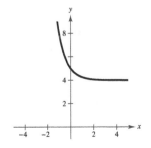

9. $f(x) = 5^{x-2} + 4$

Horizontal asymptote: $y = 4$

x	-1	0	1	2	3
$f(x)$	4.008	4.04	4.2	5	9

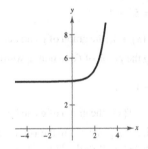

11. $f(x) = \left(\frac{1}{2}\right)^{-x} + 3 = 2^x + 3$

Horizontal asymptote: $y = 3$

x	-2	-1	0	1	2
$f(x)$	3.25	3.5	4	5	7

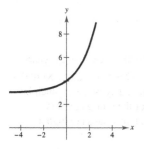

13. $\left(\frac{1}{3}\right)^{x-3} = 9$

$\left(\frac{1}{3}\right)^{x-3} = 3^2$

$\left(\frac{1}{3}\right)^{x-3} = \left(\frac{1}{3}\right)^{-2}$

$x - 3 = -2$

$x = 1$

15. $e^{3x-5} = e^7$

$3x - 5 = 7$

$3x = 12$

$x = 4$

17. $f(x) = 5^x,\ g(x) = 5^x + 1$

Because $g(x) = f(x) + 1$, the graph of g can be obtained by shifting the graph of f one unit upward.

19. $f(x) = 3^x,\ g(x) = 1 - 3^x$

Because $g(x) = 1 - f(x)$, the graph of g can be obtained by reflecting the graph of f in the x-axis and shifting the graph one unit upward. (**Note:** This is equivalent to shifting the graph of f one unit upward and then reflecting the graph in the x-axis.)

21. $f(x) = e^x$

$f(3.4) = e^{3.4} \approx 29.964$

23. $f(x) = e^x$

$f\left(\frac{3}{5}\right) = e^{3/5} \approx 1.822$

25. $h(x) = e^{-x/2}$

x	-2	-1	0	1	2
$h(x)$	2.72	1.65	1	0.61	0.37

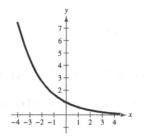

27. $f(x) = e^{x+2}$

x	-3	-2	-1	0	1
$f(x)$	0.37	1	2.72	7.39	20.09

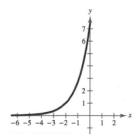

29. $F(t) = 1 - e^{-t/3}$

(a) $F(1) \approx 0.283$

(b) $F(2) \approx 0.487$

(c) $F(5) \approx 0.811$

31. $P = \$5000,\ r = 3\%,\ t = 10$ years

Compounded n times per year: $A = P\left(1 + \dfrac{r}{n}\right)^{nt} = 5000\left(1 + \dfrac{0.03}{n}\right)^{10n}$

Compounded continuously: $A = Pe^{rt} = 5000e^{0.03(10)}$

n	1	2	4	12	365	Continuous
A	\$6719.58	\$6734.28	\$6741.74	\$6746.77	\$6749.21	\$6749.29

33. $3^3 = 27$

$\log_3 27 = 3$

35. $e^{0.8} = 2.2255\ldots$

$\ln 2.2255\ldots = 0.8$

37. $f(x) = \log x$

$f(1000) = \log 1000$

$= \log 10^3 = 3$

39. $g(x) = \log_2 x$

$g\!\left(\tfrac{1}{4}\right) = \log_2 \tfrac{1}{4}$

$= \log_2 2^{-2} = -2$

41. $\log_4(x + 7) = \log_4 14$

$x + 7 = 14$

$x = 7$

43. $\ln(x + 9) = \ln 4$

$x + 9 = 4$

$x = -5$

45. $g(x) = \log_7 x \Rightarrow x = 7^y$

Domain: $(0, \infty)$

x-intercept: $(1, 0)$

Vertical asymptote: $x = 0$

x	$\tfrac{1}{7}$	1	7	49
$g(x)$	-1	0	1	2

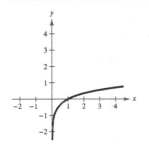

47. $f(x) = 4 - \log(x + 5)$

Domain: $(-5, \infty)$

Because

$4 - \log(x + 5) = 0 \Rightarrow \log(x + 5) = 4$

$x + 5 = 10^4$

$x = 10^4 - 5$

$= 9995.$

x-intercept: $(9995, 0)$

Vertical asymptote: $x = -5$

x	-4	-3	-2	-1	0	1
$f(x)$	4	3.70	3.52	3.40	3.30	3.22

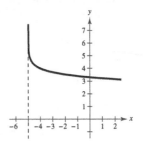

49. $f(22.6) = \ln 22.6 \approx 3.118$

51. $f\!\left(\sqrt{e}\right) = \tfrac{1}{2} \ln \sqrt{e} = 0.25$

53. $f(x) = \ln x + 6 = 6 + \ln x$

Domain: $(0, \infty)$

$\ln x + 6 = 0$

$\ln x = -6$

$x = e^{-6}$

x-intercept: $\left(e^{-6}, 0\right)$

Vertical asymptote: $x = 0$

x	$\tfrac{1}{4}$	$\tfrac{1}{2}$	1	2	3
$f(x)$	4.613	5.037	6	6.693	7.098

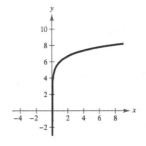

55. $f(x) = \ln(x - 6)$

Domain: $(6, \infty)$

$$\ln(x - 6) = 0$$
$$x - 6 = e^0$$
$$x - 6 = 1$$
$$x = 7$$

x-intercept: $(7, 0)$

Vertical asymptote: $x = 6$

x	6.5	7	8	9	10
$f(x)$	-0.693	0	0.693	1.099	1.386

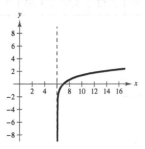

57. $M = m - 5 \log\left(\dfrac{d}{10}\right)$

Let $m = 2.08$ and $M = 1.3$ and solve for d.

$$1.3 = 2.08 - 5 \log\left(\frac{d}{10}\right)$$

$$-0.78 = -5 \log\left(\frac{d}{10}\right)$$

$$0.156 = \log\left(\frac{d}{10}\right)$$

$$10^{0.156} = 10^{\log(d/10)}$$

$$10^{0.156} = \frac{d}{10}$$

$$10 \cdot 10^{0.156} = d$$

$$d = 10^{1.156} \approx 14.32 \text{ parsecs}$$

59. (a) $\log_2 6 = \dfrac{\log 6}{\log 2} \approx 2.585$

(b) $\log_2 6 = \dfrac{\ln 6}{\ln 2} \approx 2.585$

61. (a) $\log_{1/2} 5 = \dfrac{\log 5}{\log(1/2)} \approx -2.322$

(b) $\log_{1/2} 5 = \dfrac{\ln 5}{\ln(1/2)} \approx -2.322$

63. $\log_2 \frac{5}{3} = \log_2 5 - \log_2 3$

65. $\log_2 \frac{9}{5} = \log_2 9 - \log_2 5$
$$= \log_2 3^2 - \log_2 9$$
$$= 2 \log_2 3 - \log_2 5$$

67. $\log 7x^2 = \log 7 + \log x^2$
$$= \log 7 + 2 \log x$$

69. $\log_3 \dfrac{9}{\sqrt{x}} = \log_3 9 - \log_3 \sqrt{x}$
$$= \log_3 3^2 - \log_3 x^{1/2}$$
$$= 2 - \frac{1}{2} \log_3 x$$

71. $\ln x^2 y^2 z = \ln x^2 + \ln y^2 + \ln z$
$$= 2 \ln x + 2 \ln y + \ln z$$

73. $\ln 7 + \ln x = \ln(7x)$

75. $\log x - \dfrac{1}{2} \log y = \log x - \log y^{1/2}$
$$= \log\left(\frac{x}{\sqrt{y}}\right)$$

77. $\dfrac{1}{2} \log_3 x - 2 \log_3(y + 8) = \log_3 x^{1/2} - \log_3(y + 8)^2$
$$= \log_3 \sqrt{x} - \log_3(y + 8)^2$$
$$= \log_3 \frac{\sqrt{x}}{(y + 8)^2}$$

79. $t = 50 \log \dfrac{18{,}000}{18{,}000 - h}$

(a) Domain: $0 \le h < 18{,}000$

(b)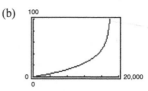

Vertical asymptote: $h = 18{,}000$

(c) As the plane approaches its absolute ceiling, it climbs at a slower rate, so the time required increases.

(d) $50 \log \dfrac{18{,}000}{18{,}000 - 4000} \approx 5.46$ minutes

81. $5^x = 125$

$\qquad 5^x = 5^3$

$\qquad x = 3$

83. $e^x = 3$

$\qquad x = \ln 3 \approx 1.099$

85. $\ln x = 4$

$\qquad x = e^4 \approx 54.598$

87. $e^{4x} = e^{x^2+3}$

$\qquad 4x = x^2 + 3$

$\qquad 0 = x^2 - 4x + 3$

$\qquad 0 = (x - 1)(x - 3)$

$\qquad x = 1, \; x = 3$

89. $2^x - 3 = 29$

$\qquad 2^x = 32$

$\qquad 2^x = 2^5$

$\qquad x = 5$

91. $\ln 3x = 8.2$

$\qquad e^{\ln 3x} = e^{8.2}$

$\qquad 3x = e^{8.2}$

$\qquad x = \dfrac{e^{8.2}}{3} \approx 1213.650$

93. $\ln x + \ln(x - 3) = 1$

$\qquad \ln\left[x(x - 3)\right] = 1$

$\qquad \ln\left(x^2 - 3x\right) = 1$

$\qquad e^{\ln\left(x^2 - 3x\right)} = e^1$

$\qquad x^2 - 3x - e = 0$

$\qquad\qquad x = \dfrac{-b \pm \sqrt{b^2 - 4ac}}{2a}$

$\qquad\qquad x = \dfrac{-(-3) \pm \sqrt{(-3)^2 - 4(1)(-e)}}{2(1)}$

$\qquad\qquad x = \dfrac{3 \pm \sqrt{9 + 4e}}{2}$

$\qquad\qquad x = \dfrac{3 + \sqrt{9 + 4e}}{2} \approx 3.729$

$\qquad\qquad x = \dfrac{3 - \sqrt{9 + 4e}}{2}$ is extraneous since the domain of the $\ln x$ term is $x > 0$.

95. $\log_8(x - 1) = \log_8(x - 2) - \log_8(x + 2)$

$\qquad \log_8(x - 1) = \log_8\left(\dfrac{x - 2}{x + 2}\right)$

$\qquad\qquad x - 1 = \dfrac{x - 2}{x + 2}$

$\qquad (x - 1)(x + 2) = x - 2$

$\qquad x^2 + x - 2 = x - 2$

$\qquad\qquad x^2 = 0$

$\qquad\qquad x = 0$

Because $x = 0$ is not in the domain of $\log_8(x - 1)$ or of $\log_8(x - 2)$, it is an extraneous solution. The equation has no solution.

97. $\log(1 - x) = -1$

$$1 - x = 10^{-1}$$

$$1 - \tfrac{1}{10} = x$$

$$x = 0.900$$

99. $25e^{-0.3x} = 12$

Graph $y_1 = 25e^{-0.3x}$ and $y_2 = 12$.

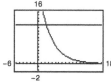

The graphs intersect at $x \approx 2.447$.

101. $2\ln(x + 3) - 3 = 0$

Graph $y_1 = 2\ln(x + 3) - 3$.

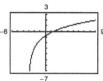

The x-intercept is at $x \approx 1.482$.

103. $P = 8500,\ A = 3(8500) = 25{,}500,\ r = 1.5\%$

$$A = Pe^{rt}$$

$$25{,}500 = 8500e^{0.015t}$$

$$3 = e^{0.015t}$$

$$\ln 3 = 0.015t$$

$$t = \frac{\ln 3}{0.015} \approx 73.2 \text{ years}$$

105. $y = 3e^{-2x/3}$

Exponential decay model

Matches graph (e).

106. $y = 4e^{2x/3}$

Exponential growth model

Matches graph (b).

107. $y = \ln(x + 3)$

Logarithmic model

Vertical asymptote: $x = -3$

Graph includes $(-2, 0)$

Matches graph (f).

108. $y = 7 - \log(x + 3)$

Logarithmic model

Vertical asymptote: $x = -3$

Matches graph (d).

109. $y = 2e^{-(x+4)^2/3}$

Gaussian model

Matches graph (a).

110. $y = \dfrac{6}{1 + 2e^{-2x}}$

Logistics growth model

Matches graph (c).

111. $y = ae^{bx}$

Using the point $(0, 2)$, you have

$$2 = ae^{b(0)}$$

$$2 = ae^0$$

$$2 = a(1)$$

$$2 = a$$

Then, using the point $(4, 3)$, you have

$$3 = 2e^{b(4)}$$

$$3 = 2e^{4b}$$

$$\tfrac{3}{2} = e^{4b}$$

$$\ln \tfrac{3}{2} = 4b$$

$$\tfrac{1}{4}\ln\left(\tfrac{3}{2}\right) = b$$

So, $y = 2e^{\frac{1}{4}\ln\left(\frac{3}{2}\right)x}$

or

$$y = 2e^{0.1014x}$$

113. $y = 0.0499e^{-(x-71)^2/128},\ 40 \le x \le 100$

Graph $y_1 = 0.0499e^{-(x-71)^2/128}$.

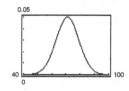

The average test score is 71.

115. $\beta = 10 \log\left(\dfrac{I}{10^{-12}}\right)$

$\dfrac{\beta}{10} = \log\left(\dfrac{I}{10^{-12}}\right)$

$10^{\beta/10} = \dfrac{I}{10^{-12}}$

$I = 10^{\beta/10-12}$

(a) $\beta = 60$

$I = 10^{60/10-12}$

$= 10^{-6} \text{ watt/m}^2$

(b) $\beta = 135$

$I = 10^{135/10-12}$

$= 10^{1.5}$

$= 10\sqrt{10} \text{ watts/m}^2$

(c) $\beta = 1$

$I = 10^{1/10-12}$

$= 10^{\frac{1}{10}} \times 10^{-12}$

$\approx 1.259 \times 10^{-12} \text{ watt/m}^2$

117. True. By the inverse properties, $\log_b b^{2x} = 2x$.

Problem Solving for Chapter 5

1. $y = a^x$

$y_1 = 0.5^x$

$y_2 = 1.2^x$

$y_3 = 2.0^x$

$y_4 = x$

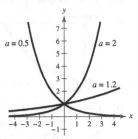

The curves $y = 0.5^x$ and $y = 1.2^x$ cross the line $y = x$.

From checking the graphs it appears that $y = x$ will cross $y = a^x$ for $0 \le a \le 1.44$.

3. The exponential function, $y = e^x$, increases at a faster rate than the polynomial $y = x^n$.

5. (a) $f(u + v) = a^{u+v} = a^u \cdot a^v = f(u) \cdot f(v)$

(b) $f(2x) = a^{2x} = \left(a^x\right)^2 = \left[f(x)\right]^2$

7. (a)
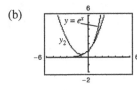

(b)
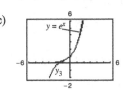

(c)

9.
$$f(x) = e^x - e^{-x}$$
$$y = e^x - e^{-x}$$
$$x = e^y - e^{-y}$$
$$x = \frac{e^{2y} - 1}{e^y}$$
$$xe^y = e^{2y} - 1$$
$$e^{2y} - xe^y - 1 = 0$$
$$e^y = \frac{x \pm \sqrt{x^2 + 4}}{2} \quad \text{Quadratic Formula}$$

Choosing the positive quantity for e^y you have

$$y = \ln\left(\frac{x + \sqrt{x^2 + 4}}{2}\right). \text{ So,}$$

$$f^{-1}(x) = \ln\left(\frac{x + \sqrt{x^2 + 4}}{2}\right).$$

11. Answer (c). $y = 6\left(1 - e^{-x^2/2}\right)$

The graph passes through $(0, 0)$ and neither (a) nor (b) pass through the origin. Also, the graph has y-axis symmetry and a horizontal asymptote at $y = 6$.

13. $y_1 = c_1\left(\frac{1}{2}\right)^{t/k_1}$ and $y_2 = c_2\left(\frac{1}{2}\right)^{t/k_2}$

$$c_1\left(\frac{1}{2}\right)^{t/k_1} = c_2\left(\frac{1}{2}\right)^{t/k_2}$$

$$\frac{c_1}{c_2} = \left(\frac{1}{2}\right)^{(t/k_2 - t/k_1)}$$

$$\ln\left(\frac{c_1}{c_2}\right) = \left(\frac{t}{k_2} - \frac{t}{k_1}\right)\ln\left(\frac{1}{2}\right)$$

$$\ln c_1 - \ln c_2 = t\left(\frac{1}{k_2} - \frac{1}{k_1}\right)\ln\left(\frac{1}{2}\right)$$

$$t = \frac{\ln c_1 - \ln c_2}{\left[(1/k_2) - (1/k_1)\right]\ln(1/2)}$$

15. (a) $y_1 \approx 252{,}606(1.0310)^t$

(b) $y_2 \approx 400.88t^2 - 1464.6t + 291{,}782$

(c)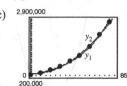

(d) The exponential model is a better fit for the data, but neither would be reliable to predict the population of the United States in 2020. The exponential model approaches infinity rapidly.

17.
$$(\ln x)^2 = \ln x^2$$
$$(\ln x)^2 - 2\ln x = 0$$
$$\ln x(\ln x - 2) = 0$$
$$\ln x = 0 \text{ or } \ln x = 2$$
$$x = 1 \text{ or } \quad x = e^2$$

19. $y_4 = (x - 1) - \frac{1}{2}(x - 1)^2 + \frac{1}{3}(x - 1)^3 - \frac{1}{4}(x - 1)^4$

The pattern implies that

$$\ln x = (x - 1) - \frac{1}{2}(x - 1)^2 + \frac{1}{3}(x - 1)^3 - \frac{1}{4}(x - 1)^4 + \dots.$$

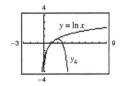

21. $y = 80.4 - 11 \ln x$

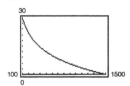

$$y(300) = 80.4 - 11 \ln 300 \approx 17.7 \text{ ft}^3/\text{min}$$

23. (a)

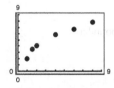

(b) The data could best be modeled by a logarithmic model.

(c) The shape of the curve looks much more logarithmic than linear or exponential.

(d) $y \approx 2.1518 + 2.7044 \ln x$

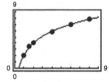

(e) The model is a good fit to the actual data.

25. (a)

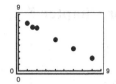

(b) The data could best be modeled by a linear model.

(c) The shape of the curve looks much more linear than exponential or logarithmic.

(d) $y \approx -0.7884x + 8.2566$

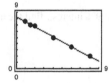

(e) The model is a good fit to the actual data.

Practice Test for Chapter 5

1. Solve for x: $x^{3/5} = 8$.

2. Solve for x: $3^{x-1} = \frac{1}{81}$.

3. Graph $f(x) = 2^{-x}$.

4. Graph $g(x) = e^x + 1$.

5. If $5000 is invested at 9% interest, find the amount after three years if the interest is compounded
   (a) monthly.
   (b) quarterly.
   (c) continuously.

6. Write the equation in logarithmic form: $7^{-2} = \frac{1}{49}$.

7. Solve for x: $x - 4 = \log_2 \frac{1}{64}$.

8. Given $\log_b 2 = 0.3562$ and $\log_b 5 = 0.8271$, evaluate $\log_b \sqrt[4]{8/25}$.

9. Write $5 \ln x - \frac{1}{2} \ln y + 6 \ln z$ as a single logarithm.

10. Using your calculator and the change of base formula, evaluate $\log_9 28$.

11. Use your calculator to solve for N: $\log_{10} N = 0.6646$

12. Graph $y = \log_4 x$.

13. Determine the domain of $f(x) = \log_3(x^2 - 9)$.

14. Graph $y = \ln(x - 2)$.

15. True or false: $\dfrac{\ln x}{\ln y} = \ln(x - y)$

16. Solve for x: $5^x = 41$

17. Solve for x: $x - x^2 = \log_5 \frac{1}{25}$

18. Solve for x: $\log_2 x + \log_2(x - 3) = 2$

19. Solve for x: $\dfrac{e^x + e^{-x}}{3} = 4$

20. Six thousand dollars is deposited into a fund at an annual interest rate of 13%. Find the time required for the investment to double if the interest is compounded continuously.

CHAPTER 6
Topics in Analytic Geometry

CHAPTER 6
Topics in Analytic Geometry

Section 6.1   Lines

1. inclination

3. $\left| \dfrac{m_2 - m_1}{1 + m_1 m_2} \right|$

5. $m = \tan \dfrac{\pi}{6} = \dfrac{\sqrt{3}}{3}$

7. $m = \tan \dfrac{3\pi}{4} = -1$

9. $m = \tan \dfrac{\pi}{3} = \sqrt{3}$

11. $m = \tan 0.39 \approx 0.4111$

13. $m = \tan 1.27 \approx 3.2236$

15. $m = \tan 1.81 \approx -4.1005$

17. $m = 1$

$1 = \tan \theta$

$\theta = \dfrac{\pi}{4}$ radian $= 45°$

19. $m = \dfrac{2}{3}$

$\dfrac{2}{3} = \tan \theta$

$\theta = \arctan\left(\dfrac{2}{3}\right)$

≈ 0.5880 radian $\approx 33.7°$

21. $m = -1$

$-1 = \tan \theta$

$\theta = 180° + \arctan(-1)$

$= \dfrac{3\pi}{4}$ radians $= 135°$

23. $m = -\dfrac{3}{2}$

$-\dfrac{3}{2} = \tan \theta$

$\theta = \tan^{-1}\left(-\dfrac{3}{2}\right) + \pi$

≈ 2.1588 radians $\approx 123.7°$

25. $\left(\sqrt{3}, 2\right), (0, 1)$

$m = \dfrac{1 - 2}{0 - \sqrt{3}} = \dfrac{-1}{-\sqrt{3}} = \dfrac{1}{\sqrt{3}}$

$\dfrac{1}{\sqrt{3}} = \tan \theta$

$\theta = \arctan \dfrac{1}{\sqrt{3}}$

$= \dfrac{\pi}{6}$ radian $= 30°$

27. $\left(-\sqrt{3}, -1\right), (0, -2)$

$m = \dfrac{-2 - (-1)}{0 - \left(-\sqrt{3}\right)} = \dfrac{-1}{\sqrt{3}}$

$-\dfrac{1}{\sqrt{3}} = \tan \theta$

$\theta = \arctan\left(-\dfrac{1}{\sqrt{3}}\right) = \dfrac{5\pi}{6}$ radians $= 150°$

29. $(6, 1), (10, 8)$

$m = \dfrac{8 - 1}{10 - 6} = \dfrac{7}{4}$

$\dfrac{7}{4} = \tan \theta$

$\theta = \arctan \dfrac{7}{4} \approx 1.0517$ radians $\approx 60.3°$

31. $(-2, 20), (10, 0)$

$m = \dfrac{0 - 20}{10 - (-2)} = -\dfrac{20}{12} = -\dfrac{5}{3}$

$-\dfrac{5}{3} = \tan \theta$

$\theta = \pi + \arctan\left(-\dfrac{5}{3}\right) \approx 2.1112$ radians $\approx 121.0°$

33. $\left(\dfrac{1}{4}, \dfrac{3}{2}\right), \left(\dfrac{1}{3}, \dfrac{1}{2}\right)$

$m = \dfrac{\frac{1}{2} - \frac{3}{2}}{\frac{1}{3} - \frac{1}{4}} = -\dfrac{1}{\frac{1}{12}} = -12$

$-12 = \tan \theta$

$\theta = \arctan(-12) + \pi \approx 1.6539$ radians

$\approx 94.8°$

35. $2x + 2y - 5 = 0$

$$y = -x + \frac{5}{2} \Rightarrow m = -1$$

$$-1 = \tan \theta$$

$$\theta = \arctan(-1) = \frac{3\pi}{4} \text{ radians} = 135°$$

37. $3x - 3y + 1 = 0$

$$y = x + \frac{1}{3} \Rightarrow m = 1$$

$$1 = \tan \theta$$

$$\theta = \arctan 1 = \frac{\pi}{4} \text{ radian} = 45°$$

39. $x + \sqrt{3}y + 2 = 0$

$$y = -\frac{1}{\sqrt{3}}x - \frac{2}{\sqrt{3}} \Rightarrow m = -\frac{1}{\sqrt{3}}$$

$$-\frac{1}{\sqrt{3}} = \tan \theta$$

$$\theta = \arctan\left(-\frac{1}{\sqrt{3}}\right) = \frac{5\pi}{6} \text{ radians} = 150°$$

41. $6x - 2y + 8 = 0$

$$y = 3x + 4 \Rightarrow m = 3$$

$$3 = \tan \theta$$

$$\theta = \arctan 3 \approx 1.2490 \text{ radians} \approx 71.6°$$

43. $4x + 5y - 9 = 0$

$$y = -\frac{4}{5}x + \frac{9}{5} \Rightarrow m = -\frac{4}{5}$$

$$-\frac{4}{5} = \tan \theta$$

$$\theta = \tan^{-1}\left(-\frac{4}{5}\right) + \pi$$

$$\approx 2.4669 \text{ radians} \approx 141.3°$$

45. $3x + y = 3 \Rightarrow y = -3x + 3 \Rightarrow m_1 = -3$

$x - y = 2 \Rightarrow y = x - 2 \Rightarrow m_2 = 1$

$$\tan \theta = \left| \frac{1 - (-3)}{1 + (-3)(1)} \right| = 2$$

$$\theta = \arctan 2 \approx 1.1071 \text{ radians} \approx 63.4°$$

47. $x - y = 0 \Rightarrow y = x \Rightarrow m_1 = 1$

$3x - 2y = -1 \Rightarrow y = \frac{3}{2}x + \frac{1}{2} \Rightarrow m_2 = \frac{3}{2}$

$$\tan \theta = \left| \frac{\frac{3}{2} - 1}{1 + \left(\frac{3}{2}\right)(1)} \right| = \frac{1}{5}$$

$$\theta = \arctan \frac{1}{5} \approx 0.1974 \text{ radian} \approx 11.3°$$

49. $x - 2y = 7 \Rightarrow y = \frac{1}{2}x - \frac{7}{2} \Rightarrow m_1 = \frac{1}{2}$

$6x + 2y = 5 \Rightarrow y = -3x + \frac{5}{2} \Rightarrow m_2 = -3$

$$\tan \theta = \left| \frac{-3 - \frac{1}{2}}{1 + \left(\frac{1}{2}\right)(-3)} \right| = 7$$

$$\theta = \arctan 7 \approx 1.4289 \text{ radians} \approx 81.9°$$

51. $x + 2y = 8 \Rightarrow y = -\frac{1}{2}x + 4 \Rightarrow m_1 = -\frac{1}{2}$

$x - 2y = 2 \Rightarrow y = \frac{1}{2}x - 1 \Rightarrow m_2 = \frac{1}{2}$

$$\tan \theta = \left| \frac{\frac{1}{2} - \left(-\frac{1}{2}\right)}{1 + \left(-\frac{1}{2}\right)\left(\frac{1}{2}\right)} \right| = \frac{4}{3}$$

$$\theta = \arctan\left(\frac{4}{3}\right) \approx 0.9273 \text{ radian} \approx 53.1°$$

53. $0.05x - 0.03y = 0.21 \Rightarrow y = \frac{5}{3}x - 7 \Rightarrow m_1 = \frac{5}{3}$

$0.07x + 0.02y = 0.16 \Rightarrow y = -\frac{7}{2}x + 8 \Rightarrow m_2 = -\frac{7}{2}$

$$\tan \theta = \left| \frac{\left(-\frac{7}{2}\right) - \left(\frac{5}{3}\right)}{1 + \left(\frac{5}{3}\right)\left(-\frac{7}{2}\right)} \right| = \frac{31}{29}$$

$$\theta = \arctan\left(\frac{31}{29}\right) \approx 0.8187 \text{ radian} \approx 46.9°$$

55. Let $A = (1, 5)$, $B = (3, 8)$, and $C = (4, 5)$.

Slope of AB: $m_1 = \dfrac{8 - 5}{3 - 1} = \dfrac{3}{2}$

Slope of BC: $m_2 = \dfrac{5 - 8}{4 - 3} = \dfrac{-3}{1} = -3$

Slope of AC: $m_3 = \dfrac{5 - 5}{4 - 1} = \dfrac{0}{3} = 0$

$\tan A = \left| \dfrac{0 - \dfrac{3}{2}}{1 + \left(\dfrac{3}{2}\right)(0)} \right| = \dfrac{\dfrac{3}{2}}{1} = \dfrac{3}{2}$

$\tan B = \left| \dfrac{\dfrac{3}{2} - (-3)}{1 + (-3)\left(\dfrac{3}{2}\right)} \right| = \dfrac{\dfrac{9}{2}}{\dfrac{7}{2}} = \dfrac{9}{7}$

$\tan C = \left| \dfrac{-3 - 0}{1 + (0)(-3)} \right| = \dfrac{3}{1} = 3$

$A = \arctan\left(\dfrac{3}{2}\right) \approx 56.3°$

$B = \arctan \dfrac{9}{7} \approx 52.1°$

$C = \arctan 3 \approx 71.6°$

57. Let $A = (-4, -1)$, $B = (3, 2)$, and $C = (1, 0)$.

Slope of AB: $m_1 = \dfrac{-1 - 2}{-4 - 3} = \dfrac{3}{7}$

Slope of BC: $m_2 = \dfrac{2 - 0}{3 - 1} = 1$

Slope of AC: $m_3 = \dfrac{-1 - 0}{-4 - 1} = \dfrac{1}{5}$

$\tan A = \left| \dfrac{\dfrac{1}{5} - \dfrac{3}{7}}{1 + \left(\dfrac{3}{7}\right)\left(\dfrac{1}{5}\right)} \right| = \dfrac{\dfrac{8}{35}}{\dfrac{38}{35}} = \dfrac{4}{9}$

$A = \arctan\left(\dfrac{4}{19}\right) \approx 11.9°$

$\tan B = \left| \dfrac{1 - \dfrac{3}{7}}{1 + \left(\dfrac{3}{7}\right)(1)} \right| = \dfrac{\dfrac{4}{7}}{\dfrac{10}{7}} = \dfrac{2}{5}$

$B = \arctan\left(\dfrac{2}{5}\right) \approx 21.8°$

$C = 180° - A - B$

$\approx 180° - 11.9° - 21.8° = 146.3°$

59. $(x_1, y_1) = (1, 2)$

$y = x + 2 \Rightarrow x - y + 2 = 0$

$d = \dfrac{|(1)(1) + (-1)(2) + 2|}{\sqrt{1^2 + (-1)^2}} = \dfrac{1}{\sqrt{2}} = \dfrac{\sqrt{2}}{2} \approx 0.7071$

61. $(x_1, y_1) = (2, 3)$

$y = 2x - 3 \Rightarrow 2x - y - 3 = 0$

$d = \dfrac{|2(2) + (-1)(3) + (-3)|}{\sqrt{2^2 + (-1)^2}} = \dfrac{2}{\sqrt{5}} = \dfrac{2\sqrt{5}}{5} \approx 0.8944$

63. $(x_1, y_1) = (-2, 4)$

$y = -x + 6 \Rightarrow x + y - 6 = 0$

$d = \dfrac{|(1)(-2) + (1)(4) + (-6)|}{\sqrt{1^2 + 1^2}} = \dfrac{4}{\sqrt{2}} = 2\sqrt{2} \approx 2.8284$

65. $(x_1, y_1) = (1, -2)$

$y = 3x - 6 \Rightarrow 3x - y - 6 = 0$

$d = \dfrac{|(3)(1) + (-1)(-2) + (-6)|}{\sqrt{3^2 + (-1)^2}} = \dfrac{1}{\sqrt{10}} = \dfrac{\sqrt{10}}{10} \approx 0.3162$

67. $(x_1, y_1) = (2, 3)$

$3x + y = 1 \Rightarrow 3x + y - 1 = 0$

$$d = \frac{|3(2) + (1)(3) + (-1)|}{\sqrt{3^2 + 1^2}} = \frac{8}{\sqrt{10}} = \frac{8\sqrt{10}}{10} = \frac{4\sqrt{10}}{5} \approx 2.5298$$

69. $(x_1, y_1) = (6, 2)$

$-3x + 4y = -5 \Rightarrow -3x + 4y + 5 = 0$

$$d = \frac{|(-3)(6) + (4)(2) + (5)|}{\sqrt{(-3)^2 + (4)^2}} = \frac{5}{\sqrt{25}} = 1$$

71. $(x_1, y_1) = (-2, 4)$

$4x + 3y = 5 \Rightarrow 4x + 3y - 5 = 0$

$$d = \frac{|(4)(-2) + (3)(4) + (-5)|}{\sqrt{4^2 + 3^2}} = \frac{1}{\sqrt{25}} = \frac{1}{5}$$

73. (a)

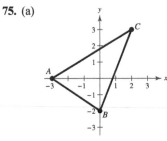

(b) Slope of the line AC: $m = \dfrac{1 - 0}{3 - (-1)} = \dfrac{1}{4}$

Equation of the line AC: $y - 0 = \dfrac{1}{4}(x + 1)$

$$x - 4y + 1 = 0$$

Altitude from

$B = (0, 3)$: $h = \dfrac{|(1)(0) + (-4)(3) + (1)|}{\sqrt{1^2 + (-4)^2}} = \dfrac{11}{\sqrt{17}} = \dfrac{11\sqrt{17}}{17}$

(c) Length of the base AC: $b = \sqrt{(3 + 1)^2 + (1 - 0)^2} = \sqrt{17}$

Area of the triangle: $A = \dfrac{1}{2}bh$

$$= \frac{1}{2}(\sqrt{17})\left(\frac{11}{\sqrt{17}}\right) = \frac{11}{2} \text{ units}^2$$

75. (a)

(b) Slope of the line AC: $m = \dfrac{3 - 0}{2 + 3} = \dfrac{3}{5}$

Equation of the line AC: $y - 0 = \dfrac{3}{5}(x + 3)$

$$3x - 5y + 9 = 0$$

Altitude from $B = (0, -2)$: $h = \dfrac{|3(0) + (-5)(-2) + (9)|}{\sqrt{3^2 + (-5)^2}} = \dfrac{19}{\sqrt{34}} = \dfrac{19\sqrt{34}}{34}$

(c) Length of the base AC: $b = \sqrt{(2 + 3)^2 + (3 - 0)^2} = \sqrt{34}$

Area of the triangle: $A = \dfrac{1}{2}bh$

$$= \frac{1}{2}(\sqrt{34})\left(\frac{19}{\sqrt{34}}\right) = \frac{19}{2} \text{ units}^2$$

77. (a)

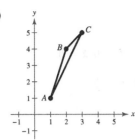

(b) Slope of the line AC: $b = \sqrt{(5+2)^2 + (1-0)^2} = \sqrt{50} = 5\sqrt{2}$

Equation of the line AC: $y - 1 = 2(x - 1)$

$$2x - y - 1 = 0$$

Altitude from

$$B = (2, 4): h = \frac{|(2)(2) + (-1)(4) + (-1)|}{\sqrt{2^2 + (-1)^2}} = \frac{1}{\sqrt{5}} = \frac{\sqrt{5}}{5}$$

(c) Length of the base AC: $b = \sqrt{(3-1)^2 + (5-1)^2} = \sqrt{20} = 2\sqrt{5}$

Area of the triangle: $A = \frac{1}{2}bh$

$$= \frac{1}{2}(2\sqrt{5})\left(\frac{\sqrt{5}}{5}\right) = 1 \text{ unit}^2$$

79. $x + y = 1 \Rightarrow (0, 1)$ is a point on the line $\Rightarrow x_1 = 0$
and $y_1 = 1$

$x + y = 5 \Rightarrow A = 1, B = 1,$ and $C = -5$

$$d = \frac{|1(0) + 1(1) + (-5)|}{\sqrt{1^2 + 1^2}} = \frac{4}{\sqrt{2}} = 2\sqrt{2}$$

81. Slope: $m = \tan 0.1 \approx 0.1003$

Change in elevation: $\sin 0.1 = \dfrac{x}{2(5280)}$

$$x \approx 1054 \text{ feet}$$

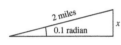

Not drawn to scale

83. Slope $= \frac{3}{5}$

Inclination $= \tan^{-1} \frac{3}{5} \approx 31.0°$

85. $\tan \gamma = \frac{6}{9}$

$\gamma = \arctan\left(\frac{2}{3}\right) \approx 33.69°$

$\beta = 90 - \gamma \approx 56.31°$

Also, because the right triangles containing α and β are equal, $\alpha = \gamma \approx 33.69°$

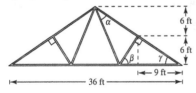

87. True. The inclination of a line is related to its slope by $m = \tan \theta$. If the line has an inclination of 0 radians, then the slope is 0 radians.

89. False. Substitute $m_1 = \tan \theta_1$ and $m_2 = \tan \theta_2$ into the formula for the angle between two lines.

91. False. By definition, the inclination of a nonhorizontal line is the positive angle θ measured counter clockwise from the x-axis to the line. So, the angle θ can be acute, right or obtuse. The angle θ between two lines is less than $\pi/2$ because, if $\theta > \dfrac{\pi}{2}$, then $\tan \theta < 0$.

Because the formula for the angle between two lines involves absolute value, then $\tan \theta$ will always be positive. So, θ cannot be larger than $\pi/2$.

93. (a) $(0, 0) \Rightarrow x_1 = 0$ and $y_1 = 0$

$y = mx + 4 \Rightarrow 0 = mx - y + 4$

$$d = \frac{|m(0) + (-1)(0) + 4|}{\sqrt{m^2 + (-1)^2}} = \frac{4}{\sqrt{m^2 + 1}}$$

(b)

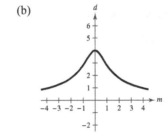

(c) The maximum distance of 4 occurs when the slope m is 0 and the line through $(0, 4)$ is horizontal.

(d) The graph has a horizontal asymptote at $d = 0$. As the slope becomes larger, the distance between the origin and the line, $y = mx + 4$, becomes smaller and approaches 0.

Section 6.2 Introduction to Conics: Parabolas

1. conic

3. locus

5. axis

7. focal chord

9. $y^2 = 4x$

Vertex: $(0, 0)$

$p = 1 > 0$

The graph opens to the right because p is positive.
So, the equation matches graph (c).

10. $x^2 = 2y$

Vertex: $(0, 0)$

$p = \dfrac{1}{2} > 0$

The graph opens upward because p is positive. So, the
equation matches graph (a).

11. $x^2 = -8y$

Vertex: $(0, 0)$

$p = -2 < 0$

The graph opens downward because p is negative. So,
the equation matches graph (b).

12. $y^2 = -12x$

Vertex: $(0, 0)$

$p = -3 < 0$

The graph opens to the left because p is negative. So, the
equation matches graph (d).

13. Vertex: $(0, 0) \Rightarrow h = 0, k = 0$

Graph opens upward.

$x^2 = 4py$

Focus: $(0, 1)$

$p = 1$

$x^2 = 4(1)y$

$x^2 = 4y$

15. Vertex: $(0, 0) \Rightarrow h = 0, k = 0$

Focus: $\left(0, \frac{1}{2}\right) \Rightarrow p = \frac{1}{2}$

$x^2 = 4py$

$x^2 = 4\left(\frac{1}{2}\right)y$

$x^2 = 2y$

17. Focus: $(-2, 0) \Rightarrow p = -2$

$y^2 = 4px$

$y^2 = 4(-2)x$

$y^2 = -8x$

19. Vertex: $(0, 0) \Rightarrow h = 0, k = 0$

Directrix: $y = 2 \Rightarrow p = -2$

$x^2 = 4py$

$x^2 = 4(-2)y$

$x^2 = -8y$

21. Vertex: $(0, 0) \Rightarrow h = 0, k = 0$

Directrix: $x = -1 \Rightarrow p = 1$

$y^2 = 4px$

$y^2 = 4(1)x$

$y^2 = 4x$

23. Vertex: $(0, 0) \Rightarrow h = 0, k = 0$

Vertical axis

Passes through: $(4, 6)$

$x^2 = 4py$

$4^2 = 4p(6)$

$16 = 24p$

$p = \frac{2}{3}$

$x^2 = 4\left(\frac{2}{3}\right)y$

$x^2 = \frac{8}{3}y$

25. Vertex: $(0, 0) \Rightarrow h = 0, k = 0$

Horizontal axis

Passes through: $(-2, 5)$

$$y^2 = 4px$$
$$5^2 = 4p(-2)$$
$$25 = -8p$$
$$p = -\frac{25}{8}$$
$$y^2 = 4\left(-\frac{25}{8}\right)x$$
$$y^2 = -\frac{25}{2}x$$

27. Vertex: $(2, 6) \Rightarrow h = 2, k = 6$

Focus: $(2, 4) \Rightarrow p = -2$

$$(x - h)^2 = 4p(y - k)$$
$$(x - 2)^2 = 4(-2)(y - 6)$$
$$(x - 2)^2 = -8(y - 6)$$

29. Vertex: $(6, 3) \Rightarrow h = 6, k = 3$

Focus: $(4, 3) \Rightarrow p = -2$

$$(y - k)^2 = 4p(x - h)$$
$$(y - 3)^2 = 4(-2)(x - 6)$$
$$(y - 3)^2 = -8(x - 6)$$

31. Vertex: $(0, 2)$

Directrix: $y = 4$

Vertical axis

$$p = 2 - 4 = -2$$
$$(x - 0)^2 = 4(-2)(y - 2)$$
$$x^2 = -8(y - 2)$$

33. Focus: $(2, 2)$

Directrix: $x = -2$

Horizontal axis

Vertex: $(0, 2)$

$$p = 2 - 0 = 2$$
$$(y - 2)^2 = 4(2)(x - 0)$$
$$(y - 2)^2 = 8x$$

35. Vertex: $(3, -3) \Rightarrow h = 3, k = -3$

Vertical Axis; Passes through $(0, 0)$

$$(x - h)^2 = 4p(y - k)$$
$$(x - 3)^2 = 4p(y + 3)$$
$$(0 - 3)^2 = 4p(0 + 3)$$
$$9 = 12p$$
$$p = \frac{3}{4}$$
$$(x - 3)^2 = 3(y + 3)$$

37. $y = \frac{1}{2}x^2$

$$x^2 = 2y$$
$$x^2 = 4\left(\frac{1}{2}\right)y \Rightarrow h = 0, k = 0, p = \frac{1}{2}$$

Vertex: $(0, 0)$

Focus: $\left(0, \frac{1}{2}\right)$

Directrix: $y = -\frac{1}{2}$

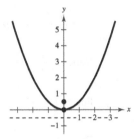

39. $y^2 = -6x$

$$y^2 = 4\left(-\frac{3}{2}\right)x \Rightarrow h = 0, k = 0, p = -\frac{3}{2}$$

Vertex: $(0, 0)$

Focus: $\left(-\frac{3}{2}, 0\right)$

Directrix: $x = \frac{3}{2}$

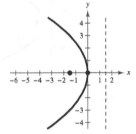

41. $x^2 + 12y = 0$

$$x^2 = -12y = 4(-3)y \Rightarrow h = 0, k = 0, p = -3$$

Vertex: $(0, 0)$

Focus: $(0, -3)$

Directrix: $y = 3$

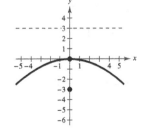

43. $(x - 1)^2 + 8(y + 2) = 0$

$$(x - 1)^2 = 4(-2)(y + 2)$$

$h = 1, k = -2, p = -2$

Vertex: $(1, -2)$

Focus: $(1, -4)$

Directrix: $y = 0$

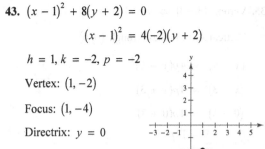

45. $(y + 7)^2 = 4\left(x - \dfrac{3}{2}\right)$

$$(y + 7)^2 = 4(1)\left(x - \dfrac{3}{2}\right)$$

$h = \dfrac{3}{2}, k = -7, p = 1$

Vertex: $\left(\dfrac{3}{2}, -7\right)$

Focus: $\left(\dfrac{5}{2}, -7\right)$

Directrix: $x = \dfrac{1}{2}$

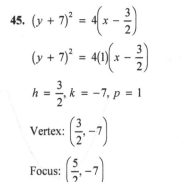

47. $\qquad y = \dfrac{1}{4}(x^2 - 2x + 5)$

$$4y = x^2 - 2x + 5$$

$$4y - 5 + 1 = x^2 - 2x + 1$$

$$4y - 4 = (x - 1)^2$$

$$(x - 1)^2 = 4(1)(y - 1)$$

$h = 1, k = 1, p = 1$

Vertex: $(1, 1)$

Focus: $(1, 2)$

Directrix: $y = 0$

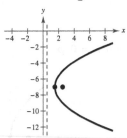

49. $y^2 + 6y + 8x + 25 = 0$

$$y^2 + 6y + 9 = -8x - 25 + 9$$

$$(y + 3)^2 = 4(-2)(x + 2)$$

$h = -2, k = -3, p = -2$

Vertex: $(-2, -3)$

Focus: $(-4, -3)$

Directrix: $x = 0$

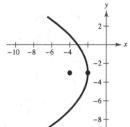

51. $x^2 + 4x - 6y = -10$

$$x^2 + 4x + 4 = 6y - 10 + 4$$

$$x^2 + 4x + 4 = 6y - 6$$

$$(x + 2)^2 = 6(y - 1)$$

$$(x + 2)^2 = 4\left(\dfrac{3}{2}\right)(y - 1)$$

$h = -2, k = 1, p = \dfrac{3}{2}$

Vertex: $(-2, 1)$

Focus: $\left(-2, \dfrac{5}{2}\right)$

Directrix: $y = -\dfrac{1}{2}$

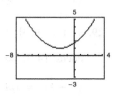

53. $y^2 + x + y = 0$

$$y^2 + y + \dfrac{1}{4} = -x + \dfrac{1}{4}$$

$$\left(y + \dfrac{1}{2}\right)^2 = 4\left(-\dfrac{1}{4}\right)\left(x - \dfrac{1}{4}\right)$$

$h = \dfrac{1}{4}, k = -\dfrac{1}{2}, p = -\dfrac{1}{4}$

Vertex: $\left(\dfrac{1}{4}, -\dfrac{1}{2}\right)$

Focus: $\left(0, -\dfrac{1}{2}\right)$

Directrix: $x = \dfrac{1}{2}$

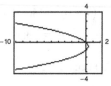

55. $x^2 = 8y$

$x^2 = 4(2)y \Rightarrow p = 2$

Focus: $(0, 2)$

$d_1 = 2 - b$

$d_2 = \sqrt{(6-0)^2 + \left(\dfrac{9}{2} - 2\right)^2} = \sqrt{36 + \dfrac{25}{4}}$

$2 - b = \dfrac{13}{2}$

$b = -\dfrac{9}{2}$

$m = \dfrac{-(9/2) - (9/2)}{0 - 6} = \dfrac{3}{2}$

Tangent line: $y = \dfrac{3}{2}x - \dfrac{9}{2}$

57. $x^2 = 2y \Rightarrow p = \dfrac{1}{2}$

Point: $(4, 8)$

Focus: $\left(0, \dfrac{1}{2}\right)$

$d_1 = \dfrac{1}{2} - b$

$d_2 = \sqrt{(4-0)^2 + \left(8 - \dfrac{1}{2}\right)^2}$

$= \dfrac{17}{2}$

$d_1 = d_2 \Rightarrow b = -8$

$m = \dfrac{8 - (-8)}{4 - 0} = 4$

Tangent line: $y = 4x - 8$

59. $y = -2x^2$

$x^2 = -\dfrac{1}{2}y \Rightarrow p = -\dfrac{1}{8}$

Point: $(-1, -2)$

Focus: $\left(0, -\dfrac{1}{8}\right)$

$d_1 = b - \left(-\dfrac{1}{8}\right) = b + \dfrac{1}{8}$

$d_2 = \sqrt{(-1-0)^2 + \left(-2 - \left(-\dfrac{1}{8}\right)\right)^2}$

$= \dfrac{17}{8}$

$d_1 = d_2 \Rightarrow b = 2$

$m = \dfrac{-2 - 2}{-1 - 0} = 4$

$y = 4x + 2$

61. $y^2 = 4px, \ p = 1.5$

$y^2 = 4(1.5)x$

$y^2 = 6x$

63. Vertex: $(0, 0)$

$(y - 0)^2 = 4p(x - 0)$

$y^2 = 4px$

At $(1000, 800): 800^2 = 4p(1000) \Rightarrow p = 160$

$y^2 = 4(160)x$

$y^2 = 640x$

65. (a) $\quad x^2 = 4py$

$32^2 = 4p\left(\dfrac{1}{12}\right)$

$1024 = \dfrac{1}{3}p$

$3072 = p$

$x^2 = 4(3072)y$

$x^2 = 12{,}288y \ \text{(in feet)}$

(b) $\dfrac{1}{24} = \dfrac{x^2}{12{,}288}$

$\dfrac{12{,}288}{24} = x^2$

$512 = x^2$

$x \approx 22.6 \text{ feet}$

67. Vertex: $(0, 48) \Rightarrow h = 0, k = 48$

Passes through $\left(10\sqrt{3}, 0\right)$

Vertical axis

$$(x - 0)^2 = 4p(y - 48)$$

$$\left(10\sqrt{3} - 0\right)^2 = 4p(0 - 48)$$

$$300 = -192p$$

$$-\frac{25}{16} = p$$

$$x^2 = 4\left(-\frac{25}{16}\right)(y - 48)$$

$$x^2 = -\frac{25}{4}(y - 48)$$

69. $x^2 = 4p(y - 12)$

$(4, 10)$ on curve:

$$16 = 4p(10 - 12) = -8p \Rightarrow p = -2$$

$$x^2 = 4(-2)(y - 12) = -8y + 96$$

$$y = \frac{-x^2 + 96}{8}$$

$$y = 0 \text{ if } x^2 = 96 \Rightarrow x = 4\sqrt{6}$$

So, the width is about $2\left(4\sqrt{6}\right) \approx 19.6$ meters.

71. (a) $x^2 = 4py$

$$60^2 = 4p(20) \Rightarrow p = 45$$

Focus: $(0, 45)$

(b) $x^2 = 4(45)y$ or $y = \frac{1}{180}x^2$

73. (a) $V = 17,500\sqrt{2}$ mi/h

$\approx 24,750$ mi/h

(b) $p = -4100, (h, k) = (0, 4100)$

$$(x - 0)^2 = 4(-4100)(y - 4100)$$

$$x^2 = -16,400(y - 4100)$$

75. (a) $x^2 = -\frac{v^2}{16}(y - s)$

$$x^2 = -\frac{(28)^2}{16}(y - 100)$$

$$x^2 = -49(y - 100)$$

(b) The ball hits the ground when $y = 0$.

$$x^2 = -49(0 - 100)$$

$$x^2 = 4900$$

$$x = 70$$

The ball travels 70 feet.

77. False. It is not possible for a parabola to intersect its directrix. If the graph crossed the directrix there would exist points closer to the directrix than the focus.

79. True. If the axis (line connecting the vertex and focus) is horizontal, then the directrix must be vertical.

81. Both (a) and (b) are parabolas with vertical axes, while (c) is a parabola with a horizontal axis.

So, equations (a) and (b) are equivalent when $p = \dfrac{1}{4a}$.

(a) $y = a(x - h)^2 + k$

(b)
$$(x - h)^2 = 4p(y - k)$$
$$(x - h)^2 = 4py - 4pk$$
$$(1/4p)\left((x - h)^2 + 4pk\right) = 4py(1/4p)$$
$$(1/4p)(x - h)^2 + k = y = a(x - h)^2 + k$$
$$a = \left(\frac{1}{4p}\right)$$
$$4a = (1/p)$$
$$p = \frac{1}{4a}$$

83. The graph of $x^2 + y^2 = 0$ is a single point, $(0, 0)$.

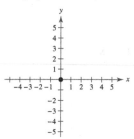

The plane intersects the double-napped cone at the vertices of the cones.

85. (a) $A = \dfrac{8}{3}(2)^{1/2}(4)^{3/2}$

$$= \frac{8}{3}\left(\sqrt{2}\right)(8)$$

$$= \frac{64\sqrt{2}}{3} \text{ square units}$$

(b) As p approaches zero, the parabola becomes narrower and narrower, so the area becomes smaller and smaller.

Section 6.3   Ellipses

1. ellipse; foci

3. minor axis

5. $\dfrac{x^2}{4} + \dfrac{y^2}{9} = 1$

   Center: $(0, 0)$

   $a = 3, b = 2$

   Vertical major axis
   Matches graph (b).

6. $\dfrac{x^2}{9} + \dfrac{y^2}{4} = 1$

   Center: $(0, 0)$

   $a = 3, b = 2$

   Horizontal major axis
   Matches graph (c).

7. $\dfrac{(x - 2)^2}{16} + (y + 1)^2 = 1$

   Center: $(2, -1)$

   $a = 4, b = 1$

   Horizontal major axis
   Matches graph (a).

8. $\dfrac{(x + 2)^2}{9} + \dfrac{(y + 2)^2}{4} = 1$

   Center: $(-2, -2)$

   $a = 3, b = 2$

   Horizontal major axis
   Matches graph (d).

9. Center: $(0, 0)$

   $a = 4, b = 2$

   Vertical major axis

   $\dfrac{(x - h)^2}{b^2} + \dfrac{(y - k)^2}{a^2} = 1$

   $\dfrac{x^2}{4} + \dfrac{y^2}{16} = 1$

11. Center: $(0, 0)$

    Vertices: $(\pm 7, 0) \Rightarrow a = 7$

    Foci: $(\pm 2, 0) \Rightarrow c = 2$

    $b^2 = a^2 - c^2 = 49 - 4 = 45$

    $\dfrac{x^2}{a^2} + \dfrac{y^2}{b^2} = 1$

    $\dfrac{x^2}{49} + \dfrac{y^2}{45} = 1$

13. Center: $(0, 0)$

    Foci: $(\pm 4, 0) \Rightarrow c = 4$

    Length of horizontal major axis: $10 \Rightarrow a = 5$

    $b^2 = a^2 - c^2 = 25 - 16 = 9$

    $\dfrac{x^2}{a^2} + \dfrac{y^2}{b^2} = 1$

    $\dfrac{x^2}{25} + \dfrac{y^2}{9} = 1$

15. Major axis vertical

    Passes through: $(0, 6)$ and $(3, 0)$

    $a = 6, b = 3$

    $\dfrac{x^2}{b^2} + \dfrac{y^2}{a^2} = 1$

    $\dfrac{x^2}{9} + \dfrac{y^2}{36} = 1$

17. Vertices: $(\pm 6, 0) \Rightarrow a = 6$

    Major axis horizontal

    Passes through: $(4, 1)$

    $\dfrac{x^2}{36} + \dfrac{y^2}{b^2} = 1$

    $\dfrac{4^2}{36} + \dfrac{1^2}{b^2} = 1$

    $16b^2 + 36 = 36b^2$

    $36 = 20b^2$

    $\dfrac{9}{5} = b^2$

    $\dfrac{x^2}{36} + \dfrac{y^2}{\frac{9}{5}} = 1$ or $\dfrac{x^2}{36} + \dfrac{5y^2}{9} = 1$

19. Center: $(2, 3)$

$a = 3, b = 1$

Vertical major axis

$$\frac{(x-h)^2}{b^2} + \frac{(y-k)^2}{a^2} = 1$$

$$\frac{(x-2)^2}{1} + \frac{(y-3)^2}{9} = 1$$

21. Vertices: $(2, 0), (10, 0) \Rightarrow a = 4$

Horizontal major axis

Length of minor axis: $4 \Rightarrow b = 2$

Center: $(6, 0) \Rightarrow h = 6, k = 0$

$$\frac{(x-h)^2}{a^2} + \frac{(y-k)^2}{b^2} = 1$$

$$\frac{(x-6)^2}{16} + \frac{(y-0)^2}{4} = 1$$

$$\frac{(x-6)^2}{16} + \frac{y^2}{4} = 1$$

23. Foci: $(0, 0), (4, 0) \Rightarrow c = 2$

Length of major axis: $6 \Rightarrow a = 3$

Center: $(2, 0) = (h, k)$

$b^2 = a^2 - c^2 = 9 - 4 = 5$

$$\frac{(x-h)^2}{a^2} + \frac{(y-k)^2}{b^2} = 1$$

$$\frac{(x-2)^2}{9} + \frac{y^5}{5} = 1$$

25. Center: $(1, 3)$

Vertex: $(-2, 3) \Rightarrow a = 3$

Major axis horizontal

Length of minor axis: $4 \Rightarrow b = 2$

$$\frac{(x-h)^2}{a^2} + \frac{(y-k)^2}{b^2} = 1$$

$$\frac{(x-1)^2}{9} + \frac{(y-3)^2}{4} = 1$$

27. Center: $(1, 4)$

Vertices: $(1, 0)$ and $(1, 8) \Rightarrow a = 4$

Major axis vertical

$$a = 2c$$
$$4 = 2c$$
$$c = 2$$
$$b^2 = a^2 - c^2$$
$$b^2 = 16 - 4 = 12$$

$$\frac{(x-h)^2}{b^2} + \frac{(y-k)^2}{a^2} = 1$$

$$\frac{(x-1)^2}{12} + \frac{(y-4)^2}{16} = 1$$

29. Vertices: $(0, 2), (4, 2) \Rightarrow a = 2$

Center: $(2, 2)$

Endpoints of the minor axis: $(2, 3), (2, 1) \Rightarrow b = 1$

Horizontal major axis:

$$\frac{(x-h)^2}{a^2} + \frac{(y-k)^2}{b^2} = 1$$

$$\frac{(x-2)^2}{4} + \frac{(y-2)^2}{1} = 1$$

31. $\dfrac{x^2}{25} + \dfrac{y^2}{16} = 1$

$a = 5, b = 4, c = 3$

Center: $(0, 0)$

Vertices: $(\pm 5, 0)$

Foci: $(\pm 3, 0)$

Eccentricity: $e = \dfrac{3}{5}$

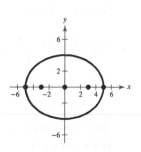

33. $9x^2 + y^2 = 36$

$$\frac{x^2}{4} + \frac{y^2}{36} = 1$$

$a = 6, b = 2$

$c^2 = a^2 - b^2$

$\quad = 36 - 4 = 32$

Center: $(0, 0)$

Vertices: $(0, \pm 6)$

Foci: $\left(0, \pm 4\sqrt{2}\right)$

Eccentricity: $e = \dfrac{4\sqrt{2}}{6} = \dfrac{2\sqrt{2}}{3}$

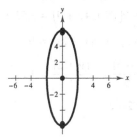

35. $\dfrac{(x-4)^2}{16} + \dfrac{(y+1)^2}{25} = 1$

$a = 5, b = 4$

$c^2 = a^2 - b^2 = 25 - 16 = 9 \Rightarrow c = 3$

Center: $(4, -1)$

Vertices: $(4, 4), (4, -6)$

Foci: $(4, 2), (4, -4)$

Eccentricity: $e = \dfrac{3}{5}$

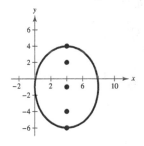

37. $\dfrac{(x+5)^2}{9/4} + (y-1)^2 = 1$

$a = \dfrac{3}{2}, b = 1, c = \dfrac{\sqrt{5}}{2}$

Center: $(-5, 1)$

Vertices: $\left(-\dfrac{7}{2}, 1\right), \left(-\dfrac{13}{2}, 1\right)$

Foci: $\left(-5 + \dfrac{\sqrt{5}}{2}, 1\right), \left(-5 - \dfrac{\sqrt{5}}{2}, 1\right)$

Eccentricity: $e = \dfrac{\sqrt{5}}{3}$

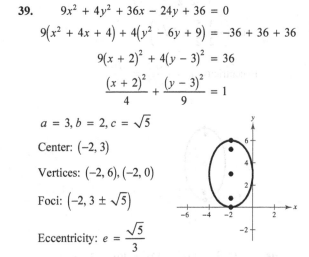

39. $9x^2 + 4y^2 + 36x - 24y + 36 = 0$

$9(x^2 + 4x + 4) + 4(y^2 - 6y + 9) = -36 + 36 + 36$

$9(x + 2)^2 + 4(y - 3)^2 = 36$

$$\frac{(x+2)^2}{4} + \frac{(y-3)^2}{9} = 1$$

$a = 3, b = 2, c = \sqrt{5}$

Center: $(-2, 3)$

Vertices: $(-2, 6), (-2, 0)$

Foci: $\left(-2, 3 \pm \sqrt{5}\right)$

Eccentricity: $e = \dfrac{\sqrt{5}}{3}$

41. $x^2 + 5y^2 - 8x - 30y - 39 = 0$

$(x^2 - 8x + 16) + 5(y^2 - 6y + 9) = 39 + 16 + 45$

$(x - 4)^2 + 5(y - 3)^2 = 100$

$$\frac{(x-4)^2}{100} + \frac{(y-3)^2}{20} = 1$$

$a = 10, b = \sqrt{20} = 2\sqrt{5},$

$c = \sqrt{80} = 4\sqrt{5}$

Center: $(4, 3)$

Foci: $\left(4 \pm 4\sqrt{5}, 3\right)$

Vertices: $(14, 3), (-6, 3)$

Eccentricity: $e = \dfrac{4\sqrt{5}}{10} = \dfrac{2\sqrt{5}}{5}$

43.
$$6x^2 + 2y^2 + 18x - 10y + 2 = 0$$

$$6\left(x^2 + 3x + \frac{9}{4}\right) + 2\left(y^2 - 5y + \frac{25}{4}\right) = -2 + \frac{27}{2} + \frac{25}{2}$$

$$6\left(x + \frac{3}{2}\right)^2 + 2\left(y - \frac{5}{2}\right)^2 = 24$$

$$\frac{\left(x + \frac{3}{2}\right)^2}{4} + \frac{\left(y - \frac{5}{2}\right)^2}{12} = 1$$

$$a = \sqrt{12} = 2\sqrt{3}, b = 2, c = \sqrt{8} = 2\sqrt{2}$$

Center: $\left(-\frac{3}{2}, \frac{5}{2}\right)$

Vertices: $\left(-\frac{3}{2}, \frac{5}{2} \pm 2\sqrt{3}\right)$

Foci: $\left(-\frac{3}{2}, \frac{5}{2} \pm 2\sqrt{2}\right)$

Eccentricity: $e = \frac{2\sqrt{2}}{2\sqrt{3}} = \frac{\sqrt{6}}{3}$

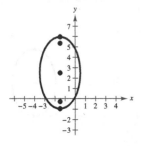

45.
$$12x^2 + 20y^2 - 12x + 40y - 37 = 0$$

$$12\left(x^2 - x + \frac{1}{4}\right) + 20\left(y^2 + 2y + 1\right) = 37 + 3 + 20$$

$$12\left(x - \frac{1}{2}\right)^2 + 20(y + 1)^2 = 60$$

$$\frac{\left(x - \frac{1}{2}\right)^2}{5} + \frac{(y + 1)^2}{3} = 1$$

$$a^2 = 5, b^2 = 3$$

$$c^2 = 5 - 3 = 2$$

Center: $\left(\frac{1}{2}, -1\right)$

Vertices: $\left(\frac{1}{2} \pm \sqrt{5}, -1\right)$

Foci: $\left(\frac{1}{2} \pm \sqrt{2}, -1\right)$

Eccentricity: $e = \frac{\sqrt{2}}{\sqrt{5}} = \frac{\sqrt{10}}{5}$

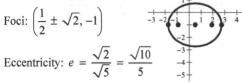

47. $5x^2 + 3y^2 = 15$

$$\frac{x^2}{3} + \frac{y^2}{5} = 1$$

$$a = \sqrt{5}, b = \sqrt{3}, c = \sqrt{2}$$

Center: $(0, 0)$

Vertices: $\left(0, \pm\sqrt{5}\right)$

Foci: $\left(0, \pm\sqrt{2}\right)$

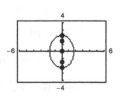

Eccentricity: $e = \frac{\sqrt{10}}{5}$

49.
$$x^2 + 9y^2 - 10x + 36y + 52 = 0$$

$$\left(x^2 - 10x + 25\right) + 9\left(y^2 + 4y + 4\right) = -52 + 25 + 36$$

$$(x - 5)^2 + 9(y + 2)^2 = 9$$

$$\frac{(x - 5)^2}{9} + \frac{(y + 2)^2}{1} = 1$$

$$a = 3, b = 1, c = 2\sqrt{2}$$

Center: $(5, -2)$

Vertices: $(8, -2), (2, -2)$

Foci: $\left(5 \pm 2\sqrt{2}, -2\right)$

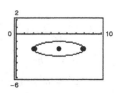

Eccentricity: $e = \frac{2\sqrt{2}}{3}$

51. Vertices: $(\pm 5, 0) \Rightarrow a = 5$

$$e = \frac{3}{5} \Rightarrow c = \frac{3}{5}a = 3$$

$$b^2 = a^2 - c^2 = 25 - 9 = 16$$

Center: $(0, 0) = (h, k)$

$$\frac{(x - h)^2}{a^2} + \frac{(y - k)^2}{b^2} = 1$$

$$\frac{x^2}{25} + \frac{y^2}{16} = 1$$

53. (a)

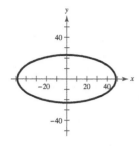

$$\frac{x^2}{2352.25} + \frac{y^2}{23^2} = 1 \text{ or } \frac{x^2}{529} + \frac{y^2}{2352.25} = 1$$

$$a = \frac{97}{2}, b = 23, c = \sqrt{\left(\frac{97}{2}\right)^2 - (23)^2} \approx 4.7$$

(b) Distance between foci: $2(4.7) \approx 85.4$ feet

55. The length of the major axis and minor axis are 280 millimeters and 160 millimeters, respectively.
Therefore,

$$2a = 280 \Rightarrow a = 140 \text{ and } 2b = 160 \Rightarrow b = 80.$$

$$a^2 = b^2 + c^2$$

$$140^2 = 80^2 + c^2$$

$$13{,}200 = c^2$$

$$\sqrt{13{,}200} = c$$

$$20\sqrt{33} = c$$

The kidney stone and spark plug are each located at a focus, therefore they are $2c$ millimeters apart, or
$2\left(20\sqrt{33}\right) = 40\sqrt{33} \approx 229.8$ millimeters apart.

57. $a + c = 6378 + 939 = 7317$

$a - c = 6378 + 215 = 6593$

Solving this system for a and c yields

$a + c = 7317$

$a - c = 6593$

$\qquad 2a = 13{,}910$

$\qquad a = 6955$

$6955 + c = 7317$

$\qquad c = 362$

Eccentricity: $e = \dfrac{c}{a} = \dfrac{362}{6955} \approx 0.0520$

59. $\dfrac{x^2}{9} + \dfrac{y^2}{16} = 1$

$a = 4, b = 3, c = \sqrt{7}$

Points on the ellipse: $(\pm 3, 0), (0, \pm 4)$

Length of latus recta: $\dfrac{2b^2}{a} = \dfrac{2(3)^2}{4} = \dfrac{9}{2}$

Additional points: $\left(\pm\dfrac{9}{4}, -\sqrt{7}\right), \left(\pm\dfrac{9}{4}, \sqrt{7}\right)$

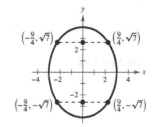

61. $5x^2 + 3y^2 = 15$

$$\frac{x^2}{3} + \frac{y^2}{5} = 1$$

$a = \sqrt{5}, b = \sqrt{3}, c = \sqrt{2}$

Points on the ellipse: $\left(\pm\sqrt{3}, 0\right), \left(0, \pm\sqrt{5}\right)$

Length of latus recta: $\dfrac{2b^2}{a} = \dfrac{2 \cdot 3}{\sqrt{5}} = \dfrac{6\sqrt{5}}{5}$

Additional points: $\left(\pm\dfrac{3\sqrt{5}}{5}, \pm\sqrt{2}\right)$

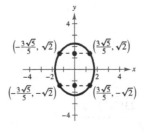

63. False. The graph of $\dfrac{x^2}{4} + y^4 = 1$ is not an ellipse.

The degree of y is 4, not 2.

65. *Sample answer:* Foci: $(2, 2), (10, 2) \Rightarrow c = 4$

Center: $(6, 2)$

Let $a^2 = 324$ and $b^2 = 308$

So that $c^2 = a^2 - b^2$.

$$\frac{(x-6)^2}{324} + \frac{(y-2)^2}{308} = 1$$

67. $\dfrac{x^2}{a^2} + \dfrac{y^2}{b^2} = 1$

(a) $a + b = 20 \Rightarrow b = 20 - a$

$A = \pi ab = \pi a(20 - a)$

(b) $264 = \pi a(20 - a)$

$0 = -\pi a^2 + 20\pi a - 264$

$0 = \pi a^2 - 20\pi a + 264$

By the Quadratic Formula: $a \approx 14$ or $a \approx 6$.
Choosing the larger value of a, you have $a \approx 14$ and $b \approx 6$.

The equation of an ellipse with an area of 264 is

$$\frac{x^2}{196} + \frac{y^2}{36} = 1.$$

69.

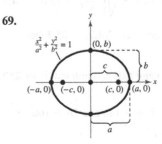

The length of half the major axis is a and the length of half the minor axis is b. Find the distance between $(0, b)$ and $(c, 0)$ and $(0, b)$ and $(-c, 0)$.

$$d_1 = \sqrt{(0-c)^2 + (b-0)^2} = \sqrt{c^2 + b^2}$$

$$d_2 = \sqrt{(0-(-c))^2 + (b-0)^2} = \sqrt{c^2 + b^2}$$

The sum of the distances from any point on the ellipse to the two foci is constant. Using the vertex $(a, 0)$, the constant sum is $(a + c) + (a - c) = 2a$.

So, the sum of the distances from $(0, b)$ to the two foci is

$$\sqrt{c^2 + b^2} + \sqrt{c^2 + b^2} = 2a$$
$$2\sqrt{c^2 + b^2} = 2a$$
$$\sqrt{c^2 + b^2} = a$$
$$c^2 + b^2 = a^2$$

So, $a^2 = b^2 + c^2$ for the ellipse $\dfrac{x^2}{a^2} + \dfrac{y^2}{b^2} = 1$,

where $a > 0, b > 0$.

Section 6.4   Hyperbolas

1. hyperbola; foci

3. transverse axis; center

5. $\dfrac{y^2}{9} - \dfrac{x^2}{25} = 1$

Center: $(0, 0)$

$a = 3, b = 5$

Vertical transverse axis

Matches graph (b).

6. $\dfrac{x^2}{9} - \dfrac{y^2}{25} = 1$

Center: $(0, 0)$

$a = 3, b = 5$

Horizontal transverse axis

Matches graph (d).

7. $\dfrac{x^2}{25} - \dfrac{(y + 2)^2}{9} = 1$

Center: $(0, -2)$

$a = 5, b = 3$

Horizontal transverse axis

Matches graph (c).

8. $\dfrac{(y + 4)^2}{25} - \dfrac{(x - 2)^2}{9} = 1$

Center: $(2, -4)$

$a = 5, b = 3$

Vertical transverse axis

Matches graph (a).

9. Vertices: $(0, \pm 2) \Rightarrow a = 2$

Foci: $(0, \pm 4) \Rightarrow c = 4$

$b^2 = c^2 - a^2 = 16 - 4 = 12$

Center: $(0, 0) = (h, k)$

$\dfrac{(y - k)^2}{a^2} - \dfrac{(x - h)^2}{b^2} = 1$

$\dfrac{y^2}{4} - \dfrac{x^2}{12} = 1$

11. Vertices: $(2, 0), (6, 0) \Rightarrow a = 2$

Foci: $(0, 0), (8, 0) \Rightarrow c = 4$

$b^2 = c^2 - a^2 = 16 - 4 = 12$

Center: $(4, 0) = (h, k)$

$\dfrac{(x - h)^2}{a^2} - \dfrac{(y - k)^2}{b^2} = 1$

$\dfrac{(x - 4)^2}{4} - \dfrac{y^2}{12} = 1$

13. Vertices: $(4, 1), (4, 9) \Rightarrow a = 4$

Foci: $(4, 0), (4, 10) \Rightarrow c = 5$

$b^2 = c^2 - a^2 = 25 - 16 = 9$

Center: $(4, 5) = (h, k)$

$\dfrac{(y - k)^2}{a^2} - \dfrac{(x - h)^2}{b^2} = 1$

$\dfrac{(y - 5)^2}{16} - \dfrac{(x - 4)^2}{9} = 1$

15. Vertices: $(2, 3), (2, -3) \Rightarrow a = 3$

Passes through the point: $(0, 5)$

Center: $(2, 0) = (h, k)$

$\dfrac{(y - k)^2}{a^2} - \dfrac{(x - h)^2}{b^2} = 1$

$\dfrac{y^2}{9} - \dfrac{(x - 2)^2}{b^2} = 1$

$\dfrac{(x - 2)^2}{b^2} = \dfrac{y^2}{9} - 1 = \dfrac{y^2 - 9}{9}$

$b^2 = \dfrac{9(x - 2)^2}{y^2 - 9} = \dfrac{9(-2)^2}{25 - 9}$

$= \dfrac{36}{16} = \dfrac{9}{4}$

$\dfrac{y^2}{9} - \dfrac{(x - 2)^2}{9/4} = 1$

$\dfrac{y^2}{9} - \dfrac{4(x - 2)^2}{9} = 1$

17. Vertices: $(0, -3), (4, -3) \Rightarrow a = 2$

Center: $(2, -3)$

Passes through: $(-4, 5)$

$\dfrac{(x - h)^2}{a^2} - \dfrac{(y - k)^2}{b^2} = 1$

$\dfrac{(x - 2)^2}{4} - \dfrac{(y + 3)^2}{b^2} = 1$

$\dfrac{(-4 - 2)^2}{4} - \dfrac{(5 + 3)^2}{b^2} = 1$

$9 - \dfrac{64}{b^2} = 1 \Rightarrow b^2 = 8$

$\dfrac{(x - 2)^2}{4} - \dfrac{(y + 3)^2}{8} = 1$

19. $x^2 - y^2 = 1$

$a = 1, b = 1, c = \sqrt{2}$

Center: $(0, 0)$

Vertices: $(\pm 1, 0)$

Foci: $\left(\pm\sqrt{2}, 0\right)$

Asymptotes: $y = \pm x$

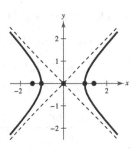

21. $\dfrac{y^2}{36} - \dfrac{x^2}{100} = 1$

$a = 6, b = 10$

$c^2 = a^2 + b^2 = 136 \Rightarrow c = 2\sqrt{34}$

Center: $(0, 0)$

Vertices: $(0, \pm 6)$

Foci: $\left(0, \pm 2\sqrt{34}\right)$

Asymptotes: $y = \pm\dfrac{3}{5}x$

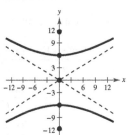

23. $2y^2 - \dfrac{x^2}{2} = 2$

$y^2 - \dfrac{x^2}{4} = 1$

$a = 1, b = 2,$

$c^2 = a^2 + b^2 = 5 \Rightarrow c = \sqrt{5}$

Center: $(0, 0)$

Vertices: $(0, \pm 1)$

Foci: $\left(0, \pm\sqrt{5}\right)$

Asymptotes: $y = \pm\dfrac{1}{2}x$

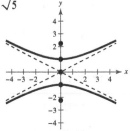

25. $\dfrac{(x-1)^2}{4} - \dfrac{(y+2)^2}{1} = 1$

$a = 2, b = 1, c = \sqrt{5}$

Center: $(1, -2)$

Vertices: $(-1, -2), (3, -2)$

Foci: $\left(1 \pm \sqrt{5}, -2\right)$

Asymptotes: $y = -2 \pm \dfrac{1}{2}(x - 1)$

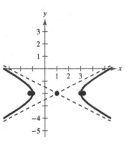

27. $\dfrac{(y+6)^2}{1/9} - \dfrac{(x-2)^2}{1/4} = 1$

$a = \dfrac{1}{3}, b = \dfrac{1}{2},$

$c = \dfrac{\sqrt{13}}{6}$

Center: $(2, -6)$

Vertices: $\left(2, -\dfrac{17}{3}\right), \left(2, -\dfrac{19}{3}\right)$

Foci: $\left(2, -6 \pm \dfrac{\sqrt{13}}{6}\right)$

Asymptotes: $y = -6 \pm \dfrac{2}{3}(x - 2)$

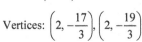

29. $9x^2 - y^2 - 36x - 6y + 18 = 0$

$9(x^2 - 4x + 4) - (y^2 + 6y + 9) = -18 + 36 - 9$

$9(x - 2)^2 - (y + 3)^2 = 9$

$\dfrac{(x-2)^2}{1} - \dfrac{(y+3)^2}{9} = 1$

$a = 1, b = 3, c = \sqrt{10}$

Center: $(2, -3)$

Vertices: $(1, -3), (3, -3)$

Foci: $\left(2 \pm \sqrt{10}, -3\right)$

Asymptotes:

$y = -3 \pm 3(x - 2)$

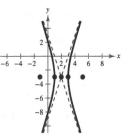

31. $4x^2 - y^2 + 8x + 2y - 1 = 0$

$4(x^2 + 2x + 1) - (y^2 - 2y + 1) = 1 + 4 - 1$

$4(x + 1)^2 - (y - 1)^2 = 4$

$\dfrac{(x+1)^2}{1} - \dfrac{(y-1)^2}{4} = 1$

$a = 1, b = 2, c = \sqrt{5}$

Center: $(-1, 1)$

Vertices: $(-2, 1), (0, 1)$

Foci: $\left(-1 \pm \sqrt{5}, 1\right)$

Asymptotes: $y = 1 \pm 2(x + 1)$

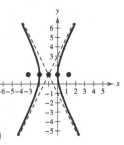

33. $2x^2 - 3y^2 = 6$

$$\frac{x^2}{3} - \frac{y^2}{2} = 1$$

$a = \sqrt{3}, b = \sqrt{2}, c = \sqrt{5}$

Center: $(0, 0)$

Vertices: $\left(\pm\sqrt{3}, 0\right)$

Foci: $\left(\pm\sqrt{5}, 0\right)$

Asymptotes: $y = \pm\sqrt{\frac{2}{3}}x = \pm\frac{\sqrt{6}}{3}x$

To use a graphing utility, solve for y first.

$$y^2 = \frac{2x^2 - 6}{3}$$

$y_1 = \sqrt{\dfrac{2x^2 - 6}{3}}$

$y_2 = -\sqrt{\dfrac{2x^2 - 6}{3}}$ $\Big\}$ Hyperbola

$y_3 = \dfrac{\sqrt{6}}{3}x$

$y_4 = -\dfrac{\sqrt{6}}{3}x$ $\Big\}$ Asymptotes

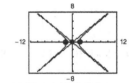

35. $25y^2 - 9x^2 = 225$

$$\frac{y^2}{9} - \frac{x^2}{25} = 1$$

$a = 3, b = 5,$

$c^2 = a^2 + b^2 = 9 + 25 = 34 \Rightarrow c = \sqrt{34}$

Center: $(0, 0)$

Vertices: $(0, \pm3)$

Foci: $\left(0, \pm\sqrt{34}\right)$

Asymptotes: $y = \pm\dfrac{3}{5}x$

To use a graphing utility, solve for y first.

$$y^2 = \frac{225 + 9x^2}{25}$$

$y_1 = \sqrt{\dfrac{9x^2 + 225}{25}}$

$y_2 = -\sqrt{\dfrac{9x^2 + 225}{25}}$ $\Big\}$ Hyperbola

$y_3 = \dfrac{3}{5}x$

$y_4 = -\dfrac{3}{5}x$ $\Big\}$ Asymptotes

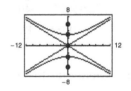

37. $9y^2 - x^2 + 2x + 54y + 62 = 0$

$9(y^2 + 6y + 9) - (x^2 - 2x + 1) = -62 - 1 + 81$

$$9(y + 3)^2 - (x - 1)^2 = 18$$

$$\frac{(y + 3)^2}{2} - \frac{(x - 1)^2}{18} = 1$$

$a = \sqrt{2}, b = 3\sqrt{2}, c = 2\sqrt{5}$

Center: $(1, -3)$

Vertices: $\left(1, -3 \pm \sqrt{2}\right)$

Foci: $\left(1, -3 \pm 2\sqrt{5}\right)$

Asymptotes: $y = -3 \pm \dfrac{1}{3}(x - 1)$

To use a graphing utility, solve for y first.

$$9(y + 3)^2 = 18 + (x - 1)^2$$

$$y = -3 \pm \sqrt{\frac{18 + (x - 1)^2}{9}}$$

$y_1 = -3 + \dfrac{1}{3}\sqrt{18 + (x - 1)^2}$

$y_2 = -3 - \dfrac{1}{3}\sqrt{18 + (x - 1)^2}$ $\Big\}$ Hyperbola

$y_3 = -3 + \dfrac{1}{3}(x - 1)$

$y_4 = -3 - \dfrac{1}{3}(x - 1)$ $\Big\}$ Asymptotes

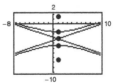

39. Vertices: $(\pm 1, 0) \Rightarrow a = 1$

Asymptotes: $y = \pm 5x \Rightarrow \dfrac{b}{a} = 5, b = 5$

Center: $(0, 0) = (h, k)$

$$\dfrac{(x - h)^2}{a^2} - \dfrac{(y - k)^2}{b^2} = 1$$

$$\dfrac{x^2}{1} - \dfrac{y^2}{25} = 1$$

41. Foci: $(0, \pm 8) \Rightarrow c = 8$

Asymptotes: $y = \pm 4x \Rightarrow \dfrac{a}{b} = 4 \Rightarrow a = 4b$

Center: $(0, 0) = (h, k)$

$$c^2 = a^2 + b^2 \Rightarrow 64 = 16b^2 + b^2$$

$$\dfrac{64}{17} = b^2 \Rightarrow a^2 = \dfrac{1024}{17}$$

$$\dfrac{(y - k)^2}{a^2} - \dfrac{(x - h)^2}{b^2} = 1$$

$$\dfrac{y^2}{1024/17} - \dfrac{x^2}{64/17} = 1$$

$$\dfrac{17y^2}{1024} - \dfrac{17x^2}{64} = 1$$

43. Vertices: $(1, 2), (3, 2) \Rightarrow a = 1$

Asymptotes: $y = x, y = 4 - x$

$$\dfrac{b}{a} = 1 \Rightarrow \dfrac{b}{1} = 1 \Rightarrow b = 1$$

Center: $(2, 2) = (h, k)$

$$\dfrac{(x - h)^2}{a^2} - \dfrac{(y - k)^2}{b^2} = 1$$

$$\dfrac{(x - 2)^2}{1} - \dfrac{(y - 2)^2}{1} = 1$$

45. Vertices: $(3, 0), (3, 4) \Rightarrow a = 2$

Asymptotes: $y = \dfrac{2}{3}x, y = 4 - \dfrac{2}{3}x$

$$\dfrac{a}{b} = \dfrac{2}{3} \Rightarrow b = 3$$

Center: $(3, 2) = (h, k)$

$$\dfrac{(y - k)^2}{a^2} - \dfrac{(x - h)^2}{b^2} = 1$$

$$\dfrac{(y - 2)^2}{4} - \dfrac{(x - 3)^2}{9} = 1$$

47. Foci: $(-1, -1), (9, -1) \Rightarrow c = 5$

Asymptotes: $y = \dfrac{3}{4}x - 4, y = -\dfrac{3}{4}x + 2$

$$\dfrac{b}{a} = \dfrac{3}{4} \Rightarrow b = 3, a = 4$$

Center: $(4, -1) = (h, k)$

$$\dfrac{(x - h)^2}{a^2} - \dfrac{(y - k)^2}{b^2} = 1$$

$$\dfrac{(x - 4)^2}{16} - \dfrac{(y + 1)^2}{9} = 1$$

49. (a) Vertices: $(\pm 1, 0) \Rightarrow a = 1$

Horizontal transverse axis

Center: $(0, 0)$

$$\dfrac{x^2}{a^2} - \dfrac{y^2}{b^2} = 1$$

Point on the graph: $(2, 13)$

$$\dfrac{2^2}{1^2} - \dfrac{13^2}{b^2} = 1$$

$$4 - \dfrac{169}{b^2} = 1$$

$$3b^2 = 169$$

$$b^2 = \dfrac{169}{3}$$

So, $\dfrac{x^2}{1} - \dfrac{y^2}{169/3} = 1.$

(b) When $y = 5$: $x^2 = 1 + \dfrac{5^2}{56.33}$

$$x = \sqrt{1 + \dfrac{25}{56.33}} \approx 1.2016$$

So, the width is about $2x \approx 2.403$ feet.

51.
$$2c = 4 \text{ mi} = 21{,}120 \text{ ft}$$
$$c = 10{,}560 \text{ ft}$$
$$(1100 \text{ ft/s})(18 \text{ s}) = 19{,}800 \text{ ft}$$

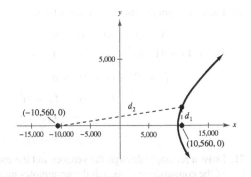

The lightning occurred 19,800 feet further from B than from A:

$$d_2 - d_1 = 2a = 19{,}800 \text{ ft}$$
$$a = 9900 \text{ ft}$$
$$b^2 = c^2 - a^2 = (10{,}560)^2 - (9900)^2$$
$$b^2 = 13{,}503{,}600$$

$$\frac{x^2}{(9900)^2} - \frac{y^2}{13{,}503{,}600} = 1$$

$$\frac{x^2}{98{,}010{,}000} - \frac{y^2}{13{,}503{,}600} = 1$$

53. (a) Foci: $(\pm 150, 0) \Rightarrow c = 150$

Center: $(0, 0) = (h, k)$

$$\frac{d_2}{186{,}000} - \frac{d_1}{186{,}000} = 0.001 \Rightarrow 2a = 186, \ a = 93$$

$$b^2 = c^2 - a^2 = 150^2 - 93^2 = 13{,}851$$

$$\frac{x^2}{93^2} - \frac{y^2}{13{,}851} = 1$$

$$x^2 = 93^2\left(1 + \frac{75^2}{13{,}851}\right) \approx 12{,}161$$

$$x \approx 110.3 \text{ miles}$$

(c) Using the asymptote with positive slope,

$$y = k \pm \frac{b}{a}(x - h)$$

$$y = \frac{\sqrt{13{,}851}}{\sqrt{8694}} x$$

$$y = \frac{27\sqrt{19}}{93} x$$

(b) $c - a = 150 - 93 = 57$ miles

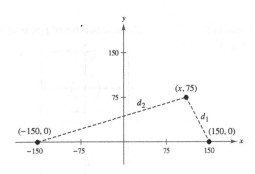

55. $9x^2 + 4y^2 - 18x + 16y - 119 = 0$
$A = 9, C = 4$
$AC = (9)(4) = 36 > 0 \Rightarrow$ Ellipse

57. $4x^2 - y^2 - 4x - 3 = 0$
$A = 4, C = -1$
$AC = (4)(-1) = -4 < 0 \Rightarrow$ Hyperbola

59. $y^2 - 4x^2 + 4x - 2y - 4 = 0$
$A = -4, C = 1$
$AC = (-4)(1) = -4 < 0 \Rightarrow$ Hyperbola

61. $4x^2 + 25y^2 + 16x + 250y + 541 = 0$
$A = 4, C = 25$
$AC = (4)(25) = 100 > 0 \Rightarrow$ Ellipse

63. $25x^2 - 10x - 200y - 119 = 0$
$A = 25, C = 0$
$AC = 25(0) = 0 \Rightarrow$ Parabola

65. $100x^2 + 100y^2 - 100x + 400y + 409 = 0$
$A = 100, C = 100$
$A = C \Rightarrow$ Circle

67. True. For a hyperbola, $c^2 = a^2 + b^2$ or

$$e^2 = \frac{c^2}{a^2} = 1 + \frac{b^2}{a^2}.$$

The larger the ratio of b to a, the larger the eccentricity $e = c/a$ of the hyperbola.

69. False. The graph is two intersecting lines.

$$x^2 - y^2 + 4x - 4y = 0$$

$$\left(x^2 + 4x + 4\right) - \left(y^2 + 4y + 4\right) = 4 - 4$$

$$(x - 2)^2 - (y + 2)^2 = 0$$

$$(x - 2)^2 = (y + 2)^2$$

$$x - 2 = \pm(y - 2)$$

$$y = x \text{ and } y = -x + 4$$

71. Draw a rectangle through the vertices and the endpoints of the conjugate axis. Sketch the asymptotes by drawing lines through the opposite corners of the rectangle.

73. Because the transverse axis is vertical,

$$\frac{(y + 5)^2}{9} - \frac{(x - 3)^2}{4} = 1, \text{ where } a = 3, b = 2, h = 3,$$

and $k = -5$ the equations of the asymptotes should be

$$y = k \pm \frac{a}{b}(x - h)$$

$$y = -5 \pm \frac{3}{2}(x - 3)$$

$$y = \frac{3}{2}x - \frac{19}{2} \text{ and } y = -\frac{3}{2}x - \frac{1}{2}.$$

75.

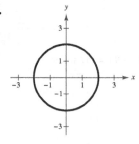

Value of C	Possible number of points of intersection
$C > 2$	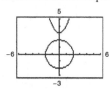
$C = 2$	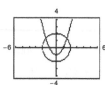
$-2 < C < 2$	
$C = -2$	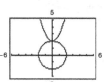

$C < -2$ 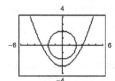   or   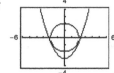   or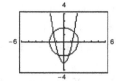

For $C \le -2$, analyze the two curves to determine the number of points of intersection.

$C = -2$: $x^2 + y^2 = 4$ and $y = x^2 - 2$

$$x^2 = y + 2$$

Substitute: $(y + 2) + y^2 = 4$

$$y^2 + y - 2 = 0$$

$$(y + 2)(y - 1) = 0$$

$$y = -2, 1$$

$x^2 = y + 2$	$x^2 = y + 2$
$x^2 = -2 + 2$	$x^2 = 1 + 2$
$x^2 = 0$	$x^2 = 3$
$x = 0$	$x = \pm\sqrt{3}$
$(0, -2)$	$\left(-\sqrt{3}, 1\right), \left(\sqrt{3}, 1\right)$

There are three points of intersection when $C = -2$.

$C < -2$: $x^2 + y^2 = 4$ and $y = x^2 + C$

$$x^2 = y - C$$

Substitute: $(y - C) + y^2 = 4$

$$y^2 + y - 4 - C = 0$$

$$y = \frac{-1 \pm \sqrt{(1)^2 - (4)(1)(-C - 4)}}{2}$$

$$y = \frac{-1 \pm \sqrt{1 + 4(C + 4)}}{2}$$

If $1 + 4(C + 4) < 0$, there are no real solutions (no points of intersection):

$$1 + 4C + 16 < 0$$

$$4C < -17$$

$$C < \frac{-17}{4}, \text{ no points of intersection}$$

If $1 + 4(C + 4) = 0$, there is one real solution (two points of intersection):

$$1 + 4C + 16 = 0$$

$$4C = -17$$

$$C = \frac{-17}{4}, \text{ two points of intersection}$$

If $1 + 4(C + 4) > 0$, there are two real solutions (four points of intersection):

$$1 + 4C + 16 > 0$$

$$4C > -17$$

$$C > \frac{-17}{4}, \text{ (but } C < -2\text{), four points of intersection}$$

Summary:

a. no points of intersection: $C > 2$ or $C < \dfrac{-17}{4}$

b. one point of intersection: $C = 2$

c. two points of intersection: $-2 < C < 2$ or $C = \dfrac{-17}{4}$

d. three points of intersection: $C = -2$

e. four points of intersection: $\dfrac{-17}{4} < C < -2$

Section 6.5   Rotation of Conics

1. rotation; axes

3. invariant under rotation

5. $\theta = 90°$; Point: $(2, 0)$

$x = x' \cos \theta - y' \sin \theta$ $\qquad$ $y = x' \sin \theta + y' \cos \theta$

$2 = x' \cos 90° - y' \sin 90°$ $\quad$ $0 = x' \sin 90° + y' \cos 90°$

$2 = -y'$ $\qquad\qquad\qquad\qquad$ $0 = x'$

$y' = -2$

So, $(x', y') = (0, -2)$.

7. $\theta = 30°$; Point: $(1, 3)$

$\begin{aligned} x &= x' \cos \theta - y' \sin \theta \\ y &= x' \sin \theta + y' \cos \theta \end{aligned} \Rightarrow \begin{cases} 1 = x' \cos 30° - y' \sin 30° \\ 3 = x' \sin 30° + y' \cos 30° \end{cases}$

Solving the system yields $(x', y') = \left(\dfrac{3 + \sqrt{3}}{2}, \dfrac{3\sqrt{3} - 1}{2} \right)$.

9. $\theta = 45°$; Point: $(2, 1)$

$\begin{aligned} x &= x' \cos \theta - y' \sin \theta \\ y &= x' \sin \theta + y' \cos \theta \end{aligned} \Rightarrow \begin{cases} 2 = x' \cos 45° - y' \sin 45° \\ 1 = x' \sin 45° + y' \cos 45° \end{cases}$

Solving the system yields $(x', y') = \left(\dfrac{3\sqrt{2}}{2}, -\dfrac{\sqrt{2}}{2} \right)$.

11. $\theta = 60°$; Point: $(1, 2)$

$\begin{aligned} x &= x' \cos \theta - y' \sin \theta \\ y &= x' \sin \theta + y' \cos \theta \end{aligned} \Rightarrow \begin{cases} 1 = x' \cos 60° - y' \sin 60° \\ 2 = x' \sin 60° + y' \cos 60° \end{cases}$

Solving the system yields $(x', y') = \left(\dfrac{1}{2} + \sqrt{3}, 1 - \dfrac{\sqrt{3}}{2} \right) = \left(\dfrac{1 + 2\sqrt{3}}{2}, \dfrac{2 - \sqrt{3}}{2} \right)$.

13. $xy + 3 = 0$, $A = 0$, $B = 1$, $C = 0$

$\cot 2\theta = \dfrac{A - C}{B} = 0 \Rightarrow 2\theta = \dfrac{\pi}{2} \Rightarrow \theta = \dfrac{\pi}{4}$

$x = x' \cos \dfrac{\pi}{4} - y' \sin \dfrac{\pi}{4}$ $\qquad$ $y = x' \sin \dfrac{\pi}{4} + y' \cos \dfrac{\pi}{4}$ $\qquad\qquad$ $xy + 3 = 0$

$ = x'\left(\dfrac{\sqrt{2}}{2} \right) - y'\left(\dfrac{\sqrt{2}}{2} \right)$ $\qquad$ $ = x'\left(\dfrac{\sqrt{2}}{2} \right) + y'\left(\dfrac{\sqrt{2}}{2} \right)$ $\qquad$ $\left(\dfrac{x' - y'}{\sqrt{2}} \right)\left(\dfrac{x' + y'}{\sqrt{2}} \right) + 3 = 0$

$ = \dfrac{x' - y'}{\sqrt{2}}$ $\qquad\qquad\qquad$ $ = \dfrac{x' + y'}{\sqrt{2}}$ $\qquad\qquad$ $\dfrac{(x')^2}{2} - \dfrac{(y')^2}{2} = -3$

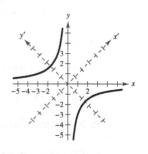

$\dfrac{(y')^2}{6} - \dfrac{(x')^2}{6} = 1$

15. $xy + 2x - y + 4 = 0$

$A = 0, B = 1, C = 0$

$\cot 2\theta = \dfrac{A - C}{B} = 0 \Rightarrow 2\theta = \dfrac{\pi}{2} \Rightarrow \theta = \dfrac{\pi}{4}$

$x = x' \cos \dfrac{\pi}{4} - y' \sin \dfrac{\pi}{4}$ $\qquad$ $y = x' \sin \dfrac{\pi}{4} + y' \cos \dfrac{\pi}{4}$

$\quad = \dfrac{x' - y'}{\sqrt{2}}$ $\qquad\qquad\qquad$ $= \dfrac{x' + y'}{\sqrt{2}}$

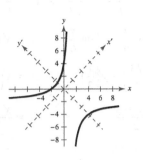

$xy + 2x - y + 4 = 0$

$\left(\dfrac{x' - y'}{\sqrt{2}}\right)\left(\dfrac{x' + y'}{\sqrt{2}}\right) + 2\left(\dfrac{x' - y'}{\sqrt{2}}\right) - \left(\dfrac{x' + y'}{\sqrt{2}}\right) + 4 = 0$

$\dfrac{(x')^2}{2} - \dfrac{(y')^2}{2} + \dfrac{2x'}{\sqrt{2}} - \dfrac{2y'}{\sqrt{2}} - \dfrac{x'}{\sqrt{2}} - \dfrac{y'}{\sqrt{2}} + 4 = 0$

$\left[(x')^2 + \sqrt{2}x' + \left(\dfrac{\sqrt{2}}{2}\right)^2\right] - \left[(y')^2 + 3\sqrt{2}y' + \left(\dfrac{3\sqrt{2}}{2}\right)^2\right] = -8 + \left(\dfrac{\sqrt{2}}{2}\right)^2 - \left(\dfrac{3\sqrt{2}}{2}\right)^2$

$\left(x' + \dfrac{\sqrt{2}}{2}\right)^2 - \left(y' + \dfrac{3\sqrt{2}}{2}\right)^2 = -12$

$\dfrac{\left(y' + \dfrac{3\sqrt{2}}{2}\right)^2}{12} - \dfrac{\left(x' + \dfrac{\sqrt{2}}{2}\right)^2}{12} = 1$

17. $5x^2 - 6xy + 5y^2 - 12 = 0$

$A = 5, B = -6, C = 5$

$\cot 2\theta = \dfrac{A - C}{B} = 0 \Rightarrow 2\theta = \dfrac{\pi}{2} \Rightarrow \theta = \dfrac{\pi}{4}$

$x = x' \cos \dfrac{\pi}{4} - y' \sin \dfrac{\pi}{4}$ $\qquad$ $y = x' \sin \dfrac{\pi}{4} + y' \cos \dfrac{\pi}{4}$

$\quad = x'\left(\dfrac{\sqrt{2}}{2}\right) - y'\left(\dfrac{\sqrt{2}}{2}\right)$ $\qquad$ $= x'\left(\dfrac{\sqrt{2}}{2}\right) + y'\left(\dfrac{\sqrt{2}}{2}\right)$

$\quad = \dfrac{x' - y'}{\sqrt{2}}$ $\qquad\qquad\qquad$ $= \dfrac{x' + y'}{\sqrt{2}}$

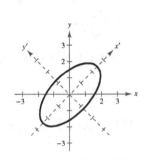

$5x^2 - 6xy + 5y^2 - 12 = 0$

$5\left(\dfrac{x' - y'}{\sqrt{2}}\right)^2 - 6\left(\dfrac{x' - y'}{\sqrt{2}}\right)\left(\dfrac{x' + y'}{\sqrt{2}}\right) + 5\left(\dfrac{x' + y'}{\sqrt{2}}\right)^2 - 12 = 0$

$\dfrac{5(x')^2}{2} - 5x'y' + \dfrac{5(y')^2}{2} - 3(x')^2 + 3(y')^2 + \dfrac{5(x')^2}{2} + 5x'y' + \dfrac{5(y')^2}{2} - 12 = 0$

$2(x')^2 + 8(y')^2 = 12$

$\dfrac{(x')^2}{6} + \dfrac{(y')^2}{3/2} = 1$

19. $13x^2 + 6\sqrt{3}xy + 7y^2 - 16 = 0$

$A = 13, B = 6\sqrt{3}, C = 7$

$\cot 2\theta = \dfrac{A - C}{B} = \dfrac{1}{\sqrt{3}} \Rightarrow 2\theta = \dfrac{\pi}{3} \Rightarrow \theta = \dfrac{\pi}{6}$

$x = x' \cos \dfrac{\pi}{6} - y' \sin \dfrac{\pi}{6} \qquad y = x' \sin \dfrac{\pi}{6} + y' \cos \dfrac{\pi}{6}$

$\quad = x'\left(\dfrac{\sqrt{3}}{2}\right) - y'\left(\dfrac{1}{2}\right) \qquad\quad = x'\left(\dfrac{1}{2}\right) + y'\left(\dfrac{\sqrt{3}}{2}\right)$

$\quad = \dfrac{\sqrt{3}x' - y'}{2} \qquad\qquad\quad = \dfrac{x' + \sqrt{3}y'}{2}$

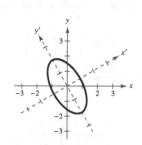

$$13x^2 + 6\sqrt{3}xy + 7y^2 - 16 = 0$$

$$13\left(\dfrac{\sqrt{3}x' - y'}{2}\right)^2 + 6\sqrt{3}\left(\dfrac{\sqrt{3}x' - y'}{2}\right)\left(\dfrac{x' + \sqrt{3}y'}{2}\right) + 7\left(\dfrac{x' + \sqrt{3}y'}{2}\right)^2 - 16 = 0$$

$$\dfrac{39(x')^2}{4} - \dfrac{13\sqrt{3}x'y'}{2} + \dfrac{13(y')^2}{4} + \dfrac{18(x')^2}{4} + \dfrac{18\sqrt{3}x'y'}{4} - \dfrac{6\sqrt{3}x'y'}{4}$$

$$-\dfrac{18(y')^2}{4} + \dfrac{7(x')^2}{4} + \dfrac{7\sqrt{3}x'y'}{2} + \dfrac{21(y')^2}{4} - 16 = 0$$

$$16(x')^2 + 4(y')^2 = 16$$

$$\dfrac{(x')^2}{1} + \dfrac{(y')^2}{4} = 1$$

21. $x^2 + 2xy + y^2 + \sqrt{2}x - \sqrt{2}y = 0, A = 1, B = 2, C = 1$

$\cot 2\theta = \dfrac{A - C}{B} = \dfrac{1 - 1}{2} = 0 \Rightarrow 2\theta = \dfrac{\pi}{2} \Rightarrow \theta = \dfrac{\pi}{4}$

$x = x' \cos \dfrac{\pi}{4} - y' \sin \dfrac{\pi}{4} \qquad y = x' \sin \dfrac{\pi}{4} + y' \cos \dfrac{\pi}{4}$

$\quad = \dfrac{x' - y'}{\sqrt{2}} \qquad\qquad\qquad = \dfrac{x' + y'}{\sqrt{2}}$

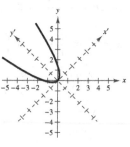

$x^2 + 2xy + y^2 + \sqrt{2}x - \sqrt{2}y = 0$

$$\left(\dfrac{x' - y'}{\sqrt{2}}\right)^2 + 2\left(\dfrac{x' - y'}{\sqrt{2}}\right)\left(\dfrac{x' + y'}{\sqrt{2}}\right) + \left(\dfrac{x' + y'}{\sqrt{2}}\right)^2 + \sqrt{2}\left(\dfrac{x' - y'}{\sqrt{2}}\right) - \sqrt{2}\left(\dfrac{x' + y'}{\sqrt{2}}\right) = 0$$

$$\dfrac{(x')^2}{2} - x'y' + \dfrac{(y')^2}{2} + (x')^2 - (y')^2 + \dfrac{(x')^2}{2} + x'y' + \dfrac{(y')^2}{2} + x' - y' - x' - y' = 0$$

$$2(x')^2 - 2y' = 0$$

$$2(x')^2 = 2y'$$

$$(x')^2 = y'$$

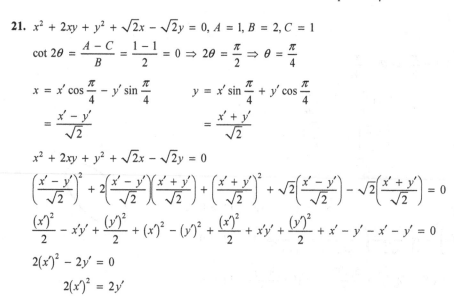

23. $9x^2 + 24xy + 16y^2 + 19x - 130y = 0$

$A = 9, B = 24, C = 16$

$\cot 2\theta = \dfrac{A - C}{B} = -\dfrac{7}{24} \Rightarrow \theta \approx 53.13°$

$\cos 2\theta = -\dfrac{7}{25}$

$\sin \theta = \sqrt{\dfrac{1 - \cos 2\theta}{2}} = \sqrt{\dfrac{1 - \left(-\dfrac{7}{25}\right)}{2}} = \dfrac{4}{5}$

$\cos \theta = \sqrt{\dfrac{1 + \cos 2\theta}{2}} = \sqrt{\dfrac{1 + \left(-\dfrac{7}{25}\right)}{2}} = \dfrac{3}{5}$

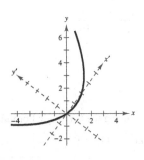

$x = x' \cos \theta - y' \sin \theta = x'\left(\dfrac{3}{5}\right) - y'\left(\dfrac{4}{5}\right) = \dfrac{3x' - 4y'}{5}$ $y = x' \sin \theta + y' \cos \theta = x'\left(\dfrac{4}{5}\right) + y'\left(\dfrac{3}{5}\right) = \dfrac{4x' + 3y'}{5}$

$9x^2 + 24xy + 16y^2 + 90x - 130y = 0$

$9\left(\dfrac{3x' - 4y'}{5}\right)^2 + 24\left(\dfrac{3x' - 4y'}{5}\right)\left(\dfrac{4x' + 3y'}{5}\right) + 16\left(\dfrac{4x' + 3y'}{5}\right)^2 + 90\left(\dfrac{3x' - 4y'}{5}\right) - 130\left(\dfrac{4x' + 3y'}{5}\right) = 0$

$\dfrac{81(x')^2}{25} - \dfrac{216x'y'}{25} + \dfrac{144(y')^2}{25} + \dfrac{288(x')^2}{25} - \dfrac{168x'y'}{25} - \dfrac{288(y')^2}{25} + \dfrac{256(x')^2}{25} + \dfrac{384x'y'}{25} + \dfrac{144(y')^2}{25}$

$+ 54x' - 72y' - 104x' - 78y' = 0$

$25(x')^2 - 50x' - 150y' = 0$

$(x')^2 - 2x' = 6y'$

$(x')^2 - 2x' + 1 = 6y' + 1$

$(x' - 1)^2 = 6\left(y' + \dfrac{1}{6}\right)$

25. $x^2 - 4xy + 2y^2 = 6$

$A = 1, B = -4, C = 2$

$\cot 2\theta = \dfrac{A - C}{B} = \dfrac{1 - 2}{-4} = \dfrac{1}{4}$

$\dfrac{1}{\tan 2\theta} = \dfrac{1}{4}$

$\tan 2\theta = 4$

$2\theta \approx 75.96$

$\theta \approx 37.98°$

To graph conic with a graphing calculator, solve for y in terms of x.

$x^2 - 4xy + 2y^2 = 6$

$y^2 - 2xy + x^2 = 3 - \dfrac{x^2}{2} + x^2$

$(y - x)^2 = 3 + \dfrac{x^2}{2}$

$y - x = \pm\sqrt{3 + \dfrac{x^2}{2}}$

$y = x \pm \sqrt{3 + \dfrac{x^2}{2}}$

Enter $y_1 = x + \sqrt{3 + \dfrac{x^2}{2}}$ and $y_2 = x - \sqrt{3 + \dfrac{x^2}{2}}$.

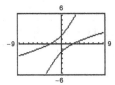

27. $14x^2 + 16xy + 9y^2 = 44$

$A = 14, B = 16, C = 9$

$\cot 2\theta = \dfrac{A - C}{B} = \dfrac{14 - 9}{16} = \dfrac{5}{16}$

$\tan 2\theta = \dfrac{16}{5}$

$2\theta \approx 72.65°$

$\theta \approx 36.32°$

Solve for y in terms of x using the Quadratic Formula.

$(9)y^2 + (16x)y + \left(14x^2 - 44\right) = 0$

$$y = \frac{-b \pm \sqrt{b^2 - 4ac}}{2a}$$

$$y = \frac{-(16x) \pm \sqrt{(16x)^2 - 4(9)\left(14x^2 - 44\right)}}{2(9)}$$

$$y = \frac{-16x \pm \sqrt{-248x^2 + 1548}}{18}$$

Use $y_1 = \dfrac{-16x + \sqrt{-248x^2 + 1548}}{18}$

and $y_2 = \dfrac{-16x - \sqrt{-248x^2 + 1548}}{18}$

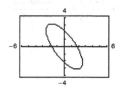

29. $2x^2 + 4xy + 2y^2 + \sqrt{26}x + 3y = -15$

$A = 2, B = 4, C = 2$

$\cot 2\theta = \dfrac{A - C}{B} = 0 \Rightarrow 2\theta = \dfrac{\pi}{2} \Rightarrow \theta = \dfrac{\pi}{4}$ or $45°$

Solve for y in terms of x using the Quadratic Formula.

$2y^2 + (4x + 3)y + \left(2x^2 + \sqrt{26}x + 15\right) = 0$

$$y = \frac{-(4x + 3) \pm \sqrt{(4x + 3)^2 - 4(2)\left(2x^2 + \sqrt{26}x + 15\right)}}{2(2)}$$

$$= \frac{-(4x + 3) \pm \sqrt{(4x + 3)^2 - 8\left(2x^2 + \sqrt{26}x + 15\right)}}{4}$$

Enter $y_1 = \dfrac{-(4x + 3) + \sqrt{(4x + 3)^2 - 8\left(2x^2 + \sqrt{26}x + 15\right)}}{4}$ and

$y_2 = \dfrac{-(4x + 3) - \sqrt{(4x + 3)^2 - 8\left(2x^2 + \sqrt{26}x + 15\right)}}{4}$.

31. $xy + 2 = 0$

$B^2 - 4AC = 1 \Rightarrow$ The graph is a hyperbola.

$\cot 2\theta = \dfrac{A - C}{B} = 0 \Rightarrow \theta = 45°$

Matches graph (e).

32. $x^2 - xy + 3y^2 - 5 = 0$

$A = 1, B = -1, C = 3$

$B^2 - 4AC = (-1)^2 - 4(1)(3) = -11$

The graph is an ellipse.

$\cot 2\theta = \dfrac{A - C}{B} = \dfrac{1 - 3}{-1} = 2 \Rightarrow \theta \approx 13.28°$

Matches graph (a).

33. $3x^2 + 2xy + y^2 - 10 = 0$

$B^2 - 4AC = (2)^2 - 4(3)(1) = -8 \Rightarrow$

The graph is an ellipse or circle.

$\cot 2\theta = \dfrac{A - C}{B} = 1 \Rightarrow \theta = 22.5°$

Matches graph (d).

34. $x^2 - 4xy + 4y^2 + 10x - 30 = 0$

$A = 1, B = -4, C = 4$

$B^2 - 4AC = (-4)^2 - 4(1)(4) = 0$

The graph is a parabola.

$\cot 2\theta = \dfrac{A - C}{B} = \dfrac{1 - 4}{-4} = \dfrac{3}{4} \Rightarrow \theta \approx 26.57°$

Matches graph (c).

35. $x^2 + 2xy + y^2 = 0$

$(x + y)^2 = 0$

$x + y = 0$

$y = -x$

The graph is a line. Matches graph (f).

36. $-2x^2 + 3xy + 2y^2 + 3 = 0$

$B^2 - 4AC = (3)^2 - 4(-2)(2) = 25 \Rightarrow$

The graph is a hyperbola.

$\cot 2\theta = \dfrac{A - C}{B} = -\dfrac{4}{3} \Rightarrow \theta \approx -18.43°$

Matches graph (b).

37. (a) $16x^2 - 8xy + y^2 - 10x + 5y = 0$

$B^2 - 4AC = (-8)^2 - 4(16)(1) = 0$

The graph is a parabola.

(b) $y^2 + (-8x + 5)y + (16x^2 - 10x) = 0$

$y = \dfrac{-(-8x + 5) \pm \sqrt{(-8x + 5)^2 - 4(1)(16x^2 - 10x)}}{2(1)}$

$= \dfrac{(8x - 5) \pm \sqrt{(8x - 5)^2 - 4(16x^2 - 10x)}}{2}$

(c)
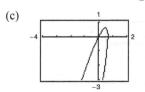

39. (a) $12x^2 - 6xy + 7y^2 - 45 = 0$

$B^2 - 4AC = (-6)^2 - 4(12)(7) = -300 < 0$

The graph is an ellipse.

(b) $7y^2 + (-6x)y + (12x^2 - 45) = 0$

$y = \dfrac{-(-6x) \pm \sqrt{(-6x)^2 - 4(7)(12x^2 - 45)}}{2(7)}$

$= \dfrac{6x \pm \sqrt{36x^2 - 28(12x^2 - 45)}}{14}$

(c)
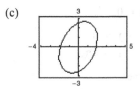

41. (a) $x^2 - 6xy - 5y^2 + 4x - 22 = 0$

$B^2 - 4AC = (-6)^2 - 4(1)(-5) = 56 > 0$

The graph is a hyperbola.

(b) $-5y^2 + (-6x)y + (x^2 + 4x - 22) = 0$

$y = \dfrac{-(-6x) \pm \sqrt{(-6x)^2 - 4(-5)(x^2 + 4x - 22)}}{2(-5)}$

$= \dfrac{6x \pm \sqrt{36x^2 + 20(x^2 + 4x - 22)}}{-10}$

$= \dfrac{-6x \pm \sqrt{36x^2 + 20(x^2 + 4x - 22)}}{10}$

(c)
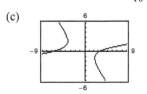

43. (a) $x^2 + 4xy + 4y^2 - 5x - y - 3 = 0$

$$B^2 - 4AC = (4)^2 - 4(1)(4) = 0$$

The graph is a parabola.

(b) $4y^2 + (4x - 1)y + (x^2 - 5x - 3) = 0$

$$y = \frac{-(4x - 1) \pm \sqrt{(4x - 1)^2 - 4(4)(x^2 - 5x - 3)}}{2(4)}$$

$$= \frac{-(4x - 1) \pm \sqrt{(4x - 1)^2 - 16(x^2 - 5x - 3)}}{8}$$

(c)

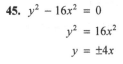

45. $y^2 - 16x^2 = 0$

$$y^2 = 16x^2$$

$$y = \pm 4x$$

Two intersecting lines

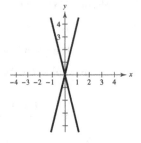

47. $15x^2 - 2xy - y^2 = 0$

$(5x - y)(3x + y) = 0$

$5x - y = 0 \quad 3x + y = 0$

$\quad y = 5x \quad\quad y = -3x$

Two intersecting lines

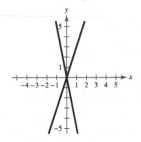

49. $x^2 - 2xy + y^2 = 0$

$$y^2 - 2xy + x^2 = x^2 - x^2$$

$$(y - x)^2 = 0$$

$$y - x = 0$$

$$y = x$$

Line

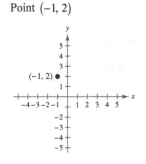

51. $\quad x^2 + y^2 + 2x - 4y + 5 = 0$

$$x^2 + 2x + 1 + y^2 - 4y + 4 = -5 + 1 + 4$$

$$(x + 1)^2 + (y - 2)^2 = 0$$

Point $(-1, 2)$

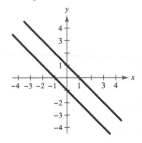

53. $x^2 + 2xy + y^2 - 1 = 0$

$$(x + y)^2 - 1 = 0$$

$$(x + y)^2 = 1$$

$$x + y = \pm 1$$

$$y = -x \pm 1$$

Two parallel lines

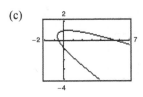

55. $x^2 - 4y^2 - 20x - 64y - 172 = 0 \;\Rightarrow\; (x - 10)^2 - 4(y + 8)^2 = 16$

<u>$16x^2 + 4y^2 - 320x + 64y + 1600 = 0$</u> $\Rightarrow 16(x - 10)^2 + 4(y + 8)^2 = 256$

$17x^2 \qquad\;\; - 340x \qquad\qquad 1428 = 0$

$\qquad\qquad (17x - 238)(x - 6) = 0$

$\qquad\qquad\qquad\qquad\quad x = 6 \text{ or } x = 14$

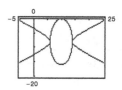

When $x = 6$: $6^2 - 4y^2 - 20(6) - 64y - 172 = 0$

$\qquad\qquad\qquad -4y^2 - 64y - 256 = 0$

$\qquad\qquad\qquad\qquad y^2 + 16y + 64 = 0$

$\qquad\qquad\qquad\qquad\quad (y + 8)^2 = 0$

$\qquad\qquad\qquad\qquad\qquad\quad y = -8$

When $x = 14$: $14^2 - 4y^2 - 20(14) - 64y - 172 = 0$

$\qquad\qquad\qquad -4y^2 - 64y - 256 = 0$

$\qquad\qquad\qquad\qquad y^2 + 16y + 64 = 0$

$\qquad\qquad\qquad\qquad\quad (y + 8)^2 = 0$

$\qquad\qquad\qquad\qquad\qquad\quad y = -8$

Points of intersection: $(6, -8), (14, -8)$

57. $x^2 + 4y^2 - 2x - 8y + 1 = 0 \;\Rightarrow\; (x - 1)^2 + 4(y - 1)^2 = 4$

<u>$-x^2 \qquad\quad + 2x - 4y - 1 = 0$</u> $\Rightarrow y = -\frac{1}{4}(x - 1)^2$

$\qquad 4y^2 \qquad - 12y \qquad = 0$

$\qquad\quad 4y(y - 3) = 0$

$\qquad\qquad\qquad y = 0 \text{ or } y = 3$

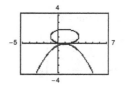

When $y = 0$: $x^2 + 4(0)^2 - 2x - 8(0) + 1 = 0$

$\qquad\qquad\qquad\quad x^2 - 2x + 1 = 0$

$\qquad\qquad\qquad\quad (x - 1)^2 = 0$

$\qquad\qquad\qquad\qquad\quad x = 1$

When $y = 3$: $-x^2 + 2x - 4(3) - 1 = 0$

$\qquad\qquad\qquad x^2 - 2x + 13 = 0$

No real solution

Point of intersection: $(1, 0)$

59.

$$\begin{array}{r} x^2 + y^2 - 4 = 0 \\ \underline{3x - y^2 = 0} \\ x^2 + 3x - 4 = 0 \end{array}$$

$$(x + 4)(x - 1) = 0$$

$$x = -4 \text{ or } x = 1$$

When $x = -4$: $\;3(-4) - y^2 = 0$

$$y^2 = -12$$

No real solution

When $x = 1$: $\;3(1) - y^2 = 0$

$$y^2 = 3$$

$$y = \pm\sqrt{3}$$

The points of intersection are $\left(1, \sqrt{3}\right)$ and $\left(1, -\sqrt{3}\right)$.

The standard forms of the equations are:

$$x^2 + y^2 = 4$$

$$y^2 = 3x$$

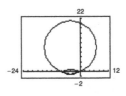

61. $-x^2 - y^2 - 8x + 20y - 7 = 0 \;\Rightarrow\; (x + 4)^2 + (y - 10)^2 = 123$: Circle

$$\underline{x^2 + 9y^2 + 8x + 4y + 7 = 0} \;\Rightarrow\; (x + 4)^2 + 9\left(y + \tfrac{2}{9}\right)^2 = \tfrac{85}{9} \;\Rightarrow\; \dfrac{(x + 4)}{\frac{85}{9}} + \dfrac{\left(y + \frac{2}{9}\right)^2}{\frac{85}{81}} = 1\text{: Ellipse}$$

$$8y^2 + 24y = 0$$

$$8y(y + 3) = 0$$

$$y = 0 \text{ or } y = -3$$

When $y = 0$: $x^2 + 9(0)^2 + 8x + 4(0) + 7 = 0$

$$x^2 + 8x + 7 = 0$$

$$(x + 7)(x + 1) = 0$$

$$x = -7 \text{ or } x = -1$$

When $y = -3$: $x^2 + 9(-3)^2 + 8x + 4(-3) + 7 = 0$

$$x^2 + 8x + 76 = 0$$

No real solutions

The points of intersection are $(-7, 0)$ and $(-1, 0)$.

63. (a) Because $A = 1$, $B = -2$ and $c = 1$ you have $\cot 2\theta = \dfrac{A - C}{B} = \dfrac{1 - 1}{-2} = 0 \Rightarrow 2\theta = \dfrac{\pi}{2} \Rightarrow \theta = \dfrac{\pi}{4}$

which implies that $x = x'\cos\dfrac{\pi}{4} - y'\sin\dfrac{\pi}{4}$

$$= x'\left(\dfrac{1}{\sqrt{2}}\right) - y'\left(\dfrac{1}{\sqrt{2}}\right)$$

$$= \dfrac{x' - y'}{\sqrt{2}}$$

and $y = x'\sin\dfrac{\pi}{4} + y'\cos\dfrac{\pi}{4}$

$$= x'\left(\dfrac{1}{\sqrt{2}}\right) + y'\left(\dfrac{1}{\sqrt{2}}\right)$$

$$= \dfrac{x' + y'}{\sqrt{2}}.$$

The equation in the $x'y'$-system is obtained by $x^2 - 2xy - 27\sqrt{2}x + y^2 + 9\sqrt{2}y + 378 = 0.$

$$\left(\dfrac{x' - y'}{\sqrt{2}}\right)^2 - 2\left(\dfrac{x' - y'}{\sqrt{2}}\right)\left(\dfrac{x' + y'}{\sqrt{2}}\right) - 27\sqrt{2}\left(\dfrac{x' - y'}{\sqrt{2}}\right) + \left(\dfrac{x' + y'}{\sqrt{2}}\right)^2 + 9\sqrt{2}\left(\dfrac{x' + y'}{\sqrt{2}}\right) + 378 = 0$$

$$\dfrac{(x')^2}{2} - \dfrac{2x'y'}{\sqrt{2}} + \dfrac{(y')^2}{2} - (x')^2 + (y')^2 - 27x' + 27y' + \dfrac{(x')^2}{2} + \dfrac{2x'y'}{\sqrt{2}} + \dfrac{(y')^2}{2} + 9x' + 9y' + 378 = 0$$

$$2(y')^2 + 36y' - 18x' + 378 = 0$$

$$(y')^2 + 18y' + 81 = 9x' - 189 + 81$$

$$(y' + 9)^2 = 9(x' - 12)$$

$$(y' + 9)^2 = 4(9/4)(x' - 12)$$

(b) Since $p = 9/4 = 2.25$, the distance from the vertex to the receiver is 2.25 feet.

65. $x^2 + xy + ky^2 + 6x + 10 = 0$

$B^2 - 4AC = 1^2 - 4(1)(k) = 1 - 4k > 0 \Rightarrow -4k > -1 \Rightarrow k < \frac{1}{4}$

True. For the graph to be a hyperbola, the discriminant must be greater than zero.

67. $r^2 = x^2 + y^2 = (x'\cos\theta - y'\sin\theta)^2 + (y'\cos\theta + x'\sin\theta)^2$

$$= (x')^2\cos^2\theta - 2x'y'\cos\theta\sin\theta + (y')^2\sin^2\theta + (y')^2\cos^2\theta + 2x'y'\cos\theta\sin\theta + (x')^2\sin^2\theta$$

$$= (x')^2\left(\cos^2\theta + \sin^2\theta\right) + (y')^2\left(\sin^2\theta + \cos^2\theta\right) = (x')^2 + (y')^2$$

So, $(x')^2 + (y')^2 = r^2.$

Section 6.6 Parametric Equations

1. plane curve

3. eliminating; parameter

5. (a) $x = \sqrt{t}, y = 3 - t$

t	0	1	2	3	4
x	0	1	$\sqrt{2}$	$\sqrt{3}$	2
y	3	2	1	0	-1

(b)

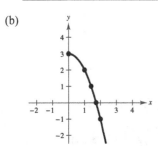

(c) $x = \sqrt{t} \;\;\Rightarrow x^2 = t$

$y = 3 - t \;\;\Rightarrow y = 3 - x^2$

The graph of the parametric equations only shows the right half of the parabola, whereas the rectangular equation yields the entire parabola.

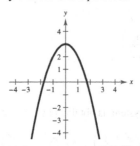

7. $x = t, y = -5t$

t	-2	-1	0	1	2
x	-2	-1	0	1	2
y	10	5	0	-5	-10

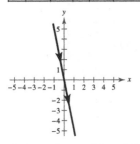

The curve is traced from left to right.

9. $x = t^2, y = 3t$

t	-2	-1	0	1	2
x	4	1	0	1	4
y	-6	-3	0	3	6

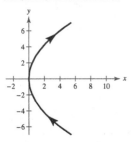

The curve is traced clockwise.

11. $x = 3 \cos \theta, y = 2 \sin^2 \theta, 0 \le \theta \le \pi$

θ	0	$\dfrac{\pi}{4}$	$\dfrac{\pi}{2}$	$\dfrac{3\pi}{4}$	π
x	3	$\dfrac{3\sqrt{2}}{2}$	0	$-\dfrac{3\sqrt{2}}{2}$	-3
y	0	1	2	1	0

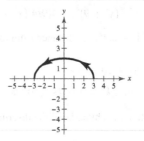

The curve is traced from right to left.

13. (a) $x = t, y = 4t$

t	-2	-1	0	1	2
x	-2	-1	0	1	2
y	-8	-4	0	4	8

(b) $x = t \;\;\Rightarrow t = x$

$y = 4t \;\;\Rightarrow y = 4x$

15. (a) $x = -t + 1, \ y = -3t$

t	-2	-1	0	1	2
x	3	2	1	0	-1
y	6	3	0	-3	-6

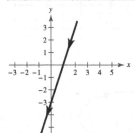

(b) $x = -t + 1 \Rightarrow t = 1 - x$

$y = -3t$

$y = -3(1 - x)$

$y = -3 + 3x$

$y = 3x - 3$

17. (a) $x = \frac{1}{4}t, \ y = t^2$

t	-2	-1	0	1	2
x	$-\frac{1}{2}$	$-\frac{1}{4}$	0	$\frac{1}{4}$	$\frac{1}{2}$
y	4	1	0	1	4

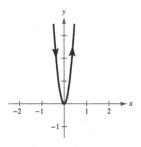

(b) $x = \frac{1}{4}t \Rightarrow t = 4x$

$y = t^2 \Rightarrow y = 16x^2$

19. (a) $x = t^2, \ y = -2t$

t	-2	-1	0	1	2
x	4	1	0	1	4
y	4	2	0	-2	-4

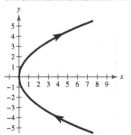

(b) $x = t^2 \Rightarrow t = \pm\sqrt{x}$

$y = -2t$

$\ \ = \pm 2\sqrt{x}$

21. (a) $x = \sqrt{t}, \ y = 1 - t$

t	0	1	2	3
x	0	1	$\sqrt{2}$	$\sqrt{3}$
y	1	0	-1	-2

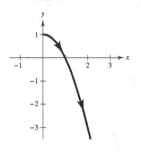

(b) $x = \sqrt{t} \Rightarrow x^2 = t, \ t \geq 0$

$y = 1 - t$

$\ \ = 1 - x^2, \ x \geq 0$

23. (a) $x = \sqrt{t} - 3, y = t^3$

t	0	1	2	3	4
x	-3	-2	$\sqrt{2} - 3$	$\sqrt{3} - 3$	-1
y	0	1	8	27	64

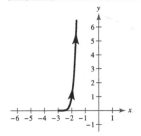

(b) $x = \sqrt{t} - 3 \Rightarrow t = (x + 3)^2$

$y = t^3 \Rightarrow y = \left[(x + 3)^2\right]^3 = (x + 3)^6, x \geq -3$

25. (a) $x = t + 1, y = \dfrac{t}{t + 1}$

t	-3	-2	0	1	2
x	-2	-1	1	2	3
y	$\dfrac{3}{2}$	2	0	$\dfrac{1}{2}$	$\dfrac{2}{3}$

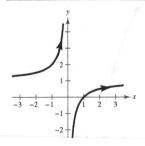

(b) $x = t + 1 \Rightarrow t = x - 1$

$y = \dfrac{t}{t + 1} \Rightarrow y = \dfrac{x - 1}{x}$

27. (a) $x = 4 \cos \theta, y = 2 \sin \theta$

θ	0	$\dfrac{\pi}{2}$	π	$\dfrac{3\pi}{2}$	2π
x	4	0	-4	0	4
y	0	2	0	-2	0

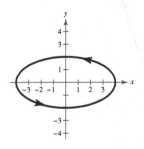

(b) $x = 4 \cos \theta \Rightarrow \left(\dfrac{x}{4}\right)^2 = \cos^2 \theta$

$y = 2 \sin \theta \Rightarrow \left(\dfrac{y}{2}\right)^2 = \sin^2 \theta$

$\left(\dfrac{x}{4}\right)^2 + \left(\dfrac{y}{2}\right)^2 = 1$

$\dfrac{x^2}{16} + \dfrac{y^2}{4} = 1$

29. (a) $x = 1 + \cos\theta$, $y = 1 + 2\sin\theta$

θ	0	$\dfrac{\pi}{2}$	π	$\dfrac{3\pi}{2}$	2π
x	2	1	0	1	2
y	1	3	1	−1	1

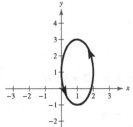

(b) $x = 1 + \cos\theta \Rightarrow (x-1)^2 = \cos^2\theta$

$$y = 1 + 2\sin\theta \Rightarrow \left(\frac{y-1}{2}\right)^2 = \sin^2\theta$$

$$\frac{(x-1)^2}{1} + \frac{(y-1)^2}{4} = 1$$

31. (a) $x = 2\sec\theta$, $y = \tan\theta$, $\dfrac{\pi}{2} \le \theta \le \dfrac{3\pi}{2}$

θ	$\pi/2$	$3\pi/4$	π	$5\pi/4$	$3\pi/2$
x	undefined	$-2\sqrt{2}$	-2	$-2\sqrt{2}$	undefined
y	undefined	-1	0	1	undefined

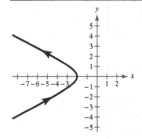

(b) $x = 2\sec\theta \Rightarrow \sec\theta = \dfrac{x}{2}$

$$y = \tan\theta$$

$$1 + \tan^2\theta = \sec^2\theta$$

$$1 + y^2 = \left(\frac{x}{2}\right)^2$$

$$1 + y^2 = \frac{x^2}{4}$$

$$\frac{x^2}{4} - y^2 = 1, \; x \le -2$$

33. (a) $x = 3\cos\theta$, $y = 3\sin\theta$

θ	0	$\pi/4$	$\pi/2$	$3\pi/4$	π	$5\pi/4$	$3\pi/2$	$7\pi/4$	2π
x	3	$3\sqrt{2}/2$	0	$-3\sqrt{2}/2$	-3	$-3\sqrt{2}/2$	0	$3\sqrt{2}/2$	3
y	0	$3\sqrt{2}/2$	3	$3\sqrt{2}/2$	0	$-3\sqrt{2}/2$	-3	$-3\sqrt{2}/2$	0

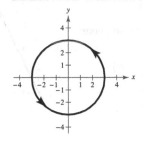

(b) $x = 3\cos\theta \Rightarrow \cos\theta = \dfrac{x}{3}$

$$y = 3\sin\theta \Rightarrow \sin\theta = \frac{x}{3}$$

$$\sin^2\theta + \cos^2\theta = 1$$

$$\left(\frac{x}{3}\right)^2 + \left(\frac{x}{3}\right)^2 = 1$$

$$x^2 + y^2 = 9$$

35. (a) $x = e^t, y = e^{3t}$

t	-3	-2	-1	0	1
x	0.0498	0.1353	0.3679	1	2.7183
y	0.0001	0.0024	0.0498	1	20.0855

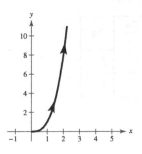

(b) $x = e^t$

   $y = e^{3t} = \left(e^t\right)^3 \Rightarrow y = x^3, x > 0$

37. (a) $x = t^3, y = 3 \ln t$

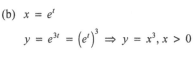

t	$\frac{1}{2}$	1	2	3	4
x	$\frac{1}{8}$	1	8	27	64
y	-2.0794	0	2.0794	3.2958	4.1589

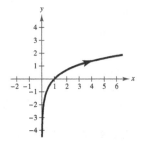

(b) $x = t^3 \qquad \Rightarrow x^{1/3} = t$

   $y = 3 \ln t \Rightarrow y = \ln t^3$

   $y = \ln\left(x^{1/3}\right)^3$

   $y = \ln x$

39. $x = t, y = \sqrt{t}$

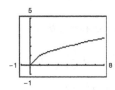

41. $x = 2t, y = |t + 1|$

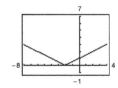

43. $x = 4 + 3 \cos \theta, y = -2 + \sin \theta$

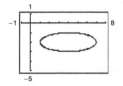

45. $x = 2 \csc \theta, y = 4 \cot \theta$

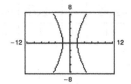

47. $x = \dfrac{t}{2}, y = \ln\left(t^2 + 1\right)$

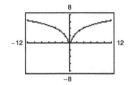

49. By eliminating the parameter, each curve becomes

   $y = 2x + 1.$

(a) $x = t$

   $y = 2t + 1$

   There are no restrictions
   on x and y.

   Domain: $(-\infty, \infty)$

   Orientation:

   Left to right

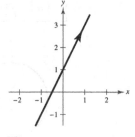

(b) $x = \cos \theta$ $\Rightarrow -1 \le x \le 1$

  $y = 2 \cos \theta + 1 \Rightarrow -1 \le y \le 3$

  The graph oscillates.

  Domain: $[-1, 1]$

  Orientation:

  Depends on θ

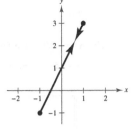

(c) $x = e^{-t}$ $\Rightarrow x > 0$

  $y = 2e^{-t} + 1 \Rightarrow y > 1$

  Domain: $(0, \infty)$

  Orientation: Downward or right to left

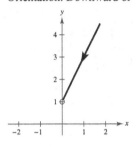

(d) $x = e^{t}$ $\Rightarrow x > 0$

  $y = 2e^{t} + 1 \Rightarrow y > 1$

  Domain: $(0, \infty)$

  Orientation: Upward or left to right

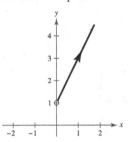

51. $x = x_1 + t(x_2 - x_1),\ y = y_1 + t(y_2 - y_1)$

$$\frac{x - x_1}{x_2 - x_1} = t$$

$$y = y_1 + \left(\frac{x - x_1}{x_2 - x_1}\right)(y_2 - y_1)$$

$$y - y_1 = \frac{y_2 - y_1}{x_2 - x_1}(x - x_1) = m(x - x_1)$$

53. $x = h + a \cos \theta,\ y = k + b \sin \theta$

$$\frac{x - h}{a} = \cos \theta, \frac{y - k}{b} = \sin \theta$$

$$\frac{(x - h)^2}{a^2} + \frac{(y - k)^2}{b^2} = 1$$

55. Line through $(0, 0)$ and $(3, 6)$

  From Exercise 51:

  $x = x_1 + t(x_2 - x_1)$ $y = y_1 + t(y_2 - y_1)$

  $\quad = 0 + t(3 - 0)$ $\quad = 0 + t(6 - 0)$

  $\quad = 3t$ $\quad = 6t$

57. Circle with center $(3, 2)$; radius 4

  From Exercise 52:

  $x = 3 + 4 \cos \theta$

  $y = 2 + 4 \sin \theta$

59. Ellipse

  Vertices: $(\pm 5, 0) \Rightarrow (h, k) = (0, 0)$ and $a = 5$

  Foci: $(\pm 4, 0) \Rightarrow c = 4$

  $b^2 = a^2 - c^2 \Rightarrow 25 - 16 = 9 \Rightarrow b = 3$

  From Exercise 53:

  $x = h + a \cos \theta = 5 \cos \theta$

  $y = k + b \sin \theta = 3 \sin \theta$

61. Hyperbola

  Vertices: $(1, 0), (9, 0) \Rightarrow (h, k) = (5, 0)$ and $a = 4$

  Foci: $(0, 0), (10, 0) \Rightarrow c = 5$

  $c^2 = a^2 + b^2 \Rightarrow 25 = 16 + b^2 \Rightarrow b = 3$

  From Exercise 54:

  $x = 5 + 4 \sec \theta$

  $y = 3 \tan \theta$

63. Line segment between $(0, 0)$ and $(-5, 2)$.

  $x = x_1 + t(x_2 - x_1)$ and $y = y_1 + t(y_2 - y_1)$

  $x = 0 + t(-5 - 0)$ and $y = 0_1 + t(2 - 0)$

  $x = -5t$

  $y = 2t$

  $0 \le t \le 1$

65. Left branch of the hyperbola with vertices $(\pm 3, 0)$ and foci $(\pm 5, 0)$.

$a = 3, c = 5$

Center: $(0, 0), h = 0, k = 0$.

$b^2 = c^2 - a^2 = 25 - 9 = 16$

$x = h + a \sec \theta \Rightarrow x = 0 + 3 \sec \theta$

$y = k + b \tan \theta \Rightarrow y = 0 + 4 \tan \theta$

$x = 3 \sec \theta$

$y = 4 \tan \theta$

$\dfrac{\pi}{2} < \theta < \dfrac{3\pi}{2}$

67. $y = 3x - 2$

  (a) $t = x \Rightarrow x = t$ and $y = 3t - 2$

  (b) $t = 2 - x \Rightarrow x = -t + 2$ and

     $y = 3(-t + 2) - 2 = -3t + 4$

69. $x = 2y + 1$

  (a) $t = x \Rightarrow x = t$ and $t = 2y + 1 \Rightarrow y = \dfrac{1}{2}t - \dfrac{1}{2}$

  (b) $t = 2 - x \Rightarrow x = -t + 2$ and

     $-t + 2 = 2y + 1 \Rightarrow y = -\dfrac{1}{2}t + \dfrac{1}{2}$

71. $y = x^2 + 1$

  (a) $t = x \Rightarrow x = t$ and $y = t^2 + 1$

  (b) $t = 2 - x \Rightarrow x = -t + 2$ and

     $y = (-t + 2)^2 + 1 = t^2 - 4t + 5$

73. $y = 1 - 2x^2$

  (a) $t = x \Rightarrow x = t$

     $\Rightarrow y = 1 - 2t^2$

  (b) $t = 2 - x \Rightarrow x = -t + 2$

     $\Rightarrow y = 1 - 2(-t + 2)^2 = -2t^2 + 8t - 7$

75. $y = \dfrac{1}{x}$

  (a) $t = x \Rightarrow x = t$ and $y = \dfrac{1}{t}$

  (b) $t = 2 - x \Rightarrow x = -t + 2$ and

     $y = \dfrac{1}{-t + 2} = \dfrac{-1}{t - 2}$

77. $y = e^x$

  (a) $t = x \Rightarrow x = t$ and $y = e^t$

  (b) $t = 2 - x \Rightarrow x = -t + 2$ and $y = e^{-t+2}$

79. $x = 4(\theta - \sin \theta)$

  $y = 4(1 - \cos \theta)$

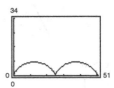

81. $x = 2\theta - 4 \sin \theta$

  $y = 2 - 4 \cos \theta$

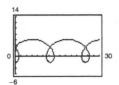

83. $x = 3 \cos^3 \theta$

  $y = 3 \sin^3 \theta$

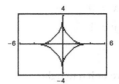

85. $x = 2 \cot \theta$

  $y = 2 \sin^2 \theta$

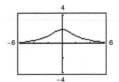

87. $x = 2 \cos \theta \Rightarrow -2 \le x \le 2$

  $y = \sin 2\theta \Rightarrow -1 \le y \le 1$

  Matches graph (b).

  Domain: $[-2, 2]$

  Range: $[-1, 1]$

88. $x = 4\cos^3 \theta \Rightarrow -4 \le x \le 4$

   $y = 6\sin^3 \theta \Rightarrow -6 \le y \le 6$

   Matches graph (c).

   Domain: $[-4, 4]$

   Range: $[-6, 6]$

89. $x = \frac{1}{2}(\cos \theta + \theta \sin \theta)$

   $y = \frac{1}{2}(\sin \theta - \theta \cos \theta)$

   Matches graph (d).

   Domain: $(-\infty, \infty)$

   Range: $(-\infty, \infty)$

90. $x = \frac{1}{2}\cot \theta \Rightarrow -\infty < x < \infty$

   $y = 4\sin \theta \cos \theta \Rightarrow -2 \le y \le 2$

   Matches graph (a).

   Domain: $(-\infty, \infty)$

   Range: $[-2, 2]$

91. $x = (v_0 \cos \theta)t$ and $y = h + (v_0 \sin \theta)t - 16t^2$

   (a) $\theta = 60°$, $v_0 = 88$ ft/sec

   $x = (88 \cos 60°)t$ and $y = (88 \sin 60°)t - 16t^2$

   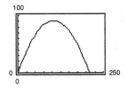

   Maximum height: 90.7 feet

   Range: 209.6 feet

   (b) $\theta = 60°$, $v_0 = 132$ ft/sec

   $x = (132 \cos 60°)t$ and $y = (132 \sin 60°)t - 16t^2$

   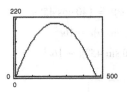

   Maximum height: 204.2 feet

   Range: 471.6 feet

   (c) $\theta = 45°$, $v_0 = 88$ ft/sec

   $x = (88 \cos 45°)t$ and $y = (88 \sin 45°)t - 16t^2$

   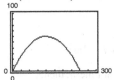

   Maximum height: 60.5 ft

   Range: 242.0 ft

   (d) $\theta = 45°$, $v_0 = 132$ ft/sec

   $x = (132 \cos 45°)t$ and $y = (132 \sin 45°)t - 16t^2$

   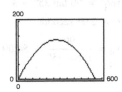

   Maximum height: 136.1 ft

   Range: 544.5 ft

93. (a) 100 miles per hour $= 100\left(\frac{5280}{3600}\right)$ ft/sec $= \frac{440}{3}$ ft/sec

$x = \left(\frac{440}{3}\cos\theta\right)t \approx (146.67\cos\theta)t$

$y = 3 + \left(\frac{440}{3}\sin\theta\right)t - 16t^2 \approx 3 + (146.67\sin\theta)t - 16t^2$

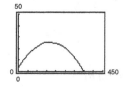

(b) For $\theta = 15°$:

$x = \left(\frac{440}{3}\cos 15°\right)t \approx 141.7t$

$y = 3 + \left(\frac{440}{3}\sin 15°\right)t - 16t^2 \approx 3 + 38.0t - 16t^2$

The ball hits the ground inside the ballpark, so it is not a home run.

(c) For $\theta = 23°$:

$x = \left(\frac{440}{3}\cos 23°\right)t \approx 135.0t$

$y = 3 + \left(\frac{440}{3}\sin 23°\right)t - 16t^2 \approx 3 + 57.3t - 16t^2$

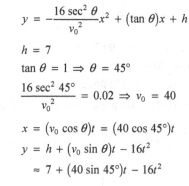

The ball easily clears the 7-foot fence at 408 feet so it is a home run.

(d) Find θ so that $y = 7$ when $x = 408$ by graphing the parametric equations for θ values between $15°$ and $23°$. This occurs when $\theta \approx 19.3°$.

95. (a) $x = (\cos 35°)v_0 t$

$y = 7 + (\sin 35°)v_0 t - 16t^2$

(b) If the ball is caught at time t_1, then:

$90 = (\cos 35°)v_0 t_1$

$4 = 7 + (\sin 35°)v_0 t_1 - 16t_1^2$

$v_0 t_1 = \dfrac{90}{\cos 35°} \Rightarrow -3 = (\sin 35°)\dfrac{90}{\cos 35°} - 16t_1^2$

$\Rightarrow 16t_1^2 = 90\tan 35° + 3$

$\Rightarrow t_1 \approx 2.03$ seconds

$\Rightarrow v_0 = \dfrac{90}{t_1\cos 35°} \approx 54.09$ ft/sec

(c)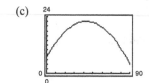

Maximum height ≈ 22 feet

(d) From part (b), $t_1 \approx 2.03$ seconds.

97. $y = 7 + x - 0.02x^2$

(a) Exercise 98 result:

$y = -\dfrac{16\sec^2\theta}{v_0^2}x^2 + (\tan\theta)x + h$

$h = 7$

$\tan\theta = 1 \Rightarrow \theta = 45°$

$\dfrac{16\sec^2 45°}{v_0^2} = 0.02 \Rightarrow v_0 = 40$

$x = (v_0\cos\theta)t = (40\cos 45°)t$

$y = h + (v_0\sin\theta)t - 16t^2$

$\approx 7 + (40\sin 45°)t - 16t^2$

(b)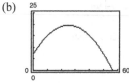

(c) Maximum height: 19.5 feet

Range: 56.2 feet

99. When the circle has rolled θ radians, the center is at $(a\theta, a)$.

$$\sin \theta = \sin(180° - \theta)$$

$$= \frac{|AC|}{b} = \frac{|BD|}{b} \Rightarrow |BD| = b \sin \theta$$

$$\cos \theta = -\cos(180° - \theta)$$

$$= \frac{|AP|}{-b} \Rightarrow |AP| = -b \cos \theta$$

So, $x = a\theta - b \sin \theta$ and $y = a - b \cos \theta$.

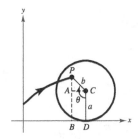

101. True

$$x = t$$
$$y = t^2 + 1 \Rightarrow y = x^2 + 1$$
$$x = 3t$$
$$y = 9t^2 + 1 \Rightarrow y = x^2 + 1$$

103. False. It is possible for both x and y to be functions of t, but y cannot be a function of x. For example, consider the parametric equations $x = 3 \cos t$ and $y = 3 \sin t$. Both x and y are functions of t. However, after eliminating the parameter and finding the rectangular equation $\frac{x^2}{9} + \frac{y^2}{9} = 1$, you can see that y is not a function of x.

105. The use of parametric equations is useful when graphing two functions simultaneously on the same coordinate system. For example, this is useful when tracking the path of an object so the position and the time associated with that position can be determined.

107.

$$x = \sqrt{t - 1} \Rightarrow x^2 = t - 1$$
$$x^2 + 1 = t$$
$$y = 2t$$
$$y = 2(x^2 + 1)$$
$$y = 2x^2 + 2, \, x \geq 0$$

The parametric equation for x is defined only when $t \geq 1$, so the domain of the rectangular equation is $x \geq 0$.

109. The graph is the same, but the orientation is reversed.

Section 6.7   Polar Coordinates

1. pole

3. polar

5. Polar coordinates: $\left(2, \dfrac{5\pi}{6}\right)$

Additional representations:

$$\left(-2, \frac{\pi}{6} - \pi\right) = \left(-2, -\frac{5\pi}{6}\right)$$

$$\left(2, \frac{\pi}{6} - 2\pi\right) = \left(2, -\frac{11\pi}{6}\right)$$

$$\left(-2, \frac{\pi}{6} + \pi\right) = \left(-2, \frac{7\pi}{6}\right)$$

7. Polar coordinates: $\left(4, -\dfrac{\pi}{3}\right)$

Additional representations:

$$\left(4, -\frac{\pi}{3} + 2\pi\right) = \left(4, \frac{5\pi}{3}\right)$$

$$\left(-4, -\frac{\pi}{3} - \pi\right) = \left(-4, -\frac{4\pi}{3}\right)$$

$$\left(-4, -\frac{\pi}{3} + \pi\right) = \left(-4, \frac{2\pi}{3}\right)$$

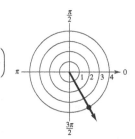

9. Polar coordinates: $(2, 3\pi)$

Additional representations:

$$(2, 3\pi - 2\pi) = (2, \pi)$$

$$(-2, 3\pi - 2\pi) = (-2, \pi)$$

$$(-2, 3\pi - 3\pi) = (-2, 0)$$

11. Polar coordinates: $\left(-2, \dfrac{2\pi}{3}\right)$

Additional representations:

$$\left(-2, \frac{2\pi}{3} - 2\pi\right) = \left(-2, -\frac{4\pi}{3}\right)$$

$$\left(2, \frac{2\pi}{3} + \pi\right) = \left(2, \frac{5\pi}{3}\right)$$

$$\left(-2, \frac{2\pi}{3} - \pi\right) = \left(-2, -\frac{\pi}{3}\right)$$

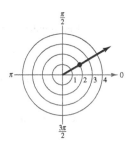

13. Polar coordinates: $\left(0, \dfrac{7\pi}{6}\right)$

Additional representations:

$\left(0, \dfrac{7\pi}{6} - \pi\right) = \left(0, \dfrac{\pi}{6}\right)$

$\left(0, \dfrac{7\pi}{6} - 2\pi\right) = \left(0, -\dfrac{5\pi}{6}\right)$

$\left(0, \dfrac{7\pi}{6} - 3\pi\right) = \left(0, -\dfrac{11\pi}{6}\right)$

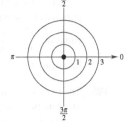

or $(0, \theta)$ for any θ,

$-2\pi < \theta < 2\pi$

15. Polar coordinates: $\left(\sqrt{2}, 2.36\right)$

Additional representations:

$\left(\sqrt{2}, 2.36 - 2\pi\right) \approx \left(\sqrt{2}, -3.92\right)$

$\left(-\sqrt{2}, 2.36 - \pi\right) \approx \left(-\sqrt{2}, -0.78\right)$

$\left(-\sqrt{2}, 2.36 + \pi\right) \approx \left(-\sqrt{2}, 5.50\right)$

17. Polar coordinates: $(-3, -1.57)$

Additional representations:

$(3, 1.57)$

$(-3, 4.71)$

$(3, -4.71)$

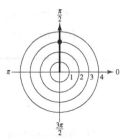

19. Polar coordinates: $(0, \pi) = (r, \theta)$

$x = 0 \cos \pi = (0)(-1) = 0$

$y = 0 \sin \pi = (0)(0) = 0$

Rectangular coordinates: $(0, 0)$

21. Polar coordinates: $\left(3, \dfrac{\pi}{2}\right)$

$x = 3 \cos \dfrac{\pi}{2} = 0$

$y = 3 \sin \dfrac{\pi}{2} = 3$

Rectangular coordinates: $(0, 3)$

23. Polar coordinates: $\left(2, \dfrac{3\pi}{4}\right)$

$x = 2 \cos \dfrac{3\pi}{4} = -\sqrt{2}$

$y = 2 \sin \dfrac{3\pi}{4} = \sqrt{2}$

Rectangular coordinates: $\left(-\sqrt{2}, \sqrt{2}\right)$

25. Polar coordinates: $\left(-2, \dfrac{7\pi}{6}\right) = (r, \theta)$

$x = r \cos \theta = -2 \cos \dfrac{7\pi}{6} = \sqrt{3}$

$y = r \sin \theta = -2 \sin \dfrac{7\pi}{6} = 1$

Rectangular coordinates: $\left(\sqrt{3}, 1\right)$

27. Polar coordinates: $\left(-3, -\dfrac{\pi}{3}\right)$

$x = r \cos \theta = (-3) \cos\left(-\dfrac{\pi}{3}\right) = (-3)\left(\dfrac{1}{2}\right) = -\dfrac{3}{2}$

$y = r \sin \theta = (-3) \sin\left(-\dfrac{\pi}{3}\right) = (-3)\left(-\dfrac{\sqrt{3}}{2}\right) = \dfrac{3\sqrt{3}}{2}$

Rectangular coordinates: $\left(-\dfrac{3}{2}, \dfrac{3\sqrt{3}}{2}\right)$

29. Polar coordinates: $\left(2, \dfrac{7\pi}{8}\right) = (r, \theta)$

$x = r \cos \theta = 2 \cos \dfrac{7\pi}{8} \approx -1.85$

$y = r \sin \theta = 2 \sin \dfrac{7\pi}{8} \approx 0.77$

Regular coordinates: $(-1.85, 0.77)$

31. Polar coordinates: $\left(1, \dfrac{5\pi}{12}\right)$

$x = \cos\dfrac{5\pi}{12} \approx 0.26$

$y = \sin\dfrac{5\pi}{12} \approx 0.97$

Rectangular coordinates: $(0.26, 0.97)$

33. Polar coordinates: $(-2.5, 1.1) = (r, \theta)$

$x = r\cos\theta = -2.5\cos 1.1 \approx -1.13$

$y = r\sin\theta = -2.5\sin 1.1 \approx -2.23$

Rectangular coordinates: $(-1.13, -2.23)$

35. Polar coordinates: $(2.5, -2.9) = (r, \theta)$

$x = r\cos\theta = 2.5\cos(-2.9) \approx -2.43$

$y = r\sin\theta = 2.5\sin(-2.9) \approx -0.60$

Rectangular coordinates: $(-2.43, -0.60)$

37. Polar coordinates: $(-3.1, 7.92) = (r, \theta)$

$x = r\cos\theta = -3.1\cos 7.92 \approx 0.20$

$y = r\sin\theta = -3.1\sin 7.92 \approx -3.09$

Rectangular coordinates: $(0.20, -3.09)$

39. Rectangular coordinates: $(1, 1)$

$r = \pm\sqrt{2},\ \tan\theta = 1,\ \theta = \dfrac{\pi}{4}$

Polar coordinates: $\left(\sqrt{2}, \dfrac{\pi}{4}\right)$

41. Rectangular coordinates: $(-3, -3)$

$r = 3\sqrt{2},\ \tan\theta = 1,\ \theta = \dfrac{5\pi}{4}$

Polar coordinates: $\left(3\sqrt{2}, \dfrac{5\pi}{4}\right)$

43. Rectangular coordinates: $(3, 0)$

$r = \sqrt{9 + 0} = 3,\ \tan\theta = 0,\ \theta = 0$

Polar coordinates: $(3, 0)$

45. Rectangular coordinates: $(0, -5)$

$r = 5,\ \tan\theta$ undefined, $\theta = \dfrac{\pi}{2}$

Polar coordinates: $\left(5, \dfrac{3\pi}{2}\right)$

47. Rectangular coordinates: $\left(-\sqrt{3}, -\sqrt{3}\right)$

$r = \pm\sqrt{3 + 3} = \pm\sqrt{6},\ \tan\theta = 1,\ \theta = \dfrac{5\pi}{4}$

Polar coordinates: $\left(\sqrt{6}, \dfrac{5\pi}{4}\right)$

49. Rectangular coordinates: $\left(\sqrt{3}, -1\right)$

$r = \sqrt{3 + 1} = 2,\ \tan\theta = -\dfrac{1}{\sqrt{3}},\ \theta = \dfrac{11\pi}{6}$

Polar coordinates: $\left(2, \dfrac{11\pi}{6}\right)$

51. Rectangular coordinates: $(3, -2)$

$R \blacktriangleright Pr(3, -2) \approx 3.61 = r$

$R \blacktriangleright P\theta(3, -2) \approx 5.70 = \theta$

Polar coordinates: $(3.61, 5.70)$

53. Rectangular coordinates: $(-5, 2)$

$R \blacktriangleright Pr(-5, -2) \approx 5.39 = r$

$R \blacktriangleright P\theta(-5, -2) \approx 2.76 = \theta$

Polar coordinates: $(5.39, 2.76)$

55. Rectangular coordinates: $\left(-\sqrt{3}, -4\right)$

$R \blacktriangleright Pr\left(-\sqrt{3}, -4\right) \approx 4.36 = r$

$R \blacktriangleright P\theta\left(-\sqrt{3}, -4\right) \approx 4.30 = \theta$

Polar coordinates: $(4.36, 4.30)$

57. Rectangular coordinates: $\left(\dfrac{5}{2}, \dfrac{4}{3}\right)$

$R \blacktriangleright Pr\left(\dfrac{5}{2}, \dfrac{4}{3}\right) \approx 2.83 = r$

$R \blacktriangleright P\theta\left(\dfrac{5}{2}, \dfrac{4}{3}\right) \approx 0.49 = \theta$

Polar coordinates: $(2.83, 0.49)$

59. $x^2 + y^2 = 9$

$r = 3$

61. $y = x$

$r\cos\theta = r\sin\theta$

$1 = \tan\theta$

$\theta = \dfrac{\pi}{4}$

63.
$$x = 10$$
$$r \cos \theta = 10$$
$$r = 10 \sec \theta$$

65.
$$3x - y + 2 = 0$$
$$3r \cos \theta - r \sin \theta + 2 = 0$$
$$r(3 \cos \theta - \sin \theta) = -2$$
$$r = \frac{-2}{3 \cos \theta - \sin \theta}$$

67.
$$xy = 16$$
$$(r \cos \theta)(r \sin \theta) = 16$$
$$r^2 = 16 \sec \theta \csc \theta = 32 \csc 2\theta$$

69.
$$x = a$$
$$r \cos \theta = a$$
$$r = a \sec \theta$$

71.
$$x^2 + y^2 = a^2$$
$$r^2 = a^2$$
$$r = a$$

73.
$$x^2 + y^2 - 2ax = 0$$
$$r^2 - 2a \, r \cos \theta = 0$$
$$r(r - 2a \cos \theta) = 0$$
$$r - 2a \cos \theta = 0$$
$$r = 2a \cos \theta$$

75.
$$\left(x^2 + y^2\right)^2 = x^2 - y^2$$
$$\left(r^2\right)^2 = x^2 - y^2$$
$$r^4 = x^2 - y^2$$
$$r^2 = \frac{x^2}{r^2} - \frac{y^2}{r^2}$$
$$r^2 = \left(\frac{x}{r}\right)^2 - \left(\frac{y}{r}\right)^2$$
$$r^2 = \cos^2 \theta - \sin^2 \theta$$
$$r^2 = \cos 2\theta$$

77.
$$y^3 = x^2$$
$$(r \sin \theta)^3 = (r \cos \theta)^2$$
$$r^3 \sin^3 \theta = r^2 \cos^2 \theta$$
$$\frac{r \sin^3 \theta}{\cos^2 \theta} = 1$$
$$r \sin \theta \tan^2 \theta = 1$$
$$r = \csc \theta \cot^2 \theta$$

79.
$$r = 5$$
$$r^2 = 25$$
$$x^2 + y^2 = 25$$

81.
$$\theta = \frac{2\pi}{3}$$
$$\tan \theta = \tan \frac{2\pi}{3}$$
$$\frac{y}{x} = -\sqrt{3}$$
$$y = -\sqrt{3}x$$
$$\sqrt{3}x + y = 0$$

83. $\tan\left(\dfrac{\pi}{2}\right)$ is undefined, therefore it is the vertical line
$$x = 0.$$

85.
$$r = 4 \csc \theta$$
$$r \sin \theta = 4$$
$$y = 4$$

87.
$$r = -3 \sec \theta$$
$$\frac{r}{\sec \theta} = -3$$
$$r \cos \theta = -3$$
$$x = -3$$

89.
$$r = -2 \cos \theta$$
$$r^2 = -2r \cos \theta$$
$$x^2 + y^2 = -2x$$
$$x^2 + y^2 + 2x = 0$$

91.
$$r^2 = \cos \theta$$
$$r^3 = r \cos \theta$$
$$\left(\pm\sqrt{x^2 + y^2}\right)^3 = x$$
$$\pm\left(x^2 + y^2\right)^{3/2} = x$$
$$\left(x^2 + y^2\right)^3 = x^2$$
$$x^2 + y^2 = x^{2/3}$$
$$x^2 + y^2 - x^{2/3} = 0$$

93.
$$r^2 = \sin 2\theta = 2 \sin \theta \cos \theta$$
$$r^2 = 2\left(\frac{y}{r}\right)\left(\frac{x}{r}\right) = \frac{2xy}{r^2}$$
$$r^4 = 2xy$$
$$\left(x^2 + y^2\right)^2 = 2xy$$

95.
$$r = 2 \sin 3\theta$$
$$r = 2 \sin(\theta + 2\theta)$$
$$r = 2[\sin \theta \cos 2\theta + \cos \theta \sin 2\theta]$$
$$r = 2\left[\sin \theta(1 - 2 \sin^2 \theta) + \cos \theta(2 \sin \theta \cos \theta)\right]$$
$$r = 2\left[\sin \theta - 2 \sin^3 \theta + 2 \sin \theta \cos^2 \theta\right]$$
$$r = 2\left[\sin \theta - 2 \sin^3 \theta + 2 \sin \theta(1 - \sin^2 \theta)\right]$$
$$r = 2(3 \sin \theta - 4 \sin^3 \theta)$$
$$r^4 = 6r^3 \sin \theta - 8r^3 \sin^3 \theta$$
$$(x^2 + y^2)^2 = 6(x^2 + y^2)y - 8y^3$$
$$(x^2 + y^2)^2 = 6x^2y - 2y^3$$

97.
$$r = \frac{2}{1 + \sin \theta}$$
$$r(1 + \sin \theta) = 2$$
$$r + r \sin \theta = 2$$
$$r = 2 - r \sin \theta$$
$$\pm\sqrt{x^2 + y^2} = 2 - y$$
$$x^2 + y^2 = (2 - y)^2$$
$$x^2 + y^2 = 4 - 4y + y^2$$
$$x^2 + 4y - 4 = 0$$

99.
$$r = \frac{6}{2 - 3 \sin \theta}$$
$$r(2 - 3 \sin \theta) = 6$$
$$2r = 6 + 3r \sin \theta$$
$$2\left(\pm\sqrt{x^2 + y^2}\right) = 6 + 3y$$
$$4(x^2 + y^2) = (6 + 3y)^2$$
$$4x^2 + 4y^2 = 36 + 36y + 9y^2$$
$$4x^2 - 5y^2 - 36y - 36 = 0$$

101. The graph of the polar equation consists of all points that are six units from the pole.
$$r = 6$$
$$r^2 = 36$$
$$x^2 + y^2 = 36$$

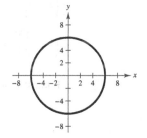

103. The graph of the polar equation consists of all points that make an angle of $\pi/6$ with the polar axis.
$$\theta = \frac{\pi}{6}$$
$$\tan \theta = \tan \frac{\pi}{6}$$
$$\frac{y}{x} = \frac{\sqrt{3}}{3}$$
$$y = \frac{\sqrt{3}}{3}x$$
$$3y = \sqrt{3}x$$
$$-\sqrt{3}x + 3y = 0$$

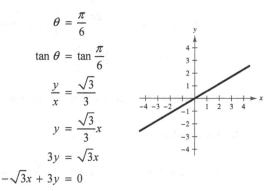

105. The graph of the polar equation is not evident by simple inspection. Convert to rectangular form first.
$$r = 3 \sec \theta$$
$$r \cos \theta = 3$$
$$x = 3$$
$$x - 3 = 0$$

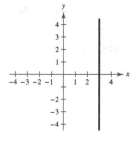

107. The graph of the polar equation consists of all points on the circle with radius 1 and center $(0, 1)$.
$$r = 2 \sin \theta$$
$$r^2 = 2r \sin \theta$$
$$x^2 + y^2 = 2y$$
$$x^2 + y^2 - 2y = 0$$
$$x^2 + y^2 - 2y + 1 = 1$$
$$x^2 + (y - 1)^2 = 1$$

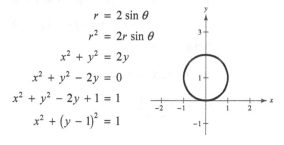

109. (a)

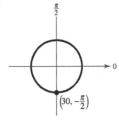

Since the passengers enter a car at the point

$(r, \theta) = \left(30, -\dfrac{\pi}{2}\right)$, $r = 30$ is the polar equation for

the model.

(b) Since it takes 45 seconds for the Ferris wheel to complete one revolution clockwise, after 15 seconds a passenger car makes one-third of one revolution or an angle of $\dfrac{2\pi}{3}$ radians.

Because $\theta = -\dfrac{\pi}{2} - \dfrac{2\pi}{3} = -\dfrac{7\pi}{6}$, the passenger car

is at $\left(30, -\dfrac{7\pi}{6}\right) = \left(30, \dfrac{5\pi}{6}\right)$.

(c) Polar coordinates: $\left(30, \dfrac{5\pi}{6}\right)$

$x = r \cos \theta = 30 \cos \dfrac{5\pi}{6} = 15\sqrt{3} \approx 25.98$

$y = r \sin \theta = 30 \sin \dfrac{5\pi}{6} = 15$

Rectangular coordinates: $(25.98, 15)$

The car is about 25.98 feet to the left of the center and 15 feet above the center.

111. True. Because r is a directed distance, then the point (r, θ) can be represented as $(r, \theta \pm 2n\pi)$.

113. Rectangular coordinates: $\left(1, -\sqrt{3}\right)$

$r = \sqrt{1 + 3} = \sqrt{4} = 2$, $\tan \theta = \dfrac{-\sqrt{3}}{1}$, $\theta = -\dfrac{\pi}{3}$, the point lies in Quadrant IV.

Polar coordinates: $(r, \theta)\left(2, -\dfrac{\pi}{3}\right)$ or $\left(2, \dfrac{5\pi}{3}\right)$

115. (a)

$$r = 2(h \cos \theta + k \sin \theta)$$
$$r = 2\left(h\left(\dfrac{x}{r}\right) + k\left(\dfrac{y}{r}\right)\right)$$
$$r = \dfrac{2hx + 2ky}{r}$$
$$r^2 = 2hx + 2ky$$
$$x^2 + y^2 = 2hx + 2ky$$
$$x^2 - 2hx + y^2 - 2ky = 0$$
$$\left(x^2 - 2hx + h^2\right) + \left(y^2 - 2ky + k^2\right) = h^2 + k^2$$
$$(x - h)^2 + (y - k)^2 = h^2 + k^2$$

Center: (h, k)

Radius: $\sqrt{h^2 + k^2}$

Center: $\left(\dfrac{1}{2}, \dfrac{3}{2}\right)$

(b) $r = \cos \theta + 3 \sin \theta = 2\left(\dfrac{1}{2} \cos \theta + \dfrac{3}{2} \sin \theta\right)$

$h = \dfrac{1}{2}$, $k = \dfrac{3}{2}$

$$\left(x - \dfrac{1}{2}\right)^2 + \left(y - \dfrac{3}{2}\right)^2 = \left(\dfrac{1}{2}\right)^2 + \left(\dfrac{3}{2}\right)^2$$

$$\left(x - \dfrac{1}{2}\right)^2 + \left(y - \dfrac{3}{2}\right)^2 = \dfrac{5}{2}$$

Radius: $r = \sqrt{\dfrac{5}{2}} = \dfrac{\sqrt{10}}{2}$

Section 6.8   Graphs of Polar Equations

1. $\theta = \dfrac{\pi}{2}$

3. convex limaçon

5. lemniscate

7. $r = 3 \cos \theta$

Circle

9. $r = 3(1 - 2 \cos \theta)$

Limaçon with inner loop

11. $r = 6 \cos 2\theta$

Rose curve with 4 petals

13. $r = 6 + 3 \cos \theta$

$\theta = \dfrac{\pi}{2}$: $-r = 6 + 3 \cos(-\theta)$

$-r = 6 + 3 \cos \theta$

Not an equivalent equation

Polar axis: $r = 6 + 3 \cos (-\theta)$

$r = 6 + 3 \cos \theta$

Equivalent equation

Pole: $-r = 6 + 3 \cos \theta$

Not an equivalent equation

Answer: Symmetric with respect to polar axis.

15. $r = \dfrac{2}{1 + \sin \theta}$

$\theta = \dfrac{\pi}{2}$: $r = \dfrac{2}{1 + \sin(\pi - \theta)}$

$r = \dfrac{2}{1 + \sin \pi \cos \theta - \cos \pi \sin \theta}$

$r = \dfrac{2}{1 + \sin \theta}$

Equivalent equation

Polar axis: $r = \dfrac{2}{1 + \sin(-\theta)}$

$r = \dfrac{2}{1 - \sin \theta}$

Not an equivalent equation

Pole: $-r = \dfrac{2}{1 + \sin \theta}$

Answer: Symmetric with respect to $\theta = \pi/2$

17. $r^2 = 36 \cos 2\theta$

$\theta = \dfrac{\pi}{2}$: $(-r)^2 = 36 \cos 2(-\theta)$

$r^2 = 36 \cos 2\theta$

Equivalent equation

Polar axis: $r^2 = 36 \cos 2(-\theta)$

$r^2 = 36 \cos 2\theta$

Equivalent equation

Pole: $(-r)^2 = 36 \cos 2\theta$

$r^2 = 36 \cos 2\theta$

Equivalent equation

Answer: Symmetric with respect to $\theta = \dfrac{\pi}{2}$, the polar axis, and the pole

19. $|r| = |10 - 10 \sin \theta| = 10|1 - \sin \theta| \le 10(2) = 20$

$|1 - \sin \theta| = 2$

$1 - \sin \theta = 2$ or $1 - \sin \theta = -2$

$\sin \theta = -1$ $\sin \theta = 3$

$\theta = \dfrac{3\pi}{2}$ Not possible

Maximum: $|r| = 20$ when $\theta = \dfrac{3\pi}{2}$

$0 = 10(1 - \sin \theta)$

$\sin \theta = 1$

$\theta = \dfrac{\pi}{2}$

Zero: $r = 0$ when $\theta = \dfrac{\pi}{2}$

21. $|r| = |4 \cos 3\theta| = 4|\cos 3\theta| \le 4$

$|\cos 3\theta| = 1$

$\cos 3\theta = \pm 1$

$\theta = 0, \dfrac{\pi}{3}, \dfrac{2\pi}{3}$

Maximum: $|r| = 4$ when $\theta = 0, \dfrac{\pi}{3}, \dfrac{2\pi}{3}$

$0 = 4 \cos 3\theta$

$\cos 3\theta = 0$

$\theta = \dfrac{\pi}{6}, \dfrac{\pi}{2}, \dfrac{5\pi}{6}$

Zero: $r = 0$ when $\theta = \dfrac{\pi}{6}, \dfrac{\pi}{2}, \dfrac{5\pi}{6}$

23. $r = 5$

Symmetric with respect to $\theta = \dfrac{\pi}{2}$, polar axis, pole

Circle with radius 5

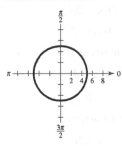

25. $r = \dfrac{\pi}{4}$

Symmetric with respect to $\theta = \dfrac{\pi}{2}$, polar axis, pole

Circle with radius $\dfrac{\pi}{4}$

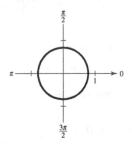

27. $r = 3 \sin \theta$

Symmetric with respect to $\theta = \dfrac{\pi}{2}$

Circle with radius $\dfrac{3}{2}$

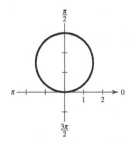

29. $r = 3(1 - \cos \theta)$

Symmetric with respect
to the polar axis

$\dfrac{a}{b} = \dfrac{3}{3} = 1 \Rightarrow$ Cardioid

$|r| = 6$ when $\theta = \pi$

$r = 0$ when $\theta = 0$

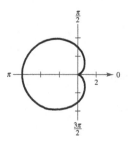

31. $r = 4(1 + \sin \theta)$

Symmetric with respect

to $\theta = \dfrac{\pi}{2}$

$\dfrac{a}{b} = \dfrac{4}{4} = 1 \Rightarrow$ Cardioid

$|r| = 8$ when $\theta = \dfrac{\pi}{2}$

$r = 0$ when $\theta = \dfrac{3\pi}{2}$

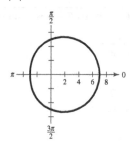

33. $r = 5 + 2 \cos \theta$

Symmetric with respect to the polar axis

$\dfrac{a}{b} = \dfrac{5}{2} > 1 \Rightarrow$ Dimpled limaçon

$|r| = 7$ when $\theta = 0$

35. $r = 1 - 3 \sin \theta$

Symmetric with respect to $\theta = \dfrac{\pi}{2}$

$\dfrac{a}{b} = \dfrac{1}{3} < 1 \Rightarrow$ Limaçon with inner loop

$|r| = 4$ when $\theta = \dfrac{3\pi}{2}$

$r = 0$ when $\theta = 0.3398, 2.802$

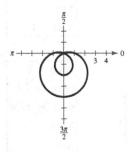

37. $r = 3 - 6 \cos \theta$

Symmetric with respect to the polar axis

$\dfrac{a}{b} = \dfrac{1}{2} < 1 \Rightarrow$ Limaçon with inner loop

$|r| = 9$ when $\theta = \pi$

$r = 0$ when $\theta = \dfrac{\pi}{3}, \dfrac{5\pi}{3}$

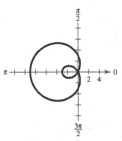

39. $r = 5 \sin 2\theta$

Symmetric with respect to $\theta = \pi/2$, the polar axis, and the pole

Rose curve $(n = 2)$ with 4 petals

$|r| = 5$ when $\theta = \dfrac{\pi}{4}, \dfrac{3\pi}{4}, \dfrac{5\pi}{4}, \dfrac{7\pi}{4}$

$r = 0$ when $\theta = 0, \dfrac{\pi}{2}, \pi$

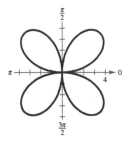

41. $r = 6 \cos 3\theta$

Symmetric with respect to polar axis

Rose curve $(n = 3)$ with three petals

$|r| = 6$ when $\theta = 0, \dfrac{\pi}{3}, \dfrac{2\pi}{3}, \pi$

$r = 0$ when $\theta = \dfrac{\pi}{6}, \dfrac{\pi}{2}, \dfrac{5\pi}{6}$

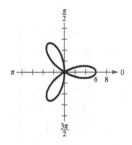

43. $r = 2 \sec \theta$

$r = \dfrac{2}{\cos \theta}$

$r \cos \theta = 2$

$x = 2 \Rightarrow$ Line

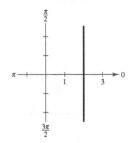

45. $r = \dfrac{3}{\sin \theta - 2 \cos \theta}$

$r(\sin \theta - 2 \cos \theta) = 3$

$y - 2x = 3$

$y = 2x + 3 \Rightarrow$ Line

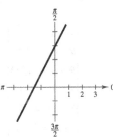

47. $r^2 = 9 \cos 2\theta$

Symmetric with respect
to the polar axis, $\theta = \pi/2$,
and the pole

Lemniscate

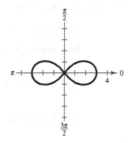

49. $r = \dfrac{9}{4}$

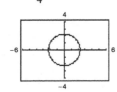

$0 \le \theta \le 2\pi$

θmin $= 0$	
θmax $= 2\pi$	
θstep $= \pi/24$	
Xmin $= -6$	
Xmax $= 6$	
Xscl $= 1$	
Ymin $= -4$	
Ymax $= 4$	
Yscl $= 1$	

51. $r = \dfrac{5\pi}{8}$

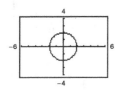

$0 \le \theta \le 2\pi$

θmin $= 0$	
θmax $= 2\pi$	
θstep $= \pi/24$	
Xmin $= -6$	
Xmax $= 6$	
Xscl $= 1$	
Ymin $= -4$	
Ymax $= 4$	
Yscl $= 1$	

53. $r = 8 \cos \theta$

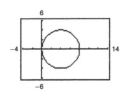

$0 \le \theta \le 2\pi$

θmin $= 0$	
θmax $= 2\pi$	
θstep $= \pi/24$	
Xmin $= -4$	
Xmax $= 14$	
Xscl $= 2$	
Ymin $= -6$	
Ymax $= 6$	
Yscl $= 2$	

55. $r = 3(2 - \sin \theta)$

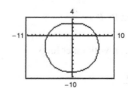

$0 \le \theta \le 2\pi$

θmin $= 0$	
θmax $= 2\pi$	
θstep $= \pi/24$	
Xmin $= -11$	
Xmax $= 10$	
Xscl $= 1$	
Ymin $= -10$	
Ymax $= 4$	
Yscl $= 1$	

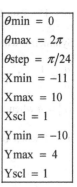

57. $r = 8 \sin \theta \cos^2 \theta$

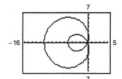

$0 \le \theta \le 2\pi$

θmin $= 0$	
θmax $= 2\pi$	
θstep $= \pi/24$	
Xmin $= -4$	
Xmax $= 5$	
Xscl $= 1$	
Ymin $= -3$	
Ymax $= 3$	
Yscl $= 1$	

59. $r = 3 - 8 \cos \theta$

$0 \le \theta < 2\pi$

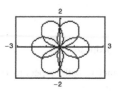

61. $r = 2 \cos\left(\dfrac{3\theta}{2}\right)$

$0 \le \theta < 4\pi$

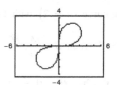

63. $r^2 = 9 \sin 2\theta$

$0 \le \theta < \pi$

65.
$$r = 2 - \sec \theta = 2 - \frac{1}{\cos \theta}$$
$$r \cos \theta = 2 \cos \theta - 1$$
$$r(r \cos \theta) = 2r \cos \theta - r$$
$$\left(\pm\sqrt{x^2 + y^2}\right)x = 2x - \left(\pm\sqrt{x^2 + y^2}\right)$$
$$\left(\pm\sqrt{x^2 + y^2}\right)(x + 1) = 2x$$
$$\left(\pm\sqrt{x^2 + y^2}\right) = \frac{2x}{x + 1}$$
$$x^2 + y^2 = \frac{4x^2}{(x + 1)^2}$$
$$y^2 = \frac{4x^2}{(x + 1)^2} - x^2 = \frac{4x^2 - x^2(x + 1)^2}{(x + 1)^2} = \frac{4x^2 - x^2(x^2 + 2x + 1)}{(x + 1)^2}$$
$$= \frac{-x^4 - 2x^3 + 3x^2}{(x + 1)^2} = \frac{-x^2(x^2 + 2x - 3)}{(x + 1)^2}$$
$$y = \pm\sqrt{\frac{x^2(3 - 2x - x^2)}{(x + 1)^2}} = \pm\left|\frac{x}{x + 1}\right|\sqrt{3 - 2x - x^2}$$

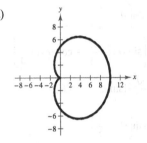

The graph has an asymptote at $x = -1$.

67. $r = \dfrac{3}{\theta}$

$$\theta = \frac{3}{r} = \frac{3 \sin \theta}{r \sin \theta} = \frac{3 \sin \theta}{y}$$

$$y = \frac{3 \sin \theta}{\theta}$$

As $\theta \to 0, y \to 3$

69. (a)

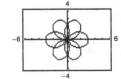

The graph is a cardioid.

(b) Since $|r|$ is at a maximum when $r = 10$ at $\theta = 0$ radians, the microphone is most sensitive to sound when $\theta = 0$.

71. True. The equation is of the form $r = a \sin n\theta$, where n is odd, so it has five petals.

73. $r = 3 \sin k\theta$

(a) $r = 3 \sin 1.5\theta$
    $0 \le \theta < 4\pi$

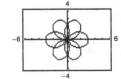

(b) $r = 3 \sin 2.5\theta$
    $0 \le \theta < 4\pi$

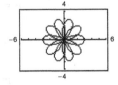

(c) Yes. $r = 3 \sin(k\theta)$.

Find the minimum value of $\theta, (\theta > 0)$, that is a multiple of 2π that makes $k\theta$ a multiple of 2π.

75. $r = 10 \cos \theta$

(a) $0 \le \theta \le \dfrac{\pi}{2}$

Upper half of circle

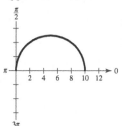

(b) $\dfrac{\pi}{2} \le \theta \le \pi$

Lower half of circle

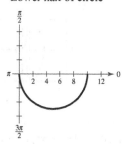

(c) $-\dfrac{\pi}{2} \le \theta \le \dfrac{\pi}{2}$

Entire circle

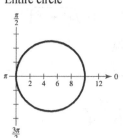

(d) $\dfrac{\pi}{4} \le \theta \le \dfrac{3\pi}{4}$

Left half of circle

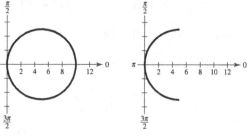

77. Let the curve $r = f(\theta)$ be rotated by ϕ to form the curve $r = g(\theta)$.

If (r_1, θ_1) is a point on $r = f(\theta)$, then $(r_1, \theta_1 + \phi)$ is on $r = g(\theta)$.

That is, $g(\theta_1 + \phi) = r_1 = f(\theta_1)$. Letting $\theta = \theta_1 + \phi$, or $\theta_1 = \theta - \phi$,

you see that $g(\theta) = g(\theta_1 + \phi) = f(\theta_1) = f(\theta - \phi)$.

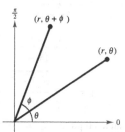

79. (a) $r = 2 - \sin\left(\theta - \dfrac{\pi}{4}\right)$

$\qquad = 2 - \left[\sin\theta\cos\dfrac{\pi}{4} - \cos\theta\sin\dfrac{\pi}{4}\right]$

$\qquad = 2 - \dfrac{\sqrt{2}}{2}(\sin\theta - \cos\theta)$

(b) $r = 2 - \sin\left(\theta - \dfrac{\pi}{2}\right)$

$\qquad = 2 - \left[\sin\theta\cos\dfrac{\pi}{2} - \cos\theta\sin\dfrac{\pi}{2}\right]$

$\qquad = 2 + \cos\theta$

(c) $r = 2 - \sin(\theta - \pi)$

$\qquad = 2 - \left[\sin\theta\cos\pi - \cos\theta\sin\pi\right]$

$\qquad = 2 + \sin\theta$

(d) $r = 2 - \sin\left(\theta - \dfrac{3\pi}{2}\right)$

$\qquad = 2 - \left[\sin\theta\cos\dfrac{3\pi}{2} - \cos\theta\sin\dfrac{3\pi}{2}\right]$

$\qquad = 2 - \cos\theta$

Section 6.9   Polar Equations of Conics

1. conic

3. vertical; left

5. $r = \dfrac{2e}{1 + e\cos\theta}$

$e = 1\!: r = \dfrac{2}{1 + \cos\theta} \Rightarrow$ parabola

$e = 0.5\!: r = \dfrac{1}{1 + 0.5\cos\theta} \Rightarrow$ ellipse

$e = 1.5\!: r = \dfrac{3}{1 + 1.5\cos\theta} \Rightarrow$ hyperbola

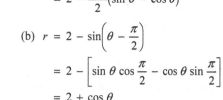

7. $r = \dfrac{2e}{1 - e\sin\theta}$

$e = 1\!: r = \dfrac{2}{1 - \sin\theta} \Rightarrow$ parabola

$e = 0.5\!: r = \dfrac{1}{1 - 0.5\sin\theta} \Rightarrow$ ellipse

$e = 1.5\!: r = \dfrac{3}{1 - 1.5\sin\theta} \Rightarrow$ hyperbola

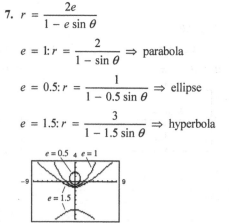

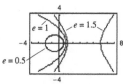

9. $r = \dfrac{4}{1 - \cos \theta}$

$e = 1 \Rightarrow$ Parabola

Vertical directrix to the left of the pole

Matches graph (c).

10. $r = \dfrac{3}{2 + \cos \theta}$

$e = \dfrac{1}{2} \Rightarrow$ Ellipse

Vertical directrix to the right of the pole

Matches graph (d).

11. $r = \dfrac{4}{1 + \sin \theta}$

$e = 1 \Rightarrow$ Parabola

Horizontal directrix above the pole

Matches graph (a).

12. $r = \dfrac{4}{1 - 3 \sin \theta}$

$e = 3 \Rightarrow$ Hyperbola

Horizontal directrix below pole

Matches graph (b).

13. $r = \dfrac{3}{1 - \cos \theta}$

$e = 1 \Rightarrow$ Parabola

Vertex: $\left(\dfrac{3}{2}, \pi \right)$

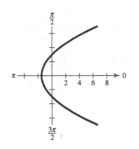

15. $r = \dfrac{5}{1 - \sin \theta}$

$e = 1$, the graph is a parabola.

Vertex: $\left(\dfrac{5}{2}, -\dfrac{\pi}{2} \right)$

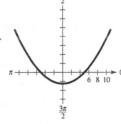

17. $r = \dfrac{2}{2 - \cos \theta} = \dfrac{1}{1 - (1/2) \cos \theta}$

$e = \dfrac{1}{2} < 1$, the graph is an ellipse.

Vertices: $(2, 0), \left(\dfrac{2}{3}, \pi \right)$

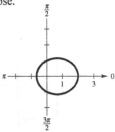

19. $r = \dfrac{6}{2 + \sin \theta} = \dfrac{3}{1 + (1/2) \sin \theta}$

$e = \dfrac{1}{2} < 1$, the graph is an ellipse.

Vertices: $\left(2, \dfrac{\pi}{2} \right), \left(6, \dfrac{3\pi}{2} \right)$

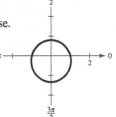

21. $r = \dfrac{3}{2 + 4 \sin \theta} = \dfrac{3/2}{1 + 2 \sin \theta}$

$e = 2 > 1$, the graph is a hyperbola.

Vertices: $\left(\dfrac{1}{2}, \dfrac{\pi}{2} \right), \left(-\dfrac{3}{2}, \dfrac{3\pi}{2} \right)$

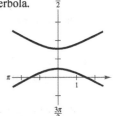

23. $r = \dfrac{3}{2 - 6 \cos \theta} = \dfrac{3/2}{1 - 3 \cos \theta}$

$e = 3 > 1$, the graph is a hyperbola.

Vertices: $\left(-\dfrac{3}{4}, 0 \right), \left(\dfrac{3}{8}, \pi \right)$

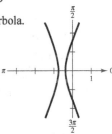

25. $r = \dfrac{-1}{1 - \sin \theta}$

$e = 1 \Rightarrow$ Parabola

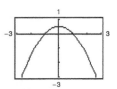

27. $r = \dfrac{3}{-4 + 2\cos\theta}$

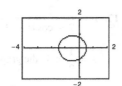

$e = \dfrac{1}{2} \Rightarrow$ Ellipse

29.

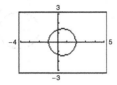

Ellipse

31. $r = \dfrac{14}{14 + 17\sin\theta} = \dfrac{1}{1 + (17/14)\sin\theta}$

$e = \dfrac{17}{14} > 1 \Rightarrow$ Hyperbola

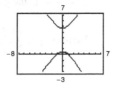

33. $r = \dfrac{3}{1 - \cos(\theta - \pi/4)}$

Rotate the graph in Exercise 13 through the angle $\pi/4$.

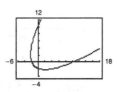

35. $r = \dfrac{6}{2 + \sin\left(\theta + \dfrac{\pi}{6}\right)}$

Rotate the graph in Exercise 19 through the angle $-\pi/6$.

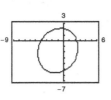

37. Parabola: $e = 1$

Directrix: $x = -1$

Vertical directrix to the left of the pole

$r = \dfrac{1(1)}{1 - 1\cos\theta} = \dfrac{1}{1 - \cos\theta}$

39. Ellipse: $e = \dfrac{1}{2}$

Directrix: $x = 3$

$p = 3$

Vertical directrix to the right of the pole

$r = \dfrac{(1/2)(3)}{1 + (1/2)\cos\theta} = \dfrac{\frac{3}{2}}{1 + \frac{1}{2}\cos\theta} = \dfrac{3}{2 + \cos\theta}$

41. Hyperbola: $e = 2$

Directrix: $x = 1$

$p = 1$

Vertical directrix to the right of the pole

$r = \dfrac{2(1)}{1 + 2\cos\theta} = \dfrac{2}{1 + 2\cos\theta}$

43. Parabola

Vertex: $(2, 0) \Rightarrow e = 1, p = 4$

Vertical directrix to the right of the pole

$r = \dfrac{1(4)}{1 + 1\cos\theta} = \dfrac{4}{1 + \cos\theta}$

45. Parabola

Vertex: $(5, \pi) \Rightarrow e = 1, p = 10$

Vertical directrix to the left of the pole

$r = \dfrac{1(10)}{1 - 1\cos\theta} = \dfrac{10}{1 - \cos\theta}$

47. Ellipse: Vertices $(2, 0), (10, \pi)$

Center: $(4, \pi); c = 4, a = 6, e = \dfrac{2}{3}$

Vertical directrix to the right of the pole

$r = \dfrac{(2/3)p}{1 + (2/3)\cos\theta} = \dfrac{2p}{3 + 2\cos\theta}$

$2 = \dfrac{2p}{3 + 2\cos 0}$

$p = 5$

$r = \dfrac{2(5)}{3 + 2\cos\theta} = \dfrac{10}{3 + 2\cos\theta}$

49. Ellipse: Vertices $(20, 0), (4, \pi)$

Center: $(8, 0); c = 8, a = 12, e = \dfrac{2}{3}$

Vertical directrix to the left of the pole

$r = \dfrac{(2/3)p}{1 - (2/3)\cos\theta} = \dfrac{2p}{3 - 2\cos\theta}$

$20 = \dfrac{2p}{3 - 2\cos 0}$

$p = 10$

$r = \dfrac{2(10)}{3 - 2\cos\theta} = \dfrac{20}{3 - 2\cos\theta}$

51. Hyperbola: Vertices $\left(1, \dfrac{3\pi}{2}\right), \left(9, \dfrac{3\pi}{2}\right)$

Center: $\left(5, \dfrac{3\pi}{2}\right)$; $c = 5$, $a = 4$, $e = \dfrac{5}{4}$

Horizontal directrix below the pole

$$r = \frac{(5/4)p}{1 - (5/4)\sin\theta} = \frac{5p}{4 - 5\sin\theta}$$

$$1 = \frac{5p}{4 - 5\sin(3\pi/2)}$$

$$p = \frac{9}{5}$$

$$r = \frac{5(9/5)}{4 - 5\sin\theta} = \frac{9}{4 - 5\sin\theta}$$

53. When $\theta = 0$, $r = c + a = ea + a = a(1 + e)$.

Therefore,

$$a(1 + e) = \frac{ep}{1 - e\cos 0}$$

$$a(1 + e)(1 - e) = ep$$

$$a(1 - e^2) = ep.$$

So, $r = \dfrac{ep}{1 - e\cos\theta} = \dfrac{(1 - e^2)a}{1 - e\cos\theta}$

55. Earth:

(a) $r = \dfrac{\left[1 - (0.0167)^2(9.2957 \times 10^7)\right]}{1 - 0.0167\cos\theta} \approx \dfrac{9.2931 \times 10^7}{1 - 0.0167\cos\theta}$

(b) Perihelion distance: $r = 9.2957 \times 10^7(1 - 0.0167) \approx 9.1405 \times 10^7$ miles

Aphelion distance: $r = 9.2957 \times 10^7(1 + 0.0167) \approx 9.4509 \times 10^7$ miles

57. Venus:

(a) $r = \dfrac{\left[1 - (0.0067)^2(1.0821 \times 10^8)\right]}{1 - 0.0067\cos\theta} \approx \dfrac{1.0821 \times 10^8}{1 - 0.0067\cos\theta}$

(b) Perihelion distance: $r = 1.0821 \times 10^8(1 - 0.0067) \approx 1.0748 \times 10^8$ kilometers

Aphelion distance: $r = 1.0821 \times 10^8(1 + 0.0067) \approx 1.0894 \times 10^8$ kilometers

59. Mars:

(a) $r = \dfrac{\left[1 - (0.0935)^2(1.4162 \times 10^8)\right]}{1 - 0.0935\cos\theta} \approx \dfrac{1.4038 \times 10^8}{1 - 0.0935\cos\theta}$

(b) Perihelion distance: $r = 1.4162 \times 10^8(1 - 0.0935) \approx 1.2838 \times 10^8$ miles

Aphelion distance: $r = 1.4162 \times 10^8(1 + 0.0935) \approx 1.5486 \times 10^8$ miles

61. $r = \dfrac{3}{2 + \sin\theta}$

$= \dfrac{3/2}{1 + 1/2\sin\theta}$

Because $e = \dfrac{1}{2}$, the equation represents an ellipse.

63. True. The graphs represent the same hyperbola, although the graphs are not traced out in the same order as θ goes from 0 to 2π.

65. True. The conic is an ellipse because the eccentricity is less than 1.

$e = \frac{2}{3} < 1$

67.
$$\frac{x^2}{a^2} + \frac{y^2}{b^2} = 1$$

$$\frac{r^2 \cos^2 \theta}{a^2} + \frac{r^2 \sin^2 \theta}{b^2} = 1$$

$$\frac{r^2 \cos^2 \theta}{a^2} + \frac{r^2(1 - \cos^2 \theta)}{b^2} = 1$$

$$r^2 b^2 \cos^2 \theta + r^2 a^2 - r^2 a^2 \cos^2 \theta = a^2 b^2$$

$$r^2(b^2 - a^2)\cos^2 \theta + r^2 a^2 = a^2 b^2$$

Since $b^2 - a^2 = -c^2$, we have:

$$-r^2 c^2 \cos^2 \theta + r^2 a^2 = a^2 b^2$$

$$-r^2\left(\frac{c}{a}\right)^2 \cos^2 \theta + r^2 = b^2, \, e = \frac{c}{a}$$

$$-r^2 e^2 \cos^2 \theta + r^2 = b^2$$

$$r^2(1 - e^2 \cos^2 \theta) = b^2$$

$$r^2 = \frac{b^2}{1 - e^2 \cos^2 \theta}$$

69. $\dfrac{x^2}{169} + \dfrac{y^2}{144} = 1$

$a = 13, b = 12, c = 5, e = \dfrac{5}{13}$

$$r^2 = \frac{144}{1 - (25/169) \cos^2 \theta} = \frac{24{,}336}{169 - 25 \cos^2 \theta}$$

71. $\dfrac{x^2}{9} - \dfrac{y^2}{16} = 1$

$a = 3, b = 4, c = 5, e = \dfrac{5}{3}$

$$r^2 = \frac{-16}{1 - (25/9) \cos^2 \theta} = \frac{144}{25 \cos^2 \theta - 9}$$

73. One focus: $(5, 0)$

Vertices: $(4, 0), (4, \pi)$

$a = 4, c = 5 \Rightarrow b = 3$ and $e = \dfrac{5}{4}$

$$\frac{x^2}{16} - \frac{y^2}{9} = 1$$

$$r^2 = \frac{-9}{1 - (25/16) \cos^2 \theta} = \frac{-144}{16 - 25 \cos^2 \theta} = \frac{144}{25 \cos^2 \theta}$$

75. The graph of $r = \dfrac{5}{1 - \sin \theta}$ is a parabola with a horizontal directrix below the pole and opens upward.

(a) The graph of $r = \dfrac{5}{1 - \cos \theta}$ is a parabola with a vertical directrix to the left of the pole and opens to the right.

(b) The graph of $r = \dfrac{5}{1 + \sin \theta}$ is a parabola with a horizontal directrix above the pole and opens downward.

(c) The graph of $r = \dfrac{5}{1 + \cos \theta}$ is a parabola with a vertical directrix to the right of the pole and opens to the left.

(d) The graph of $r = \dfrac{5}{1 - \sin[\theta = (\pi/4)]}$ is the graph of $r = \dfrac{5}{1 - \sin \theta}$ rotated through the angle $\pi/4$.

77. $r = \dfrac{4}{1 - 0.4 \cos \theta}$

(a) Because $e < 1$, the conic is an ellipse.

(b) $r = \dfrac{4}{1 + 0.4 \cos \theta}$ has a vertical directrix to the

right of the pole and $r = \dfrac{4}{1 - 0.4 \sin \theta}$ has a

horizontal directrix below the pole. The given polar

equation, $r = \dfrac{4}{1 - 0.4 \cos \theta}$, has a vertical directrix

to the left of the pole.

(c)

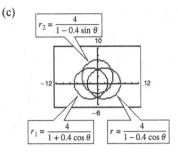

Review Exercises for Chapter 6

1. Points: $(-1, 2)$ and $(2, 5)$

$$m = \frac{5 - 2}{2 - (-1)} = \frac{3}{3} = 1$$

$$\tan \theta = 1 \Rightarrow \theta = \frac{\pi}{4} \text{ radian} = 45°$$

3. $5x + 2y + 4 = 0$

$$2y = -5x - 4$$

$$y = -\frac{5}{2}x - 2$$

$$m = -\frac{5}{2}$$

$$\tan \theta = -\frac{5}{2}$$

$$\theta = \arctan\left(-\frac{5}{2}\right) \approx \pi - 1.1902 = 1.9513 \text{ radians or } 111.8°$$

5. $4x + y = 2 \Rightarrow y = -4x + 2 \Rightarrow m_1 = -4$

$-5x + y = -1 \Rightarrow y = 5x - 1 \Rightarrow m_2 = 5$

$$\tan \theta = \left| \frac{5 - (-4)}{1 + (-4)(5)} \right| = \frac{9}{19}$$

$$\theta = \arctan \frac{9}{19} \approx 0.4424 \text{ radian} \approx 25.35°$$

7. $2x - 7y = 8 \Rightarrow y = \frac{2}{7}x - \frac{8}{7} \Rightarrow m_1 = \frac{2}{7}$

$0.4x + y = 0 \Rightarrow y = -0.4x \Rightarrow m_2 = -0.4$

$$\tan \theta = \left| \frac{-0.4 - (2/7)}{1 + (2/7)(-0.4)} \right| = \frac{24}{31}$$

$$\theta = \arctan\left(\frac{24}{31}\right) \approx 0.6588 \text{ radian} \approx 37.7°$$

9. $(4, 3) \Rightarrow x_1 = 4, y_1 = 3$

$$2x - y - 1 = 0 \Rightarrow A = 2, B = -1, C = -1$$

$$d = \frac{|(2)(4) + (-1)(3) + (-1)|}{\sqrt{(2)^2 + (-1)^2}} = \frac{4}{\sqrt{5}} = \frac{4\sqrt{5}}{5}$$

11. Hyperbola

13. Vertex: $(0, 0) = (h, k)$

Focus: $(0, 3) \Rightarrow p = 3$

$$(x - h)^2 = 4p(y - k)$$

$$(x - 0)^2 = 4(3)(y - 0)$$

$$x^2 = 12y$$

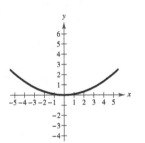

15. Vertex: $(0, 2) = (h, k)$

Focus: $x = -3 \Rightarrow p = 3$

$$(y - k)^2 = 4p(x - h)$$

$$(y - 2)^2 = 12x$$

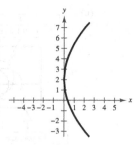

17. $y = 2x^2 \Rightarrow x^2 = \dfrac{1}{2}y \Rightarrow p = \dfrac{1}{8}$

Focus: $\left(0, \dfrac{1}{8}\right)$

$$d_1 = b + \dfrac{1}{8}$$

$$d_2 = \sqrt{(-1 - 0)^2 + \left(2 - \dfrac{1}{8}\right)^2}$$

$$= \sqrt{1 + \dfrac{225}{64}} = \dfrac{17}{8}$$

$$d_1 = d_2$$

$$b + \dfrac{1}{8} = \dfrac{17}{8}$$

$$b = 2$$

slope $m = \dfrac{-2 - 2}{0 + 1} = -4.$

Point-slope: $y - 2 = -4(x + 1)$

Tangent line: $y = -4x - 2$

19. Parabola

Opens downward

Vertex: $(0, 10)$

$$(x - h)^2 = 4p(y - k)$$

$$x^2 = 4p(y - 10)$$

Solution points: $(\pm 3, 8)$

$$9 = 4p(8 - 10)$$

$$9 = -8p$$

$$-\dfrac{9}{8} = p$$

$$x^2 = 4\left(-\dfrac{9}{8}\right)(y - 10)$$

$$x^2 = -\dfrac{9}{2}(y - 10)$$

To find the x-intercepts, let $y = 0$.

$$x^2 = 45$$

$$x = \pm\sqrt{45} = \pm 3\sqrt{5}$$

At the base, the archway is $2\left(3\sqrt{5}\right) = 6\sqrt{5} \approx 13.4$
meters wide.

21. Vertices: $(2, 0), (2, 16) \Rightarrow a = 8$

Center: $(2, 8) = (h, k)$

Minor axis of length $6 \Rightarrow b = 3$

Ends of minor axis: $(2, 3), (2, 13)$

$$\dfrac{(x - h)^2}{b^2} + \dfrac{(y - k)^2}{a^2} = 1$$

$$\dfrac{(x - 2)^2}{9} + \dfrac{(y - 8)^2}{64} = 1$$

23. Vertices: $(0, 1), (4, 1) \Rightarrow a = 2, (h, k) = (2, 1)$

Endpoints of minor axis: $(2, 0), (2, 2) \Rightarrow b = 1$

$$\dfrac{(x - h)^2}{a^2} + \dfrac{(y - k)^2}{b^2} = 1$$

$$\dfrac{(x - 2)^2}{4} + (y - 1)^2 = 1$$

25. $2a = 10 \Rightarrow a = 5$

$b = 4$

$c^2 = a^2 - b^2 = 25 - 16 = 9 \Rightarrow c = 3$

The foci occur 3 feet from the center of the arch on a line connecting the tops of the pillars.

27. $\dfrac{(x + 2)^2}{64} + \dfrac{(y - 5)^2}{36} = 1$

$a = 8, b = 6$

$c^2 = a^2 - b^2 = 64 - 36 = 28 \Rightarrow c = 2\sqrt{7}$

Center: $(-2, 5)$

Vertices: $(-10, 5), (6, 5)$

Foci: $\left(-2 \pm 2\sqrt{7}, 5\right)$

Eccentricity: $e = \dfrac{2\sqrt{7}}{8} = \dfrac{\sqrt{7}}{4}$

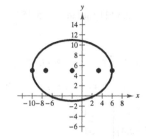

29.
$$16x^2 + 9y^2 - 32x + 72y + 16 = 0$$
$$16(x^2 - 2x + 1) + 9(y^2 + 8y + 16) = -16 + 16 + 144$$
$$16(x - 1)^2 + 9(y + 4)^2 = 144$$
$$\frac{(x-1)^2}{9} + \frac{(y+4)^2}{16} = 1$$

$a = 4, b = 3, c = \sqrt{7}$

Center: $(1, -4)$

Vertices: $(1, 0)$, and $(1, -8)$

Foci: $\left(1, -4 \pm \sqrt{7}\right)$

Eccentricity: $e = \dfrac{\sqrt{7}}{4}$

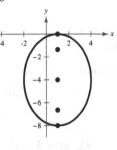

31. Vertices: $(0, \pm 6) \Rightarrow a = 6, (h, k) = (0, 0)$

Foci: $(0, \pm 8) \Rightarrow c = 8$

$$b^2 = c^2 - a^2 = 64 - 36 = 28$$
$$\frac{(y-k)^2}{a^2} - \frac{(x-h)^2}{b^2} = 1$$
$$\frac{y^2}{36} - \frac{x^2}{28} = 1$$

33. Foci: $(\pm 5, 0) \Rightarrow c = 5, (h, k) = (0, 0)$

Asymptotes:
$$y = \pm\frac{3}{4}x \Rightarrow y = \pm\frac{b}{a}x \Rightarrow b = 3, a = 4$$
$$\frac{(x-h)^2}{a^2} - \frac{(y-k)^2}{b^2} = 1$$
$$\frac{x^2}{4^2} - \frac{y^2}{3^2} = 1 \Rightarrow \frac{x^2}{16} - \frac{y^2}{9} = 1$$

35. $\dfrac{(x-4)^2}{49} - \dfrac{(y+2)^2}{25} = 1$

$a = 7, b = 5$

$c^2 = a^2 + b^2 = 49 + 25 = 74 \Rightarrow c = \sqrt{74}$

Center: $(4, -2)$

Vertices: $(11, -2), (-3, -2)$

Foci: $\left(4 \pm \sqrt{74}, -2\right)$

Asymptotes: $y = -2 \pm \dfrac{5}{7}(x - 4)$

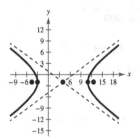

37.
$$9x^2 - 16y^2 - 18x - 32y - 151 = 0$$
$$9(x^2 - 2x + 1) - 16(y^2 + 2y + 1) = 151 + 9 - 16$$
$$9(x - 1)^2 - 16(y + 1)^2 = 144$$
$$\frac{(x-1)^2}{16} - \frac{(y+1)^2}{9} = 1$$

$a = 4, b = 3, c = 5$

Center: $(1, -1)$

Vertices: $(5, -1)$ and $(-3, -1)$

Foci: $(6, -1)$ and $(-4, -1)$

Asymptotes: $y = -1 \pm \dfrac{3}{4}(x - 1)$

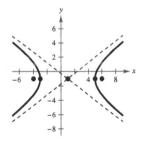

39. Since the microphones were two miles apart, 2 miles $\cdot$ 5280 $\Rightarrow c = 5280$. Since sound travels at 1100 feet per second, $\left| d_1 - d_2 \right| = 2a = 6600 \Rightarrow a = 3300$.

$b^2 = c^2 - a^2$

$\qquad = 5280^2 - 3300^2$

$\qquad = 27{,}878{,}400 - 10{,}890{,}000$

$\qquad = 16{,}988{,}400$

Since the explosion took place 6600 feet farther from B than from A. The locus of all points that are 6600 feet closer to A than to B is one

branch of the hyperbola of the form $\dfrac{x^2}{a^2} - \dfrac{y^2}{b^2} = 1$, so it follows that

$$\dfrac{x^2}{10{,}890{,}000} - \dfrac{y^2}{16{,}988{,}400} = 1.$$

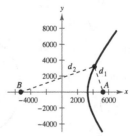

41. $5x^2 - 2y^2 + 10x - 4y + 17 = 0$

$AC = 5(-2) = -10 < 0$

Hyperbola

43. $3x^2 + 2y^2 - 12x + 12y + 29 = 0$

$A = 3, C = 2$

$AC = 3(2) = 6 > 0$

Ellipse

45. $xy + 5 = 0$

$A = C = 0, B = 1$

$B^2 - 4AC = 1^2 - 4(0)(0) = 1 > 0 \Rightarrow$ Hyperbola

$\cot 2\theta = \dfrac{A - C}{B} = \dfrac{0 - 0}{1} = 0 \Rightarrow 2\theta = \dfrac{\pi}{2} \Rightarrow \theta = \dfrac{\pi}{4}$

$x = x' \cos \dfrac{\pi}{4} - y' \sin \dfrac{\pi}{4} = \dfrac{x' - y'}{\sqrt{2}}$

$y = x' \sin \dfrac{\pi}{4} + y' \cos \dfrac{\pi}{4} = \dfrac{x' + y'}{\sqrt{2}}$

$\left(\dfrac{x' - y'}{\sqrt{2}} \right)\left(\dfrac{x' + y'}{\sqrt{2}} \right) + 5 = 0$

$\dfrac{(x')^2 - (y')^2}{2} = -5$

$\dfrac{(y')^2}{10} - \dfrac{(x')^2}{10} = 1$

47. $5x^2 - 2xy + 5y^2 - 12 = 0$

$A = C = 5, B = -2$

$B^2 - 4AC = (-2)^2 - 4(5)(5) = -96 < 0$

The graph is an ellipse.

$\cot 2\theta = 0 \Rightarrow 2\theta = \dfrac{\pi}{2} \Rightarrow \theta = \dfrac{\pi}{4}$

$x = x' \cos \dfrac{\pi}{4} - y' \sin \dfrac{\pi}{4} = \dfrac{x' - y'}{\sqrt{2}}$

$y = x' \sin \dfrac{\pi}{4} + y' \cos \dfrac{\pi}{4} = \dfrac{x' + y'}{\sqrt{2}}$

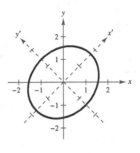

$$5\left(\dfrac{x' - y'}{\sqrt{2}}\right)^2 - 2\left(\dfrac{x' - y'}{\sqrt{2}}\right)\left(\dfrac{x' + y'}{\sqrt{2}}\right) + 5\left(\dfrac{x' + y'}{\sqrt{2}}\right)^2 - 12 = 0$$

$$\dfrac{5}{2}\left[(x')^2 - 2(x'y') + (y')^2\right] - \left[(x')^2 - (y')^2\right] + \dfrac{5}{2}\left[(x')^2 + 2(x'y') + (y')^2\right] = 12$$

$$4(x')^2 + 6(y')^2 = 12$$

$$\dfrac{(x')^2}{3} + \dfrac{(y')^2}{2} = 1$$

49. (a) $16x^2 - 24xy + 9y^2 - 30x - 40y = 0$

$B^2 - 4AC = (-24)^2 - 4(16)(9) = 0$

The graph is a parabola.

(b) To use a graphing utility, we need to solve for y in terms of x.

$9y^2 + (-24x - 40)y + (16x^2 - 30x) = 0$

$$y = \dfrac{-(-24x - 40) \pm \sqrt{(-24x - 40)^2 - 4(9)(16x^2 - 30x)}}{2(9)}$$

$$= \dfrac{(24x + 40) \pm \sqrt{(24x + 40)^2 - 36(16x^2 - 30x)}}{18}$$

(c)

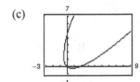

51. (a) $x^2 - 10xy + y^2 + 1 = 0$

Since $B^2 - 4AC = (-10)^2 - 4(1)(1) > 0 \Rightarrow$ Hyperbola

(b) Use the Quadratic Formula to solve for y in terms of x: $y^2 - 10xy + x^2 + 1 = 0$

$$y = \dfrac{10x \pm \sqrt{100x^2 - 4(x^2 + 1)}}{2}$$

(c)

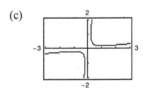

53. $x = 3t - 2, y = 7 - 4t$

(a)

t	-2	-1	0	1	2
x	-8	-5	-2	1	4
y	15	11	7	3	-1

(b)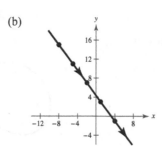

55. (a)

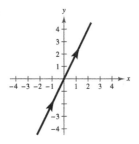

(b) $x = 2t \Rightarrow \dfrac{x}{2} = t$

$y = 4t \Rightarrow y = 4\left(\dfrac{x}{2}\right) = 2x$

57. (a)

(b) $x = t^2, x \geq 0$

$y = \sqrt{t} \Rightarrow y^2 = t$

$x = \left(y^2\right)^2 \Rightarrow x$

$= y^4 \Rightarrow y = \sqrt[4]{x}$

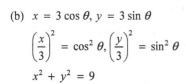

59. (a)

(b) $x = 3\cos\theta, y = 3\sin\theta$

$\left(\dfrac{x}{3}\right)^2 = \cos^2\theta, \left(\dfrac{y}{3}\right)^2 = \sin^2\theta$

$x^2 + y^2 = 9$

61. $y = 2x + 3$

(a) $t = x \Rightarrow x = t$

$y = 2x + 3 = 2t + 3$

(b) $t = x + 1 \Rightarrow x = t - 1$

$y = 2x + 3 = 2(t - 1) + 3 = 2t + 1$

(c) $t = 3 - x \Rightarrow x = 3 - t$

$y = 2x + 3 = 2(3 - t) + 3 = 9 - 2t$

63. $y = x^2 + 3$

(a) $t = x \Rightarrow x = t$

$y = x^2 + 3 = t^2 + 3$

(b) $t = x + 1 \Rightarrow x = t - 1$

$y = x^2 + 3 = (t - 1)^2 + 3 = t^2 - 2t + 4$

(c) $t = 3 - x \Rightarrow x = 3 - t$

$y = x^2 + 3 = (3 - t)^2 + 3 = t^2 - 6t + 12$

65. $y = 1 - 4x^2$

(a) $t = x \Rightarrow x = t$

$y = 1 - 4x^2 = 1 - 4t^2$

(b) $t = x + 1 \Rightarrow x = t - 1$

$y = 1 - 4x^2 = 1 - 4(t - 1)^2 = -4t^2 + 8t - 3$

(c) $t = 3 - x \Rightarrow x = 3 - t$

$y = 1 - 4x^2 = 1 - 4(3 - t)^2 = -4t^2 + 24t - 35$

67. Polar coordinates: $\left(4, \dfrac{5\pi}{6}\right)$

Additional polar representations:

$\left(4, -\dfrac{7\pi}{6}\right), \left(-4, -\dfrac{\pi}{6}\right), \left(-4, \dfrac{11\pi}{6}\right)$

69. Polar coordinates: $(-7, 4.19)$

Additional polar representations: $(7, 1.05), (-7, -2.09)$
$(7, -5.23)$

71. Polar coordinates: $\left(0, \dfrac{\pi}{2}\right) = (r, \theta)$

$x = r \cos \theta = 0 \cos \dfrac{\pi}{2} = 0$

$y = r \sin \theta = 0 \sin \dfrac{\pi}{2} = 0$

Rectangular coordinates: $(0, 0)$

73. Polar coordinates: $\left(-1, \dfrac{\pi}{3}\right)$

$x = -1 \cos \dfrac{\pi}{3} = -\dfrac{1}{2}$

$y = -1 \sin \dfrac{\pi}{3} = -\dfrac{\sqrt{3}}{2}$

Rectangular coordinates: $\left(-\dfrac{1}{2}, -\dfrac{\sqrt{3}}{2}\right)$

75. Rectangular coordinates: $(3, 3)$

$r = \sqrt{(3)^2 + (3)^2} = \sqrt{18} = 3\sqrt{2}$

$\tan \theta = 1, \theta = \dfrac{\pi}{4}$

Polar coordinates: $\left(3\sqrt{2}, \dfrac{\pi}{4}\right)$

77. Rectangular coordinates: $\left(-\sqrt{5}, \sqrt{5}\right)$

$r = \sqrt{\left(-\sqrt{5}\right)^2 + \left(\sqrt{5}\right)^2} = \sqrt{10}$

$\tan \theta = -1, \theta = \dfrac{3\pi}{4}$

Polar coordinates: $\left(\sqrt{10}, \dfrac{3\pi}{4}\right)$

79. $x^2 + y^2 = 81$

$r^2 = 81$

$r = 9$

81. $\qquad x = 5$

$r \cos \theta = 5$

$r = \dfrac{5}{\cos \theta}$

$r = 5 \sec \theta$

83. $\qquad\qquad xy = 5$

$(r \cos \theta)(r \sin \theta) = 5$

$r^2 = \dfrac{5}{\sin \theta \cos \theta}$

$\qquad = \dfrac{10}{\sin 2\theta} = 10 \csc 2\theta$

85. $\qquad r = 4$

$r^2 = 16$

$x^2 + y^2 = 16$

87. $\qquad r = 3 \cos \theta$

$r^2 = 3r \cos \theta$

$x^2 + y^2 = 3x$

89. $\qquad\qquad r^2 = \sin \theta$

$r^3 = r \sin \theta$

$\left(\pm\sqrt{x^2 + y^2}\right)^3 = y$

$\left(x^2 + y^2\right)^3 = y^2$

$x^2 + y^2 = y^{2/3}$

91. $r = 6$

Circle of radius 6 centered at the pole

Symmetric with respect to $\theta = \pi/2$, the polar axis and the pole

Maximum value of $|r| = 6$, for all values of θ

Zeros: None

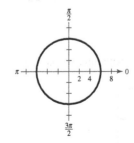

93. $r = -2(1 + \cos\theta)$

Symmetric with respect to the polar axis

Maximum value of $|r| = 4$ when $\theta = 0$

Zeros: $r = 0$ when $\theta = \pi$

$\dfrac{a}{b} = \dfrac{2}{2} = 1 \Rightarrow$ Cardioid

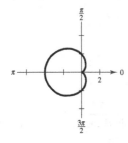

95. $r = 4\sin 2\theta$

Rose curve $(n = 2)$ with 4 petals

Symmetric with respect to $\theta = \pi/2$, the polar axis, and the pole

Maximum value of $|r| = 4$ when $\theta = \dfrac{\pi}{4}, \dfrac{3\pi}{4}, \dfrac{5\pi}{4}, \dfrac{7\pi}{4}$

Zeros: $r = 0$ when $\theta = 0, \dfrac{\pi}{2}, \pi, \dfrac{3\pi}{2}$

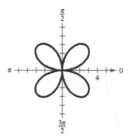

97. $r = 2 + 6\sin\theta$

Limaçon with inner loop

$r = f(\sin\theta) \Rightarrow \theta = \dfrac{\pi}{2}$ symmetry

Maximum value: $|r| = 8$ when $\theta = \dfrac{\pi}{2}$

Zeros: $2 + 6\sin\theta = 0 \Rightarrow \sin\theta = -\dfrac{1}{3} \Rightarrow \theta \approx 3.4814, 5.9433$

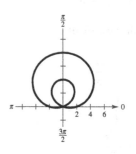

99. $r^2 = 9\sin\theta$

$r = \pm 3\sqrt{\sin\theta}$

Symmetric with respect to polar axis, $\theta = \dfrac{\pi}{2}$, and the pole.

Maximum value of $|r| = 3$ when $\theta = \dfrac{\pi}{2}$

Zeros: $r = 0$ when $\theta = 0$ and π

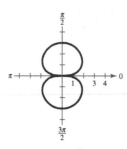

101. $r = 3(2 - \cos\theta)$

$\quad = 6 - 3\cos\theta$

$\dfrac{a}{b} = \dfrac{6}{3} = 2$

Convex limaçon

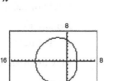

103. $r = 8\cos 3\theta$

Rose curve $(n = 3)$ with three petals

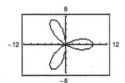

105. $r = \dfrac{1}{1 + 2\sin\theta}, e = 2$

Hyperbola symmetric with respect to $\theta = \dfrac{\pi}{2}$ and

having vertices at $\left(\dfrac{1}{3}, \dfrac{\pi}{2}\right)$ and $\left(-1, \dfrac{3\pi}{2}\right)$

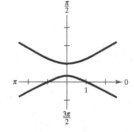

107. $r = \dfrac{4}{5 - 3\cos\theta}$

$r = \dfrac{4/5}{1 - (3/5)\cos\theta},\ e = \dfrac{3}{5}$

Ellipse symmetric with respect to the polar axis and having vertices at $(2, 0)$ and $(1/2, \pi)$.

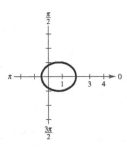

109. Parabola: $r = \dfrac{ep}{1 - e\cos\theta},\ e = 1$

Vertex: $(2, \pi)$

Focus: $(0, 0) \Rightarrow p = 4$

$r = \dfrac{4}{1 - \cos\theta}$

111. Ellipse: $r = \dfrac{ep}{1 - e\cos\theta}$

Vertices: $(5, 0), (1, \pi) \Rightarrow a = 3$

One focus: $(0, 0) \Rightarrow c = 2$

$e = \dfrac{c}{a} = \dfrac{2}{3},\ p = \dfrac{5}{2}$

$r = \dfrac{(2/3)(5/2)}{1 - (2/3)\cos\theta} = \dfrac{5/3}{1 - (2/3)\cos\theta} = \dfrac{5}{3 - 2\cos\theta}$

113. $a + c = 122{,}800 + 4000 \Rightarrow a + c = 126{,}800$

$a - c = 119 + 4000 \quad\Rightarrow a - c = 4{,}119$

$\overline{ 2a = 130{,}919}$

$ a = 65{,}459.5$

$ c = 61{,}340.5$

$e = \dfrac{c}{a} = \dfrac{61{,}340.5}{65{,}459.5} \approx 0.937$

$r = \dfrac{ep}{1 - e\cos\theta} \approx \dfrac{0.937p}{1 - 0.937\cos\theta}$

$r = 126{,}800$ when $\theta = 0$

$126{,}800 = \dfrac{ep}{1 - e\cos 0}$

$ep = 126{,}800\left(1 - \dfrac{61{,}340.5}{65{,}459.5}\right) \approx 7978.81$

So, $r \approx \dfrac{7978.81}{1 - 0.937\cos\theta}$.

When

$\theta = \dfrac{\pi}{3},\ r \approx \dfrac{7978.81}{1 - 0.937\cos(\pi/3)} \approx 15{,}011.87$ miles.

The distance from the surface of Earth and the satellite is $15{,}011.87 - 4000 \approx 11{,}011.87$ miles.

115. False.

$\dfrac{x^2}{4} - y^4 = 1$ is a fourth-degree equation.

The equation of a hyperbola is a second degree equation.

117. False.

$(r, \theta), (r, \theta + 2\pi), (-r, \theta + \pi)$, etc.

All represent the same point.

119. (a) $x^2 + y^2 = 25$

$r = 5$

The graphs are the same. They are both circles centered at $(0, 0)$ with a radius of 5.

(b) $x - y = 0 \Rightarrow y = x$

$\theta = \dfrac{\pi}{4}$

The graphs are the same. They are both lines with slope 1 and intercept $(0, 0)$.

Problem Solving for Chapter 6

1. (a) $\theta = \pi - 1.10 - 0.84 \approx 1.2016$ radians

(b) $\sin 0.84 = \dfrac{x}{3250} \Rightarrow x = 3250\sin 0.84 \approx 2420$ feet

$\sin 1.10 = \dfrac{y}{6700} \Rightarrow y = 6700\sin 1.10 \approx 5971$ feet

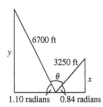

3. Let (x, x) be the corner of the square in Quadrant I.

$A = 4x^2$

$\dfrac{x^2}{a^2} + \dfrac{x^2}{b^2} = 1 \Rightarrow x^2 = \dfrac{a^2 b^2}{a^2 + b^2}$

So, $A = \dfrac{4a^2 b^2}{a^2 + b^2}$.

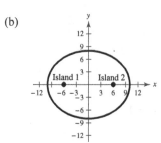

5. (a)

Because $d_1 + d_2 \le 20$, by definition, the outer bound that the boat can travel is an ellipse. The islands are the foci.

(b)

Island 1 is located at $(-6, 0)$ and Island 2 is located at $(6, 0)$.

(c) $d_1 + d_2 = 2a = 20 \Rightarrow a = 10$

The boat traveled 20 miles.

The vertex is $(10, 0)$.

(d) $c = 6, a = 10 \Rightarrow b^2 = a^2 - c^2 = 64$

$\dfrac{x^2}{100} + \dfrac{y^2}{64} = 1$

7. $Ax^2 + Cy^2 + Dx + Ey + F = 0$

Assume that the conic is *not* degenerate.

(a) $A = C, A \ne 0$. Complete the square with respect to x and y, to write the standard equation of a circle.

$$Ax^2 + Ay^2 + Dx + Ey + F = 0$$

$$x^2 + y^2 + \frac{D}{A}x + \frac{E}{A}y + \frac{F}{A} = 0$$

$$\left(x^2 + \frac{D}{A}x + \frac{D^2}{4A^2}\right) + \left(y^2 + \frac{E}{A}y + \frac{E^2}{4A^2}\right) = -\frac{F}{A} + \frac{D^2}{4A^2} + \frac{E^2}{4A^2}$$

$$\left(x + \frac{D}{2A}\right)^2 + \left(y + \frac{E}{2A}\right)^2 = \frac{D^2 + E^2 - 4AF}{4A^2}$$

$$(x - h)^2 + (y - k)^2 = r^2$$

This is a circle with center $\left(-\dfrac{D}{2A}, -\dfrac{E}{2A}\right)$ and radius $\dfrac{\sqrt{D^2 + E^2 - 4AF}}{2|A|}$.

(b) $A = 0$ or $C = 0$ (but not both).

Case 1: Let $C = 0$. Complete the square with respect to x to write the standard equation of a parabola with horizontal axis.

$$Ax^2 + Dx + Ey + F = 0$$

$$x^2 + \frac{D}{A}x = -\frac{E}{A}y - \frac{F}{A}$$

$$x^2 + \frac{D}{A}x + \frac{D^2}{4A^2} = -\frac{E}{A}y - \frac{F}{A} + \frac{D^2}{4A^2}$$

$$\left(x + \frac{D}{2A}\right)^2 = -\frac{E}{A}\left(y + \frac{F}{E} - \frac{D^2}{4AE}\right)$$

$$\left(x + \frac{D}{2A}\right)^2 = -\frac{E}{A}\left(y + \left(\frac{4AF}{E} - \frac{D^2}{4AE}\right)\right)$$

$$(x - h)^2 = 4p(y - k)$$

This is a parabola with vertex $\left(-\dfrac{D}{2A}, \dfrac{D^2 - 4AF}{4AE}\right)$.

Case 2: $A = 0$ yields a similar result when you complete the square with respect to y to have a parabola with vertical axis.

(c) $AC > 0 \Rightarrow A$ and C are either both positive or are both negative, if that is the case, move the terms to the other side of the equation so that they are both positive.

Complete the square with respect to x and to y to write the standard equation of an ellipse.

$$Ax^2 + Cy^2 + Dx + Ey + F = 0$$

$$A\left(x^2 + \frac{D}{A}x + \frac{D^2}{4A^2}\right) + C\left(y^2 + \frac{E}{C}y + \frac{E^2}{4C^2}\right) = -F + \frac{D^2}{4A} + \frac{E^2}{4C}$$

$$A\left(x + \frac{D}{2A}\right)^2 + C\left(y + \frac{E}{2C}\right)^2 = \frac{CD^2 + AE^2 - 4ACF}{4AC}$$

$$\frac{\left(x + \frac{D}{2A}\right)^2}{\left(\frac{CD^2 + AE^2 - 4ACF}{4A^2C}\right)} + \frac{\left(y + \frac{E}{2C}\right)^2}{\left(\frac{CD^2 + AE^2 - 4ACF}{4AC^2}\right)} = 1$$

$$\frac{(x - h)^2}{a^2} + \frac{(y - k)^2}{b^2} = 1$$

Because A and C are both positive, $4A^2C$ and $4AC^2$ are both positive. $CD^2 + AE^2 - 4ACF$ must be positive or the conic is degenerate. So, we have an ellipse with center $\left(-\dfrac{D}{2A}, -\dfrac{E}{2C}\right)$. The values of $\dfrac{CD^2 + AE^2 - 4ACF}{4A^2C}$ and

$\dfrac{CD^2 + AE^2 - 4ACF}{4AC^2}$ will determine if the major axis is vertical or horizontal.

(d) $AC < 0 \Rightarrow A$ and C have opposite signs. Let's assume that A is positive and C is negative.

If A is negative and C is positive, move the terms to the other side of the equation. From part (c) above completing the square with respect to x and to y yields the standard equation of the hyperbola.

$$\frac{\left(x + \frac{D}{2A}\right)^2}{\left(\frac{CD^2 + AE^2 - 4ACF}{4A^2C}\right)} + \frac{\left(y + \frac{E}{2C}\right)^2}{\left(\frac{CD^2 + AE^2 - 4ACF}{4AC^2}\right)} = 1$$

$$\frac{(x - h)^2}{a^2} + \frac{(y - k)^2}{b^2} = 1.$$

Because $A > 0$ and $C < 0$, the first denominator is positive if $CD^2 + AE^2 - 4ACF < 0$ and is negative if

$CD^2 + AE^2 - 4ACF > 0$, since $4A^2C$ is negative. Recall in the first sentence we assumed A is positive and C is

negative. The second denominator would have the *opposite* sign because $4AC^2 > 0$. So, we have a hyperbola with center

$\left(-\dfrac{D}{2A}, -\dfrac{E}{2C}\right)$.

9. At the point $(a, 0)$, the difference of the distances to the foci $(\pm c, 0)$ is $(c + a) - (c - a) = 2a$. Let (x, y) be a point on the hyperbola.

$$2a = \sqrt{(x + c)^2 + y^2} - \sqrt{(x - c)^2 + y^2}$$

$$2a + \sqrt{(x - c)^2 + y^2} = \sqrt{(x + c)^2 + y^2}$$

$$4a^2 + 4a\sqrt{(x - c)^2 + y^2} + (x - c)^2 + y^2 = (x + c)^2 + y^2$$

$$4a\sqrt{(x - c)^2 + y^2} = 4cx - 4a^2$$

$$a\sqrt{(x - c)^2 + y^2} = cx - a^2$$

$$a^2\left(x^2 - 2cx + c^2 + y^2\right) = c^2x^2 - 2a^2cx + a^4$$

$$a^2\left(c^2 - a^2\right) = \left(c^2 - a^2\right)x^2 - a^2y^2$$

$$1 = \frac{x^2}{a^2} - \frac{y^2}{c^2 - a^2}$$

Thus, $c^2 = a^2 + b^2$.

11. To change the orientation, replace t with $-t$.

$$x = \cos(-t) = \cos t$$

$$y = 2 \sin(-t) = -2 \sin t$$

13. (a)
$$r = 2 \cos 2\theta \sec \theta$$

$$r = \frac{2\left(\cos^2 \theta - \sin^2 \theta\right)}{\cos \theta}$$

$$\sqrt{x^2 + y^2} = \frac{2\left[\left(\dfrac{x}{\sqrt{x^2+y^2}}\right)^2 - \left(\dfrac{y}{\sqrt{x^2+y^2}}\right)^2\right]}{\dfrac{x}{\sqrt{x^2+y^2}}}$$

$$\frac{x}{\sqrt{x^2 + y^2}} \cdot \sqrt{x^2 + y^2} = 2\left[\frac{x^2}{x^2 + y^2} - \frac{y^2}{x^2 + y^2}\right]$$

$$x = 2\left[\frac{x^2 - y^2}{x^2 + y^2}\right]$$

$$x\left(x^2 + y^2\right) = 2\left(x^2 - y^2\right)$$

$$x^3 + xy^2 = 2x^2 - 2y^2$$

$$2y^2 + xy^2 = 2x^2 - x^3$$

$$y^2(2 + x) = x^2(2 - x)$$

$$y^2 = x^2\left(\frac{2 - x}{2 + x}\right)$$

(b) $x = \dfrac{2 - 2t^2}{1 + t^2}$ and $y = \dfrac{t\left(2 - 2t^2\right)}{1 + t^2}$

(c)

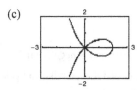

15. $x = (a - b)\cos t + b \cos\left(\dfrac{a - b}{b}t\right)$

$y = (a - b)\sin t - b \sin\left(\dfrac{a - b}{b}t\right)$

(a) $a = 2, b = 1$

$x = \cos t + \cos t = 2\cos t$

$y = \sin t - \sin t = 0$

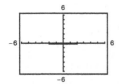

The graph oscillates between -2 and 2 on the x-axis.

(b) $a = 3, b = 1$

$x = 2\cos t + \cos 2t$

$y = 2\sin t - \sin 2t$

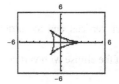

The graph is a three-sided figure with counterclockwise orientation.

(c) $a = 4, b = 1$

$x = 3\cos t + \cos 3t$

$y = 3\sin t - \sin 3t$

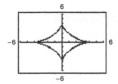

The graph is a four-sided figure with counterclockwise orientation.

(d) $a = 10, b = 1$

$x = 9\cos t + \cos 9t$

$y = 9\sin t - \sin 9t$

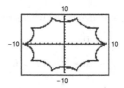

The graph is a ten-sided figure with counterclockwise orientation.

(e) $a = 3, b = 2$

$x = \cos t + 2\cos\dfrac{t}{2}$

$y = \sin t - 2\sin\dfrac{t}{2}$

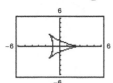

The graph looks the same as the graph in part (b), but is oriented clockwise instead of counterclockwise.

(f) $a = 4, b = 3$

$x = \cos t + 3\cos\dfrac{t}{3}$

$y = \sin t - 3\sin\dfrac{t}{3}$

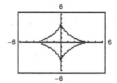

The graph is the same as the graph in part (c), but is oriented clockwise instead of counterclockwise.

17. *Sample answer:*

$r = 2\cos\left(\dfrac{1}{2}\theta\right)$

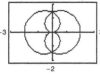

$r = 3\sin\left(\dfrac{5\theta}{2}\right)$

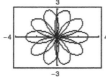

$r = -\cos\left(\sqrt{2}\theta\right), -2\pi \le \theta \le 2\pi$

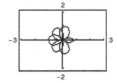

$r = -2\sin\left(\dfrac{4\theta}{7}\right)$

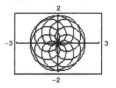

If n is a rational number, then the curve has a finite number of petals. If n is an irrational number, then the curve has an infinite number of petals.

Chapter 6 Practice Test

1. Find the angle, θ, between the lines $3x + 4y = 12$ and $4x - 3y = 12$.

2. Find the distance between the point $(5, -9)$ and the line $3x - 7y = 21$.

3. Find the vertex, focus and directrix of the parabola $x^2 - 6x - 4y + 1 = 0$.

4. Find an equation of the parabola with its vertex at $(2, -5)$ and focus at $(2, -6)$.

5. Find the center, foci, vertices, and eccentricity of the ellipse $x^2 + 4y^2 - 2x + 32y + 61 = 0$.

6. Find an equation of the ellipse with vertices $(0, \pm 6)$ and eccentricity $e = \frac{1}{2}$.

7. Find the center, vertices, foci, and asymptotes of the hyperbola $16y^2 - x^2 - 6x - 128y + 231 = 0$.

8. Find an equation of the hyperbola with vertices at $(\pm 3, 2)$ and foci at $(\pm 5, 2)$.

9. Rotate the axes to eliminate the xy-term. Sketch the graph of the resulting equation, showing both sets of axes.
   $5x^2 + 2xy + 5y^2 - 10 = 0$

10. Use the discriminant to determine whether the graph of the equation is a parabola, ellipse, or hyperbola.
    (a) $6x^2 - 2xy + y^2 = 0$
    (b) $x^2 + 4xy + 4y^2 - x - y + 17 = 0$

11. Convert the polar point $\left(\sqrt{2}, \dfrac{3\pi}{4}\right)$ to rectangular coordinates.

12. Convert the rectangular point $\left(\sqrt{3}, -1\right)$ to polar coordinates.

13. Convert the rectangular equation $4x - 3y = 12$ to polar form.

14. Convert the polar equation $r = 5 \cos \theta$ to rectangular form.

15. Sketch the graph of $r = 1 - \cos \theta$.

16. Sketch the graph of $r = 5 \sin 2\theta$.

17. Sketch the graph of $r = \dfrac{3}{6 - \cos \theta}$.

18. Find a polar equation of the parabola with its vertex at $\left(6, \dfrac{\pi}{2}\right)$ and focus at $(0, 0)$.

For Exercises 19 and 20, eliminate the parameter and write the corresponding rectangular equation.

19. $x = 3 - 2 \sin \theta$, $y = 1 + 5 \cos \theta$

20. $x = e^{2t}$, $y = e^{4t}$

CHECKPOINTS

CHECKPOINTS
Chapter P

Checkpoints for Section P.1

1. (a) Natural numbers: $\left\{\frac{6}{3}, 8\right\}$

(b) Whole numbers: $\left\{\frac{6}{3}, 8\right\}$

(c) Integers: $\left\{-22, -1, \frac{6}{3}, 8\right\}$

(d) Rational numbers: $\left\{-22, -7.5, -1 - \frac{1}{4}, \frac{6}{3}, 8\right\}$

(e) Irrational numbers: $\left\{-\pi, \frac{1}{2}\sqrt{2}\right\}$

2.

(a) The point representing the real number $\frac{5}{2} = 2.5$ lies halfway between 2 and 3, on the real number line.

(b) The point representing the real number -1.6 lies between -2 and -1 but closer to -2, on the real number line.

(c) The point representing the real number lies $-\frac{3}{4}$ between -1 and 0 but closer to -1, on the real number line.

(d) The point representing the real number 0.7 lies between 0 and 1 but closer to 1, on the real number line.

3. (a) Because -5 lies to the left of 1 on the real number line, you can say that -5 is *less than 1*, and write $-5 < 1$.

(b) Because $\frac{3}{2}$ lies to the left of 7 on the real number line, you can say that $\frac{3}{2}$ is *less than 7*, and write $\frac{3}{2} < 7$.

(c) Because $-\frac{2}{3}$ lies to the right of $-\frac{3}{4}$ on the real number line, you can say that $-\frac{2}{3}$ is *greater than* $-\frac{3}{4}$, and write $-\frac{2}{3} > -\frac{3}{4}$.

4. (a) The inequality $x > -3$ denotes all real numbers greater than -3.

(b) The inequality $0 < x \leq 4$ means that $x > 0$ and $x \leq 4$. This double inequality denotes all real numbers between 0 and 4, including 4 but not including 0.

5. The interval consists of real numbers greater than or equal to -2 and less than 5.

6. The inequality $-2 \leq x < 4$ can represent the statement "x is less than 4 and at least -2."

7. (a) $|1| = 1$

(b) $-\left|\frac{3}{4}\right| = -\left(\frac{3}{4}\right) = -\frac{3}{4}$

(c) $\frac{2}{|-3|} = \frac{2}{3}$

(d) $-|0.7| = -(0.7) = -0.7$

8. (a) If $x > -3$, then $\dfrac{|x+3|}{x+3} = \dfrac{x+3}{x+3} = 1$.

(b) If $x < -3$, then $\dfrac{|x+3|}{x+3} = \dfrac{-(x+3)}{x+3} = -1$.

9. (a) $|-3| < |4|$ because $|-3| = 3$ and $|4| = 4$, and 3 is less than 4.

(b) $-|-4| = -|-4|$ because $-|-4| = -4$ and $-|4| = -4$.

(c) $|-3| > -|-3|$ because $|-3| = 3$ and $-|-3| = -3$, and 3 is greater than -3.

10. (a) The distance between 35 and -23 is
$$|35 - (-23)| = |58| = 58.$$

(b) The distance between -35 and -23 is
$$|-35 - (-23)| = |-12| = 12.$$

(c) The distance between 35 and 23 is
$$|35 - 23| = |12| = 12.$$

11.

Algebraic Expression	Terms	Coefficients
$-2x + 4$	$-2x, 4$	$-2, 4$

12.

Expression	Value of Variable	Substitute	Value of Expression
$4x - 5$	$x - 0$	$4(0) - 5$	$0 - 5 = -5$

13. (a) $x + 9 = 9 + x$: This statement illustrates the Commutative Property of Addition. In other words, you obtain the same result whether you add x and 9, or 9 and x.

(b) $5(x^3 \cdot 2) = (5x^3)2$: This statement illustrates the Associative Property of Multiplication. In other words, to form the product $5 \cdot x^3 \cdot 2$, it does not matter whether 5 and $(x^3 \cdot 2)$, or $5x^3$ and 2 are multiplied first.

(c) $(2 + 5x^2)y^2 = 2y^2 + 5x^2 \cdot y^2$: This statement illustrates the Distributive Property. In other words, the terms 2 and $5x^2$ are multiplied by y^2.

14. (a) $\dfrac{3}{5} \cdot \dfrac{x}{6} = \dfrac{3x}{30} = \dfrac{3x + 3}{30 \div 3} = \dfrac{x}{10}$

(b) $\dfrac{x}{10} + \dfrac{2x}{5} = \dfrac{x}{10} + \dfrac{2x}{5} \cdot \dfrac{2}{2}$

$= \dfrac{x}{10} + \dfrac{2x}{5} = \dfrac{x}{10} + \dfrac{2x}{5} \cdot \dfrac{2}{2}$

$= \dfrac{x}{10} + \dfrac{4x}{10}$

$= \dfrac{x + 4x}{10}$

$= \dfrac{5x + 5}{10 \div 5}$

$= \dfrac{x}{2}$

Checkpoints for Section P.2

1. (a)

$7 - 2x = 15$	Write original equation.
$-2x = 8$	Subtract 7 from each side.
$x = -4$	Divide each side by -2.

Check: $7 - 2x = 15$

$7 - 2(-4) \overset{?}{=} 15$

$7 + 8 \overset{?}{=} 15$

$15 = 15$

(b)

$7x - 9 = 5x + 7$	Write original equation.
$2x - 9 = 7$	Subtract $5x$ from each side.
$2x = 16$	Add 9 from each side.
$x = 8$	Divide each side by 2.

Check: $7x - 9 = 5x + 7$

$7(8) - 9 = 5(8 + 7)$

$56 - 9 = 40 + 7$

$47 = 47 \checkmark$

2.

$\dfrac{4x}{9} - \dfrac{1}{3} = x + \dfrac{5}{3}$	Write original equation.
$(9)\left(\dfrac{4x}{9}\right) - (9)\left(\dfrac{1}{3}\right) = (9)x + 9\left(\dfrac{5}{3}\right)$	Multiply each term by the LCD.
$4x - 3 = 9x + 15$	Simplify.
$-5x = 18$	Combine like terms.
$x = -\dfrac{18}{5}$	Divide each side by -5.

3.

$\dfrac{3x}{x - 4} = 5 + \dfrac{12}{x - 4}$	Write original equation.
$(x - 4)\left(\dfrac{3x}{x - 4}\right) = (x - 4)5 + (x - 4)\left(\dfrac{12}{x - 4}\right)$	Multiply each term by LCD.
$3x = 5x - 20 + 12, \; x \neq 4$	Simplify.
$-2x = -8$	Divide each side by -2.
$x = 4$	Extraneous solution

In the original equation, $x = 4$ yields a denominator of zero. So, $x = 4$ is an extraneous solution, and the original equation has no solution.

4. $\quad 2x^2 - 3x + 1 = 6 \qquad\qquad$ Write original equation.

$\quad\quad 2x^2 - 3x - 5 = 0 \qquad\qquad$ Write in general form.

$\quad (2x - 5)(x + 1) = 0 \qquad\qquad$ Factor.

$\quad\quad\quad 2x - 5 = 0 \Rightarrow x = \frac{5}{2} \quad$ Set 1st factor equal to 0.

$\quad\quad\quad x + 1 = 0 \Rightarrow x = -1 \quad$ Set 2nd factor equal to 0.

The solutions are $x = -1$ and $x = \frac{5}{2}$.

Check: $x = -1$

$$2x^2 - 3x + 1 = 6$$

$$2(-1)^2 - 3(-1) + 1 \overset{?}{=} 6$$

$$2(1) + 3 + 1 \overset{?}{=} 6$$

$$6 = 6 \checkmark$$

$x = \frac{5}{2}$

$$2x^2 - 3x + 1 = 6$$

$$2\left(\tfrac{5}{2}\right)^2 - 3\left(\tfrac{5}{2}\right) + 1 \overset{?}{=} 6$$

$$2\left(\tfrac{25}{4}\right) - \tfrac{15}{2} + 1 \overset{?}{=} 6$$

$$6 = 6 \checkmark$$

5. (a) $\quad 3x^2 = 36 \qquad\qquad$ Write original equation.

$\quad\quad x^2 = 12 \qquad\qquad$ Divide each side by 3.

$\quad\quad x = \pm\sqrt{12} \qquad$ Extract square roots.

$\quad\quad x = \pm 2\sqrt{3}$

The solutions are $x = \pm 2\sqrt{3}$.

Check: $\quad x = -2\sqrt{3}$

$$3x^2 = 36$$

$$3\left(-2\sqrt{3}\right)^2 \overset{?}{=} 36$$

$$3(12) \overset{?}{=} 36$$

$$36 = 36 \checkmark$$

$x = 2\sqrt{3}$

$$3x^2 = 36$$

$$3\left(2\sqrt{3}\right)^2 \overset{?}{=} 36$$

$$3(12) \overset{?}{=} 36$$

$$36 = 36 \checkmark$$

(b) $(x - 1)^2 = 10$

$$x - 1 = \pm\sqrt{10}$$

$$x = 1 \pm \sqrt{10}$$

The solutions are $x = 1 \pm \sqrt{10}$.

Check: $\quad x = 1 - \sqrt{10}$

$$(x - 1)^2 = 10$$

$$\left[\left(1 - \sqrt{10}\right) - 1\right]^2 \overset{?}{=} 10$$

$$\left(-\sqrt{10}\right)^2 \overset{?}{=} 10$$

$$10 = 10 \checkmark$$

$x = 1 + \sqrt{10}$

$$(x - 1)^2 = 10$$

$$\left[\left(1 + \sqrt{10}\right) - 1\right]^2 \overset{?}{=} 10$$

$$\left(\sqrt{10}\right)^2 \overset{?}{=} 10$$

$$10 = 10 \checkmark$$

6.

$$x^2 - 4x - 1 = 0 \qquad \text{Write original equation.}$$

$$x^2 - 4x = 1 \qquad \text{Add 1 to each side.}$$

$$x^2 - 4x + (2)^2 = 1 + (2)^2 \qquad \text{Add } 2^2 \text{ to each side.}$$

$$(\text{half of } 4)^2$$

$$(x - 2)^2 = 5 \qquad \text{Simplify.}$$

$$x - 2 = \pm\sqrt{5} \qquad \text{Extract square roots.}$$

$$x = 2 \pm \sqrt{5} \qquad \text{Add 2 to each side.}$$

The solutions are $x = 2 \pm \sqrt{5}$.

Check: $x = 2 - \sqrt{5}$

$$x^2 - 4x - 1 = 0$$

$$\left(2 - \sqrt{5}\right)^2 - 4\left(2 - \sqrt{5}\right) - 1 \overset{?}{=} 0$$

$$\left(4 - 4\sqrt{5} + 5\right) - 8 + 4\sqrt{5} - 1 \overset{?}{=} 0$$

$$4 + 5 - 8 - 1 \overset{?}{=} 0$$

$$0 = 0 \checkmark$$

$x = 2 + \sqrt{5}$ also checks. $\checkmark$

7.

$$3x^2 - 10x - 2 = 0 \qquad \text{Original equation}$$

$$3x^2 - 10x = 2 \qquad \text{Add 2 to each side.}$$

$$x^2 - \frac{10}{3}x = \frac{2}{3} \qquad \text{Divide each side by 3.}$$

$$x^2 - \frac{10}{3}x + \left(\frac{5}{3}\right)^2 = \frac{2}{3} + \left(\frac{5}{3}\right)^2 \qquad \text{Add } \left(\frac{5}{3}\right)^2 \text{ to each side.}$$

$$\left(x - \frac{5}{3}\right)^2 = \frac{31}{9} \qquad \text{Simplify.}$$

$$x - \frac{5}{3} = \pm\frac{\sqrt{31}}{3} \qquad \text{Extract square roots.}$$

$$x = \frac{5}{3} \pm \frac{\sqrt{31}}{3} \qquad \text{Add } \frac{5}{3} \text{ to each side.}$$

The solutions are $\frac{5}{3} \pm \frac{\sqrt{31}}{3}$.

8. $3x^2 + 2x - 10 = 0$ Write original equation.

$$x = \frac{-6 \pm \sqrt{6^2 - 4ac}}{2a}$$ Quadratic Formula

$$x = \frac{-2 \pm \sqrt{(2)^2 - 4(3)(-10)}}{2(3)}$$ Substitute $a = 3$, $b = 2$ and $c = -10$.

$$x = \frac{-2 \pm \sqrt{4 + 120}}{6}$$ Simplify.

$$x = \frac{-2 \pm \sqrt{124}}{6}$$ Simplify.

$$x = \frac{-2 \pm 2\sqrt{31}}{6}$$ Simplify.

$$x = \frac{2\left(-1 \pm \sqrt{31}\right)}{6}$$ Factor our common factor.

$$x = \frac{-1 \pm \sqrt{31}}{3}$$ Simplify.

The solutions are $\dfrac{-1 \pm \sqrt{31}}{3}$.

Check: $x = \dfrac{-1 \pm \sqrt{31}}{3}$

$$3x^2 + 2x - 10 = 0$$

$$3\left(\frac{-1 + \sqrt{31}}{3}\right)^2 + 2\left(\frac{-1 + \sqrt{31}}{3}\right) - 10 \overset{?}{=} 0$$

$$\frac{1}{3} + \frac{2\sqrt{31}}{3} + \frac{31}{3} + \frac{2\sqrt{31}}{3} - 10 \overset{?}{=} 0$$

$$10 - 10 \overset{?}{=} 0$$

$$0 = 0 \checkmark$$

The solution $x = \dfrac{-1 - \sqrt{31}}{3}$ also checks. $\checkmark$

9. $18x^2 - 48x + 32 = 0$

$9x^2 - 24x + 16 = 0$

$$x = \frac{-b \pm \sqrt{b^2 - 4ac}}{2a}$$

$$x = \frac{-(-24) \pm \sqrt{(-24)^2 - 4(9)(16)}}{2(9)}$$

$$x = \frac{24 \pm \sqrt{0}}{18}$$

$$x = \frac{4}{3}$$

The quadratic equation has only one solution: $x = \dfrac{4}{3}$.

10.

$9x^4 - 12x^2 = 0$	Write original equation.
$3x^2(3x^2 - 4) = 0$	Factor out common factor.
$3x^2 = 0 \implies x = 0$	Set 1st factor equal to 0.
$3x^2 - 4 = 0 \implies 3x^2 = 4$	Set 2nd factor equal to 0.

$$x^2 = \frac{4}{3}$$

$$x = \pm\sqrt{\frac{4}{3}}$$

$$x = \pm\frac{2\sqrt{3}}{3}$$

Check: $x = 0$

$$9x^4 - 12x^2 = 0$$

$$9(0)^4 - 12(0)^2 \overset{?}{=} 0$$

$$0 = 0 \checkmark$$

$$x = \frac{2\sqrt{3}}{3}$$

$$9x^4 - 12x^2 = 0$$

$$9\left(\frac{2\sqrt{3}}{3}\right)^4 - 12\left(\frac{2\sqrt{3}}{3}\right)^2 \overset{?}{=} 0$$

$$9\left(\frac{16}{9}\right) - 12\left(\frac{4}{3}\right) \overset{?}{=} 0$$

$$16 - 16 \overset{?}{=} 0$$

$$0 = 0 \checkmark$$

The solution $x = \dfrac{-2\sqrt{3}}{3}$ also checks. $\checkmark$

So, the solutions are $x = 0$ and $x = \pm\dfrac{2\sqrt{3}}{3}$.

11. (a)

$x^3 - 5x^2 - 2x + 10 = 0$	Write original equation.
$x^2(x - 5) - 2(x - 5) = 0$	Factor by grouping.
$(x - 5)(x^2 - 2) = 0$	Distributive Property
$x - 5 = 0 \Rightarrow x = 5$	Set 1st factor equal to 0.
$x^2 - 2 = 0 \Rightarrow x^2 = 2$	Set 2nd factor equal to 0.
$\quad\quad\quad\quad = \pm\sqrt{2}$	

Check: $x = 5$

$$x^3 - 5x^2 - 2x + 10 = 0$$

$$(5)^3 - 5(5)^2 - 2(5) + 10 \overset{?}{=} 0$$

$$125 - 125 - 10 + 10 \overset{?}{=} 0$$

$$0 = 0 \checkmark$$

$x = \sqrt{2}$

$$x^3 - 5x^2 - 2x + 10 = 0$$

$$(\sqrt{2})^3 - 5(\sqrt{2})^2 - 2(\sqrt{2}) + 10 \overset{?}{=} 0$$

$$2\sqrt{2} - 10 - 2\sqrt{2} + 10 \overset{?}{=} 0$$

$$0 = 0 \checkmark$$

The solution $x = -\sqrt{2}$ also checks. $\checkmark$

So, the solutions are $x = 5$ and $x = \pm\sqrt{2}$.

(b) $6x^3 - 27x^2 - 54x = 0$

$3x(2x^2 - 9x - 18) = 0$	Factor out common factor.
$3x(2x + 3)(x - 6) = 0$	Factor quadratic factor.
$3x = 0 \Rightarrow x = 0$	Set 1st factor equal to 0.
$2x + 3 = 0 \Rightarrow x = -\frac{3}{2}$	Set 2nd factor equal to 0.
$x - 6 = 0 \Rightarrow x = 6$	Set 3rd factor equal to 0.

Check: $x = 0$

$$6x^3 - 27x^2 - 54x = 0$$

$$6(0)^3 - 27(0)^2 - 54(0) \overset{?}{=} 0$$

$$0 = 0 \checkmark$$

$x = -\frac{3}{2}$

$$6x^3 - 27x^2 - 54x = 0$$

$$6\left(-\tfrac{3}{2}\right)^3 - 27\left(-\tfrac{3}{2}\right)^2 - 54\left(-\tfrac{3}{2}\right) \overset{?}{=} 0$$

$$6\left(-\tfrac{27}{8}\right)^3 - 27\left(\tfrac{9}{4}\right)^2 + 27(3) \overset{?}{=} 0$$

$$-\tfrac{81}{4} - \tfrac{243}{4} + 81 \overset{?}{=} 0$$

$$0 = 0 \checkmark$$

$x = 6$

$$6x^3 - 27x^2 - 54x = 0$$

$$6(6)^3 - 27(6)^2 - 54(6) \overset{?}{=} 0$$

$$6(216) - 27(36) - 324 \overset{?}{=} 0$$

$$0 = 0 \checkmark$$

So, the solutions are $x = 0$, $x = -\frac{3}{2}$, and $x = 6$.

12. $-\sqrt{40 - 9x} + 2 = x$ ⠀⠀⠀⠀⠀⠀Write original equation.

⠀⠀$-\sqrt{40 - 9x} = x - 2$ ⠀⠀⠀⠀⠀Isolated radical.

⠀⠀$\left(-\sqrt{40 - 9x}\right)^2 = (x - 2)^2$ ⠀⠀Square each side.

⠀⠀⠀⠀$40 - 9x = x^2 - 4x + 4$ ⠀⠀Simplify.

⠀⠀⠀⠀⠀⠀$0 = x^2 + 5x - 36$ ⠀⠀Write in general form.

⠀⠀⠀⠀⠀⠀$0 = (x - 4)(x + 9)$ ⠀⠀Factor.

⠀⠀⠀$x - 4 = 0 \Rightarrow x = 4$ ⠀⠀Set 1st factor equal to 0.

⠀⠀⠀$x + 9 = 0 \Rightarrow x = -9$ ⠀⠀Set 2nd factor equal to 0.

Check: $x = 4$

$$-\sqrt{40 - 9x} + 2 = x$$

$$-\sqrt{40 - 9(4)} + 2 \overset{?}{=} 4$$

$$-\sqrt{4} + 2 \overset{?}{=} 4$$

$$-2 + 2 \overset{?}{=} 4$$

$$0 \neq 4 \ \text{✗}$$

$x = 4$ is an extraneous solution.

$x = -9$

$$-\sqrt{40 - 9(-9)} + 2 \overset{?}{=} -9$$

$$-\sqrt{121} + 2 \overset{?}{=} -9$$

$$-11 + 2 \overset{?}{=} -9$$

$$-9 \overset{?}{=} -9 \ \checkmark$$

So, the only solution is $x = -9$.

13. $(x - 5)^{2/3} = 16$ ⠀⠀⠀⠀⠀⠀Write original equation.

⠀$\sqrt[3]{(x - 5)^2} = 16$ ⠀⠀⠀⠀⠀Rewrite in radical form.

⠀⠀$(x - 5)^2 = 4096$ ⠀⠀⠀⠀Cube each side.

⠀⠀⠀$x - 5 = \pm 64$ ⠀⠀⠀⠀Extract square roots.

⠀⠀⠀⠀$x = 5 \pm 64$ ⠀⠀⠀⠀Add 5 to each side.

⠀⠀⠀⠀$x = -59, x = 69$

Check: $x = -59$ ⠀⠀⠀⠀⠀⠀⠀⠀$x = 69$

$$(x - 5)^{2/3} = 16 \qquad\qquad (x - 5)^{2/3} = 16$$

$$(-59 - 5)^{2/3} \overset{?}{=} 16 \qquad (69 - 5)^{2/3} \overset{?}{=} 16$$

$$(-64)^{2/3} \overset{?}{=} 16 \qquad\qquad (64)^{2/3} \overset{?}{=} 16$$

$$(-4)^2 \overset{?}{=} 16 \qquad\qquad\quad (4)^2 \overset{?}{=} 16$$

$$16 = 16 \ \checkmark \qquad\qquad\quad 16 = 16 \ \checkmark$$

So, the solutions are $x = -59$ and $x = 69$.

14. $\left|x^2 + 4x\right| = 7x + 18$

First Equation

$$x^2 + 4x = 7x + 18 \qquad \text{Use positive expression.}$$
$$x^2 - 3x - 18 = 0 \qquad \text{Write in general form.}$$
$$(x + 3)(x - 6) = 0 \qquad \text{Factor.}$$
$$x + 3 = 0 \Rightarrow x = -3 \qquad \text{Set 1st factor equal to 0.}$$
$$x - 6 = 0 \Rightarrow x = 6 \qquad \text{Set 2nd factor equal to 0.}$$

Second Equation

$$-\left(x^2 + 4x\right) = 7x + 18 \qquad \text{Use negative expression.}$$
$$-x^2 - 4x = 7x + 18 \qquad \text{Distributive Property}$$
$$0 = x^2 + 11x + 18 \qquad \text{Write in general form.}$$

Use the Quadratic equation to solve the equation $0 = x^2 + 11x + 18$.

$$x = \frac{-b \pm \sqrt{b^2 - 4ac}}{2a}$$

$$x = \frac{-11 \pm \sqrt{11^2 - 4(1)(18)}}{2(1)}$$

$$x = \frac{-11 \pm \sqrt{49}}{2}$$

$$x = -2, -9$$

The possible solutions are $x = -9$, $x = -3$, $x = -2$, and $x = 6$.

Because $x = -9$ and $x = -3$ are extraneous, the only solutions are $x = -2$ and $x = 6$.

Checkpoints for Section P.3

1.

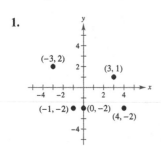

2. To sketch a scatter plot of the data shown in the table, first draw a vertical axis to represent the number of employees E (in thousands) and a horizontal axis to represent the year. Then plot the resulting points. Note that the break in the *t*-axis indicates that the numbers through 2004 have been omitted.

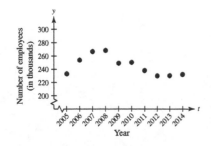

3. Let $(x_1, y_1) = (3, 1)$ and $(x_2, y_2) = (-3, 0)$.

Then apply the Distance Formula.

$$d = \sqrt{(x_2 - x_1)^2 + (y_2 - y_1)^2}$$
$$= \sqrt{(-3 - 3)^2 + (0 - 1)^2}$$
$$= \sqrt{(-6)^2 + (-1)^2}$$
$$= \sqrt{36 + 1}$$
$$= \sqrt{37}$$
$$\approx 6.08$$

So, the distance between the points is about 6.08 units.

4. The three points are plotted in the figure.

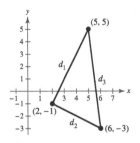

Using the Distance Formula, the lengths of the three sides are as follows.

$$d_1 = \sqrt{(5-2)^2 + (5-(-1))^2}$$

$$= \sqrt{3^2 + 6^2}$$

$$= \sqrt{9 + 36}$$

$$= \sqrt{45}$$

$$d_2 = \sqrt{(6-2)^2 + (-3-(-1))^2}$$

$$= \sqrt{4^2 + (-2)^2}$$

$$= \sqrt{16 + 4}$$

$$= \sqrt{20}$$

$$d_3 = \sqrt{(6-5)^2 + (-3-5)^2}$$

$$= \sqrt{(1)^2 + (-8)^2}$$

$$= \sqrt{1 + 64}$$

$$= \sqrt{65}$$

Because $(d_1)^2 + (d_2)^2 = 45 + 20 = \left(\sqrt{65}\right)^2 = (d_3)^2$ you can conclude by the Pythagorean Theorem that the triangle must be a right triangle.

5. Let $(x_1, y_1) = (-2, 8)$ and $(x_2, y_2) = (-4, -0)$.

$$\text{Midpoint} = \left(\frac{x_1 + x_2}{2}, \frac{y_1 + y_2}{2}\right)$$

$$= \left(\frac{-2 + 4}{2}, \frac{8 + (-10)}{2}\right)$$

$$= \left(\frac{2}{2}, -\frac{2}{2}\right)$$

$$= 1, -1$$

The midpoint of the line segment is $(1, -1)$.

6. You can find the length of the pass by finding the distance between the points $(10, 10)$ and $(25, 32)$.

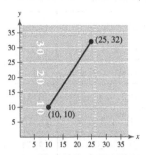

$$d = \sqrt{(x_2 - x_1)^2 + (y_2 - y_1)^2}$$

$$= \sqrt{(25 - 10)^2 + (32 - 10)^2}$$

$$= \sqrt{15^2 + 22^2}$$

$$= \sqrt{225 + 484}$$

$$= \sqrt{709}$$

$$\approx 26.6 \text{ years}$$

So, the pass is about 26.6 yards long.

7. Assuming that the annual revenue from Yahoo! Inc. followed a linear pattern, you can estimate the 2013 annual revenue by finding the midpoint of the line segment connecting the points $(2012, 5.0)$ and $(2014, 4.6)$.

$$\text{Midpoint} = \left(\frac{x_1 + x_2}{2}, \frac{y_1 + y_2}{2}\right)$$

$$= \left(\frac{2012 + 2014}{2}, \frac{5.0 + 4.6}{2}\right)$$

$$= (2013, 4.8)$$

Yahoo! Inc
Annual Revenue

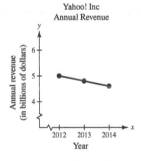

So, you can estimate the annual revenue for Yahoo! Inc. was $4.8 billion in 2013.

8. (a) To graph $y = x^2 + 3$, construct a table of values that consists of several solution points. Then plot the points and connect them with a smooth curve.

x	$y = x^2 + 3$	(x, y)
-2	$y = (-2)^2 + 3 = 7$	$(-2, 7)$
-1	$y = (-1)^2 + 3 = 4$	$(-1, 4)$
0	$y = (0)^2 + 3 = 3$	$(0, 3)$
1	$y = (1)^2 + 3 = 4$	$(1, 4)$
2	$y = 2(2) + 3 = 7$	$(2, 7)$

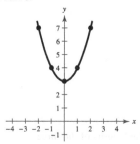

(b) To graph $y = 1 - x^2$, construct a table of values that consists of several solution points. Then plot the points and connect them with a smooth curve.

x	$y = 1 - x^2$	(x, y)
-2	$y = 1 - (-2)^2 = -3$	$(-2, -3)$
-1	$y = 1 - (-1)^2 = 0$	$(-1, 0)$
0	$y = 1 - (0)^2 = 1$	$(0, 1)$
1	$y = 1 - (1)^2 = 0$	$(1, 0)$
2	$y = 1 - (2)^2 = -3$	$(2, -3)$

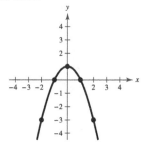

9. To find the x-intercepts, let $y = 0$ and solve for x.

$$y = -x^2 - 5x$$
$$0 = -x^2 - 5x$$
$$0 = -x(x + 5)$$
$$x = 0 \text{ and } x = -5$$

x-intercepts: $(0, 0), (-5, 0)$

To find the y-intercept, let $x = 0$ and solve for y.

$$y = -x^2 - 5x$$
$$y = -(0)^2 - 5(0)$$
$$y = 0$$

y-intercept: $(0, 0)$

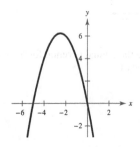

10. x-Axis:

$$y^2 = 6 - x \quad \text{Write original equation.}$$
$$(-y)^2 = 6 - x \quad \text{Replace } y \text{ with } -y.$$
$$y^2 = 6 - x \quad \text{Result is the original equation.}$$

y-Axis:

$$y^2 = 6 - x \qquad \text{Write original equation.}$$
$$y^2 = 6 - (-x) \quad \text{Replace } x \text{ with } -x.$$
$$y^2 = 6 + x \qquad \text{Result is } not \text{ an equivalent equation.}$$

Origin:

$$y^2 = 6 - x \qquad \text{Write original equation.}$$
$$(-y)^2 = 6 - (-x) \quad \text{Replace } y \text{ with } -y \text{ and } x \text{ with } -x.$$
$$y^2 = 6 + x \qquad \text{Result is } not \text{ an equivalent equation.}$$

Of the three tests for symmetry, the only one that is satisfied is the test for x-axis symmetry.

11. Of the three test of symmetry, the only one that is satisfied is the test for y-axis symmetry because

$y = (-x)^2 - 4$ is equivalent to $y = x^2 - 4$. Using

symmetry, you only need to find solution points to the right of the y-axis and then reflect them about the y-axis to obtain the graph.

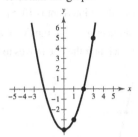

12. The equation $y = |x - 2|$ fails all three tests for

symmetry and consequently its graph is not symmetric with respect to either axis or to the origin. So, construct a table of values. Then plot and connect the points.

x	$y = |x - 2|$	(x, y)
-2	$y = |(-2) - 2| = 4$	$(-2, 4)$
-1	$y = |(-1) - 2| = 3$	$(-1, 3)$
0	$y = |(0) - 2| = 2$	$(0, 2)$
1	$y = |(1) - 2| = 1$	$(1, 1)$
2	$y = |(2) - 2| = 0$	$(3, 0)$
3	$y = |(2) - 2| = 0$	$(3, 1)$
4	$y = |(2) - 2| = 0$	$(4, 2)$

From the table, you can see that the x-intercept is $(2, 0)$

and the y-intercept is $(0, 2)$.

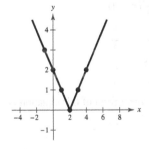

13. The radius of the circle is the distance between

$(1, -2)$ and $(-3, -5)$.

$$r = \sqrt{(x - h)^2 + (y - k)^2}$$
$$= \sqrt{[1 - (-3)]^2 + [-2 - (-5)]^2}$$
$$= \sqrt{4^2 + 3^2}$$
$$= \sqrt{16 + 9}$$
$$= \sqrt{25}$$
$$= 5$$

Using $(h, k) = (-3, -5)$ and $r = 5$, the equation of the circle is

$$(x - h)^2 + (y - k)^2 = r^2$$
$$[x - (-3)]^2 + [y - (-5)]^2 = (5)^2$$
$$(x + 3)^2 + (y + 5)^2 = 25.$$

Checkpoints for Section P.4

1. (a) $y = 3x + 2$: Because $b = 2$, the y-intercept is $(0, 2)$. Because the slope is $m = 3$, the line rises three units for each unit the line moves to the right.

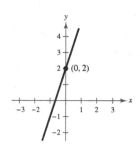

(c) $4x + y = 5$: By writing this equation in slope-intercept form

$$4x + y = 5$$
$$y = -4x + 5$$

you can see that the y-intercept is $(0, 5)$. Because the slope is $m = -4$, the line falls four units for each unit the line moves to the right.

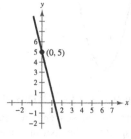

(b) $y = -3$: By writing this equation in the form $y = (0)x - 3$, you can see that the y-intercept is $(0, -3)$ and the slope is $m = 0$. A zero slope implies that the line is horizontal.

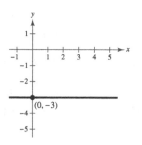

2. (a) The slope of the line passing through $(-5, -6)$ and $(2, 8)$ is $m = \dfrac{8 - (-6)}{2 - (-5)} = \dfrac{14}{7} = 2$.

(b) The slope of the line passing through $(4, 2)$ and $(2, 5)$ is $m = \dfrac{5 - 2}{2 - 4} = \dfrac{3}{-2} = -\dfrac{3}{2}$.

(c) The slope of the line passing through $(0, 0)$ and $(0, -6)$ is $m = \dfrac{-6 - 0}{0 - 0} = \dfrac{-6}{0}$. Because division by 0 is undefined, the slope is undefined and the line is vertical.

(d) The slope of the line passing through $(0, -1)$ and $(3, -1)$ is $m = \dfrac{-1 - (-1)}{3 - 0} = \dfrac{0}{3} = 0$.

3. (a) Use the point-slope form with $m = 2$ and $(x_1, y_1) = (3, -7)$.

$$y - y_1 = m(x - x_1)$$
$$y - (-7) = 2(x - 3)$$
$$y + 7 = 2x - 6$$
$$y = 2x - 13$$

The slope-intercept form of this equation is $y = 2x - 13$.

(b) Use the point-slope form with $m = \dfrac{-2}{3}$ and $(x_1, y_1) = (1, 1)$

$$y - y_1 = m(x - x_1)$$
$$y - 1 = \dfrac{-2}{3}(x - 1)$$
$$y - 1 = \dfrac{-2}{3}x + \dfrac{2}{3}$$
$$y = \dfrac{-2}{3}x + \dfrac{5}{3}$$

The slope-intercept form of this equation is $y = -\dfrac{2}{3}x + \dfrac{5}{3}$.

(c) Use the point-slope form with $m = 0$ and $(x_1, y_1) = (1, 1)$.

$$y - y_1 = m(x - x_1)$$
$$y - 1 = 0(x - 1)$$
$$y - 1 = 0$$
$$y = 1$$

The slope-intercept of the equation is the line $y = 1$.

4. By writing the equation of the given line in slope-intercept form

$$5x - 3y = 8$$
$$-3y = -5x + 8$$
$$y = \frac{5}{3}x - \frac{8}{3}$$

You can see that it has a slope of $m = \frac{5}{3}$.

(a) Any line parallel to the given line must also have a slope of $m = \frac{5}{3}$. So, the line through $(-4, 1)$ that is parallel to the given line has the following equation.

$$y - y_1 = m(x - x_1)$$
$$y - 1 = \frac{5}{3}(x - (-4))$$
$$y - 1 = \frac{5}{3}(x + 4)$$
$$y - 1 = \frac{5}{3}x + \frac{20}{3}$$
$$y = \frac{5}{3}x + \frac{23}{3}$$

(b) Any line perpendicular to the given line must also have a slope of $m = -\frac{3}{5}$ because $-\frac{3}{5}$ is the negative reciprocal of $\frac{5}{3}$. So, the line through $(-4, 1)$ that is perpendicular to the given line has the following equation.

$$y - y_1 = m(x - x_1)$$
$$y - 1 = -\frac{3}{5}(x - (-4))$$
$$y - 1 = -\frac{3}{5}(x + 4)$$
$$y - 1 = -\frac{3}{5}x - \frac{12}{5}$$
$$y = -\frac{3}{5}x - \frac{7}{5}$$

5. The horizontal length of the ramp is 32 feet or $12(32) = 384$ inches.

So, the slope of the ramp is

$$\text{Slope} = \frac{\text{vertical change}}{\text{horizontal change}} = \frac{36 \text{ in.}}{384 \text{ in.}} \approx 0.094.$$

Because $\frac{1}{12} \approx 0.083$, the slope of the ramp is steeper than recommended.

6. The y-intercept $(0, 1500)$ tells you that the value of the copier when it was purchased $(t = 0)$ was \$1500. The slope of $m = -300$ tells you that the value of the copier decreases \$300 each year after it was purchased.

7. Let V represent the value of the machine at the end of year t. The initial value of the machine can be represented by the data point $(0, 24{,}750)$ and the salvage value of the machine can be represented by the data point $(6, 0)$. The slope of the line is

$$m = \frac{0 - 24{,}750}{6 - 0}$$
$$m = -\$4125$$

The slope represents the annual depreciation in dollars per year. Using the point-slope form, you can write the equation of the line as follows

$$V - 24{,}750 = -4125(t - 0)$$
$$V - 24{,}750 = -4125t$$
$$V = -4125t + 24{,}750$$

The equation $V = -4125t + 24{,}750$ represents the book value of the machine each year.

8. Let $t = 3$ represent 2013. Then the two given values are represented by the data points $(3, 6.5)$ and $(4, 7.2)$.

The slope of the line through these points is

$$m = \frac{7.2 - 6.5}{4 - 3} = -0.7.$$

You can find the equation that relates the sales y and the year t to be,

$$y - 6.5 = -0.7(t - 3)$$
$$y - 6.5 = -0.7t + 2.1$$
$$y = -0.7t + 4.4$$

According to this equation, the sales for Foot Locker in 2017 will be

$$y = -0.7(7) + 4.4$$
$$= 4.9 + 4.4$$
$$= \$9.3 \text{ billion.}$$

Checkpoints for Section P.5

1. (a) This mapping *does not* describe y as a function of x. The input value of -1 is assigned or matched to two different y-values.

   (b) The table *does* describe y as a function of x. Each input value is matched with exactly one output value.

2. (a) The given value of $x = -1$ corresponds to two values of y. So y is not a function of x.

   (b) To each value of x there corresponds exactly one value of y. So, y is a function of x.

3. (a) Replacing x with 2 in $f(x) = 10 - 3x^2$ yields the following.

   $$f(2) = 10 - 3(2)^2$$
   $$= 10 - 12$$
   $$= -2$$

   (b) Replacing x with -4 yields the following.

   $$f(-4) = 10 - 3(-4)^2$$
   $$= 10 - 48$$
   $$= -38$$

   (c) Replacing x with $x - 1$ yields the following.

   $$f(x - 1) = 10 - 3(x - 1)^2$$
   $$= 10 - 3(x^2 - 2x + 1)$$
   $$= 10 - 3x^2 + 6x - 3$$
   $$= -3x^2 + 6x + 7$$

4. Because $x = -2$ is less than 0, use $f(x) = x^2 + 1$ to obtain $f(-2) = (-2)^2 + 1 = 4 + 1 = 5$.

   Because $x = 2$ is greater than or equal to 0, use $f(x) = x - 1$ to obtain $f(2) = 2 - 1 = 1$.

   For $x = 3$, use $f(x) = x - 1$ to obtain $f(3) = 3 - 1 = 2$.

5. Set $f(x) = 0$ and solve for x.

   $$f(x) = 0$$
   $$x^2 - 16 = 0$$
   $$(x + 4)(x - 4) = 0$$
   $$x + 4 = 0 \Rightarrow x = -4$$
   $$x - 4 = 0 \Rightarrow x = 4$$

   So, $f(x) = 0$ when $x = -4$ or $x = 4$.

6.
$x^2 + 6x - 24 = 4x - x^2$	Set $f(x)$ equal to $g(x)$.
$2x^2 + 2x - 24 = 0$	Write in general form.
$2(x^2 + x - 12) = 0$	Factor out common factor.
$x^2 + x - 12 = 0$	Divide each side by 2.
$(x + 4)(x - 3) = 0$	Factor.
$x + 4 = 0 \Rightarrow x = -4$	Set 1st factor equal to 0.
$x - 3 = 0 \Rightarrow x = 3$	Set 2nd factor equal to 0.

   So, $f(x) = g(x)$, when $x = -4$ or $x = 3$.

7. (a) The domain of f consists of all first coordinates in the set of ordered pairs.

   Domain $= \{-2, -1, 0, 1, 2\}$

   (b) Excluding x-values that yield zero in the denominator, the domain of g is the set of all real numbers x except $x = 3$.

   (c) Because the function represents the circumference of a circle, the values of the radius r must be positive. So, the domain is the set of real numbers r such that $r > 0$.

   (d) This function is defined only for x-values for which $x - 16 \geq 0$. You can conclude that $x \geq 16$. So, the domain is the interval $[16, \infty)$.

8. Use the formula for surface area of a cylinder,
$$s = 2\pi r^2 + 2\pi rh.$$

(a) $s(r) = 2\pi r^2 + 2\pi r(4r)$

$\qquad = 2\pi r^2 + 8\pi r^2$

$\qquad = 10\pi r^2$

(b) $s(h) = 2\pi\left(\dfrac{h}{4}\right)^2 + 2\pi\left(\dfrac{h}{4}\right)h$

$\qquad = 2\pi\left(\dfrac{h^2}{16}\right) + \dfrac{\pi h^2}{2}$

$\qquad = \dfrac{1}{8}\pi h^2 + \dfrac{1}{2}\pi h^2$

$\qquad = \dfrac{5}{8}\pi h^2$

9. When $x = 60$, you can find the height of the baseball as follows

$f(x) = -0.004x^2 + 0.3x + 6$ $\qquad$ Write original function.

$f(60) = -0.004(60)^2 + 0.3(60) + 6$ $\quad$ Substitute 60 for x.

$\qquad = 9.6$ $\qquad\qquad\qquad\qquad$ Simplify.

When $x = 60$, the height of the ball thrown from the second baseman is 9.6 feet. So, the first baseman cannot catch the baseball without jumping.

10. From 2009 through 2011, use $S(t) = 69t + 151$.

2009: $S(t) = 69(9) + 151 = 772$ fuel stations.

2010: $S(t) = 69(10) + 151 = 841$ fuel stations.

2011: $S(t) = 69(11) + 151 = 910$ fuel stations.

From 2012 through 2015, use $S(t) = 160t - 803$.

2012: $S(12) = 160(12) - 803 = 1117$ fuel stations.

2013: $S(13) = 160(13) - 803 = 1277$ fuel stations.

2014: $S(14) = 160(14) - 803 = 1437$ fuel stations.

2015: $S(15) = 160(15) - 803 = 1597$ fuel stations.

11. $\dfrac{f(x+h) - f(x)}{h} = \dfrac{\left[(x+h)^2 + 2(x+h) - 3\right] - (x^2 + 2x - 3)}{h}$

$\qquad = \dfrac{x^2 + 2xh + h^2 + 2x + 2h - 3 - x^2 - 2x + 3}{h}$

$\qquad = \dfrac{2xh + h^2 + 2h}{h}$

$\qquad = \dfrac{2(2x + h + 2)}{h}$

$\qquad = 2x + h + 2, \quad h \neq 0$

Checkpoints for Section P.6

1. (a) The open dot at $(-3, -6)$ indicates that $x = -3$ is not in the domain of f. So, the domain of f is all real numbers, except $x \neq -3$, or $(-\infty, -3) \cup (-3, \infty)$.

(b) Because $(0, 3)$ is a point on the graph of f, it follows that $f(0) = 3$. Similarly, because the point $(3, -6)$ is a point on the graph of f, it follows that $f(3) = -6$.

(c) Because the graph of f does not extend above $f(0) = 3$, the range of f is the interval $(-\infty, 3]$.

2.

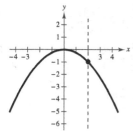

This *is* a graph of y as a function of x, because every vertical line intersects the graph at most once. That is, for a particular input x, there is at most one output y.

3. To find the zeros of a function, set the function equal to zero, and solve for the independent variable.

(a) $2x^2 + 13x - 24 = 0$ Set $f(x)$ equal to 0.

      $(2x - 3)(x + 8) = 0$ Factor.

             $2x - 3 = 0 \Rightarrow x = \dfrac{3}{2}$ Set 1st factor equal to 0.

             $x + 8 = 0 \Rightarrow x = -8$ Set 2nd factor equal to 0.

The zeros of f are $x = \dfrac{3}{2}$ and $x = -8$. The graph of f has $\left(\dfrac{3}{2}, 0\right)$ and $(-8, 0)$ as its x-intercepts.

(b) $\sqrt{t - 25} = 0$ Set $g(t)$ equal to 0.

    $\left(\sqrt{t - 25}\right)^2 = (0)^2$ Square each side.

        $t - 25 = 0$ Simplify.

             $t = 25$ Add 25 to each side.

The zero of g is $t = 25$. The graph of g has $(25, 0)$ as its t-intercept.

(c) $\dfrac{x^2 - 2}{x - 1} = 0$ Set $h(x)$ equal to zero.

    $(x - 1)\left(\dfrac{x^2 - 2}{x - 1}\right) = (x - 1)(0)$ Multiply each side by $x - 1$.

          $x^2 - 2 = 0$ Simplify.

               $x^2 = 2$ Add 2 to each side.

                $x = \pm\sqrt{2}$ Extract square roots.

The zeros of h are $x = \pm\sqrt{2}$. The graph of h has $\left(\sqrt{2}, 0\right)$ and $\left(-\sqrt{2}, 0\right)$ as its x-intercepts.

4.

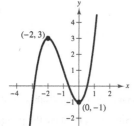

This function is increasing on the interval $(-\infty, -2)$, decreasing on the interval $(-2, 0)$, and increasing on the interval $(0, \infty)$.

5.

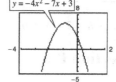

By using the zoom and the trace features or the maximum feature of a graphing utility, you can determine that the function has a relative maximum at the point $\left(-\dfrac{7}{8}, \dfrac{97}{16}\right)$ or $(-0.875, 6.0625)$.

6. (a) The average rate of change of f from
$x_1 = -3$ to $x_2 = -2$ is

$$\frac{f(x_2) - f(x_1)}{x_2 - x_1} = \frac{f(-2) - f(-3)}{-2 - (-3)}$$

$$= \frac{0 - 3}{1} = -3.$$

(b) The average rate of change of f from
$x_1 = -2$ to $x_2 = 0$ is

$$\frac{f(x_2) - f(x_1)}{x_2 - x_1} = \frac{f(0) - f(-2)}{0 - (-2)}$$

$$= \frac{0 - 0}{2} = 0.$$

7. (a) The average speed of the car from $t_1 = 0$ to
$t_2 = 1$ second is

$$\frac{s(t_2) - s(t_1)}{t_2 - t_1} = \frac{20 - 0}{1 - 0} = 20 \text{ feet per second.}$$

(b) The average speed of the car from $t_1 = 1$ to $t_2 = 4$
seconds is

$$\frac{s(t_2) - s(t_1)}{t_2 - t_1} = \frac{160 - 20}{4 - 1}$$

$$= \frac{140}{3}$$

$$\approx 46.7 \text{ feet per second.}$$

8. (a) The function $f(x) = 5 - 3x$ is neither odd nor even
because $f(-x) \neq -f(x)$ and $f(-x) \neq f(x)$ as
follows.

$$f(-x) = 5 - 3(-x)$$

$$= 5 + 3x \neq -f(x) \qquad \text{not odd}$$

$$\neq f(x) \qquad \text{not even}$$

So, the graph of f is not symmetric to the origin nor
the y-axis.

(b) The function $g(x) = x^4 - x^2 - 1$ is even because
$g(-x) = g(x)$ as follows.

$$g(-x) = (-x)^4 - (-x)^2 - 1$$

$$= x^4 - x^2 - 1$$

$$= g(x)$$

So, the graph of g is symmetric to the y-axis.

(c) The function $h(x) = 2x^3 + 3x$ is odd because
$h(-x) = -h(x)$,

$$h(-x) = 2(-x)^3 + 3(-x)$$

$$= -2x^3 - 3x$$

$$= -(2x^3 + 3x)$$

$$= -h(x)$$

So, the graph of h is symmetric to the origin.

Checkpoints for Section P.7

1. To find the equation of the line that passes through the
points $(x_1, y_1) = (-2, 6)$ and $(x_2, y_2) = (4, -4)$, first
find the slope of the line.

$$m = \frac{y_2 - y_1}{x_2 - x_1} = \frac{-9 - 6}{4 - (-2)} = \frac{-15}{6} = \frac{-5}{2}$$

Next, use the point-slope form of the equation of the line.

$$y - y_1 = m(x - x_1) \qquad \text{Point-slope form}$$

$$y - 6 = -\frac{5}{2}\left[x - (-2)\right] \qquad \text{Substitute } x_1, y_1 \text{ and } m.$$

$$y - 6 = -\frac{5}{2}(x + 2) \qquad \text{Simplify.}$$

$$y - 6 = -\frac{5}{2}x - 5 \qquad \text{Simplify.}$$

$$y = -\frac{5}{2}x + 1 \qquad \text{Simplify.}$$

$$f(x) = -\frac{5}{2}x + 1 \qquad \text{Function notation}$$

2. For $x = -\frac{3}{2}$, $f\left(-\frac{3}{2}\right) = \left[\!\left[-\frac{3}{2} + 2\right]\!\right]$

$$= \left[\!\left[\frac{1}{2}\right]\!\right]$$

$$= 0$$

Since the greatest integer $\leq \frac{1}{2}$ is 0, $f\left(-\frac{3}{2}\right) = 0$.

For $x = 1$, $f(1) = \left[\!\left[1 + 2\right]\!\right]$

$$= \left[\!\left[3\right]\!\right]$$

$$= 3$$

Since the greatest integer ≤ 3 is 3, $f(1) = 3$.

For $x = -\frac{5}{2}$, $f\left(-\frac{5}{2}\right) = \left[\!\left[-\frac{5}{2} + 2\right]\!\right]$

$$= \left[\!\left[-\frac{1}{2}\right]\!\right]$$

$$= -1$$

Since the greatest integer $\leq -\frac{1}{2}$ is -1, $f\left(-\frac{5}{2}\right) = -1$.

3. This piecewise-defined function consists of two linear functions. At $x = -4$ and to the left of $x = -4$, the graph is the line $y = -\frac{1}{2}x - 6$, and to the right of $x = -4$ the graph is the line $y = x + 5$. Notice that the point $(-4, -2)$ is a solid dot and $(-4, 1)$ is an open dot. This is because $f(-4) = -2$.

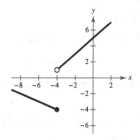

Checkpoints for Section P.8

1. (a) Relative to the graph of $f(x) = x^3$, the graph of
    $h(x) = x^3 + 5$ is an upward shift of five units.

    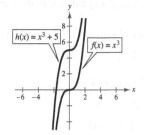

   (b) Relative to the graph of $f(x) = x^3$, the graph of
    $g(x) = (x - 3)^3 + 2$ involves a right shift of three
    units and an upward shift of two units.

    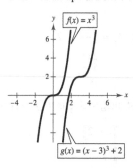

2. The graph of j is a horizontal shift of three units to the
   left *followed by* a reflection in the x-axis of the graph of
   $f(x) = x^4$. So, the equation for j is $j(x) = -(x + 3)^4$.

3. (a) **Algebraic Solution:**

    The graph of g is a reflection of the graph of f in the
    x-axis because

    $$g(x) = -\sqrt{x - 1}$$
    $$= -f(x).$$

    Graphical Solution:

    Graph f and g on the same set of coordinate axes.
    From the graph, you can see that the graph of g is a
    reflection of the graph of f in the x-axis.

    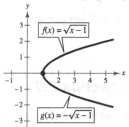

   (b) **Algebraic Solution:**

    The graph of h is a reflection of the graph of f in the
    y-axis because

    $$h(x) = \sqrt{-x - 1}$$
    $$= f(-x).$$

    Graphical Solution:

    Graph f and h on the same set of coordinate axes.
    From the graph, you can see that the graph h is a
    reflection of the graph of f, in the y-axis.

    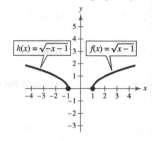

4. (a) Relative to the graph of $f(x) = x^2$, the graph of
$g(x) = 4x^2 = 4f(x)$ is a vertical stretch (each
y-value is multiplied by 4) of the graph of f.

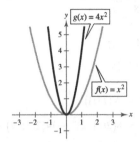

(b) Relative to the graph of $f(x) = x^2$, the graph of
$h(x) = \frac{1}{4}x^2 = \frac{1}{4}f(x)$ is a vertical shrink (each
y-value is multiplied by $\frac{1}{4}$) of the graph of f.

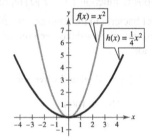

5. (a) Relative to the graph of $f(x) = x^2 + 3$, the graph
of $g(x) = f(2x) = (2x)^2 + 3 = 4x^2 + 3$ is a
horizontal shrink $(c > 1)$ of the graph of f.

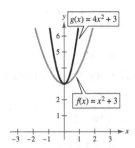

(b) Relative to the graph of $f(x) = x^2 + 3$, the graph
of $h(x) = f\left(\frac{1}{2}x\right) = \left(\frac{1}{2}x\right)^2 + 3 = \frac{1}{4}x^2 + 3$ is a
horizontal stretch $(0 < c < 1)$ of the graph of f.

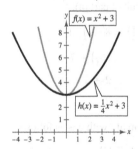

Checkpoints for Section P.9

1. The sum of f and g is
$$(f + g)(x) = f(x) + g(x)$$
$$= (x^2) + (1 - x)$$
$$= x^2 - x + 1.$$
When $x = 2$, the value of this sum is
$$(f + g)(2) = (2)^2 - (2) + 1$$
$$= 3.$$

2. The difference of f and g is
$$(f - g)(x) = f(x) - g(x)$$
$$= (x^2) - (1 - x)$$
$$= x^2 + x - 1.$$
When $x = 3$, the value of the difference is
$$(f - g)(3) = (3)^2 + (3) - 1$$
$$= 11.$$

3. The product of f and g is
$$(f\ g) = f(x)g(x)$$
$$= (x^2)(1 - x)$$
$$= x^2 - x^3$$
$$= -x^3 + x^2.$$
When $x = 3$, the value of the product is
$$(f\ g)(3) = -(3)^3 - (3)^2$$
$$= -27 + 9$$
$$= -18.$$

4. The quotient of f and g is
$$\left(\frac{f}{g}\right)(x) = \frac{f(x)}{g(x)} = \frac{\sqrt{x - 3}}{\sqrt{16 - x^2}}.$$
The quotient of g and f is
$$\left(\frac{g}{f}\right)(x) = \frac{g(x)}{f(x)} = \frac{\sqrt{16 - x^2}}{\sqrt{x - 3}}.$$
The domain of f is $[3, \infty)$ and the domain of g is $[-4, 4]$.
The intersection of these two domains is $[3, 4]$. So, the
domain of f/g is $[3, 4)$ and the domain of g/f is $(3, 4]$.

5. (a) The composition of f with g is as follows.

$$(f \circ g)(x) = f(g(x))$$
$$= f(4x^2 + 1)$$
$$= 2(4x^2 + 1) + 5$$
$$= 8x^2 + 2 + 5$$
$$= 8x^2 + 7$$

(b) The composition of g with f is as follows.

$$(g \circ f)(x) = g(f(x))$$
$$= g(2x + 5)$$
$$= 4(2x + 5)^2 + 1$$
$$= 4(4x^2 + 20x + 25) + 1$$
$$= 16x^2 + 80x + 100 + 1$$
$$= 16x^2 + 80x + 101$$

(c) Use the result of part (a).

$$(f \circ g)\left(-\tfrac{1}{2}\right) = 8\left(-\tfrac{1}{2}\right)^2 + 7$$
$$= 8\left(\tfrac{1}{4}\right) + 7$$
$$= 2 + 7$$
$$= 9$$

6. The composition of f with g is as follows.

$$(f \circ g)(x) = f(g(x))$$
$$= f(x^2 + 4)$$
$$= \sqrt{x^2 + 4}$$

The domain of f is $[0, \infty)$ and the domain of g is the set of all real numbers. The range of g is $[4, \infty)$, which is in the range of f, $[0, \infty)$. Therefore the domain of $f \circ g$ is all real numbers.

7. Let the inner function to be $g(x) = 8 - x$ and the outer function to be $f(x) = \dfrac{\sqrt[3]{x}}{5}$.

$$h(x) = \frac{\sqrt[3]{8 - x}}{5}$$
$$= f(8 - x)$$
$$= f(g(x))$$

8. (a) $(N \circ T)(t) = N(T(t))$

$$= 8(2t + 2)^2 - 14(2t + 2) + 200$$
$$= 8(4t^2 + 8t + 4) - 28t - 28 + 200$$
$$= 32t^2 + 64t + 32 - 28t - 28 + 200$$
$$= 32t^2 + 36t + 204$$

The composite function $(N \circ T)(t)$ represents the number of bacteria in the food as a function of the amount of time the food has been out of refrigeration.

(b) Let $(N \circ T)(t) = 1000$ and solve for t.

$$32t^2 + 36t + 204 = 1000$$
$$32t^2 + 36t - 796 = 0$$
$$4(8t^2 + 9t - 199) = 0$$
$$8t^2 + 9t - 199 = 0$$

Use the quadratic formula:

$$t = \frac{-9 \pm \sqrt{(9)^2 - 4(8)(-199)}}{2(8)}$$
$$= \frac{-9 \pm \sqrt{6449}}{16}$$

$t \approx 4.5$ and $t \approx -5.6$.

Using $t \approx 4.5$ hours, the bacteria count reaches approximately 1000 about 4.5 hours after the food is removed from the refrigerator.

Checkpoints for Section P.10

1. The function f multiplies each input by $\tfrac{1}{5}$. To "undo" this function, you need to multiply each input by 5.

So, the inverse function of $f(x) = \tfrac{1}{5}x$ is $f^{-1}(x) = 5x$.

To verify this, show that

$$f(f^{-1}(x)) = x \text{ and } f^{-1}(f(x)) = x.$$

$$f(f^{-1}(x)) = f(5x) = \tfrac{1}{5}(5x) = x$$

$$f^{-1}(f(x)) = f^{-1}\left(\tfrac{1}{5}x\right) = 5\left(\tfrac{1}{5}x\right) = x$$

So, the inverse function of $f(x) = \tfrac{1}{5}x$ is $f^{-1}(x) = 5x$.

2. By forming the composition of f and g, you have

$$f\big(g(x)\big) = f(7x + 4) = \frac{(7x + 4) - 4}{7} = \frac{7x}{7} = x.$$

So, it appears that g is the inverse function of f. To confirm this, form the composition of g and f.

$$g\big(f(x)\big) = g\!\left(\frac{x - 4}{7}\right) = 7\!\left(\frac{x - 4}{7}\right) + 4 = x - 4 + 4 = x$$

By forming the composition of f and h, you can see that h is *not* the inverse function of f, since the result is not the identity function x.

$$f\big(h(x)\big) = f\!\left(\frac{7}{x - 4}\right) = \frac{\left(\dfrac{7}{x - 4}\right) - 4}{7} = \frac{23 - 4x}{7(x - 4)} \neq x$$

So, g is the inverse function of f.

3. First, sketch the graphs of $f(x) = 4x - 1$ and $g(x) = \dfrac{1}{4}(x + 1)$ as shown. You can see that they appear to be reflections of

each other in the line $y = x$. This reflective property can be tested using a few points and the fact that if the point (a, b) is on

the graph of f then the point (b, a) is on the graph of g.

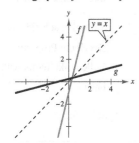

Graph of $f(x) = 4x - 1$	Graph of $g(x) = \frac{1}{4}(x + 1)$
$(-1, -5)$	$(-5, -1)$
$(0, -1)$	$(-1, 0)$
$(1, 3)$	$(3, 1)$
$(2, 7)$	$(7, 2)$

4. First, sketch the graphs of $f(x) = x^2 + 1,\ x \geq 0$ and $g(x) = \sqrt{x - 1}$ as shown. You can see that they appear to be reflections of each other in the line $y = x$. This reflective property can be tested using a few points and the fact that if the point (a, b) is on the graph of f then the point (b, a) is on the graph of g.

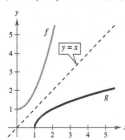

Graph of $f(x) = x^2 + 1,\ x \geq 0$	Graph of $g(x) = \sqrt{x - 1}$
$(0, 1)$	$(1, 0)$
$(1, 2)$	$(2, 1)$
$(2, 5)$	$(5, 2)$
$(3, 10)$	$(10, 3)$

5. (a) The graph of $f(x) = \frac{1}{2}(3 - x)$ is shown.

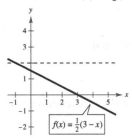

Because no horizontal line intersects the graph of f at more than one point, f is a one-to-one function and *does* have an inverse function.

(b) The graph of $f(x) = |x|$ is shown.

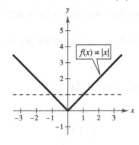

Because it is possible to find a horizontal line that intersects the graph of f at more than one point, f is *not* a one-to-one function and *does not* have an inverse function.

6. The graph of $f(x) = \dfrac{5 - 3x}{x + 2}$ is shown.

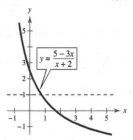

This graph passes the Horizontal Line Test. So, you know f is one-to-one and has an inverse function.

$$f(x) = \frac{5 - 3x}{x + 2} \qquad \text{Write original function.}$$

$$y = \frac{5 - 3x}{x + 2} \qquad \text{Replace } f(x) \text{ with } y.$$

$$x = \frac{5 - 3y}{y + 2} \qquad \text{Interchange } x \text{ and } y.$$

$$x(y + 2) = 5 - 3y \qquad \text{Multiply each side by } y + 2.$$

$$xy + 2x = 5 - 3y \qquad \text{Distribute Property}$$

$$xy + 3y = 5 - 2x \qquad \text{Collect like terms with } y.$$

$$y(x + 3) = 5 - 2x \qquad \text{Factor.}$$

$$y = \frac{5 - 2x}{x + 3} \qquad \text{Solve for } y.$$

$$f^{-1}(x) = \frac{5 - 2x}{x + 3} \qquad \text{Replace } y \text{ with } f^{-1}(x).$$

7. The graph of $f(x) = \sqrt[3]{10 + x}$ is shown.

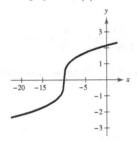

Because this graph passes the Horizontal Line Test, you know that f is one-to-one and has an inverse function.

$$f(x) = \sqrt[3]{10 + x}$$
$$y = \sqrt[3]{10 + x}$$
$$x = \sqrt[3]{10 + y}$$
$$x^3 = 10 + y$$
$$y = x^3 - 10$$
$$f^{-1}(x) = x^3 - 10$$

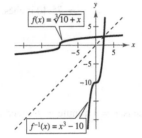

The graphs of f and f^{-1} are reflections of each other in the line $y = x$. So, the inverse of $f(x) = \sqrt[3]{10 + x}$ is $f^{-1}(x) = x^3 - 10$.

To verify, check that $f(f^{-1}(x)) = x$ and $f^{-1}(f(x)) = x$.

$$f(f^{-1}(x)) = f(x^3 - 10) \qquad\qquad f^{-1}(f(x)) = f^{-1}(\sqrt[3]{10 + x})$$
$$= \sqrt[3]{10 + (x^3 - 10)} \qquad\qquad\qquad = (\sqrt[3]{10 + x})^3 - 10$$
$$= \sqrt[3]{x^3} \qquad\qquad\qquad\qquad\qquad = 10 + x - 10$$
$$= x \qquad\qquad\qquad\qquad\qquad\qquad = x$$

Chapter 1

Checkpoints for Section 1.1

1. (a) *Sample answers:* $\dfrac{9\pi}{4} - 4\pi = -\dfrac{7\pi}{4}$

$\dfrac{9\pi}{4} - 2\pi = \dfrac{\pi}{4}$

(b) *Sample answers:* $\dfrac{-\pi}{3} + 2\pi = \dfrac{5\pi}{3}$

$\dfrac{-\pi}{3} - 2\pi = -\dfrac{7\pi}{3}$

2. (a) $\dfrac{\pi}{2} - \dfrac{\pi}{6} = \dfrac{3\pi}{6} - \dfrac{\pi}{6} = \dfrac{2\pi}{6} = \dfrac{\pi}{3}$

The complement of $\dfrac{\pi}{6}$ is $\dfrac{\pi}{3}$.

$\pi - \dfrac{\pi}{6} = \dfrac{6\pi}{6} - \dfrac{\pi}{6} = \dfrac{5\pi}{6}$

The supplement of $\dfrac{\pi}{6}$ is $\dfrac{5\pi}{6}$.

(b) Because $\dfrac{5\pi}{6}$ is greater than $\dfrac{\pi}{2}$, it has no complement.

$\pi - \dfrac{5\pi}{6} = \dfrac{6\pi}{6} - \dfrac{5\pi}{6} = \dfrac{\pi}{6}$

The supplement of $\dfrac{5\pi}{6}$ is $\dfrac{\pi}{6}$.

3. (a) $60° = \left(60 \text{ deg}\right)\left(\dfrac{\pi \text{ rad}}{180 \text{ deg}}\right) = \dfrac{\pi}{3}$ radians

(b) $320° = \left(320 \text{ deg}\right)\left(\dfrac{\pi \text{ rad}}{180 \text{ deg}}\right) = \dfrac{16\pi}{9}$ radians

4. (a) $\dfrac{\pi}{6} = \left(\dfrac{\pi}{6} \text{ rad}\right)\left(\dfrac{180 \text{ deg}}{\pi \text{ rad}}\right) = 30°$

(b) $\dfrac{5\pi}{3} = \left(\dfrac{5\pi}{3} \text{ rad}\right)\left(\dfrac{180 \text{ deg}}{\pi \text{ rad}}\right) = 300°$

5. To use the formula $s = r\theta$ first convert 160° to radian measure.

$$160° = \left(160 \text{ deg}\right)\left(\dfrac{\pi \text{ rad}}{180 \text{ deg}}\right) = \dfrac{8\pi}{9} \text{ radians}$$

Then, using a radius of $r = 27$ inches, you can find the arc length to be

$$s = r\theta$$
$$= (27)\left(\dfrac{8\pi}{9}\right)$$
$$= 24\pi$$
$$\approx 75.40 \text{ inches.}$$

6. In one revolution, the arc length traveled is

$$s = 2\pi r$$
$$= 2\pi(8)$$
$$= 16\pi \text{ centimeters.}$$

The time required for the second hand to travel this distance is $t = 1$ minute $= 60$ seconds.

So, the linear speed of the tip of the second hand is

$$\text{Linear speed} = \dfrac{s}{t}$$
$$= \dfrac{16\pi \text{ centimeters}}{60 \text{ seconds}}$$
$$\approx 0.838 \text{ centimeters per second.}$$

7. (a) Because each revolution generates 2π radians, it follows that the saw blade turns $(2400)(2\pi) = 4800\pi$ radians per minute. In other words, the angular speed is

$$\text{Angular speed} = \dfrac{\theta}{t} = \dfrac{4800\pi \text{ radians}}{1 \text{ minute}} = 4800\pi \text{ radians per minute.}$$

(b) The radius is $r = 4$. The linear speed is

$$\text{Linear speed} = \dfrac{s}{t} = \dfrac{r\theta}{t} = \dfrac{(4)(4800\pi) \text{ inches}}{60 \text{ seconds}} = 60.319 \text{ inches per minute.}$$

8. First convert 80° to radian measure as follows.

$$\theta = 80° = \left(80 \text{ deg}\right)\left(\dfrac{\pi \text{ rad}}{180 \text{ deg}}\right) = \dfrac{4\pi}{9} \text{ radians}$$

Then, using $\theta = \dfrac{4\pi}{9}$ and $r = 40$ feet, the area is

$$A = \dfrac{1}{2}r^2\theta \qquad \text{Formula for area of a sector of a circle}$$
$$= \dfrac{1}{2}(40)^2\left(\dfrac{4\pi}{9}\right) \qquad \text{Substitute for } r \text{ and } \theta$$
$$= \dfrac{3200\pi}{9} \qquad \text{Multiply}$$
$$\approx 1117 \text{ square feet.} \qquad \text{Simplify.}$$

Checkpoints for Section 1.2

1. (a) $t = \dfrac{\pi}{2}$ corresponds to the point $(x, y) = (0, 1)$

$\sin \dfrac{\pi}{2} = 1$ $\qquad\qquad$ $\csc \dfrac{\pi}{2} = 1$

$\cos \dfrac{\pi}{2} = 0$ $\qquad\qquad$ $\sec \dfrac{\pi}{2}$ is undefined.

$\tan \dfrac{\pi}{2}$ is undefined. $\qquad$ $\cot \dfrac{\pi}{2} = 0$

(b) $t = 0$ corresponds to the point $(x, y) = (1, 0)$

$\sin 0 = 0$ $\qquad\qquad$ $\csc 0$ is undefined.

$\cos 0 = 1$ $\qquad\qquad$ $\sec 0 = 1$

$\tan 0 = 0$ $\qquad\qquad$ $\cot 0$ is undefined.

(c) $t = -\dfrac{5\pi}{6}$ corresponds to the point $(x, y) = \left(-\dfrac{\sqrt{3}}{2}, -\dfrac{1}{2}\right)$

$\sin\left(-\dfrac{5\pi}{6}\right) = -\dfrac{1}{2}$ $\qquad$ $\csc\left(-\dfrac{5\pi}{6}\right) = -2$

$\cos\left(-\dfrac{5\pi}{6}\right) = -\dfrac{\sqrt{3}}{2}$ $\qquad$ $\sec\left(-\dfrac{5\pi}{6}\right) = -\dfrac{2\sqrt{3}}{3}$

$\tan\left(-\dfrac{5\pi}{6}\right) = \dfrac{\sqrt{3}}{3}$ $\qquad$ $\cot\left(-\dfrac{5\pi}{6}\right) = \sqrt{3}$

(d) $t = -\dfrac{3\pi}{4}$ corresponds to the point $(x, y) = \left(-\dfrac{\sqrt{2}}{2}, -\dfrac{\sqrt{2}}{2}\right)$

$\sin\left(-\dfrac{3\pi}{4}\right) = -\dfrac{\sqrt{2}}{2}$ $\qquad$ $\csc\left(-\dfrac{3\pi}{4}\right) = -\sqrt{2}$

$\cos\left(-\dfrac{3\pi}{4}\right) = -\dfrac{\sqrt{2}}{2}$ $\qquad$ $\sec\left(-\dfrac{3\pi}{4}\right) = -\sqrt{2}$

$\tan\left(-\dfrac{3\pi}{4}\right) = 1$ $\qquad$ $\cot\left(-\dfrac{3\pi}{4}\right) = 1$

2. (a) Because $\dfrac{9\pi}{2} = 4\pi + \dfrac{\pi}{2}$, you have $\cos \dfrac{9\pi}{2} = \cos\left(4\pi + \dfrac{\pi}{2}\right) = \cos \dfrac{\pi}{2} = 0$.

(b) Because $-\dfrac{7\pi}{3} = -2\pi - \dfrac{\pi}{3}$, you have $\sin\left(-\dfrac{7\pi}{3}\right) = \sin\left(-2\pi - \dfrac{\pi}{3}\right) = \sin\left(-\dfrac{\pi}{3}\right) = -\dfrac{\sqrt{3}}{2}$

(c) For $\cos(-t) = 0.3$, $\cos t = 0.3$ because the cosine function is even.

3. (a) 0.78183148

(b) 1.0997502

Checkpoints for Section 1.3

1.

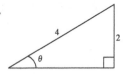

By the Pythagorean Theorem, $(\text{hyp})^2 = (\text{opp})^2 + (\text{adj})^2$, it follows that

$$\text{adj} = \sqrt{4^2 - 2^2} = \sqrt{12} = 2\sqrt{3}.$$

So, the six trigonometric functions of θ are

$$\sin \theta = \frac{\text{opp}}{\text{hyp}} = \frac{2}{4} = \frac{1}{2} \qquad \csc \theta = \frac{\text{hyp}}{\text{opp}} = \frac{4}{2} = 2$$

$$\cos \theta = \frac{\text{adj}}{\text{hyp}} = \frac{2\sqrt{3}}{4} = \frac{\sqrt{3}}{2} \qquad \sec \theta = \frac{\text{hyp}}{\text{adj}} = \frac{4}{2\sqrt{3}} = \frac{2}{\sqrt{3}} = \frac{2\sqrt{3}}{3}$$

$$\tan \theta = \frac{\text{opp}}{\text{adj}} = \frac{2}{2\sqrt{3}} = \frac{1}{\sqrt{3}} = \frac{\sqrt{3}}{3} \qquad \cot \theta = \frac{\text{adj}}{\text{opp}} = \frac{2\sqrt{3}}{2} = \sqrt{3}$$

2.

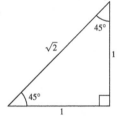

$$\cot 45° = \frac{\text{adj}}{\text{opp}} = \frac{1}{1} = 1$$

$$\sec 45° = \frac{\text{hyp}}{\text{adj}} = \frac{\sqrt{2}}{1} = \sqrt{2}$$

$$\csc 45° = \frac{\text{hyp}}{\text{opp}} = \frac{\sqrt{2}}{1} = \sqrt{2}$$

3.

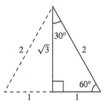

For $\theta = 60°$, you have $\text{adj} = 1$, $\text{opp} = \sqrt{3}$ and $\text{hyp} = 2$.

So, $\tan 60° = \frac{\text{opp}}{\text{adj}} = \frac{\sqrt{3}}{1} = \sqrt{3}.$

For $\theta = 30°$, you have $\text{adj} = \sqrt{3}$, $\text{opp} = 1$ and $\text{hyp} = 2$.

So, $\tan 30° = \frac{\text{opp}}{\text{adj}} = \frac{1}{\sqrt{3}} = \frac{\sqrt{3}}{3}.$

4. $34° \, 30' \, 36'' = 34° + \left(\dfrac{30}{60}\right)° + \left(\dfrac{36}{3600}\right)° = 34.51°$

$\csc(34° \, 30' \, 36'') = \csc 34.51° = \dfrac{1}{\sin 34.51°} \approx 1.765069$

5. (a) To find the value of $\sin \theta$, use the Pythagorean

Identity $\sin^2 \theta + \cos^2 \theta = 1$.

So, you have

$\sin^2 \theta + (0.96)^2 = 1$

$\quad\quad \sin^2 \theta = 1 - (0.96)^2$

$\quad\quad \sin^2 \theta = 1 - 0.9216$

$\quad\quad \sin^2 \theta = 0.0784$

$\quad\quad \sin \theta = \sqrt{0.0784}$

$\quad\quad\quad\quad = 0.28$

(b) Now, knowing the sine and cosine of θ, you can find
the tangent of θ to be

$\tan \theta = \dfrac{\sin \theta}{\cos \theta} = \dfrac{0.28}{0.96} \approx 0.2917.$

6. Given $\tan \theta = 2$

$\cot \theta = \dfrac{1}{\tan \theta} \quad$ Reciprocal Identity.

$\quad\quad = \dfrac{1}{2}$

$\sec^2 \theta = 1 + \tan^2 \theta \quad$ Pythagorean Identity

$\sec^2 \theta = 1 + (2)^2$

$\sec^2 \theta = 5$

$\sec \theta = \sqrt{5}$

Use the definitions of $\cot \theta$ and
$\sec \theta$ and the triangle to check
these results.

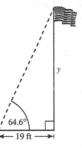

7. (a) $\tan \theta \csc \theta$

$= \left(\dfrac{\sin \theta}{\cos \theta}\right)\left(\dfrac{1}{\sin \theta}\right) \quad$ Use a Quotient Identity and a Reciprocal Identity.

$= \dfrac{1}{\cos \theta} \quad\quad\quad$ Simplify.

$= \sec \theta \quad\quad\quad\quad$ Use a Reciprocal Identity

(b) $(\csc \theta + 1)(\csc \theta - 1) = \csc^2 \theta - \csc \theta + \csc \theta - 1 \quad$ FOIL Method.

$\quad\quad\quad\quad\quad\quad\quad\quad\quad = \csc^2 \theta - 1 \quad\quad\quad\quad$ Simplify.

$\quad\quad\quad\quad\quad\quad\quad\quad\quad = \cot^2 \theta \quad\quad\quad\quad\quad$ Pythagorean identity

8. Drawing a sketch of the situation can assist you in solving the problem.

$\tan 64.6° = \dfrac{\text{opp}}{\text{adj}} = \dfrac{y}{x}$

Where $x = 19$ and y is the height of the flagpole. So, the height of the flagpole is

$y = 19 \tan 64.6°$

$\quad \approx 19(2.106)$

$\quad \approx 40$ feet.

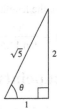

9. From the figure, you can see that the cosine of the angle θ is

$\cos \theta = \dfrac{\text{adj}}{\text{hyp}} = \dfrac{3}{6} = \dfrac{1}{2}.$

You should recognize that $\theta = 60°$.

10. From the figure, you can see that

$$\sin 11.5° = \frac{\text{opp}}{\text{hyp}} = \frac{3.5}{c}.$$

$$\sin 11.5° = \frac{3.5}{c}$$

$$c \sin 11.5° = 3.5$$

$$c = \frac{3.5}{\sin 11.5°}$$

So, the length c of the loading ramp is

$$c = \frac{3.5}{\sin 11.5} \approx \frac{3.5}{0.1994} \approx 17.6 \text{ feet.}$$

Also from the figure, you can see that

$$\tan 11.5° = \frac{\text{opp}}{\text{adj}} = \frac{3.5}{a}.$$

So, the length a of the ramp is

$$a = \frac{3.5}{\tan 11.5°} \approx \frac{3.5}{0.2034} \approx 17.2 \text{ feet.}$$

Checkpoints for Section 1.4

1. Referring to the figure shown, you can see that $x = -2$, $y = 3$ and

$$r = \sqrt{x^2 + y^2} = \sqrt{(-2)^2 + (3)^2} = \sqrt{13}.$$

So, you have the following.

$$\sin \theta = \frac{y}{r} = \frac{3}{\sqrt{13}} = \frac{3\sqrt{13}}{13}$$

$$\cos \theta = \frac{x}{r} = -\frac{2}{\sqrt{13}} = -\frac{2\sqrt{13}}{13}$$

$$\tan \theta = \frac{y}{x} = -\frac{3}{2}$$

2. Note that θ lies in Quadrant II because that is the only quadrant in which the sine is positive and the tangent is negative.

Using $\sin \theta = \frac{4}{5} = \frac{y}{r}$

and the fact that y is positive in Quadrant II, let $y = 4$ and $r = 5$.

$$r = \sqrt{x^2 + y^2}$$

$$5 = \sqrt{x^2 + 4^2}$$

$$25 = x^2 + 16$$

$$9 = x^2$$

$$\pm 3 = x$$

Since x is negative in Quadrant II, $x = -3$.

$$\cos \theta = \frac{x}{r} = -\frac{3}{5} \text{ and } \tan \theta = \frac{y}{x} = \frac{4}{-3} = -\frac{4}{3}$$

3. To begin, choose a point on the terminal side of the angle $\frac{3\pi}{2}$.

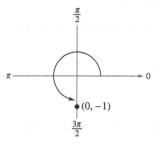

For the point $(0, -1)$, $r = 1$ and you have the following.

$$\sin \frac{3\pi}{2} = \frac{y}{r} = \frac{-1}{1} = -1$$

$$\cot \frac{3\pi}{2} = \frac{x}{y} = \frac{0}{-1} = 0$$

4. (a) Because 213° lies in Quadrant III, the angle it makes with the *x*-axis is $\theta' = 213° - 180° = 33°$.

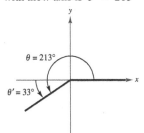

(b) Because $\dfrac{14\pi}{9}$ lies in Quadrant IV, the angle it makes with the *x*-axis is

$$\theta' = 2\pi - \frac{14\pi}{9}$$

$$= \frac{18\pi}{9} - \frac{14\pi}{9}$$

$$= \frac{4\pi}{9}.$$

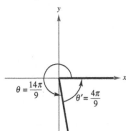

(c) Because $\dfrac{4\pi}{5}$ lies in Quadrant II, the angle it makes with the *x*-axis is

$$\theta' = \pi - \frac{4\pi}{5}$$

$$= \frac{\pi}{5}.$$

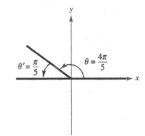

5. (a) Because $\theta = \dfrac{7\pi}{4}$ lies in Quadrant IV, the reference angle is $\theta' = 2\pi - \dfrac{7\pi}{4} = \dfrac{\pi}{4}$ as shown.

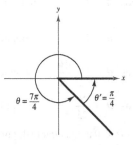

Because the sine is negative in Quadrant IV, you have $\sin \dfrac{7\pi}{4} = (-)\sin \dfrac{\pi}{7}$

$$= -\frac{\sqrt{2}}{2}.$$

(b) Because $-120° + 360° = 240°$, it follows that $-120°$ is coterminal with the third-quadrant angle 240°. So, the reference angle is $\theta' = 240° - 180° = 60°$ as shown.

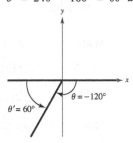

Because the cosine is negative in Quadrant III, you have $\cos(-120°) = (-)\cos 60° = -\dfrac{1}{2}$.

(c) Because $\theta = \dfrac{11\pi}{6}$ lies in Quadrant IV, the reference angle is $\theta' = 2\pi - \dfrac{11\pi}{6} = \dfrac{\pi}{6}$ as shown.

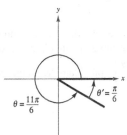

Because the tangent is negative in Quadrant IV, you have

$$\tan \frac{11\pi}{6} = (-)\tan \frac{\pi}{6} = -\frac{\sqrt{3}}{3}.$$

6. (a) Using the Pythagorean Identity $\sin^2 \theta + \cos^2 \theta = 1$, you obtain the following.

$$\sin^2 \theta + \cos^2 \theta = 1 \qquad \text{Write Identity}$$

$$\left(-\frac{4}{5}\right)^2 + \cos^2 \theta = 1 \qquad \text{Substitute } -\frac{4}{5} \text{ for } \sin \theta.$$

$$\frac{16}{25} + \cos^2 \theta = 1 \qquad \text{Simplify.}$$

$$\cos^2 \theta = 1 - \frac{16}{25} \qquad \text{Subtract } \frac{16}{25} \text{ from each side.}$$

$$\cos^2 \theta = \frac{9}{25} \qquad \text{Simplify.}$$

Because $\cos \theta < 0$ in Quadrant III, you can use the negative root to obtain

$$\cos \theta = -\sqrt{\frac{9}{25}} = -\frac{3}{5}.$$

(b) Using the trigonometric identity $\tan \theta = \dfrac{\sin \theta}{\cos \theta}$, you obtain

$$\tan \theta = \frac{-\dfrac{4}{5}}{-\dfrac{3}{5}} \qquad \text{Substitute for } \sin \theta \text{ and } \cos \theta.$$

$$= \frac{4}{3}. \qquad \text{Simplify.}$$

7.

Function	Mode	Calculator Keystrokes	Display
(a) $\tan 119°$	Degree	tan (119) ENTER	-1.8040478
(b) $\csc 5$	Radian	(sin (5)) x^{-1} ENTER	-1.0428352
(c) $\cos \dfrac{\pi}{5}$	Radian	cos (π ÷ 5) ENTER	0.8090170

Checkpoints for Section 1.5

1. Note that $y = 2 \cos x = 2(\cos x)$ indicates that the y-values for the key points will have twice the magnitude of those on the graph of $y = \cos x$. Divide the period 2π into four equal parts to get the key points.

Maximum	Intercept	Minimum	Intercept	Maximum
$(0, 2)$	$\left(\dfrac{\pi}{2}, 0\right)$	$(\pi, -2)$	$\left(\dfrac{3\pi}{2}, 0\right)$	$(2\pi, 2)$

By connecting these key points with a smooth curve and extending the curve in both directions over the interval $\left[-\dfrac{\pi}{2}, \dfrac{9\pi}{2}\right]$, you obtain the graph shown.

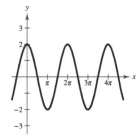

2. (a) Because the amplitude of $y = \dfrac{1}{3}\sin x$ is $\dfrac{1}{3}$, the maximum value is $\dfrac{1}{3}$ and the minimum value is $-\dfrac{1}{3}$.

Divide one cycle, $0 \le x \le 2\pi$, into for equal parts to get the key points.

Intercept	Maximum	Intercept	Minimum	Intercept
$(0, 0)$	$\left(\dfrac{\pi}{2}, \dfrac{1}{3}\right)$	$(\pi, 0)$	$\left(\dfrac{3\pi}{2}, -\dfrac{1}{3}\right)$	$(2\pi, 0)$

(b) A similar analysis shows that the amplitude of $y = 3\sin x$ is 3, and the key points are as follows.

Intercept	Maximum	Intercept	Minimum	Intercept
$(0, 0)$	$\left(\dfrac{\pi}{2}, 3\right)$	$(\pi, 0)$	$\left(\dfrac{3\pi}{2}, -3\right)$	$(2\pi, 0)$

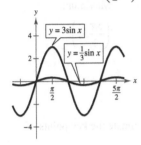

3. The amplitude is 1. Moreover, because $b = \dfrac{1}{3}$, the period is

$$\frac{2\pi}{b} = \frac{2\pi}{\frac{1}{3}} = 6\pi. \quad \text{Substitute } \frac{1}{3} \text{ for } b.$$

Now, divide the period-interval $[0, 6\pi]$ into four equal parts using the values $\dfrac{3\pi}{2}$, 3π, and $\dfrac{9\pi}{2}$ to obtain the key points.

Maximum	Intercept	Minimum	Intercept	Maximum
$(0, 1)$	$\left(\dfrac{3\pi}{2}, 0\right)$	$(3\pi, -1)$	$\left(\dfrac{9\pi}{2}, 0\right)$	$(6\pi, 1)$

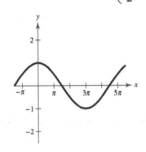

4. Algebraic Solution

The amplitude is 2 and the period is 2π.

By solving the equations

$$x - \frac{\pi}{2} = 0 \Rightarrow x = \frac{\pi}{2}$$

and

$$x - \frac{\pi}{2} = 2\pi \Rightarrow x = \frac{5\pi}{2}$$

you see that the interval $\left[\frac{\pi}{2}, \frac{5\pi}{2}\right]$ corresponds to one cycle of the graph. Dividing this interval into four equal parts produces the key points.

Maximum	Intercept	Minimum	Intercept	Maximum
$\left(\frac{\pi}{2}, 2\right)$	$(\pi, 0)$	$\left(\frac{3\pi}{2}, -2\right)$	$(2\pi, 0)$	$\left(\frac{5\pi}{2}, 2\right)$

Graphical Solution

Use a graphing utility set in *radian* mode to graph $y = 2\cos\left(x - \frac{\pi}{2}\right)$ as shown.

Use the *minimum*, *maximum*, and *zero* or *root* features of the graphing utility to approximate the key points $(1.57, 2)$, $(3.14, 0)$, $(4.71, -2)$, $(6.28, 0)$ and $(7.85, 2)$.

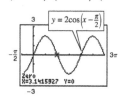

5. The amplitude is $\frac{1}{2}$ and the period is $\frac{2\pi}{b} = \frac{2\pi}{\pi} = 2$.

By solving the equations

$$\pi x + \pi = 0$$
$$\pi x = -\pi$$
$$x = -1$$

and

$$\pi x + \pi = 2\pi$$
$$\pi x = \pi$$
$$x = 1$$

you see that the interval $[-1, 1]$ corresponds to one cycle of the graph. Dividing this into four equal parts produces the key points.

Intercept	Minimum	Intercept	Maximum	Intercept
$(-1, 0)$	$\left(-\frac{1}{2}, -\frac{1}{2}\right)$	$(0, 0)$	$\left(\frac{1}{2}, \frac{1}{2}\right)$	$(1, 0)$

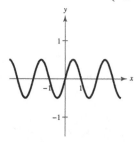

6. The amplitude is 2 and the period is 2π. The key points over the interval $[0, 2\pi]$ are

$$(0, -3), \left(\frac{\pi}{2}, -5\right), (\pi, -7), \left(\frac{3\pi}{2}, -5\right), \text{ and } (2\pi, -3).$$

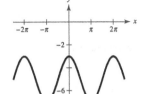

7. Use a sine model of the form $y = a \sin(bt - c) + d$.

The difference between the maximum value and minimum value is twice the amplitude of the function. So, the amplitude is

$$a = \frac{1}{2}\left[(\text{maximum depth}) - (\text{minimum depth})\right]$$

$$= \frac{1}{2}(11.3 - 0.1) = 5.6.$$

The sine function completes one half cycle between the times at which the maximum and minimum depths occur. So, the period p is

$$p = 2\left[(\text{time of min. depth}) - (\text{time of max. depth})\right]$$

$$= 2(10 - 4) = 12$$

which implies that $b = \frac{2\pi}{p} \approx 0.524.$ Because high tide occurs 4 hours after midnight, consider the maximum to be

$$bt - c = \frac{\pi}{2} \approx 1.571.$$

So, $(0.524)(4) - c \approx 1.571$

$$c \approx 0.525.$$

Because the average depth is $\frac{1}{2}(11.3 + 0.1) = 5.7$, it follows that $d = 5.7$. So, you can model the depth with the function

$$y = a \sin(bt - c) + d$$

$$= 5.6 \sin(0.524t - 0.525) + 5.7.$$

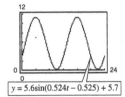

Checkpoints for Section 1.6

1. By solving the equations

$$\frac{x}{4} = -\frac{\pi}{2} \quad \text{and} \quad \frac{x}{4} = \frac{\pi}{2}$$
$$x = -2\pi \qquad\qquad x = 2\pi$$

you can see that two consecutive vertical asymptotes occur at $x = -2\pi$ and $x = 2\pi$. Between these two asymptotes, plot a few points including the x-intercept.

x	-2π	$-\pi$	0	π	2π
$f(x)$	Undef.	-1	0	1	Undef.

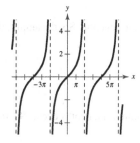

2. By solving the equations

$$2x = -\frac{\pi}{2} \quad \text{and} \quad 2x = \frac{\pi}{2}$$
$$x = -\frac{\pi}{4} \qquad\qquad x = \frac{\pi}{4}$$

you can see that two consecutive vertical asymptotes occur at $x = -\frac{\pi}{4}$ and $x = \frac{\pi}{4}$. Between these two asymptotes, plot a few points including the x-intercept.

x	$-\frac{\pi}{4}$	$-\frac{\pi}{8}$	0	$\frac{\pi}{8}$	$\frac{\pi}{4}$
$\tan 2x$	Undef.	-1	0	1	Undef.

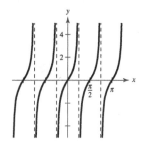

3. By solving the equations

$$\frac{x}{4} = 0 \quad \text{and} \quad \frac{x}{4} = \pi$$
$$x = 0 \qquad\qquad x = 4\pi$$

you can see that two consecutive vertical asymptotes occur at $x = 0$ and $x = 4\pi$. Between these two asymptotes, plot a few points, including the x-intercept.

x	0	π	2π	3π	4π
$\cot \frac{x}{4}$	Undef.	1	0	-1	Undef.

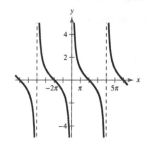

4. Begin by sketching the graph of $y = 2 \sin\left(x + \dfrac{\pi}{2}\right)$. For this function, the amplitude is 2 and the period is 2π. By solving the equations

$$x + \frac{\pi}{2} = 0 \quad \text{and} \quad x + \frac{\pi}{2} = 2\pi$$

$$x = -\frac{\pi}{2} \qquad\qquad x = \frac{3\pi}{2}$$

you can see that one cycle of the sine function corresponds to the interval from $x = -\dfrac{\pi}{2}$ to $x = \dfrac{3\pi}{2}$.

The graph of this sine function is represented by the gray curve. Because the sine function is zero at the midpoint and endpoints of this interval, the corresponding cosecant function

$$y = 2 \csc\left(x + \frac{\pi}{2}\right)$$

$$= 2\left(\cfrac{1}{\sin\left(x + \dfrac{\pi}{2}\right)}\right)$$

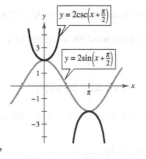

has vertical asymptotes at

$$x = -\frac{\pi}{2},\ x = \frac{\pi}{2},\ x = \frac{3\pi}{2},$$

and so on. The graph of the cosecant curve is represented by the black curve.

5. Begin by sketching the graph of $y = \cos \dfrac{x}{2}$ as indicated by the gray curve. Then, form the graph of $y = \sec \dfrac{x}{2}$ as the black curve. Note that the x-intercepts of $y = \cos \dfrac{x}{2}$, $(\pi, 0)$, $(3\pi, 0)$, $(5\pi, 0)$, … correspond to the vertical asymptotes $x = \pi$, $x = 3\pi$, $x = 5\pi$, … of the graph of $y = \sec \dfrac{x}{2}$. Moreover, notice that the period of $y = \cos \dfrac{x}{2}$ and $y = \sec \dfrac{x}{2}$ is $\dfrac{2\pi}{\frac{1}{2}} = 4\pi$.

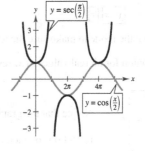

6. Consider $f(x)$ as the product of these two functions

$$y = e^x \quad \text{and} \quad y = \sin 4x$$

each of which has a set of real numbers as its domain. For any real number x, you know that $e^x |\sin 4x| \le e^x$ which means that $-e^x \le e^x \sin 4x \le e^x$.

Furthermore, because

$$f(x) = e^x \sin 4x = \pm e^x \text{ at } x = \frac{\pi}{8} \pm \frac{n\pi}{4} \text{ since}$$

$$\sin 4x = \pm 1 \text{ at } 4x = \frac{\pi}{2} + n\pi$$

and

$$f(x) = e^x \sin 4x = 0 \text{ at } x = \frac{n\pi}{4} \text{ since}$$

$$\sin 4x = 0 \text{ at } 4x = n\pi$$

the graph of f touches the curve $y = -e^x$ or $y = e^x$ at $x = \dfrac{\pi}{8} + \dfrac{n\pi}{4}$ and has x-intercepts at $x = \dfrac{n\pi}{4}$.

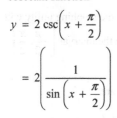

Checkpoints for Section 1.7

1. (a) Because $\sin \dfrac{\pi}{2} = 1$, and $\dfrac{\pi}{2}$ lies in $\left[-\dfrac{\pi}{2}, \dfrac{\pi}{2}\right]$, it follows that $\arcsin 1 = \dfrac{\pi}{2}$.

(b) It is not possible to evaluate $y = \sin^{-1} x$ when $x = -2$ because there is no angle whose sine is -2. Remember that the domain of the inverse sine function is $[-1, 1]$.

2. Using a graphing utility you can graph the three functions with the following keystrokes

Function	Keystroke	Display
$y = \sin x$	$\boxed{y =}$ $\boxed{\text{SIN}}$ $\boxed{(}$ $\boxed{x}$ $\boxed{)}$	$y_1 = \sin(x)$
$y = \arcsin x$	$\boxed{y =}$ $\boxed{\text{2ND}}$ $\boxed{\text{SIN}}$ $\boxed{(}$ $\boxed{x}$ $\boxed{)}$	$y_2 = \sin^{-1}(x)$
$y = x$	$\boxed{y =}$ $\boxed{x}$	$y_3 = x$

Remember to check the mode to make sure the angle measure is set to radian mode. Although the graphing utility will graph the sine function for all real values of x, restrict the viewing window to values of x to be the interval $\left[-\dfrac{\pi}{2}, \dfrac{\pi}{2}\right]$.

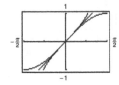

Notice that the graphs of $y_1 = \sin x, \left(-\dfrac{\pi}{2}, \dfrac{\pi}{2}\right)$ and $y_2 = \sin^{-1} x$ are reflections of each other in the line $y_3 = x$. So, g is the inverse of f.

3. Because $\cos \pi = -1$ and π lies in $[0, \pi]$, it follows that $\arccos(-1) = \cos^{-1}(-1) = \pi$.

4.

Function	Mode	Calculator Keystrokes
(a) arctan 4.84	Radian	$\boxed{\text{TAN}^{-1}}$ $\boxed{(}$ 4.84 $\boxed{)}$ $\boxed{\text{ENTER}}$

From the display, it follows that $\arctan 4.84 \approx 1.3670516$.

(b) $\arcsin(-1.1)$ Radian $\boxed{\text{SIN}^{-1}}$ $\boxed{(}$ $\boxed{(-)}$ 1.1 $\boxed{)}$ $\boxed{\text{ENTER}}$

In radian mode the calculator should display an *error* message because the domain of the inverse sine function is $[-1, 1]$.

(c) $\arccos(-0.349)$ Radian $\boxed{\text{COS}^{-1}}$ $\boxed{(}$ $\boxed{(-)}$ 0.349 $\boxed{)}$ $\boxed{\text{ENTER}}$

From the display, it follows that $\arccos(-0.349) \approx 1.9273001$.

5. (a) Because -14 lies in the domain of the arctangent function, the inverse property applies, and you have
$$\tan\left[\tan^{-1}(-14)\right] = -14.$$

(b) In this case, $\dfrac{7\pi}{4}$ does not lie in the range of the arcsine function, $-\dfrac{\pi}{2} \le y \le \dfrac{\pi}{2}$.

However, $\dfrac{7\pi}{4}$ is coterminal with $\dfrac{7\pi}{4} - 2\pi = -\dfrac{\pi}{4}$ which does lie in the range of the arcsine function, and you have

$$\sin^{-1}\left(\sin \dfrac{7\pi}{4}\right) = \sin^{-1}\left[\sin\left(-\dfrac{\pi}{4}\right)\right]$$

$$= -\dfrac{\pi}{4}.$$

(c) Because 0.54 lies in the domain of the arccosine function, the inverse property applies and you have
$$\cos(\arccos 0.54) = 0.54.$$

6. If you let $u = \arctan\left(-\dfrac{3}{4}\right)$, then $\tan u = -\dfrac{3}{4}$. Because the range of the inverse tangent function is the first and fourth quadrants and $\tan u$ is negative, u is a fourth-quadrant angle. You can sketch and label angle u.

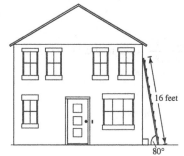

Angie whose tangent is $-\dfrac{3}{4}$.

$\sqrt{4^2 + (-3)^2} = 5$

So, $\cos\left[\arctan\left(-\dfrac{3}{4}\right)\right] = \cos u = \dfrac{4}{5}$.

7. If you let $u = \arctan x$, then $\tan u = x$, where x is any real number. Because $\tan u = \dfrac{\text{opp}}{\text{adj}} = \dfrac{x}{1}$ you can sketch a right triangle with acute angle u as shown. From this triangle, you can convert to algebraic form.

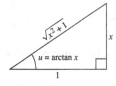

$$\sec(\arctan x) = \sec u$$
$$= \dfrac{\sqrt{x^2+1}}{1}$$
$$= \sqrt{x^2+1}$$

Checkpoints for Section 1.8

1. Because $c = 90°$, it follows that $A + B = 90°$ and $B° = 90° - 20° = 70°$.

To solve for a, use the fact that

$$\tan A = \dfrac{\text{opp}}{\text{adj}} = \dfrac{a}{b} \Rightarrow a = b \tan A.$$

So, $a = 15 \tan 20° \approx 5.46$. Similarly, to solve for c, use the fact that $\cos A = \dfrac{\text{adj}}{\text{hyp}} = \dfrac{b}{c} \Rightarrow c = \dfrac{b}{\cos A}$

So, $c = \dfrac{15}{\cos 20°} \approx 15.96$.

2.

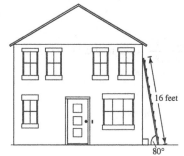

From the equation $\sin A = \dfrac{a}{c}$, it follows that

$a = c \sin A$
$= 16 \sin 80°$
≈ 15.8.

So, the height from the top of the ladder to the ground is about 15.8 feet.

3.

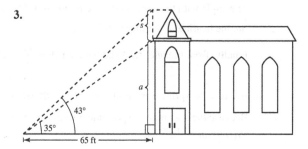

Note that this problem involves two right triangles. For the smaller right triangle, use the fact that

$\tan 35° = \dfrac{a}{65}$ to conclude that the height of the church is $a = 65 \tan 35°$.

For the larger right triangle use the equation

$\tan 43° = \dfrac{a + s}{65}$ to conclude that $a + s = 65 \tan 43°$.

So, the height of the steeple is

$s = 65 \tan 43° - a$
$= 65 \tan 43° - (65 \tan 35°)$
≈ 15.1 feet.

4.

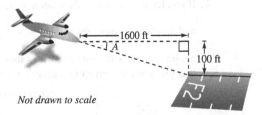

Not drawn to scale

Using the tangent function, you can see that $\tan A = \dfrac{\text{opp}}{\text{adj}} = \dfrac{100}{1600} = 0.0625$

So, the angle of depression is $A = \arctan(0.0625)$ radian

$$\approx 0.06242 \text{ radian}$$

$$\approx 3.58°.$$

5.

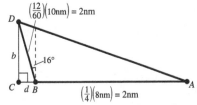

For triangle BCD, you have $B = 90° - 16° = 74°$.

The two sides of this triangle can be determined to be $b = 2 \sin 74°$ and $d = 2 \cos 74°$.

For triangle ACD, you can find angle A as follows.

$$\tan A = \frac{b}{d+2} = \frac{2 \sin 74°}{2 \cos 74° + 2} \approx 0.7535541$$

$A = \arctan A \approx \arctan 0.7535541$ radian $\approx 37°$

The angle with the north south line is $90° - 37° = 53°$.

So, the bearing of the ship is N 53° W.

Finally, from triangle ACD you have $\sin A = \dfrac{b}{c}$, which yields $c = \dfrac{b}{\sin A} = \dfrac{2 \sin 74°}{\sin 37°} \approx 3.2$ nautical miles.

6. Because the spring is at equilibrium $(d = 0)$ when $t = 0$, use the equation $d = a \sin wt$.

Because the maximum displacement from zero is 6 and the period is 3, you have the following.

Amplitude $= |a| = 6$

Period $= \dfrac{2\pi}{w} = 3 \Rightarrow w = \dfrac{2\pi}{3}$

So, an equation of motion is $d = 6 \sin \dfrac{2\pi}{3} t$.

7. Algebraic Solution

The given equation has the form $d = 4 \cos 6\pi t$, with $a \approx 4$ and $w = 6\pi$.

(a) The maximum displacement is given by the amplitude. So, the maximum displacement is 4.

(b) Frequency $= \dfrac{w}{2\pi} = \dfrac{6\pi}{2\pi} = 3$ cycles per unit of time

(c) $d = 4 \cos\big[6\pi(4)\big] = 4 \cos 24\pi = 4(1) = 4$

(d) To find the least positive value of t, for which $d = 0$, solve the equation $4 \cos 6\pi t = 0$.

First divide each side by 4 to obtain $\cos 6\pi t = 0$.

This equation is satisfied when $6\pi t = \dfrac{\pi}{2}, \dfrac{3\pi}{2}, \dfrac{5\pi}{2}, \ldots$.

Divide each of these values by 6π to obtain $t = \dfrac{1}{12}, \dfrac{1}{4}, \dfrac{5}{12}, \ldots$.

So, the least positive value of t is $t = \dfrac{1}{12}$.

Graphical Solution

(a) Use a graphing utility set in radian mode.

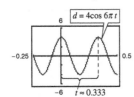

The maximum displacement is from the point of equilibrium $(d = 0)$ is 4.

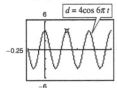

(b) The period is the time for the graph to complete one cycle, which is $t \approx 0.333$. So, the frequency is about

$\dfrac{1}{0.333} \approx 3$ per unit of time.

(c)

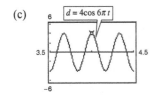

The value of d when $t = 4$ is $d = 4$

(d)

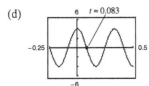

The least positive value of t for which $d = 0$ is $t = \dfrac{1}{2} \approx 0.083$.

Chapter 2

Checkpoints for Section 2.1

1. Using a reciprocal identity, you have $\cot x = \dfrac{1}{\tan x} = \dfrac{1}{\frac{1}{3}} = 3.$

   Using a Pythagorean identity, you have

   $$\sec^2 x = 1 + \tan^2 x = 1 + \left(\frac{1}{3}\right)^2 = 1 + \frac{1}{9} = \frac{10}{9}.$$

   Because $\tan x > 0$ and $\cos x < 0$, you know that the angle x lies in Quadrant III.

   Moreover, because $\sec x$ is negative when x is in Quadrant III, choose the negative root and obtain

   $$\sec x = -\sqrt{\frac{10}{9}} = -\frac{\sqrt{10}}{3}.$$

   Using a reciprocal identity, you have

   $$\cos x = \frac{1}{\sec x} = -\frac{1}{\sqrt{10}/3} = -\frac{3}{\sqrt{10}} = -\frac{3\sqrt{10}}{10}.$$

   Using a quotient identity, you have

   $$\tan x = \frac{\sin x}{\cos x} \Rightarrow \sin x = \cos x \tan x = \left(-\frac{3\sqrt{10}}{10}\right)\left(\frac{1}{3}\right) = -\frac{\sqrt{10}}{10}.$$

   Using a reciprocal identity, you have

   $$\csc x = \frac{1}{\sin x} = -\frac{1}{\sqrt{10}/10} = -\frac{10}{\sqrt{10}} = -\sqrt{10}.$$

   $\sin x = -\dfrac{\sqrt{10}}{10}$ $\qquad\qquad$ $\csc x = -\sqrt{10}$

   $\cos x = -\dfrac{3\sqrt{10}}{10}$ $\qquad\qquad$ $\sec x = -\dfrac{\sqrt{10}}{3}$

   $\tan x = \dfrac{1}{3}$ $\qquad\qquad\qquad$ $\cot x = 3$

2. First factor out a common monomial factor then use a fundamental identity.

   $$\cos^2 x \csc x - \csc x = \csc x\left(\cos^2 x - 1\right) \qquad \text{Factor out a common monomial factor.}$$

   $$= -\csc x\left(1 - \cos^2 x\right) \qquad \text{Factor out } -1.$$

   $$= -\csc x \sin^2 x \qquad\qquad \text{Pythagorean identity}$$

   $$= -\left(\frac{1}{\sin x}\right)\sin^2 x \qquad \text{Reciprocal identity}$$

   $$= -\sin x \qquad\qquad\qquad \text{Multiply.}$$

3. (a) This expression has the form $u^2 - v^2$, which is the difference of two squares. It factors as

   $$1 - \cos^2 \theta = (1 - \cos \theta)(1 + \cos \theta).$$

   (b) This expression has the polynomial form $ax^2 + bx + c$, and it factors as

   $$2\csc^2 \theta - 7\csc \theta + 6 = (2\csc \theta - 3)(\csc \theta - 2).$$

4. Use the identity $\sec^2 x = 1 + \tan^2 x$ to rewrite the expression.

$$\sec^2 x + 3\tan x + 1 = \left(1 + \tan^2 x\right) + 3\tan x + 1 \qquad \text{Pythagorean identity}$$

$$= \tan^2 x + 3\tan x + 2 \qquad \text{Combine like terms.}$$

$$= (\tan x + 2)(\tan x + 1) \qquad \text{Factor.}$$

5. $\csc x - \cos x \cot x = \dfrac{1}{\sin x} - \cos x\left(\dfrac{\cos x}{\sin x}\right) \qquad \text{Quotient and reciprocal identities}$

$$= \dfrac{1}{\sin x} - \dfrac{\cos^2 x}{\sin x} \qquad \text{Multiply.}$$

$$= \dfrac{1 - \cos^2 x}{\sin x} \qquad \text{Add fractions.}$$

$$= \dfrac{\sin^2 x}{\sin x} \qquad \text{Pythagorean identity.}$$

$$= \sin x \qquad \text{Simplify.}$$

6. $\dfrac{1}{1 + \sin \theta} + \dfrac{1}{1 - \sin \theta} = \dfrac{1 - \sin \theta + 1 + \sin \theta}{(1 + \sin \theta)(1 - \sin \theta)} \qquad \text{Add fractions.}$

$$= \dfrac{2}{1 - \sin^2 \theta} \qquad \text{Combine like terms in numerator}$$

$$\text{and multiply factors in denominator.}$$

$$= \dfrac{2}{\cos^2 \theta} \qquad \text{Pythagorean identity}$$

$$= 2\sec^2 \theta \qquad \text{Reciprocal identity}$$

7. $\dfrac{\cos^2 \theta}{1 - \sin \theta} = \dfrac{1 - \sin^2 \theta}{1 - \sin \theta} \qquad \text{Pythagorean identity}$

$$= \dfrac{(1 + \sin \theta)(1 - \sin \theta)}{1 - \sin \theta} \qquad \text{Factor the numerator as the difference of squares.}$$

$$= 1 + \sin \theta \qquad \text{Simplify.}$$

8. Begin by letting $x = 3\sin x$, then you obtain the following

$$\sqrt{9 - x^2} = \sqrt{9 - (3\sin \theta)^2} \qquad \text{Substitute } 3\sin \theta \text{ for } x.$$

$$= \sqrt{9 - 9\sin^2 \theta} \qquad \text{Rule of exponents.}$$

$$= \sqrt{9(1 - \sin^2 \theta)} \qquad \text{Factor.}$$

$$= \sqrt{9\cos^2 \theta} \qquad \text{Pythagorean identity}$$

$$= 3\cos \theta \qquad \cos \theta > 0 \text{ for } 0 < \theta = \dfrac{\pi}{2}$$

Checkpoints for Section 2.2

1. Start with the left side because it is more complicated.

$$\frac{\sin^2 \theta + \cos^2 \theta}{\cos^2 \theta \sec^2 \theta} = \frac{1}{\cos^2 \theta \sec^2 \theta} \qquad \text{Pythagorean identity}$$

$$= \frac{1}{\cos^2 \theta \left(\dfrac{1}{\cos^2 \theta}\right)} \qquad \text{Reciprocal identity}$$

$$= 1 \qquad \text{Simplify.}$$

2. Algebraic Solution

Start with the right side because it is more complicated.

$$\frac{1}{1 - \cos \beta} + \frac{1}{1 + \cos \beta} = \frac{1 + \cos \beta + 1 - \cos \beta}{(1 - \cos \beta)(1 + \cos \beta)} \qquad \text{Add fractions.}$$

$$= \frac{2}{1 - \cos^2 \beta} \qquad \text{Simplify.}$$

$$= \frac{2}{\sin^2 \beta} \qquad \text{Pythagorean identity}$$

$$= 2\csc^2 \beta \qquad \text{Reciprocal identity}$$

Numerical Solution

Use a graphing utility to create a table that shows the values of

$$y_1 = 2\csc^2 x \text{ and } y_2 = \frac{1}{1 - \cos x} + \frac{1}{1 + \cos x} \text{ for different values of } x.$$

X	Y₁	Y₂
-3	100.43	100.43
-2	2.4189	2.4189
-1	2.8246	2.8246
0	ERROR	ERROR
1	2.8246	2.8246
2	2.4189	2.4189
3	100.43	100.43

X=-3

The values for y_1 and y_2 appear to be identical, so the equation appears to be an identity.

3. Algebraic Solution

By applying identities before multiplying, you obtain the following.

$$(\sec^2 x - 1)(\sin^2 x - 1) = (\tan^2 x)(-\cos^2 x) \qquad \text{Pythagorean identities}$$

$$= \left(\frac{\sin x}{\cos x}\right)^2 (-\cos^2 x) \qquad \text{Quotient identity}$$

$$= \left(\frac{\sin^2 x}{\cos^2 x}\right)(-\cos^2 x) \qquad \text{Property of exponents}$$

$$= -\sin^2 x \qquad \text{Multiply.}$$

Graphical Solution

Using a graphing utility, let $y_1 = (\sec^2 x - 1)(\sin^2 x - 1)$ and $y_2 = -\sin^2 x$.

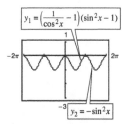

Because the graphs appear to coincide the given equation, $(\sec^2 x - 1)(\sin^2 x - 1) = -\sin^2 x$ appears to be an identity.

4. (a) $\cot x \sec x = \left(\dfrac{\cos x}{\sin x}\right)\left(\dfrac{1}{\cos x}\right)$ Convert the left into sines and cosines.

$\qquad\qquad\quad = \dfrac{1}{\sin x}$ Cancel like factors of cosines.

$\qquad\qquad\quad = \csc x$ Rewrite using reciprocal identities.

  (b) Convert the left into sines and cosines.

$\qquad \csc x - \sin x = \dfrac{1}{\sin x} - \sin x$

$\qquad\qquad\qquad\;\; = \dfrac{1 - \sin^2 x}{\sin x}$ Add fractions.

$\qquad\qquad\qquad\;\; = \dfrac{\cos^2 x}{\sin x}$ Pythagorean identity

$\qquad\qquad\qquad\;\; = \left(\dfrac{\cos x}{1}\right)\left(\dfrac{\cos x}{\sin x}\right)$ Product of fractions

$\qquad\qquad\qquad\;\; = \cos x \cot x$ Quotient identity

5. Algebraic Solution

Begin with the right side and create a monomial denominator by multiplying the numerator and denominator by $1 + \cos x$.

$\qquad \dfrac{\sin x}{1 - \cos x} = \dfrac{\sin x}{1 - \cos x}\left(\dfrac{1 + \cos x}{1 + \cos x}\right)$ Multiply numerator and denomintor by $1 + \cos x$.

$\qquad\qquad\qquad = \dfrac{\sin x + \sin x \cos x}{1 - \cos^2 x}$ Multiply.

$\qquad\qquad\qquad = \dfrac{\sin x + \sin x \cos x}{\sin^2 x}$ Pythagorean identity

$\qquad\qquad\qquad = \dfrac{\sin x}{\sin^2 x} + \dfrac{\sin x \cos x}{\sin^2 x}$ Write as separate functions.

$\qquad\qquad\qquad = \dfrac{1}{\sin x} + \dfrac{\cos x}{\sin x}$ Simplify.

$\qquad\qquad\qquad = \csc x + \cot x$ Identities

Graphical Solution

Using a graphing utility, let $y_1 = \csc x + \cot x$ and $y_2 = \dfrac{\sin x}{1 - \cos x}$.

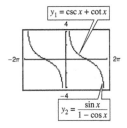

Because the graphs appear to coincide, the given equation appears to be an identity.

6. Algebraic Solution

Working with the left side, you have the following.

$$\frac{\tan^2 \theta}{1 + \sec \theta} = \frac{\sec^2 \theta - 1}{\sec \theta + 1} \qquad \text{Pythagorean identity}$$

$$= \frac{(\sec \theta + 1)(\sec \theta - 1)}{\sec \theta + 1} \qquad \text{Factor.}$$

$$= \sec \theta - 1 \qquad \text{Simplify.}$$

Now, working with the right side, you have the following.

$$\frac{1 - \cos \theta}{\cos \theta} = \frac{1}{\cos \theta} - \frac{\cos \theta}{\cos \theta} \qquad \text{Write as separate fractions.}$$

$$= \sec \theta - 1 \qquad \text{Identity and simplify.}$$

This verifies the identity because both sides are equal to $\sec \theta - 1$.

Numerical Solution

Use a graphing utility to create a table that shows the values of

$$y_1 = \frac{\tan^2 x}{1 + \sec x} \quad \text{and} \quad y_2 = \frac{1 - \cos x}{\cos x} \quad \text{for different values of } x.$$

X	Y₁	Y₂
-.75	.3667	.3667
-.5	.13949	.13949
-.25	.03205	.03205
0	0	0
.25	.03205	.03205
.5	.13949	.13949
.75	.3667	.3667

X=-.75

The values of y_1 and y_2 appear to be identical, so the equation appears to be an identity.

7. (a) $\tan x \sec^2 x - \tan x = \tan x(\sec^2 x - 1)$ Factor.

$$= \tan x \tan^2 x \qquad \text{Pythagorean identity}$$

$$= \tan^3 x \qquad \text{Multiply.}$$

(b) $(\cos^4 x - \cos^6 x)\sin x = \cos^4 x(1 - \cos^2 x)\sin x$ Factor.

$$= \cos^4 x \left(\sin^2 x\right)\sin x \qquad \text{Pythagorean identity}$$

$$= \sin^3 x \cos^4 x \qquad \text{Multiply.}$$

Checkpoints for Section 2.3

1. Begin by isolating $\sin x$ on one side of the equation.

$$\sin x - \sqrt{2} = -\sin x \qquad \text{Write original equation.}$$

$$\sin x + \sin x - \sqrt{2} = 0 \qquad \text{Add } \sin x \text{ to each side.}$$

$$\sin x + \sin x = \sqrt{2} \qquad \text{Add } \sqrt{2} \text{ to each side.}$$

$$2\sin x = \sqrt{2} \qquad \text{Combine like terms.}$$

$$\sin x = \frac{\sqrt{2}}{2} \qquad \text{Divide each side by 2.}$$

Because $\sin x$ has a period of 2π, first find all solutions in the interval $[0, 2\pi)$. These solutions are $x = \dfrac{\pi}{4}$ and $x = \dfrac{3\pi}{4}$.

Finally, add multiples of 2π to each of these solutions to obtain the general form

$$x = \frac{\pi}{4} + 2n\pi \text{ and } x = \frac{3\pi}{4} + 2n\pi \text{ where } n \text{ is an integer.}$$

2. Begin by isolating $\sin x$ on one side of the equation.

$4\sin^2 x - 3 = 0$ Write original equation.

$4\sin^2 x = 3$ Add 3 to each side.

$\sin^2 x = \dfrac{3}{4}$ Divide each side by 4.

$\sin x = \pm\sqrt{\dfrac{3}{4}}$ Extract square roots.

$\sin x = \pm\dfrac{\sqrt{3}}{2}$ Simplify.

Because $\sin x$ has a period of 2π, first find all solutions in the interval $[0, 2\pi)$. These solutions are $x = \dfrac{\pi}{3}$, $x = \dfrac{2\pi}{3}$, $x = \dfrac{4\pi}{3}$, and $x = \dfrac{5\pi}{3}$.

Finally, add multiples of 2π to each of these solutions to obtain the general form.

$x = \dfrac{\pi}{3} + 2n\pi$, $x = \dfrac{2\pi}{3} + 2n\pi$, $x = \dfrac{4\pi}{3} + 2n\pi$, and $x = \dfrac{5\pi}{3} + 2n\pi$ where n is an integer.

3. Begin by collecting all terms on one side of the equation and factoring.

$\sin^2 x = 2\sin x$ Write original equation.

$\sin^2 x - 2\sin x = 0$ Subtract $2\sin x$ from each side.

$\sin x(\sin x - 2) = 0$ Factor.

By setting each of these factors equal to zero, you obtain

$\sin x = 0$ and $\sin x - 2 = 0$

$\sin x = 2$.

In the interval $[0, 2\pi)$, the equation $\sin x = 0$ has solutions $x = 0$ and $x = \pi$. Because $\sin x$ has a period of 2π, you would obtain the general forms $x = 0 + 2n\pi$ and $x = \pi + 2n\pi$ where n is an integer by adding multiples of 2π.

No solution exists for $\sin x = 2$ because 2 is outside the range of the sine function, $[-1, 1]$, so the solutions are of the form $x = n\pi$, where n is an integer. Confirm this graphically by graphing $y = \sin^2 x - 2\sin x$.

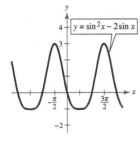

Notice that the x-intercepts occur at $-2\pi, -\pi, 0, \pi, 2\pi$ and so on.

These x-intercepts correspond to the solutions of $\sin^2 x - 2\sin x = 0$.

4. Algebraic Solution:

Treat the equation as a quadratic in $\sin x$ and factor.

$2\sin^2 x - 3\sin x + 1 = 0$ Write original equation.

$(2\sin x - 1)(\sin x - 1) = 0$ Factor.

Setting each factor equal to zero, you obtain the following solutions in the interval $[0, 2\pi)$.

$2\sin x - 1 = 0$ and $\sin x - 1 = 0$

$\sin x = \dfrac{1}{2}$ $\sin x = 1$

$x = \dfrac{\pi}{6}, \dfrac{5\pi}{6}$ $x = \dfrac{\pi}{2}$

Graphical Solution:

The x-intercepts are $x \approx 0.524$, $x = 2.618$, and $x = 1.571$.

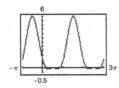

From the graph, you can conclude that the approximate solutions of $2\sin^2 x - 3\sin x + 1 = 0$ in the interval $[0, 2\pi)$ are $x \approx 0.524 = \dfrac{\pi}{6}$, $x \approx 2.618 = \dfrac{5\pi}{6}$, and $x \approx 1.571 = \dfrac{\pi}{2}$.

5. This equation contains both tangent and secant functions. You can rewrite the equation so that it has only tangent functions by using the identity $\sec^2 x = \tan^2 x + 1$.

$3\sec^2 x - 2\tan^2 x - 4 = 0$	Write original equation.
$3(\tan^2 x + 1) - 2\tan^2 x - 4 = 0$	Pythagorean identity
$3\tan^2 x + 3 - 2\tan^2 x - 4 = 0$	Distributive property
$\tan^2 x - 1 = 0$	Simplify.
$\tan^2 x = 1$	Add 1 to each side.
$\tan x = \pm 1$	Extract square roots.

Because $\tan x$ has a period of π, you can find the solutions in the interval $[0, \pi)$ to be $x = \dfrac{\pi}{4}$ and $x = \dfrac{3\pi}{4}$.

The general solution is $x = \dfrac{\pi}{4} + n\pi$ and $x = \dfrac{3\pi}{4} + n\pi$ where n is an integer.

6. Solution It is not clear how to rewrite this equation in terms of a single trigonometric function. Notice what happens when you square each side of the equation.

$\sin x + 1 = \cos x$	Write original equation.
$\sin^2 x + 2\sin x + 1 = \cos^2 x$	Square each side.
$\sin^2 x + 2\sin x + 1 = 1 - \sin^2 x$	Pythagorean identity
$\sin^2 x + \sin^2 x + 2\sin x + 1 - 1 = 0$	Rewrite equation.
$2\sin^2 x + 2\sin x = 0$	Combine like terms.
$2\sin x(\sin x + 1) = 0$	Factor.

Setting each factor equal to zero produces the following.

$2\sin x = 0$	and	$\sin x + 1 = 0$
$\sin x = 0$		$\sin x = -1$
$x = 0, \pi$		$x = \dfrac{3\pi}{2}$

Because you squared the original equation, check for extraneous solutions.

check $x = 0$	$\sin 0 + 1 \overset{?}{=} \cos 0$	Substitute 0 for x.
	$0 + 1 = 1$	Solution checks. ✓
check $x = \pi$	$\sin \pi + 1 \overset{?}{=} \cos \pi$	Substitute π for x.
	$0 + 1 \neq -1$	Solution does not check.
check $x = \dfrac{3\pi}{2}$	$\sin \dfrac{3\pi}{2} + 1 \overset{?}{=} \cos \dfrac{3\pi}{2}$	Substitute $\dfrac{3\pi}{2}$ for x.
	$-1 + 1 = 0$	Solution checks. ✓

of the three possible solutions, $x = \pi$ is extraneous. So, in the interval $[0, 2\pi)$, the two solutions are $x = 0$ and $x = \dfrac{3\pi}{2}$.

7.

$$2\sin 2t - \sqrt{3} = 0 \qquad \text{Write original equation.}$$

$$2\sin 2t = \sqrt{3} \qquad \text{Add } \sqrt{3} \text{ to each side.}$$

$$\sin 2t = \frac{\sqrt{3}}{2} \qquad \text{Divide each side by 2.}$$

In the interval $[0, 2\pi)$, you know that

$2t = \dfrac{\pi}{3}$ and $2t = \dfrac{2\pi}{3}$ are the only solutions.

So, in general you have

$$2t = \frac{\pi}{3} + 2n\pi \text{ and } 2t = \frac{2\pi}{3} + 2n\pi.$$

Dividing these results by 2, you obtain the general solution

$$t = \frac{\pi}{6} + n\pi \text{ and } t = \frac{\pi}{3} + n\pi.$$

8.

$$2\tan \frac{x}{2} - 2 = 0 \qquad \text{Write original equation.}$$

$$2\tan \frac{x}{2} = 2 \qquad \text{Add 2 to each side.}$$

$$\tan \frac{x}{2} = 1 \qquad \text{Divide each side by 2.}$$

In the interval $[0, \pi)$, you know that $\dfrac{x}{2} = \dfrac{\pi}{4}$ is the only solution. So, in general, you have

$$\frac{x}{2} = \frac{\pi}{4} + n\pi.$$

Multiplying this result by 2, you obtain the general solution

$$x = \frac{\pi}{2} + 2n\pi$$

Where n is an integer.

9.

$$4\tan^2 x + 5\tan x - 6 = 0 \qquad \text{Write original equation.}$$

$$(4\tan x - 3)(\tan x + 2) = 0 \qquad \text{Factor.}$$

$$4\tan x - 3 = 0 \text{ and } \tan x + 2 = 0 \qquad \text{Set each factor equal to zero.}$$

$$\tan x = \frac{3}{4} \qquad \tan x = -2$$

$$x = \arctan\left(\frac{3}{4}\right) \qquad x = \arctan(-2) \qquad \text{Use inverse tangent function to solve for } x.$$

These two solutions are in the interval $\left(-\dfrac{\pi}{2}, \dfrac{\pi}{2}\right)$. Recall that the range of the inverse tangent function is $\left(-\dfrac{\pi}{2}, \dfrac{\pi}{2}\right)$.

Finally, because $\tan x$ has a period of π, you add multiples of π to obtain

$$x = \arctan\left(\frac{3}{4}\right) + n\pi \text{ and } x = \arctan(-2) + n\pi$$

where n is an integer.

You can use a calculator to approximate the values of $x = \arctan\left(\dfrac{3}{4}\right) \approx 0.6435$ and $x = \arctan(-2) \approx -1.1071$.

10.

$$\sin^2 x + 2\sin x - 1 = 0 \qquad \text{Write original equation.}$$

$$\sin x = \frac{-2 \pm \sqrt{2^2 - 4(1)(-1)}}{2(1)} \qquad \text{Use the Quadratic Formula to solve for } \sin x.$$

$$\sin x = \frac{-2 \pm \sqrt{8}}{2} \qquad \text{Simplify.}$$

$$\sin x = -1 \pm \sqrt{2}$$

$$x = \arcsin\left(-1 + \sqrt{2}\right) \text{ and } x = \arcsin\left(-1 - \sqrt{2}\right) \qquad \text{Use inverse sine function to solve for } x.$$

Using the solution, $x = \arcsin\left(-1 + \sqrt{2}\right)$: $x \approx 0.4271$ and $x \approx \pi - 0.4271 = 2.7145$. These solutions lie in Quadrant I and Quadrant II.

The solution $x = \arcsin\left(-1 - \sqrt{2}\right)$ is not in the domain of the arcsine function.

11. Start with $S = 6hs + 1.5s^2\left[\left(\sqrt{3} - \cos\theta\right)\big/\sin\theta\right]$ and let $h = 3.2$ inches and $s = 0.75$ inch.

Next graph the function $S = 14.4 + 0.84375\left[\left(\sqrt{3} - \cos\theta\right)\big/\sin\theta\right]$ with a graphing utility.

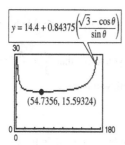

Use the minimum feature to approximate the minimum value on the graph. So, the minimum surface area of 15.6 square inches occurs when $\theta \approx 54.7356°$.

Checkpoints for Section 2.4

1. To find the exact value of $\cos\dfrac{\pi}{12}$, use the fact that

$$\frac{\pi}{12} = \frac{\pi}{3} - \frac{\pi}{4}.$$

The formula for $\cos(u - v)$ yields the following.

$$\cos\frac{\pi}{12} = \cos\left(\frac{\pi}{3} - \frac{\pi}{4}\right)$$

$$= \cos\frac{\pi}{3}\cos\frac{\pi}{4} + \sin\frac{\pi}{3}\sin\frac{\pi}{4}$$

$$= \left(\frac{1}{2}\right)\left(\frac{\sqrt{2}}{2}\right) + \left(\frac{\sqrt{3}}{2}\right)\left(\frac{\sqrt{2}}{2}\right)$$

$$= \frac{\sqrt{2}}{4} + \frac{\sqrt{6}}{4}$$

$$= \frac{\sqrt{2} + \sqrt{6}}{4}$$

2. Using the fact that $75° = 30° + 45°$, together with the formula for $\sin(u + v)$, you obtain the following.

$$\sin 75° = \sin(30° + 45°)$$

$$= \sin 30° \cos 45° + \cos 30° \sin 45°$$

$$= \left(\frac{1}{2}\right)\left(\frac{\sqrt{2}}{2}\right) + \left(\frac{\sqrt{3}}{2}\right)\left(\frac{\sqrt{2}}{2}\right)$$

$$= \frac{\sqrt{2}}{4} + \frac{\sqrt{6}}{4}$$

$$= \frac{\sqrt{2} + \sqrt{6}}{4}$$

3. Because $\sin u = \dfrac{12}{13}$ and u is in Quadrant I,

$$\cos u = \frac{5}{13} \text{ as shown.}$$

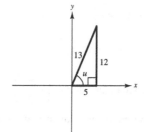

Because $\cos v = -\dfrac{3}{5}$ and v is in Quadrant II, $\sin v = \dfrac{4}{5}$ as shown.

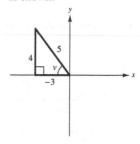

You can find $\cos(u + v)$ as follows.

$$\cos(u + v) = \cos u \cos v - \sin u \sin v$$

$$= \left(\frac{5}{13}\right)\left(-\frac{3}{5}\right) - \left(\frac{12}{13}\right)\left(\frac{4}{5}\right)$$

$$= -\frac{63}{65}$$

4. This expression fits the formula for $\sin(u + v)$. The figures show the angles $u = \arctan 1$ and $v = \arccos x$.

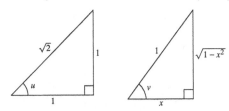

$$\sin(u + v) = \sin u \cos v + \cos u \sin v$$

$$= \sin(\arctan 1) \cos(\arccos x) + \cos(\arctan 1) \sin(\arccos x)$$

$$= \left(\frac{1}{\sqrt{2}}\right)(x) + \left(\frac{1}{\sqrt{2}}\right)\left(\sqrt{1 - x^2}\right)$$

$$= \frac{x}{\sqrt{2}} + \frac{\sqrt{1 - x^2}}{\sqrt{2}}$$

$$= \frac{x + \sqrt{1 - x^2}}{\sqrt{2}}$$

5. Using the formula for $\sin(u - v)$, you have

$$\sin\left(x - \frac{\pi}{2}\right) = \sin x \cos \frac{\pi}{2} - \cos x \sin \frac{\pi}{2}$$

$$= (\sin x)(0) - (\cos x)(1)$$

$$= -\cos x.$$

6. (a) Using the formula for $\sin(u - v) = \sin u \cos v - \cos u \sin v$, you have

$$\sin\left(\frac{3\pi}{2} - \theta\right) = \sin \frac{3\pi}{2} \cos \theta - \cos \frac{3\pi}{2} \sin\theta$$

$$= (-1)(\cos \theta) - (0)(\sin \theta)$$

$$= -\cos \theta.$$

(b) Using the formula for $\tan(u - v) = \dfrac{\tan u - \tan v}{1 + \tan u \tan v}$, you have

$$\tan\left(\theta - \frac{\pi}{4}\right) = \frac{\tan \theta - \tan \frac{\pi}{4}}{1 + \tan \theta \tan \frac{\pi}{4}}$$

$$= \frac{\tan \theta - 1}{1 + (\tan \theta)(1)}$$

$$= \frac{\tan \theta - 1}{1 + \tan \theta}.$$

7. Algebraic Solution

Using sum and difference formulas, rewrite the equation.

$$\sin\left(x + \frac{\pi}{2}\right) + \sin\left(x - \frac{3\pi}{2}\right) = 1$$

$$\sin x \cos\frac{\pi}{2} + \cos x \sin\frac{\pi}{2} + \sin x \cos\frac{3\pi}{2} - \cos x \sin\frac{3\pi}{2} = 1$$

$$(\sin x)(0) + (\cos x)(1) + (\sin x)(0) - (\cos x)(-1) = 1$$

$$\cos x + \cos x = 1$$

$$2\cos x = 1$$

$$\cos x = \frac{1}{2}$$

So, the only solutions in the interval $[0, 2\pi)$ are $x = \frac{\pi}{3}$ and $x = \frac{5\pi}{3}$.

Graphical Solution

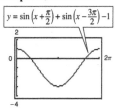

$$y = \sin\left(x + \frac{\pi}{2}\right) + \sin\left(x - \frac{3\pi}{2}\right) - 1$$

The x-intercepts are $x \approx 1.047198$ and $x \approx 5.235988$.

From the above figure, you can conclude that the approximate solutions in the interval $[0, 2\pi)$ are

$$x \approx 1.047198 = \frac{\pi}{3} \text{ and } x \approx 5.235988 = \frac{5\pi}{3}.$$

8. Using the formula for $\cos(x + h)$, you have the following.

$$\frac{\cos(x + h) - \cos x}{h} = \frac{\cos x \cos h - \sin x \sin h - \cos x}{h}$$

$$= \frac{\cos x \cos h - \cos x - \sin x \sin h}{h}$$

$$= \frac{\cos x(\cos h - 1) - \sin x \sin h}{h}$$

$$= \cos x\left(\frac{\cos h - 1}{h}\right) - \sin x\left(\frac{\sin h}{h}\right)$$

Checkpoints for Section 2.5

1. Begin by rewriting the equation so that it involves functions of x (rather than $2x$). Then factor and solve.

$\cos 2x + \cos x = 0$	Write original equation.
$2\cos^2 x - 1 + \cos x = 0$	Double-angle formula
$2\cos^2 x + \cos x - 1 = 0$	Rearrange terms
$(2\cos x - 1)(\cos x + 1) = 0$	Factor.
$2\cos x - 1 = 0 \qquad \cos x + 1 = 0$	Set factors equal to zero.
$\cos x = \frac{1}{2} \qquad\quad \cos x = -1$	Solve by cos x.
$x = \frac{\pi}{3}, \frac{5\pi}{3} \qquad\quad x = \pi$	Solutions in $[0, 2\pi)$.

So, the general solution is $x = \frac{\pi}{3} + 2n\pi$, $x = \frac{5\pi}{3} + 2n\pi$, and $x = \pi + 2n\pi$ where n is an integer.

2. Begin by drawing the angle θ, $0 < \theta < \dfrac{\pi}{2}$ given $\sin\theta = \dfrac{3}{5}$.

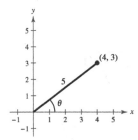

From the sketch, you know that $\sin\theta = \dfrac{y}{r} = \dfrac{3}{5}$.

Because $x = 4$, you know $\sin\theta = \dfrac{3}{5}$, $\cos\theta = \dfrac{4}{5}$, and $\tan\theta = \dfrac{3}{4}$.

Using the double angle formulas, you have the following.

$$\sin 2\theta = 2\sin\theta\cos\theta = 2\left(\frac{3}{5}\right)\left(\frac{4}{5}\right) = \frac{24}{25}$$

$$\cos 2\theta = \cos^2\theta - \sin^2\theta = \left(\frac{4}{5}\right)^2 - \left(\frac{3}{5}\right)^2 = \frac{7}{25}$$

$$\tan 2\theta = \frac{2\tan\theta}{1 - \tan^2\theta} = \frac{2\left(\frac{3}{4}\right)}{1 - \left(\frac{3}{4}\right)^2} = \frac{\frac{3}{2}}{\frac{7}{16}} = \frac{24}{7}$$

3.

$\cos 3x = \cos(2x + x)$	Rewrite $3x$ as sum of $2x$ and x.
$= \cos 2x\cos x - \sin 2x\sin x$	Sum formula
$= (2\cos^2 x - 1)(\cos x) - (2\sin x\cos x)(\sin x)$	Double-angle formulas
$= 2\cos^3 x - \cos x - 2\sin^2 x\cos x$	Distribute property and simplify.
$= 2\cos^3 x - \cos x - 2(1 - \cos^2 x)(\cos x)$	Pythagorean identity
$= 2\cos^3 x - \cos x - 2\cos x + 2\cos^3 x$	Distribute property
$= 4\cos^3 x - 3\cos x$	Simplify.

4. You can make repeated use of power-reducing formulas.

$\tan^4 x = \left(\tan^2 x\right)^2$	Property of exponets.
$= \left(\dfrac{1 - \cos 2x}{1 + \cos 2x}\right)^2$	Power-reducing formula
$= \dfrac{1 - 2\cos 2x + \cos^2 2x}{1 + 2\cos 2x + \cos^2 2x}$	Expand.
$= \dfrac{1 - 2\cos 2x + \left(\dfrac{1 + \cos 4x}{2}\right)}{1 + 2\cos 2x + \left(\dfrac{1 + \cos 4x}{2}\right)}$	Power-reducing formula
$= \dfrac{\dfrac{2 - 4\cos 2x + 1 + \cos 4x}{2}}{\dfrac{2 + 4\cos 2x + 1 + \cos 4x}{2}}$	Simplify.
$= \dfrac{3 - 4\cos 2x + \cos 4x}{3 + 4\cos 2x + \cos 4x}$	Collect like terms, invert, and multiply.

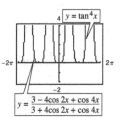

You can use a graphing utility to check this result. Notice that the graphs coincide.

5. Begin by noting 105° is one half of 210°. Then using the half-angle formula for $\cos\left(\dfrac{u}{2}\right)$ and the fact that 105° lies in Quadrant II, you have the following.

$$\cos 105° = -\sqrt{\frac{1 + \cos 210°}{2}} = -\sqrt{\frac{1 + \left(-\dfrac{\sqrt{3}}{2}\right)}{2}} = -\sqrt{\frac{2 - \sqrt{3}}{2}} = -\sqrt{\frac{2 - \sqrt{3}}{4}} = -\frac{\sqrt{2 - \sqrt{3}}}{2}$$

The negative square root is chosen because $\cos\theta$ is negative in Quadrant II.

6. Algebraic Solution

$$\cos^2 x = \sin^2\frac{x}{2} \qquad\qquad \text{Write original equation.}$$

$$\cos^2 x = \left(\pm\sqrt{\frac{1 - \cos x}{2}}\right)^2 \qquad\qquad \text{Half-angle formula}$$

$$\cos^2 x = \frac{1 - \cos x}{2} \qquad\qquad \text{Simplify.}$$

$$2\cos^2 x = 1 - \cos x \qquad\qquad \text{Multiply each side by 2.}$$

$$2\cos^2 x + \cos x - 1 = 0 \qquad\qquad \text{Simplify.}$$

$$(2\cos x - 1)(\cos x + 1) = 0 \qquad\qquad \text{Factor.}$$

$$2\cos x - 1 = 0 \qquad \cos x + 1 = 0 \qquad \text{Set each factor equal to zero.}$$

$$\cos x = \frac{1}{2} \qquad\qquad \cos x = -1 \qquad \text{Solve each equation for } \cos x.$$

$$x = \frac{\pi}{3}, \frac{5\pi}{3} \qquad\qquad x = \pi \qquad \text{Solutions in } [0, 2\pi).$$

The solutions in the interval $[0, 2\pi)$ are $x = \dfrac{\pi}{3}$, $x = \pi$, and $x = \dfrac{5\pi}{3}$.

Graphical Solution

Use a graphing utility to graph $y = \cos^2 x - \sin^2\dfrac{x}{2}$ in the interval $[0, 2\pi)$. Determine the approximate value of the x-intercepts.

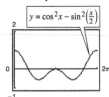

The x-intercepts are $x \approx 1.04720$, $x \approx 3.14159$, and $x \approx 5.23599$.

From the graph, you can conclude that the approximate solutions of $\cos^2 x = \sin\dfrac{2x}{2}$ in the interval $[0, 2\pi)$ are

$$x \approx 1.04720 = \frac{\pi}{3}, \ x \approx 3.14159 = \pi, \text{ and } x \approx 5.23599 = \frac{5\pi}{3}.$$

7. Using the appropriate product-to-sum formula

$$\sin u \cos v = \frac{1}{2}\left[\sin(u + v) + \sin(u - v)\right], \text{ you obtain the following.}$$

$$\sin 5x \cos 3x = \frac{1}{2}\left[\sin(5x + 3x) + \sin(5x - 3x)\right]$$

$$= \frac{1}{2}(\sin 8x + \sin 2x)$$

$$= \frac{1}{2}\sin 8x + \frac{1}{2}\sin 2x$$

8. Using the appropriate sum-to-product formula,

$$\sin u + \sin v = 2\sin\left(\frac{u+v}{2}\right)\cos\left(\frac{u-v}{2}\right),$$ you obtain the following.

$$\sin 195° + \sin 105° = 2\sin\left(\frac{195° + 105°}{2}\right)\cos\left(\frac{195° - 105°}{2}\right)$$

$$= 2\sin 150° \cos 45°$$

$$= 2\left(\frac{1}{2}\right)\left(\frac{\sqrt{2}}{2}\right)$$

$$= \frac{\sqrt{2}}{2}$$

9.

$$\sin 4x - \sin 2x = 0 \quad \text{Write orignal equation.}$$

$$2\cos\left(\frac{4x + 2x}{2}\right)\sin\left(\frac{4x - 2x}{2}\right) = 0 \quad \text{Sum-to-product formula}$$

$$2\cos 3x \sin x = 0 \quad \text{Simplify.}$$

$$\cos 3x \sin x = 0 \quad \text{Divide each side by 2.}$$

$$\cos 3x = 0 \qquad \sin x = 0 \qquad \text{Set each factor equal to zero.}$$

The solutions in the interval $[0, 2\pi)$ are $3x = \dfrac{\pi}{2}, \dfrac{3\pi}{2}$ and $x = 0, \pi$.

The general solutions for the equation $\cos 3x = 0$ are $3x = \dfrac{\pi}{2} + 2n\pi$ and $3x = \dfrac{3\pi}{2} + 2n\pi$.

So, by solving these equations for x, you have $x = \dfrac{\pi}{6} + \dfrac{2n\pi}{3}$ and $x = \dfrac{\pi}{2} + \dfrac{2n\pi}{3}$.

The general solution for the equation $\sin x = 0$ is $x = 0 + 2n\pi$ and $x = \pi + 2n\pi$.

These can be combined as $x = n\pi$.

So, the general solutions to the equation, $\sin 4x - \sin 2x = 0$ are

$$x = \frac{\pi}{6} + \frac{2n\pi}{3}, x = \frac{\pi}{2} + \frac{2n\pi}{3}, \text{ and } x = n\pi \text{ where } n \text{ is an integer.}$$

To verify these solutions you can graph $y = \sin 4x - \sin 2x$ and approximate the x-intercepts.

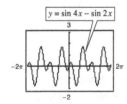

The x-intercepts occur at $0, \dfrac{\pi}{6}, \dfrac{\pi}{2}, \dfrac{5\pi}{6}, \pi, \dfrac{7\pi}{6}, \ldots$

10. Given that a football player can kick a football from ground level with an initial velocity of 80 feet per second, you have the following

$$r = \tfrac{1}{32}v_0^2 \sin 2\theta \qquad \text{Write projectile motion model.}$$

$$r = \tfrac{1}{32}(80)^2 \sin 2\theta \qquad \text{Substitute 80 for } v_0.$$

$$r = 200 \sin 2\theta \qquad \text{Simplify.}$$

Use a graphing utility to graph the model, $r = 200 \sin 2\theta$.

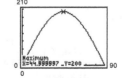

The maximum point on the graph over the interval $(0°, 90°)$ occurs at $\theta = 45°$.

So, the player must kick the football at an angle of $45°$ to yield the maximum horizontal distance of 200 feet.

Chapter 3

Checkpoints for Section 3.1

1. The third angle of the triangle is

$C = 180° - A - B = 180° - 30° - 45° = 105°.$

By the Law of Sines, you have

$$\frac{a}{\sin A} = \frac{b}{\sin B} = \frac{c}{\sin C}.$$

Using $a = 32$ centimeters produces

$b = \frac{a}{\sin A}(\sin B) = \frac{32}{\sin 30°}(\sin 45°) \approx 45.25$ centimeters

and

$c = \frac{a}{\sin A}(\sin C) = \frac{32}{\sin 30°}(\sin 105°) \approx 61.82$ centimeters

2. From the figure, note that $A = 23°$ and $C = 96°$.

So, the third angle is $B = 180° - A - C = 180° - 23° - 96° = 61°.$

By the Law of Sines, you have

$$\frac{h}{\sin A} = \frac{b}{\sin B}$$

$$h = \frac{b}{\sin B}(\sin A) = \frac{30}{\sin 61°}(\sin 23°)$$

$$\approx 13.40$$

So, the height of the tree h is approximately 13.40 meters.

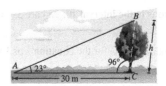

3. Sketch and label the triangle as shown.

$A = 31°, a = 12$ inches, and $b = 5$ inches.

By the Law of Sines, you have

$\frac{\sin B}{b} = \frac{\sin A}{a}$ Reciprocal form

$\sin B = b\left(\frac{\sin A}{a}\right)$ Multiply each side by b.

$\sin B = 5\left(\frac{\sin 31°}{12}\right)$ Substitute for A, a, and b.

$B \approx 12.39°$

Now, you can determine that

$C = 180° - A - B \approx 180° - 31° - 12.39° \approx 136.61°.$

Then, the remaining side is

$$\frac{c}{\sin C} = \frac{a}{\sin A}$$

$$c = \frac{a}{\sin A}(\sin C) \approx \frac{12}{\sin 31°}(\sin 136.61°)$$

$$\approx 16.01 \text{ inches.}$$

4. Sketch and label the triangle.

$a = 4$ feet, $b = 5$ feet and $A = 58°$

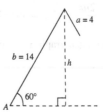

It appears that no triangle is formed.

You can verify this using the Law of Sines.

$$\frac{\sin B}{b} = \frac{\sin A}{a}$$

$$\sin B = b\left(\frac{\sin A}{a}\right)$$

$$\sin B = 14\left(\frac{\sin 60°}{4}\right) \approx 3.0311 > 1$$

This contradicts the fact that $\left|\sin B\right| \le 1$.

So, no triangle can be formed having sides $a = 4$ feet and $b = 5$ feet and angle $A = 58°$.

5. By the Law of Sines, you have

$$\frac{\sin B}{b} = \frac{\sin A}{a}$$

$$\sin B = b\left(\frac{\sin A}{a}\right) = 5\left(\frac{\sin 58°}{4.5}\right) \approx 0.9423.$$

There are two angles $B_1 \approx 70.4°$ and $B_2 \approx 180° - 70.4° = 109.6°$ between 0° and 180° whose sine is approximately 0.9423.

For $B_1 \approx 70.4°$, you obtain the following.

$$C = 180° - A - B_1 = 180° - 58° - 70.4° = 51.6°$$

$$c = \frac{a}{\sin A}(\sin C) = \frac{4.5}{\sin 58°}(\sin 51.6°) \approx 4.16 \text{ feet}$$

For $B_2 = 109.6°$, you obtain the following.

$$C = 180° - A - B_2 = 180° - 58° - 109.6° = 12.4°$$

$$c = \frac{a}{\sin A}(\sin C) = \frac{4.5}{\sin 58°}(\sin 12.4°) \approx 1.14 \text{ feet}$$

The resulting triangles are shown.

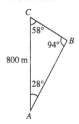

6. Consider $a = 24$ inches, $b = 18$ inches, and angle $C = 80°$ as shown. Then, the area of the triangle is

$$A = \frac{1}{2}ab \sin C = \frac{1}{2}(24)(18)\sin 80° \approx 213 \text{ square yards}.$$

7. Because lines AC and BD are parallel, it follows that $\angle ACB \cong \angle CBD$.

So, triangle ABC has the following measures as shown.

The measure of angle B is $180° - A - C = 180° - 28° - 58° = 94°$.

Using the Law of Sines, $\dfrac{a}{\sin 28°} = \dfrac{b}{\sin 94°} = \dfrac{c}{\sin 58°}$.

Because $b = 800$, $C = \dfrac{800}{\sin 94°}(\sin 58°) \approx 680.1$ meters and $a = \dfrac{800}{\sin 94°}(\sin 28°) \approx 376.5$ meters

The total distance that you swim is approximately

Distance $= 680.1 + 376.5 + 800 = 1856.6$ meters.

Checkpoints for Section 3.2

1.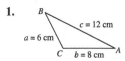

    First, find the angle opposite the longest side – side c in this case. Using the alternative form of the Law of Cosines, you find that

    $$\cos C = \frac{a^2 + b^2 - c^2}{2ab} = \frac{6^2 + 8^2 - 12^2}{2(6)(8)} \approx -0.4583.$$

    Because $\cos C$ is negative, C is an *obtuse* angle given by

    $C \approx \cos^{-1}(-0.4583) \approx 117.28°$.

    At this point, it is simpler to use the Law of Sines to determine angle B.

    $$\sin B = b\left(\frac{\sin C}{c}\right)$$

    $$\sin B = 8\left(\frac{\sin 117.28°}{12}\right) \approx 0.5925$$

    Because C is obtuse and a triangle can have at most one obtuse angle, you know that B must be acute.

    So, $B \approx \sin^{-1}(0.5925) \approx 36.34°$

    So, $A = 180° - B - C \approx 180° - 36.34° - 117.28° \approx 26.38°$.

2. $A = 80°$, $b = 16$ meters and $c = 12$ meters.

    Use the Law of Cosines to find the unknown side a in the figure.

    $a^2 = b^2 + c^2 - 2bc \cos A$

    $a^2 = 16^2 + 12^2 - 2(16)(12) \cos 80°$

    $a^2 \approx 333.3191$

    $a \approx 18.26$

    Use the Law of Sines to find angle B.

    $$\frac{\sin B}{b} = \frac{\sin A}{a}$$

    $$\sin B = b\left(\frac{\sin A}{a}\right)$$

    $$\sin B = 16\left(\frac{\sin 80°}{18.26}\right)$$

    $\sin B \approx 0.863$

    There are two angles between 0° and 180° whose sine is 0.863. The two angles are $B_1 \approx 59.66°$ and $B_2 \approx 180° - 59.66° \approx 120.34°$.

    Because side a is the longest side of the triangle, angle A must be the largest angle, therefore B must be less than 80°. So, $B \approx 59.66°$.

    Therefore, $C = 180° - A - B \approx 180° - 80° - 59.66° \approx 40.34°$.

3.

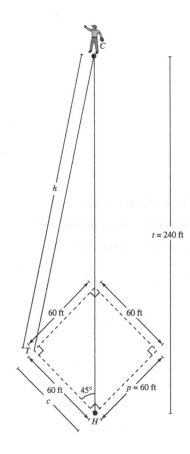

In triangle HCT, $H = 45°$ (line HC bisects the right angle at H), $t = 240$, and $c = 60$.

Using the Law of Cosines for this SAS case, you have

$$h^2 = c^2 + t^2 - 2\,ct \cos H$$

$$h^2 = 60^2 + 240^2 - 2\,(60)(240)\cos 45°$$

$$h^2 \approx 40835.3$$

$$h \approx 202.1$$

So, the center fielder is approximately 202.1 feet from the third base.

4. You have $a = 30$, $b = 56$, and $c = 40$.

So, using the alternative form of the Law of Cosines, you have

$$\cos B = \frac{a^2 + c^2 - b^2}{2ac} = \frac{30^2 + 40^2 - 56^2}{2(30)(40)} = -0.265.$$

So, $B = \cos^{-1}(-0.265) \approx 105.37°$, and thus the bearing from due north from point B to point C is $105.37° - 90° = 15.37°$, or N $15.37°$ E.

5. $a = 5$ inches, $b = 9$ inches and $c = 8$ inches.

Because $s = \dfrac{a + b + c}{2} = \dfrac{5 + 9 + 8}{2} = \dfrac{22}{2} = 11$,

Heron's Area Formula yields

$$\begin{aligned}
\text{Area} &= \sqrt{s(s - a)(s - b)(s - c)} \\
&= \sqrt{11(11 - 5)(11 - 9)(11 - 8)} \\
&= \sqrt{(11)(6)(2)(3)} \\
&= \sqrt{396} \\
&\approx 19.90 \text{ square units.}
\end{aligned}$$

Checkpoints for Section 3.3

1. From the Distance Formula, it follows that $\overrightarrow{PQ}$ and $\overrightarrow{RS}$ have the *same magnitude*.

$$\left\| \overrightarrow{PQ} \right\| = \sqrt{(3 - 0)^2 + (1 - 0)^2} = \sqrt{10}$$

$$\left\| \overrightarrow{RS} \right\| = \sqrt{(5 - 2)^2 + (3 - 2)^2} = \sqrt{10}$$

Moreover, both line segments have the *same direction* because they are both directed toward the upper right on lines having a slope of

$$\frac{1 - 0}{3 - 0} = \frac{3 - 2}{5 - 2} = \frac{1}{3}.$$

Because, $\overrightarrow{PQ}$ and $\overrightarrow{RS}$ have the same magnitude and direction, **u** and **v** are equivalent.

2. **Algebraic Solution**

   Let $P(-2, 3) = (p_1, p_2)$ and $Q(-7, 9) = (q_1, q_2)$.

   Then, the components of $\mathbf{v} = (v_1, v_2)$ are

   $v_1 = q_1 - p_1 = -7 - (-2) = -5$

   $v_2 = q_2 - p_2 = 9 - 3 = 6.$

   So, $\mathbf{v} = \langle -5, 6 \rangle$ and the magnitude of $\mathbf{v}$ is $\|\mathbf{v}\| = \sqrt{(-5)^2 + (6)^2} = \sqrt{61}.$

   Graphical Solution

   Use centimeter graph paper to plot the points $P(-2, 3)$ and $Q(-7, 9)$. Carefully sketch the vector $\mathbf{v}$.

   Use the sketch to find the components of $\mathbf{v} = \langle v_1, v_2 \rangle$. Then use a centimeter ruler to find the magnitude of $\mathbf{v}$.

   The figure shows that the components of $\mathbf{v}$ are $v_1 = -5$ and $v_2 = 6$, so $\mathbf{v} = \langle -5, 6 \rangle$. The figure also shows

   that the magnitude of $\mathbf{v}$ is $\|\mathbf{v}\| = \sqrt{61}.$

   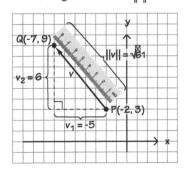

3. (a) The sum of $\mathbf{u}$ and $\mathbf{v}$ is

   $\mathbf{u} + \mathbf{v} = \langle 1, 4 \rangle + \langle 3, 2 \rangle$

   $= \langle 1 + 3, 4 + 2 \rangle$

   $= \langle 4, 6 \rangle.$

   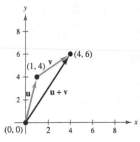

   (b) The difference of $\mathbf{u}$ and $\mathbf{v}$ is

   $\mathbf{u} + \mathbf{v} = \langle 1, 4 \rangle - \langle 3, 2 \rangle$

   $= \langle 1 - 3, 4 - 2 \rangle$

   $= \langle -2, 2 \rangle.$

   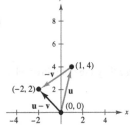

   (c) The difference of $2\mathbf{u}$ and $3\mathbf{v}$ is

   $2\mathbf{u} - 3\mathbf{v} = 2\langle 1, 4 \rangle - 3\langle 3, 2 \rangle$

   $= \langle 2, 8 \rangle - \langle 9, 6 \rangle$

   $= \langle 2 - 9, 8 - 6 \rangle$

   $= \langle -7, 2 \rangle.$

   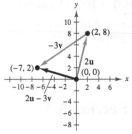

4. $u = \langle 4, -1 \rangle$ and $v = \langle 3, 2 \rangle$

   (a) $\|3u\| = |3|\|u\| = |3|\|\langle 4, -1 \rangle\| = |3|\sqrt{4^2 + (-1)^2} = |3|\sqrt{17} = 3\sqrt{17}$

   (b) $\|-2v\| = |-2|\|v\| = |-2|\|\langle 3, 2 \rangle\| = |-2|\sqrt{3^2 + 2^2} = |-2|\sqrt{13} = 2\sqrt{13}$

   (c) $\|5v\| = |5|\|v\| = |5|\|\langle 3, 2 \rangle\| = |5|\sqrt{3^2 + 2^2} = |5|\sqrt{13} = 5\sqrt{13}$

5. The unit vector in the direction of **v** is

$$\frac{v}{\|v\|} = \frac{\langle 6, -1 \rangle}{\sqrt{(6)^2 + (-1)^2}}$$

$$= \frac{1}{\sqrt{37}}\langle 6, -1 \rangle$$

$$= \left\langle \frac{6}{\sqrt{37}}, -\frac{1}{\sqrt{37}} \right\rangle.$$

This vector has a magnitude of 1 because

$$\sqrt{\left(\frac{6}{\sqrt{37}}\right)^2 + \left(-\frac{1}{\sqrt{37}}\right)^2} = \sqrt{\frac{36}{37} + \frac{1}{37}} = \sqrt{\frac{37}{37}} = 1.$$

6. $w = 6\left(\dfrac{1}{\|v\|}v\right)$

$$= 6\left(\frac{1}{\sqrt{2^2 + (-4)^2}}\langle 2, -4 \rangle\right)$$

$$= \frac{3}{\sqrt{5}}\langle 2, -4 \rangle$$

$$= \left\langle \frac{6}{\sqrt{5}}, -\frac{12}{\sqrt{5}} \right\rangle$$

7. Begin by writing the component form of vector **u**.

$$u = \langle -8 - (-2), 3 - 6 \rangle$$

$$= \langle -6, -3 \rangle$$

$$= -6i - 3j$$

The result is shown graphically.

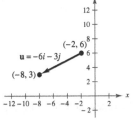

8. Perform the operations in unit vector form.

$$5u - 2v = 5(i - 2j) - 2(-3i + 2j)$$

$$= 5i - 10j + 6i - 4j$$

$$= 11i - 14j$$

9. (a) The direction angle is determined from

$$\tan \theta = \frac{b}{a} = \frac{6}{-6} = -1.$$

Because $v = -6i + 6j$ lies in Quadrant II, θ lies in Quadrant II and its reference angle is

$$\theta' = \left|\arctan(-1)\right| = \left|-\frac{\pi}{4}\right| = 45°.$$

So, it follows that the direction angle is
$$\theta = 180° - 45° = 135°.$$

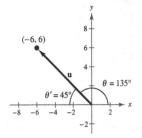

   (b) The direction angle is determined from

$$\tan \theta = \frac{b}{a} = \frac{-4}{-7} = \frac{4}{7}.$$

Because $v = -7i - 4j$ lies in Quadrant III, θ lies in Quadrant III and its reference angle is

$$\theta' = \left|\arctan\left(\frac{4}{7}\right)\right| \approx \left|0.51915 \text{ radian}\right| \approx 29.74°.$$

So, it follows that the direction angle is
$$\theta = 180° + 29.74° = 209.74°.$$

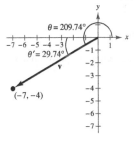

10. The velocity vector **v** has a magnitude of 100 and a direction angle of $\theta = 195°$.

$$\mathbf{v} = \|\mathbf{v}\|(\cos \theta)\mathbf{i} + \|\mathbf{v}\|(\sin \theta)\mathbf{j}$$

$$= 100(\cos 195°)\mathbf{i} + 100(\sin 195°)\mathbf{j}$$

$$\approx 100(-0.9659)\mathbf{i} + 100(-0.2588)\mathbf{j}$$

$$= -96.59\mathbf{i} - 25.88\mathbf{j}$$

$$\approx \langle -96.59, -25.88 \rangle$$

You can check that **v** has a magnitude of 100, as follows.

$$\|\mathbf{v}\| = \sqrt{(-96.59)^2 + (-25.88)^2}$$

$$\approx \sqrt{9999.4025}$$

$$\approx 100$$

11.

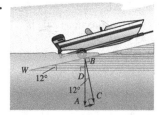

Solution

Based in the figure, you can make the following observations.

$\left\|\overline{BA}\right\|$ = force of gravity = combined weight of boat and trailer

$\left\|\overline{BC}\right\|$ = force against ramp

$\left\|\overline{AC}\right\|$ = force required to move boat up ramp = 500 pounds

By construction, triangles *BWD* and *ABC* are similar. So, angle *ABC* is 12°. In triangle *ABC*, you have

$$\sin 12° = \frac{\left\|\overline{AC}\right\|}{\left\|\overline{BA}\right\|}$$

$$\sin 12° = \frac{500}{\left\|\overline{BA}\right\|}$$

$$\left\|\overline{BA}\right\| = \frac{500}{\sin 12°}$$

$$\left\|\overline{BA}\right\| \approx 2405.$$

So, the combined weight is approximately 2405 pounds. (In the figure, note that $\overline{AC}$ is parallel to the ramp).

12. (a)

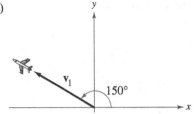

(b)

Solution

Using the figure, the velocity of the airplane (alone) is $v_1 = 450\langle\cos 150°, \sin 150°\rangle = \langle-225\sqrt{3}, 225\rangle$

and the velocity of the wind is $v_2 = 40\langle\cos 60°, \sin 60°\rangle = \langle 20, 20\sqrt{3}\rangle$.

So, the velocity of the airplane (in the wind) is $v = v_1 + v_2 = \langle-225\sqrt{3} + 20, 225 + 20\sqrt{3}\rangle \approx \langle-369.7, 259.6\rangle$

and the resultant speed of the airplane is $\|v\| \approx \sqrt{(-369.7)^2 + (259.6)^2} \approx 451.8$ miles per hour.

Finally, given that θ is the direction angle of the flight path, you have $\tan\theta \approx \dfrac{259.6}{-369.7} \approx 0.7022$

which implies that $\theta \approx 180° - 35.1° = 144.9°$.

So, the true direction of the airplane is approximately $270° + (180° - 144.9°) = 305.1°$.

Checkpoints for Section 3.4

1. (a) $\langle 3, 4\rangle \cdot \langle 2, -3\rangle = 3(2) + 4(-3)$
$= 6 - 12$
$= -6$

(b) $\langle-3, -5\rangle \cdot \langle 1, -8\rangle = (-3)(1) + (-5)(-8)$
$= -3 + 40$
$= 37$

(c) $\langle-6, 5\rangle \cdot \langle 5, 6\rangle = (-6)(5) + (5)(6)$
$= -30 + 30$
$= 0$

2. (a) Begin by finding the dot product of **u** and **v**.
$u \cdot v = \langle 3, 4\rangle \cdot \langle-2, 6\rangle$
$= 3(-2) + 4(6)$
$= -6 + 24$
$= 18$
$(u \cdot v)v = 18\langle-2, 6\rangle$
$= \langle-36, 108\rangle$

(c) Begin by finding the dot product of **v** and **v**.
$v \cdot v = \langle-2, 6\rangle \cdot \langle-2, 6\rangle$
$= -2(-2) + 6(6)$
$= 4 + 36$
$= 40$

(b) Begin by finding **u** + **v**.
$u + v = \langle 3, 4\rangle + \langle-2, 6\rangle$
$= \langle 3 + (-2), 4 + 6\rangle$
$= \langle 1, 10\rangle$
$u \cdot (u + v) = \langle 3, 4\rangle \cdot \langle 1, 10\rangle$
$= 3(1) + 4(10)$
$= 3 + 40$
$= 43$

Because $\|v\|^2 = v \cdot v = 40$, it follows that $\|v\| = \sqrt{v \cdot v}$
$= \sqrt{40}$
$= 2\sqrt{10}$.

3. $\cos \theta = \dfrac{\mathbf{u} \cdot \mathbf{v}}{\|\mathbf{u}\| \, \|\mathbf{v}\|} = \dfrac{\langle 2, 1 \rangle \cdot \langle 1, 3 \rangle}{\|\langle 2, 1 \rangle\| \, \|\langle 1, 3 \rangle\|}$

$= \dfrac{2(1) + 1(3)}{\sqrt{2^2 + 1^2} \, \sqrt{1^2 + 3^2}}$

$= \dfrac{5}{\sqrt{5} \, \sqrt{10}}$

$= \dfrac{5}{\sqrt{50}}$

$= \dfrac{5}{5\sqrt{2}}$

$= \dfrac{1}{\sqrt{2}}$

$= \dfrac{\sqrt{2}}{2}$

This implies that the angle between the two vectors is

$\theta = \cos^{-1}\left(\dfrac{\sqrt{2}}{2}\right) = \dfrac{\pi}{4} = 45°.$

4. Find the dot product of the two vectors.

$\mathbf{u} \cdot \mathbf{v} = \langle 6, 10 \rangle \cdot \left\langle -\dfrac{1}{3}, \dfrac{1}{5} \right\rangle$

$= 6\left(-\dfrac{1}{3}\right) + 10\left(\dfrac{1}{5}\right)$

$= -2 + 2$

$= 0$

Because the dot product is 0, the two vectors are orthogonal.

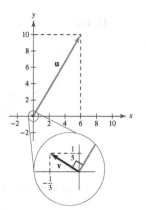

5. The projection of **u** onto **v** is

$\mathbf{w}_1 = \text{proj}_{\mathbf{v}} \, \mathbf{u} = \left(\dfrac{\mathbf{u} \cdot \mathbf{v}}{\|\mathbf{v}\|^2}\right) \mathbf{v}$

$= \left(\dfrac{\langle 3, 4 \rangle \cdot \langle 8, 2 \rangle}{\langle 8, 2 \rangle \cdot \langle 8, 2 \rangle}\right) \langle 8, 2 \rangle$

$= \left(\dfrac{3(8) + 4(2)}{8(8) + 2(2)}\right) \langle 8, 2 \rangle$

$= \left(\dfrac{32}{68}\right) \langle 8, 2 \rangle$

$= \left(\dfrac{8}{17}\right) \langle 8, 2 \rangle$

$= \left\langle \dfrac{64}{17}, \dfrac{16}{17} \right\rangle$

$= \dfrac{1}{17} \langle 64, 16 \rangle.$

The other component, $\mathbf{w}_2$ is

$\mathbf{w}_2 = \mathbf{u} - \mathbf{w}_1 = \langle 3, 4 \rangle - \left\langle \dfrac{64}{17}, \dfrac{16}{17} \right\rangle = \left\langle -\dfrac{13}{17}, \dfrac{52}{17} \right\rangle = \dfrac{1}{17} \langle -13, 52 \rangle.$

So, $\mathbf{u} = \mathbf{w}_1 + \mathbf{w}_2 = \left\langle \dfrac{64}{17}, \dfrac{16}{17} \right\rangle + \left\langle -\dfrac{13}{17}, \dfrac{52}{17} \right\rangle = \langle 3, 4 \rangle.$

6. Solution

Because the force due to gravity is vertical and downward, you can represent the gravitational force by the vector

$$\mathbf{F} = -150\mathbf{j}. \qquad \text{Force due to gravity}$$

To find the force required to keep the cart from rolling down the ramp, project $\mathbf{F}$ onto a unit vector $\mathbf{v}$ in the direction of the ramp, as follows.

$$\mathbf{v} = (\cos 15°)\mathbf{i} + (\sin 15°)\mathbf{j}$$

$$= 0.966\mathbf{i} + 0.259\mathbf{j} \qquad \text{Unit vector along ramp}$$

So, the projection of $\mathbf{F}$ onto $\mathbf{v}$ is

$$\mathbf{w}_1 = \text{proj}_{\mathbf{v}}\mathbf{F}$$

$$= \left(\frac{\mathbf{F} \cdot \mathbf{v}}{\|\mathbf{v}\|^2}\right)\mathbf{v}$$

$$= (\mathbf{F} \cdot \mathbf{v})\mathbf{v} \approx (\langle 0, -150\rangle \cdot \langle 0.966, 0.259\rangle)\mathbf{v}$$

$$\approx (-38.85)\mathbf{v}$$

$$\approx -37.5\mathbf{i} - 10.1\mathbf{j}.$$

The magnitude of this force is approximately 38.8. So, a force of approximately 38.8 pounds is required to keep the cart from rolling down the ramp.

7.

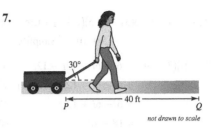

not drawn to scale

Using a projection, you can calculate the work as follows.

$$W = \left\|\text{proj}_{\overrightarrow{PQ}}\mathbf{F}\right\|\left\|\overrightarrow{PQ}\right\|$$

$$= (\cos 30°)\|\mathbf{F}\|\left\|\overrightarrow{PQ}\right\|$$

$$= \frac{\sqrt{3}}{2}(35)(40)$$

$$= 700\sqrt{3}$$

$$\approx 1212.436 \qquad \text{foot-pounds}$$

So, the work done is 1212.436 foot-pounds.

Chapter 4

Checkpoints for Section 4.1

1. (a) $(7 + 3i) + (5 - 4i) = 7 + 3i + 5 - 4i$ Remove parentheses.

                            $= (7 + 5) + (3 - 4)i$ Group like terms.

                            $= 12 - i$ Write in standard form.

    (b) $(3 + 4i) - (5 - 3i) = 3 + 4i - 5 + 3i$ Remove parentheses.

                            $= (3 - 5) + (4 + 3)i$ Group like terms.

                            $= -2 + 7i$ Write in standard form.

    (c) $2i + (-3 - 4i) - (-3 - 3i) = 2i - 3 - 4i + 3 + 3i$ Remove parentheses.

                                 $= (-3 + 3) + (2 - 4 + 3)i$ Group like terms.

                                 $= i$ Write in standard form.

    (d) $(5 - 3i) + (3 + 5i) - (8 + 2i) = 5 - 3i + 3 + 5i - 8 - 2i$ Remove parentheses.

                                    $= (5 + 3 - 8) + (-3 + 5 - 2)i$ Group like terms.

                                    $= 0 + 0i$ Simplify.

                                    $= 0$ Write in standard form.

2. (a) $-5(3 - 2i) = -5(3) - (-5)(2i)$ Distributive Property
$$= -15 + 10i \quad \text{Simplify}$$

(b) $(2 - 4i)(3 + 3i) = 6 + 6i - 12i - 12i^2$ FOIL Method
$$= 6 + 6i - 12i - 12(-1) \quad i^2 = -1$$
$$= 6 - 6i + 12 \quad \text{Simplify.}$$
$$= 18 - 6i \quad \text{Write in standard form.}$$

(c) $(4 + 5i)(4 - 5i) = 4(4 - 5i) + 5i(4 - 5i)$ Distributive Property
$$= 16 - 20i + 20i - 25i^2 \quad \text{Distributive Property}$$
$$= 16 - 20i + 20i - 25(-1) \quad i^2 = -1$$
$$= 16 + 25 \quad \text{Simplify.}$$
$$= 41 \quad \text{Write in standard form.}$$

(d) $(4 + 2i)^2 = (4 + 2i)(4 + 2i)$ Square of a binomial
$$= 4(4 + 2i) + 2i(4 + 2i) \quad \text{Distributive Property}$$
$$= 16 + 8i + 8i + 4i^2 \quad \text{Distributive Property}$$
$$= 16 + 8i + 8i + 4(-1) \quad i^2 = -1$$
$$= (16 - 4) + (8i + 8i) \quad \text{Group like terms.}$$
$$= 12 + 16i \quad \text{Write in standard form.}$$

3. (a) The complex conjugate of $3 + 6i$ is $3 - 6i$.
$$(3 + 6i)(3 - 6i) = (3)^2 - (6i)^2$$
$$= 9 - 36i^2$$
$$= 9 - 36(-1)$$
$$= 45$$

(b) The complex conjugate of $2 - 5i$ is $2 + 5i$.
$$(2 - 5i)(2 + 5i) = (2)^2 - (5i)^2$$
$$= 4 - 25i^2$$
$$= 4 - 25(-1)$$
$$= 29$$

4. $\dfrac{2 + i}{2 - i} = \dfrac{2 + i}{2 - i} \cdot \dfrac{2 + i}{2 + i}$ Multiply numerator and denominator by complex conjugate of the denominator.

$$= \frac{4 + 2i + 2i + i^2}{4 - i^2} \quad \text{Expand.}$$

$$= \frac{4 - 1 + 4i}{4 - (-1)} \quad i^2 = -1$$

$$= \frac{3 + 4i}{5} \quad \text{Simplify.}$$

$$= \frac{3}{5} + \frac{4}{5}i \quad \text{Write in standard form.}$$

5. $\sqrt{-14}\sqrt{-2} = \sqrt{14}i\sqrt{2}i = \sqrt{28}i^2 = 2\sqrt{7}(-1) = -2\sqrt{7}$

6. To solve $8x^2 + 14x + 9 = 0$, use the Quadratic formula

$$x = \frac{-b \pm \sqrt{b^2 - 4ac}}{2a}.$$

$$x = \frac{-14 \pm \sqrt{14^2 - 4(8)(9)}}{2(8)}$$ Substitute $a = 8$, $b = 14$, and $c = 9$.

$$= \frac{-14 \pm \sqrt{-92}}{16}$$ Simplify.

$$= \frac{-14 \pm 2\sqrt{23}i}{16}$$ Write $\sqrt{-92}$ in standard form.

$$= \frac{-14}{16} + \frac{2\sqrt{23}i}{16}$$ Write in standard form.

$$= \frac{-7}{8} \pm \frac{\sqrt{23}i}{8}$$ Simplify.

Checkpoints for Section 4.2

1. The third degree equation

$$x^3 + 9x = 0$$

$$x(x + 3)(x + 3) = 0$$

has exactly three solutions: $x = 0$, $x = -3$, and $x = -3$.

2. (a) $3x^2 + 2x - 1 = 0$, where $a = 3$, $b = 2$, and $c = -1$. So, the discriminant is

$$b^2 - 4ac = (2)^2 - 4(3)(-1) = 4 - (-12) = 16.$$

Because the discriminant is positive, there are two real solutions.

(b) $9x^2 + 6x + 1 = 0$, where $a = 9$, $b = 6$, and $c = 1$.

So, the discriminant is $b^2 - 4ac = (6)^2 - 4(9)(1) = 36 - 36 = 0.$

Because the discriminant is zero, there is one repeated real solution.

(c) $9x^2 + 2x + 1 = 0$, where $a = 9$, $b = 2$, and $c = 1$.

So, the discriminant is $b^2 - 4ac = (2)^2 - 4(9)(1) = 4 - 36 = -32.$

Because the discriminant is negative, there are two imaginary solutions.

3. Using $a = 1$, $b = -4$, and $c = 5$, you can apply the Quadratic Formula as follows.

$$x = \frac{-b \pm \sqrt{b^2 - 4ac}}{2a}$$

$$= \frac{-(-4) \pm \sqrt{(-4)^2 - 4(1)(5)}}{2(1)}$$

$$= \frac{4 \pm \sqrt{-4}}{2}$$

$$= \frac{4 \pm 2i}{2}$$

$$= 2 \pm i$$

4. $$x^4 + 7x^2 - 18 = 0$$

$$(x^2 + 9)(x^2 - 2) = 0$$

$$(x + 3i)(x - 3i)(x + \sqrt{2})(x - \sqrt{2}) = 0$$

Setting each factor equal to zero yields the solutions $x = -3i$, $3i$, $-\sqrt{2}$, and $\sqrt{2}$.

5. Because complex zeros occur in conjugate pairs you know that if $4i$ is a zero of f, so is $-4i$.

This means that both $(x - 4i)$ and $(x + 4i)$ are factors of f.

$$(x - 4i)(x + 4i) = x^2 - 16i^2 = x^2 + 16$$

Using long division, you can divide $x^2 + 16$ into $f(x)$ to obtain the following.

$$
\begin{array}{r}
3x - 12 \\
x^2 + 16 \overline{)3x^3 - 2x^2 + 48x - 32} \\
\underline{3x^3 \qquad\quad + 48x} \\
-2x^2 \qquad\quad - 32 \\
\underline{-2x^2 \qquad\quad - 32} \\
0
\end{array}
$$

So, you have $f(x) = (x^2 + 16)(3x - 2)$ and you can conclude that the real zeros of f are $x = -4i$, $x = 4i$, and $x = \dfrac{2}{3}$.

6. Because $-7i$ is a zero *and* the polynomial is stated to have real coefficients, you know that the conjugate $7i$ must also be a zero.

So, the four zeros are $2, -2, 7i,$ and $-7i$.

Then, using the Linear Factorization Theorem, $f(x)$ can be written as $f(x) = a(x - 2)(x + 2)(x - 7i)(x + 7i)$.

For simplicity, let $a = 1$. Then multiply the factors with real coefficients to obtain $(x + 2)(x - 2) = x^2 - 4$ and multiply the complex conjugates to obtain $(x - 7i)(x + 7i) = x^2 + 49$.

So, you obtain the following fourth-degree polynomial function.

$$f(x) = (x^2 - 4)(x^2 + 49) = x^4 + 49x^2 - 4x^2 - 196$$
$$= x^4 + 45x^2 - 196$$

7. Because $2 + i$ is a zero of f, so is $2 - i$. So,

$$
\begin{aligned}
f(x) &= a(x - 1)\big[x - (2 + i)\big]\big[x - (2 - i)\big] \\
&= a(x - 1)\big[(x - 2) - i\big]\big[(x - 2) + i\big] \\
&= a(x - 1)\big[(x - 2)^2 - i^2\big] \\
&= a(x - 1)(x^2 - 4x + 5) \\
&= a(x^3 - 5x^2 + 9x - 5)
\end{aligned}
$$

To find the value of a, use the fact that $f(2) = 2$

and obtain

$$f(2) = a(2^3 - 5(2)^2 + 9(2) - 5)$$
$$2 = a.$$

So, $a = 2$ and it follows that

$$f(x) = 2(x^3 - 5x^2 + 9x - 5)$$
$$= 2x^3 - 10x^2 + 18x - 10.$$

Checkpoints for Section 4.3

1. The number $z = 3 - 4i$ is plotted in the complex plane.

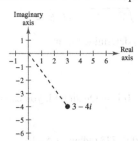

It has an absolute value of $|z| = \sqrt{3^2 + (-4)^2}$

$$= \sqrt{9 + 16}$$
$$= \sqrt{25}$$
$$= 5.$$

2. $(3 + i) + (1 + 2i) = (3 + 1) + (i + 2i)$

$$= 4 + 3i$$

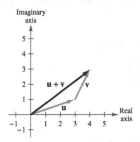

3. $(2 - 4i) - (1 + i) = (2 - 1) + (-4i - i)$

$$= 1 - 5i$$

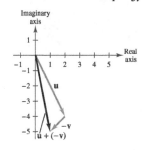

4.

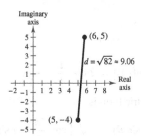

The complex conjugate of $z = 2 - 3i$ is $z = 2 + 3i$.

5. The distance between $5 - 4i$ and $6 + 5i$ is

$$d = \sqrt{(5 - 6)^2 + (-4 - 5)^2}$$
$$= \sqrt{(-1)^2 + (-9)^2}$$
$$= \sqrt{82} \approx 9.06 \text{ units}$$

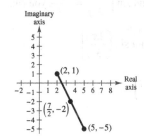

6. The midpoint of the line segment joining the points $2 + i$ and $5 - 5i$ is

$$\text{Midpoint} = \left(\frac{2 + 5}{2}, \frac{1 + (-5)}{2} \right) = \left(\frac{7}{2}, -2 \right)$$

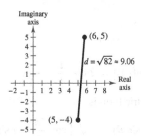

Checkpoints for Section 4.4

1. $z = 6 - 6i$

The modulus of $z = 6 - 6i$ is

$$r = \sqrt{6^2 + (-6)^2}$$
$$= \sqrt{36 + 36} = \sqrt{72} = 6\sqrt{2}$$

and the argument θ is determined from

$$\tan \theta = \frac{b}{a} = \frac{-6}{6} = -1.$$

Because $z = 6 - 6i$ lies in Quadrant IV.

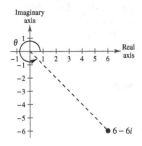

$$\theta = 2\pi - \left|\arctan(-1)\right| = 2\pi - \frac{\pi}{4} = \frac{7\pi}{4}.$$

So, the trigonometric form is

$$z = r(\cos \theta + i \sin \theta) = 6\sqrt{2}\left(\cos \frac{7\pi}{4} + i \sin \frac{7\pi}{4}\right).$$

2. The absolute value of z is

$$r = \left|3 + 4i\right| = \sqrt{3^2 + 4^2} = \sqrt{25} = 5$$

and the angle θ is determined from

$$\tan \theta = \frac{b}{a} = \frac{4}{3}.$$

Because $z = 3 + 4i$ is in Quadrant I, you can conclude that

$$\theta = \arctan \frac{4}{3} \approx 0.92728 \text{ radian} \approx 53.1°.$$

So, the trigonometric form of z is

$$z = r(\cos \theta + i \sin \theta)$$
$$= 5\left[\cos\left(\arctan \frac{4}{3}\right) + i \sin\left(\arctan \frac{4}{3}\right)\right]$$
$$\approx 5(\cos 53.1° + i \sin 53.1°)$$

3. Because $\cos 150° = -\frac{\sqrt{3}}{2}$ and $\sin 150° = \frac{1}{2}$, you can write

$$z = 2(\cos 150° + i \sin 150°)$$
$$= 2\left(-\frac{\sqrt{3}}{2} + \frac{1}{2}i\right)$$
$$= -\sqrt{3} + i$$

4. To write $z = 8\left[\cos\left(\frac{2\pi}{3}\right) + i \sin\left(\frac{2\pi}{3}\right)\right]$ in standard form, first find the trigonometric ratios. Because

$\cos\left(\frac{2\pi}{3}\right) = \frac{-1}{2}$ and $\sin\left(\frac{2\pi}{3}\right) = \frac{\sqrt{3}}{2}$, you can write

$$z = 8\left[\cos\left(\frac{2\pi}{3}\right) + i \sin\left(\frac{2\pi}{3}\right)\right]$$
$$= 8\left(-\frac{1}{2} + \frac{\sqrt{3}}{2}i\right)$$
$$= -4 + 4\sqrt{3}\, i.$$

5. $z_1z_2 = 3\left(\cos\dfrac{\pi}{3} + i\sin\dfrac{\pi}{3}\right)\cdot 4\left(\cos\dfrac{\pi}{6} + i\sin\dfrac{\pi}{6}\right)$

$\qquad = (3)(4)\left[\cos\left(\dfrac{\pi}{3} + \dfrac{\pi}{6}\right) + i\sin\left(\dfrac{\pi}{3} + \dfrac{\pi}{6}\right)\right]$

$\qquad = 12\left(\cos\dfrac{\pi}{2} + i\sin\dfrac{\pi}{2}\right)$

$\qquad = 12\left[0 + i(1)\right]$

$\qquad = 12i$

You can check this by first converting the complex numbers to their standard forms and then multiplying algebraically.

$z_1 = 3\left(\cos\dfrac{\pi}{3} + i\sin\dfrac{\pi}{3}\right) = 3\left(\dfrac{1}{2} + \dfrac{\sqrt{3}}{2}i\right) = \dfrac{3}{2} + \dfrac{3\sqrt{3}}{2}i$

$z_2 = 4\left(\cos\dfrac{\pi}{6} + i\sin\dfrac{\pi}{6}\right) = 4\left(\dfrac{\sqrt{3}}{2} + \dfrac{1}{2}i\right) = 2\sqrt{3} + 2i$

So, $z_1\, z_2 = \left(\dfrac{3}{2} + \dfrac{3\sqrt{3}}{2}i\right)\left(2\sqrt{3} + 2i\right)$

$\qquad\qquad = 3\sqrt{3} + 3i + 9i + 3\sqrt{3}\,i^2$

$\qquad\qquad = 3\sqrt{3} + 12i + 3\sqrt{3}(-1)$

$\qquad\qquad = 3\sqrt{3} + 12i - 3\sqrt{3}$

$\qquad\qquad = 12i.$

6. $\dfrac{z_1}{z_2} = \dfrac{\cos 40° + i\sin 40°}{\cos 10° + i\sin 10°}$

$\qquad = \left[\cos\left(40° - 10°\right) + i\sin\left(40° - 10°\right)\right]$

$\qquad = \cos 30° + i\sin 30°$

$\qquad = \dfrac{\sqrt{3}}{2} + \dfrac{1}{2}i$

7. $z_1 = 2\left(\cos\dfrac{\pi}{4} + i\sin\dfrac{\pi}{4}\right)$ and $z_2 = 4\left(\cos\dfrac{3\pi}{4} + i\sin\dfrac{3\pi}{4}\right)$

To find z_1z_2 in the complex plane, let

$\mathbf{u} = 2\left(\cos\dfrac{\pi}{4} + i\sin\dfrac{\pi}{4}\right) = \left\langle\sqrt{2}, \sqrt{2}\right\rangle$ and $\mathbf{v} = 4\left(\cos\dfrac{3\pi}{4} + i\sin\dfrac{3\pi}{4}\right) = \left\langle-2\sqrt{2}, 2\sqrt{2}\right\rangle$

$\|\mathbf{u}\| = \sqrt{\left(\sqrt{2}\right)^2 + \left(\sqrt{2}\right)^2} = \sqrt{4} = 2$ and $\|\mathbf{v}\| = \sqrt{\left(-2\sqrt{2}\right)^2 + \left(2\sqrt{2}\right)^2} = \sqrt{16} = 4$

So, the magnitude of the product vector is $\|\mathbf{u}\|\,\|\mathbf{v}\| = (2)(4) = 8.$ The sum of the direction angles is $\dfrac{\pi}{4} + \dfrac{3\pi}{4} = \pi.$ The product vector lies on the negative real axis and is represented in vector form as $\left\langle 0, 8\right\rangle.$ This means that $z_1z_2 = \left\langle-8, 0\right\rangle = -8.$

Checkpoints for Section 4.5

1. The absolute value of $z = -1 - i$ is $r = |-1 - i| = \sqrt{(-1)^2 + (-1)^2} = \sqrt{1 + 1} = \sqrt{2}$

and the argument θ given by $\tan \theta = \dfrac{b}{a} = \dfrac{-1}{-1} = 1$.

Because $z = -1 - i$ lies in Quadrant III, $\theta = \pi + \arctan 1 = \pi + \dfrac{\pi}{4} = \dfrac{5\pi}{4}$.

So, the trigonometric form is $z = -1 - i = \sqrt{2}\left(\cos \dfrac{5\pi}{4} + i \sin \dfrac{5\pi}{4}\right)$.

Then, by DeMoivre's Theorem, you have $(-1 - i)^4 = \left[\sqrt{2}\left(\cos \dfrac{5\pi}{4} + i \sin \dfrac{5\pi}{4}\right)\right]^4$

$$= \left(\sqrt{2}\right)^4 \left(\cos\left[\dfrac{4(5\pi)}{4}\right] + i \sin\left[\dfrac{4(5\pi)}{4}\right]\right)$$

$$= 4(\cos 5\pi + i \sin 5\pi)$$

$$= 4\left[-1 + i(0)\right]$$

$$= -4.$$

2. First, write 1 in trigonometric form $z = 1(\cos 0 + i \sin 0)$. Then, by the nth root formula, with $n = 4$ and $r = 1$, the roots are of the form

$$z_k = \sqrt[4]{1}\left(\cos \dfrac{0 + 2\pi k}{4} + i \sin \dfrac{0 + 2\pi k}{4}\right) = (1)\left(\cos \dfrac{\pi k}{2} + i \sin \dfrac{\pi k}{2}\right) = \cos \dfrac{\pi k}{2} + i \sin \dfrac{\pi k}{2}.$$

So, for $k = 0, 1, 2$ and 3, the fourth roots are as follows.

$z_0 = \cos 0 + i \sin 0 = 1 + i(0) = 1$

$z_1 = \cos \dfrac{\pi}{2} + i \sin \dfrac{\pi}{2} = 0 + i(1) = i$

$z_2 = \cos \pi + i \sin \pi = -1 + i(0) = -1$

$z_3 = \cos \dfrac{3\pi}{2} + i \sin \dfrac{3\pi}{2} = 0 + i(-1) = -i$

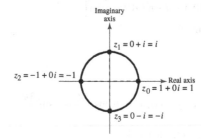

3. The absolute value of $z = -6 + 6i$ is

$$r = |-6 + 6i| = \sqrt{(-6)^2 + 6^2} = \sqrt{36 + 36} = \sqrt{72} = 6\sqrt{2}$$

and the argument θ is given by $\tan\theta = \dfrac{b}{a} = \dfrac{6}{-6} = -1.$

Because $z = -6 + 6i$ lies in Quadrant II, the trigonometric form of

z is $z = -6 + 6i = 6\sqrt{2}\left(\cos 135° + i\sin 135°\right).$

By the formula for nth roots, the cube roots have the form

$$z_k = \sqrt[3]{6\sqrt{2}}\left[\cos\left(\frac{135° + 360°k}{3}\right) + i\sin\left(\frac{135° + 360°k}{3}\right)\right].$$

Finally, for $k = 0$, 1 and 2, you obtain the roots

$$z_0 = \sqrt[3]{6\sqrt{2}}\left[\cos\left(\frac{135° + 360°(0)}{3}\right) + i\sin\left(\frac{135° + 360°(0)}{3}\right)\right].$$

$$= \sqrt[3]{6\sqrt{2}}\left(\cos 45° + i\sin 45°\right)$$

$$= \sqrt[3]{6\sqrt{2}}\left(\frac{\sqrt{2}}{2} + \frac{\sqrt{2}}{2}i\right)$$

$$\approx 1.4422 + 1.4422i$$

$$z_1 = \sqrt[3]{6\sqrt{2}}\left[\cos\left(\frac{135° + 360°(1)}{3}\right) + i\sin\left(\frac{135° + 360°(1)}{3}\right)\right].$$

$$= \sqrt[3]{6\sqrt{2}}\left(\cos 165° + i\sin 165°\right)$$

$$\approx -1.9701 + 0.5279i$$

$$z_2 = \sqrt[3]{6\sqrt{2}}\left[\cos\left(\frac{135° + 360°(2)}{3}\right) + i\sin\left(\frac{135° + 360°(2)}{3}\right)\right].$$

$$= \sqrt[3]{6\sqrt{2}}\left(\cos 285° + i\sin 285°\right)$$

$$\approx 0.5279 - 1.9701i$$

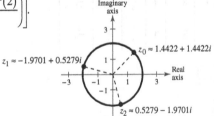

Chapter 5

Checkpoints for Section 5.1

1. Function Value

$$f\left(\sqrt{2}\right) = 8^{-\sqrt{2}}$$

Graphing Calculator Keystrokes

8 $\boxed{\wedge}$ $\boxed{(}$ $\boxed{(-)}$ $\boxed{\sqrt{}}$ 2 $\boxed{)}$ $\boxed{\text{Enter}}$

Display

0.052824803759

2. The table lists some values for each function, and the graph shows a sketch of the two functions. Note that both graphs are increasing and the graph of $g(x) = 9^x$ is increasing more rapidly than the graph of $f(x) = 3^x$.

x	-3	-2	-1	0	1	2
3^x	$\frac{1}{27}$	$\frac{1}{9}$	$\frac{1}{3}$	1	3	9
9^x	$\frac{1}{729}$	$\frac{1}{81}$	$\frac{1}{9}$	1	9	81

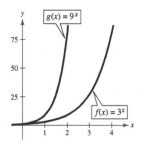

3. The table lists some values for each function and, the graph shows a sketch for each function. Note that both graphs are decreasing and the graph of $g(x) = 9^{-x}$ is decreasing more rapidly than the graph of $f(x) = 3^{-x}$

x	-2	-1	0	1	2	3
9^{-x}	64	8	1	$\frac{1}{8}$	$\frac{1}{64}$	$\frac{1}{512}$
$3^{-x}\ g(x)$	9	3	1	$\frac{1}{3}$	$\frac{1}{9}$	$\frac{1}{27}$

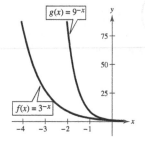

4. (a)

$$8 = 2^{2x-1} \qquad \text{Write Original equation.}$$
$$2^3 = 2^{2x-1} \qquad 8 = 2^3$$
$$3 = 2x - 1 \qquad \text{One-to-One Property}$$
$$4 = 2x$$
$$2 = x \qquad \text{Solve for } x.$$

(b)

$$\left(\tfrac{1}{3}\right)^{-x} = 27 \qquad \text{Write Original equation.}$$
$$3^x = 27 \qquad \left(\tfrac{1}{3}\right)^{-x} = 3^x$$
$$3^x = 3^3 \qquad 27 = 3^3$$
$$x = 3 \qquad \text{One-to-One Property}$$

5. (a) Because $g(x) = 4^{x-2} = f(x - 2)$, the graph of g can be obtained by shifting the graph of f two units to the right.

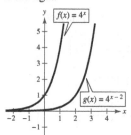

(b) Because $h(x) = 4^x + 3 = f(x) + 3$ the graph of h can be obtained by shifting the graph of f up three units.

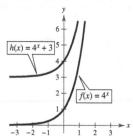

(c) Because $k(x) = 4^{-x} - 3 = f(-x) - 3$, the graph of k can be obtained by reflecting the graph of f in the y-axis and shifting the graph of f down three units.

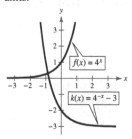

6.

Function Value	Graphing Calculator Keystrokes	Display
(a) $f(0.3) = e^{0.3}$	$\boxed{e^x}$ 0.3 $\boxed{\text{Enter}}$	1.3498588
(b) $f(-1.2) = e^{-1.2}$	$\boxed{e^x}$ $\boxed{(-)}$ 1.2 $\boxed{\text{Enter}}$	0.3011942
(c) $f(6, 2) = e^{6.2}$	$\boxed{e^x}$ 6.2 $\boxed{\text{Enter}}$	492.7490411

7. To sketch the graph of $f(x) = 5e^{0.17x}$, use a graphing utility to construct
a table of values. After constructing the table, plot the points and draw a
smooth curve.

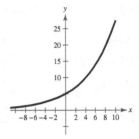

x	-3	-2	-1	0	1	2	3
$f(x)$	3.002	3.559	4.218	5.000	5.927	7.025	8.326

8. (a) For quarterly compounding, you have $n = 4$. So, in 7 years at 4%, the balance is as follows.

$$A = P\left(1 + \frac{r}{n}\right)^{nt} \qquad \text{Formula for compound interest.}$$

$$= 6000\left(1 + \frac{0.04}{4}\right)^{4(7)} \qquad \text{Substitute } P, r, n, \text{ and } t.$$

$$\approx \$7927.75 \qquad \text{Use a calculator.}$$

(b) For monthly compounding, you have $n = 12$. So in 7 years at 4%, the balance is as follows.

$$A = P\left(1 + \frac{r}{n}\right)^{nt} \qquad \text{Formula for compound interest.}$$

$$= 6000\left(1 + \frac{0.04}{12}\right)^{12(7)} \qquad \text{Substitute } P, r, n, \text{ and } t.$$

$$\approx \$7935.08 \qquad \text{Use a calculator.}$$

(c) For continuous compounding, the balance is as follows.

$$A = Pe^{rt} \qquad \text{Formula for continuous compounding.}$$

$$= 6000e^{0.04(7)} \qquad \text{Substitute } P, r, \text{ and } t.$$

$$\approx \$7938.78 \qquad \text{Use a calculator.}$$

9. Use the model for the amount of Plutonium that remains from an initial amount of 10 pounds after t years,
where $t = 0$ represents the year 1986.

$$P = 10\left(\tfrac{1}{2}\right)^{t/24,100}$$

To find the amount that remains in the year 2089, let $t = 103$.

$$P = 10\left(\tfrac{1}{2}\right)^{t/24,100} \qquad \text{Write original model.}$$

$$P = 10\left(\tfrac{1}{2}\right)^{103/24,100} \qquad \text{Substitute 103 for } t.$$

$$P \approx 9.970 \qquad \text{Use a calculator.}$$

In the year 2089, 9.970 pounds of plutonium will remain.

To find the amount that remains after 125,000 years, let $t = 125,000$.

$$P = 10\left(\tfrac{1}{2}\right)^{t/24,100} \qquad \text{Write original model.}$$

$$P = 10\left(\tfrac{1}{2}\right)^{125,000/24,100} \qquad \text{Substitute 125,000 for } t.$$

$$P \approx 0.275 \qquad \text{Use a calculator.}$$

After 125,000 years 0.275 pound of plutonium will remain.

Checkpoints for Section 5.2

1. (a) $f(1) = \log_6 1 = 0$ because $6^0 = 1$.

(b) $f\left(\frac{1}{125}\right) = \log_5 \frac{1}{125} = -3$ because $5^{-3} = \frac{1}{125}$.

(c) $f(343) = \log_7 343 = 3$ because $7^3 = 343$.

2.

Function Value	Graphing Calculator Keystrokes	Display
(a) $f(275) = \log 275$	$\boxed{\text{LOG}}$ 275 $\boxed{\text{ENTER}}$	2.4393327
(b) $f\left(-\frac{1}{2}\right) = \log -\frac{1}{2}$	$\boxed{\text{LOG}}$ $\boxed{(}$ $\boxed{(-)}$ $\boxed{(}$ 1 $\boxed{\div}$ 2 $\boxed{)}$ $\boxed{)}$ $\boxed{\text{ENTER}}$	ERROR
(c) $f\left(\frac{1}{2}\right) = \log \frac{1}{2}$	$\boxed{\text{LOG}}$ $\boxed{(}$ 1 $\boxed{\div}$ 2 $\boxed{)}$ $\boxed{\text{ENTER}}$	-0.3010300

3. (a) Using Property 2, $\log_9 9 = 1$.

(b) Using Property 3, $20^{\log_{20} 3} = 3$.

(c) Using Property 1, $\log_{\sqrt{3}} 1 = 0$.

4. $\log_5 (x^2 + 3) = \log_5 12$

$x^2 + 3 = 12$

$x^2 = 9$

$x = \pm 3$

5. (a) For $f(x) = 8^x$, construct a table of values. Then plot the points and draw a smooth curve.

x	-2	-1	0	1	2
$f(x) = 8^x$	$\frac{1}{64}$	$\frac{1}{8}$	1	8	64

(b) Because $g(x) = \log_8 x$ is the inverse function of $f(x) = 8^x$, the graph of g is obtained by plotting the points $(f(x), x)$ and connecting them with a smooth curve. The graph of g is a reflection of the graph of f in the line $y = x$.

x	$\frac{1}{64}$	$\frac{1}{8}$	1	8	64
$g(x) = \log_8 x$	-2	-1	0	1	2

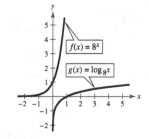

6. Begin by constructing a table of values. Note that some of the values can be obtained without a calculator by using the properties of logarithms. Then plot the points and draw a smooth curve.

x	$\frac{1}{9}$	$\frac{1}{3}$	1	3	9
$f(x) = \log_3 x$	-2	-1	0	1	2

The vertical asymptote is $x = 0$, the y-axis.

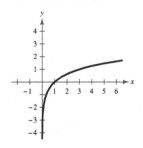

7. (a) Because $g(x) = -1 + \log_3 x = f(x) - 1$, the graph of g can be obtained by shifting the graph of f one unit down.

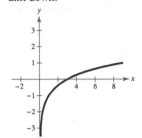

(b) Because $h(x) = \log_3(x + 3) = f(x + 3)$, the graph of h can be obtained by shifting the graph of f three units to the left.

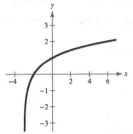

8.

Function Value	Graphing Calculator Keystrokes	Display
$f(0.01) = \ln 0.01$	LN 0.01 ENTER	-4.6051702
$f(4) = \ln 4$	LN 4 ENTER	1.3862944
$f(\sqrt{3} + 2) = \ln(\sqrt{3} + 2)$	LN (((√ 3) + 2) ENTER	1.3169579
$f(\sqrt{3} - 2) = \ln(\sqrt{3} - 2)$	LN (((√ 3) − 2) ENTER	ERROR

9. (a) $\ln e^{1/3} = \frac{1}{3}$ Inverse Property

(b) $5 \ln 1 = 5(0) = 0$ Property 1

(c) $\frac{3}{4} \ln e = \frac{3}{4}(1) = \frac{3}{4}$ Property 2

(d) $e^{\ln 7} = 7$ Inverse Property

10. Because $\ln(x + 3)$ is defined only when $x + 3 > 0$, it follows that the domain of f is $(-3, \infty)$. The graph of f is shown.

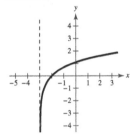

11. (a) After 1 month, the average score was the following.

$$f(1) = 75 - 6 \ln(1 + 1) \quad \text{Substitute 1 for } t.$$
$$= 75 - 6 \ln 2 \quad \text{Simplify.}$$
$$\approx 75 - 6(0.6931) \quad \text{Use a calculator.}$$
$$\approx 70.84 \quad \text{Solution}$$

(b) After 9 months, the average score was the following.

$$f(9) = 75 - 6 \ln(9 + 1) \quad \text{Substitute 9 for } t.$$
$$= 75 - 6 \ln 10 \quad \text{Simplify.}$$
$$\approx 75 - 6(2.3026) \quad \text{Use a calculator.}$$
$$\approx 61.18 \quad \text{Solution}$$

(c) After 12 months, the average score was the following.

$$f(12) = 75 - 6 \ln(12 + 1) \quad \text{Substitute 9 for } t.$$
$$= 75 - 6 \ln 13 \quad \text{Simplify.}$$
$$\approx 75 - 6(2.5649) \quad \text{Use a calculator.}$$
$$\approx 59.61 \quad \text{Solution}$$

Checkpoints for Section 5.3

1. $\log_2 12 = \dfrac{\log 12}{\log 2}$ $\log_a x = \dfrac{\log x}{\log a}$

$\approx \dfrac{1.07918}{0.30103}$ Use a calculator.

≈ 3.5850 Simplify.

2. $\log_2 12 = \dfrac{\ln 12}{\ln 2}$ $\log_a x = \dfrac{\ln x}{\ln a}$

$\approx \dfrac{2.48491}{0.69315}$ Use a calculator.

≈ 3.5850 Simplify.

3. (a) log 75 = log(3.25)   Rewrite 75 as 3 · 25.   (b) log $\frac{9}{125}$ = log 9 − log 125   Quotient Property

$\qquad\qquad$ = log 3 + log 25   Product Property $\qquad\qquad\qquad$ = log 3^2 − log 5^3 Rewrite 9 as 3^2 and 125 as 5^2.

$\qquad\qquad$ = log 3 + log 5^2 Rewrite 25 as 5^2. $\qquad\qquad\qquad$ = 2log 3 − 3 log 5   Power Property

$\qquad\qquad$ = log 3 + 2 log 5   Power Property

4. ln e^6 − ln e^2 = 6 ln e − 2ln e

$\qquad\qquad\qquad$ = 6(1) − 2(1)

$\qquad\qquad\qquad$ = 4

5. log$_3$ $\dfrac{4x^2}{\sqrt{y}}$ = log$_3$ $\dfrac{4x^2}{y^{1/2}}$ Rewrite using rational exponent.

$\qquad\qquad$ = log$_3$ $4x^2$ − log$_3$ $y^{1/2}$ Quotient Property

$\qquad\qquad$ = log$_3$ 4 + log$_3$ x^2 − log$_3$ $y^{1/2}$ Product Property

$\qquad\qquad$ = log$_3$ 4 + 2 log$_3$ x − $\dfrac{1}{2}$ log$_3$ y Power Property

6. $2\Big[\log(x + 3) − 2\log(x − 2)\Big] = 2\Big[\log(x + 3) − \log(x − 2)^2\Big]$ Power Property

$\qquad\qquad\qquad = 2\left[\log\left(\dfrac{x + 3}{(x − 2)^2}\right)\right]$ Quotient Property

$\qquad\qquad\qquad = \log\left(\dfrac{x + 3}{(x − 2)^2}\right)^2$ Power Property

$\qquad\qquad\qquad = \log\dfrac{(x + 3)^2}{(x − 2)^4}$ Simplify.

7. To solve this problem, take the natural logarithm of each of the x- and y-values of the ordered pairs.

$(\ln x, \ln y)$: $(−0.994, −0.673)$, $(0.000, 0.000)$, $(1.001, 0.668)$, $(2.000, 1.332)$, $(3.000, 2.000)$

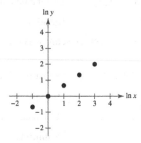

By plotting the ordered pairs, you can see that all five points appear to lie in a line. Choose any two points to determine the slope of the line. Using the points $(0, 0)$ and $(1.001, 0.668)$, the slope of the line is

$m = \dfrac{0.668 − 0}{1 − 0} = 0.668 \approx \dfrac{2}{3}$.

By the point-slope form, the equation of the line is $y = \dfrac{2}{3}x$, where $y = \ln y$ and $x = \ln x$. So, the

logarithmic equation is $\ln y = \dfrac{2}{3} \ln x$.

Checkpoints for Section 5.4

1.

Original Equation	Rewritten Equation	Solution	Property
(a) $2^x = 512$	$2^x = x^9$	$x = 9$	One-to-One
(b) $\log_6 x = 3$	$6^{\log_6 x} = 6^3$	$x = 216$	Inverse
(c) $5 - e^x = 0$	$\ln 5 = \ln e^x$	$\ln 5 = x$	Inverse
$5 = e^x$			
(d) $9^x = \frac{1}{3}$	$3^{2x} = 3^{-1}$	$2x = -1$	One-to-One
		$x = -\frac{1}{2}$	

2. (a)

$e^{2x} = e^{x^2 - 8}$	Write original equation.
$2x = x^2 - 8$	One-to-One Property
$0 = x^2 - 2x - 8$	Write in general form.
$0 = (x - 4)(x + 2)$	Factor.
$x - 4 = 0 \Rightarrow x = 4$	Set 1st factor equal to 0.
$x + 2 = 0 \Rightarrow x = -2$	Set 2nd factor equal to 0.

The solutions are $x = 4$ and $x = -2$

Check $x = -2$: $\qquad\qquad x = 4$:

$$e^{2x} = e^{x^2 - 8} \qquad\qquad e^{2(4)} \overset{?}{=} e^{(4)^2 - 8}$$

$$e^{2(-2)} \overset{?}{=} e^{(-2)^2 - 8} \qquad e^8 \overset{?}{=} e^{16 - 8}$$

$$e^{-4} \overset{?}{=} e^{4 - 8} \qquad\qquad e^8 = e^8 \ \checkmark$$

$$e^{-4} = e^{-4} \ \checkmark$$

(b)

$2(5^x) = 32$	Write original equation.
$5^x = 16$	Divide each side by 2.
$\log_5 5^x = \log_5 16$	Take log(base 5) of each side.
$x = \log_5 16$	Inverse Property
$x = \dfrac{\ln 16}{\ln 5} \approx 1.723$	Change of base formula

The solution is $x = \log_5 16 \approx 1.723$.

Check $x = \log_5 16$:

$$2(5^x) = 32$$

$$2\left[5^{(\log_5 16)}\right] \overset{?}{=} 32$$

$$2(16) \overset{?}{=} 32$$

$$32 = 32 \ \checkmark$$

3. $e^x - 7 = 23$ Write original equation.

   $e^x = 30$ Add 7 to each side.

   $\ln e^x = \ln 30$ Take natural log of each side.

   $x = \ln 30 \approx 3.401$ Inverse Property

Check $x = \ln 30$:

$e^x - 7 = 23$

$e^{(\ln 30)-7} \overset{?}{=} 23$

$30 - 7 \overset{?}{=} 23$

$23 = 23$ ✓

4. $6\left(2^{t+5}\right) + 4 = 11$ Write original equation.

   $6\left(2^{t+5}\right) = 7$ Subtract 4 from each side.

   $2^{t+5} = \dfrac{7}{6}$ Divide each side by 6.

   $\log_2 2^{t+5} = \log_2\left(\dfrac{7}{6}\right)$ Take log (base 2) of each side.

   $t + 5 = \log_2\left(\dfrac{7}{6}\right)$ Inverse Property

   $t = \log_2\left(\dfrac{7}{6}\right) - 5$ Subtract 5 from each side.

   $t = \dfrac{\ln\left(\dfrac{1}{6}\right)}{\ln 2} - 5$ Change of base formula.

   $t \approx -4.778$ Use a calculator.

The solution is $t = \log_2\left(\dfrac{7}{6}\right) - 5 \approx -4.778$.

Check $t \approx -4.778$:

   $6\left(2^{t+5}\right) + 4 = 11$

   $6\left[2^{(-4.778+5)}\right] + 4 \overset{?}{=} 11$

   $6(1.166) + 4 \overset{?}{=} 11$

   $10.998 \approx 11$ ✓

5. Algebraic Solution

$$e^{2x} - 7e^x + 12 = 0 \qquad \text{Write original equation.}$$

$$\left(e^x\right)^2 - 7e^x + 12 = 0 \qquad \text{Write in quadratic form.}$$

$$\left(e^x - 3\right)\left(e^x - 4\right) = 0 \qquad \text{Factor.}$$

$$e^x - 3 = 0 \Rightarrow e^x = 3 \qquad \text{Set 1st factor equal to 0.}$$

$$x = \ln 3 \qquad \text{Solution}$$

$$e^x - 4 = 0 \Rightarrow e^x = 4 \qquad \text{Set 2nd factor equal to 0.}$$

$$x = \ln 4 \qquad \text{Solution}$$

The solutions are $x = \ln 3 \approx 1.099$ and $x = \ln 4 \approx 1.386$.

Check $x = \ln 3$: $\qquad\qquad\qquad x = \ln 4$:

$$e^{2x} - 7e^x + 12 = 0 \qquad\qquad e^{2(\ln 4)} - 7e^{(\ln 4)} + 12 = 0$$

$$e^{2(\ln 3)} - 7e^{(\ln 3)} + 12 \overset{?}{=} 0 \qquad\qquad e^{\ln 4^2} - 7e^{\ln 4} + 12 \overset{?}{=} 0$$

$$e^{\ln\left(3^2\right)} - 7e^{\ln 3} + 12 \overset{?}{=} 0 \qquad\qquad 4^2 - 7(4) + 12 \overset{?}{=} 0$$

$$3^2 - 7(3) + 12 \overset{?}{=} 0 \qquad\qquad\qquad 0 = 0 ✓$$

$$0 = 0 ✓$$

Graphical Solution

Use a graphing utility to graph $y = e^{2x} - 7e^x + 12$ and then find the zeros.

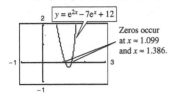

Zeros occur at $x \approx 1.099$ and $x \approx 1.386$.

So, you can conclude that the solutions are $x \approx 1.099$ and $x \approx 1.386$.

6. (a) $\ln x = \dfrac{2}{3} \qquad \text{Write original equation.}$

$$e^{\ln x} = e^{2/3} \qquad \text{Exponentiate each side.}$$

$$x = e^{2/3} \qquad \text{Inverse Property}$$

(b) $\log_2 (2x - 3) = \log_2 (x + 4) \qquad \text{Write original equation.}$

$$2x - 3 = x + 4 \qquad \text{One-to-One Property}$$

$$x = 7 \qquad \text{Solution}$$

(c) $\log 4x - \log(12 + x) = \log 2 \qquad \text{Write Original equation.}$

$$\log\left(\frac{4x}{12 + x}\right) = \log 2 \qquad \text{Quotient Property of Logarithms}$$

$$\frac{4x}{12 + x} = 2 \qquad \text{One-to-One Property}$$

$$4x = 2(12 + x) \qquad \text{Multiply each side by } (12 + x).$$

$$4x = 24 + 2x \qquad \text{Distribute.}$$

$$2x = 24 \qquad \text{Subtract } 2x \text{ from each side.}$$

$$x = 12 \qquad \text{Solution}$$

7. Algebraic Solution

$$7 + 3 \ln x = 5 \qquad \text{Write original equation.}$$

$$3 \ln x = -2 \qquad \text{Subtract 7 from each side.}$$

$$\ln x = -\frac{2}{3} \qquad \text{Divide each side by 3.}$$

$$e^{\ln x} = e^{-2/3} \qquad \text{Exponentiate each side.}$$

$$x = e^{-2/3} \qquad \text{Inverse Property}$$

$$x \approx 0.513 \qquad \text{Use a calculator.}$$

Graphical Solution

Use a graphing utility to graph $y_1 = 7 + 3 \ln x$ and $y_2 = 5$. Then find the intersection point.

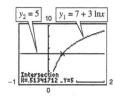

The point of intersection is about $(0.513, 5)$. So, the solution is $x \approx 0.513$.

8.
$$3 \log_4 6x = 9 \qquad \text{Write original equation.}$$

$$\log_4 6x = 3 \qquad \text{Divide each side by 3.}$$

$$4^{\log_4 6x} = 4^3 \qquad \text{Exponentiate each side (base 4).}$$

$$6x = 64 \qquad \text{Inverse Property}$$

$$x = \frac{32}{3} \qquad \text{Divide each side by 6 and simplify.}$$

Check $x = \frac{32}{3}$:

$$3 \log_4 6x = 9$$

$$3 \log_4 6\left(\frac{32}{3}\right) \overset{?}{=} 9$$

$$3 \log_4 64 \overset{?}{=} 9$$

$$3 \log_4 4^3 \overset{?}{=} 9$$

$$3 \cdot 3 \overset{?}{=} 9$$

$$9 = 9 \checkmark$$

9. Algebraic Solution

$$\log x + \log(x - 9) = 1 \qquad \text{Write original equation.}$$

$$\log[x(x - 9)] = 1 \qquad \text{Product Property of Logarithms}$$

$$10^{\log[x(x-9)]} = 10^1 \qquad \text{Exponentiate each side (base 10).}$$

$$x(x - 9) = 10 \qquad \text{Inverse Property}$$

$$x^2 - 9x - 10 = 0 \qquad \text{Write in general form.}$$

$$(x - 10)(x + 1) = 0 \qquad \text{Factor.}$$

$$x - 10 = 0 \Rightarrow x = 10 \qquad \text{Set 1st factor equal to 0.}$$

$$x + 1 = 0 \Rightarrow x = -1 \qquad \text{Set 2nd factor equal to 0.}$$

Check $x = 10$:

$$\log x + \log(x - 9) = 1$$

$$\log(10) + \log(10 - 9) \stackrel{?}{=} 1$$

$$\log 10 + \log 1 \stackrel{?}{=} 1$$

$$1 + 0 \stackrel{?}{=} 1$$

$$1 = 1 \checkmark$$

$x = -1$:

$$\log x + \log(x - 9) = 1$$

$$\log(-1) + \log(-1 - 9) \stackrel{?}{=} 1$$

$$\log(-1) + \log(-10) \stackrel{?}{=} 1$$

-1 and -10 are not in the domain of $\log x$. So, it does not check.

The solutions appear to be $x = 10$ and $x = -1$. But when you check these in the original equation, you can see that $x = 10$ is the only solution.

Graphical Solution

First, rewrite the original solution as

$$\log x + \log(x - 9) - 1 = 0.$$

Then use a graphing utility to graph the equation $y = \log x + \log(x - 9) - 1$ and find the zeros.

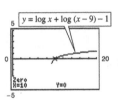

10. Using the formula for continuous compounding, the balance is

$$A = Pe^{rt}$$

$$A = 500e^{0.0525t}.$$

To find the time required for the balance to double, let $A = 1000$ and solve the resulting equation for t

$$500e^{0.0525t} = 1000 \qquad \text{Let } A = 1000.$$

$$e^{0.0525t} = 2 \qquad \text{Divide each side by 500.}$$

$$\ln e^{0.0525t} = \ln 2 \qquad \text{Take natural log of each side.}$$

$$0.0525t = \ln 2 \qquad \text{Inverse Property}$$

$$t = \frac{\ln 2}{0.0525} \qquad \text{Divide each side by 0.0525.}$$

$$t \approx 13.20 \qquad \text{Use a calculator.}$$

The balance in the account will double after approximately 13.20 years.

Because the interest rate is lower than the interest rate in Example 2, it will take more time for the account balance to double.

11. To find when sales reached \$180 billion, let $y = 80$ and solve for t.

$-614 + 342.2 \ln t = y$	Write original equation
$-614 + 342.2 \ln t = 180$	Substitute 180 for y.
$342.2 \ln t = 794$	Add 614 to each side.
$\ln t = \dfrac{794}{342.2}$	Divide each side by 342.2.
$e^{\ln t} = e^{794/342.2}$	Exponentiate each side (base e).
$t = e^{794/342.2}$	Inverse Property
$t \approx 10.2$	Use a calculator.

The solution is $t \approx 10.2$. Because $t = 9$ represents 2009, it follows that $t = 10$ represents 2010. So, sales reached \$180 billion in 2010.

Checkpoints for Section 5.5

1. Algebraic Solution

To find when the amount of U.S. online advertising spending will reach \$100 billion, let $s = 100$ and solve for t.

$0.00036e^{0.7563t} = S$	Write original model.
$0.00036e^{0.7563t} = 300$	Substitute 300 for s.
$e^{0.7563t} \approx 833{,}333.33$	Divide each side by 0.0036.
$\ln e^{0.7563t} \approx \ln 833{,}333.33$	Take natural log of each side.
$0.7563t \approx 13.6332$	Inverse Property
$t \approx 18.03$	Divide each side by 0.7563.

According to the model, the amount of U.S. online advertising spending will reach \$300 million in 2018.

Graphical Solution

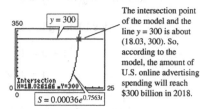

The intersection point of the model and the line $y = 300$ is about (18.03, 300). So, according to the model, the amount of U.S. online advertising spending will reach \$300 billion in 2018.

2. Let y be the number of bacteria at time t. From the given information you know that $y = 100$ when $t = 1$ and $y = 200$ when $t = 2$. Substituting this information into the model $y = ae^{bt}$ produces $100 = ae^{(1)b}$ and $200 = ae^{(2)b}$. To solve for b, solve for a in the first equation.

$100 = ae^{b}$	Write first equation.
$\dfrac{100}{e^{b}} = a$	Solve for a.

Then substitute the result into the second equation.

$200 = ae^{2b}$	Write second equation
$200 = \left(\dfrac{100}{e^{b}}\right)e^{2b}$	Substitute $\dfrac{100}{e^{b}}$ for a.
$\dfrac{200}{100} = e^{b}$	Simplify and divide each side by 100.
$2 = e^{b}$	Simplify.
$\ln 2 = \ln e^{b}$	Take natural log of each side
$\ln 2 = b$	Inverse Property

Use $b = \ln 2$ and the equation you found for a.

$a = \dfrac{100}{e^{\ln 2}}$	Substitute $\ln 2$ for b.
$= \dfrac{100}{2}$	Inverse Property
$= 50$	Simplify.

So, with $a = 50$ and $b = \ln 2$, the exponential growth model is $y = 50e^{(\ln 2)t}$.

After 3 hours, the number of bacteria will be $y = 50e^{\ln 2(3)} = 400$ bacteria.

3. Algebraic Solution

In the carbon dating model, substitute the given value of R to obtain the following.

$$\frac{1}{10^{12}}e^{-t/8223} = R \qquad \text{Write original model.}$$

$$\frac{e^{-t/8223}}{10^{12}} = \frac{1}{10^{14}} \qquad \text{Substitute } \frac{1}{10^{14}} \text{ for } R.$$

$$e^{-t/8223} = \frac{1}{10^{2}} \qquad \text{Multiply each side by } 10^{12}.$$

$$e^{-t/8223} = \frac{1}{100} \qquad \text{Simplify.}$$

$$\ln e^{-t/8223} = \ln \frac{1}{100} \qquad \text{Take natural log of each side.}$$

$$-\frac{t}{8223} \approx -4.6052 \qquad \text{Inverse Property}$$

$$t \approx 37,869 \qquad \text{Multiply each side by } -8223.$$

So, to the nearest thousand years, the age of the fossil is about 38,000 years.

Graphical Solution

Use a graphing utility to graph the formula for the ratio of carbon 14 to carbon 12 at any time t as

$$y_1 = \frac{1}{10^{12}}e^{-x/8223}.$$

In the same viewing window, graph $y_2 = \frac{1}{10^{14}}$

Use the *intersect* feature to estimate that $x \approx 18,934$ when $y = 1/10^{13}$.

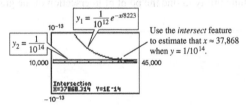

So, to the nearest thousand years, the age of fossil is about 38,000 years.

4. The graph of the function is shown below. On this bell-shaped curve, the maximum value of the curve represents the average score. From the graph, you can estimate that the average reading score for college-bound seniors in the United States in 2015 was 495.

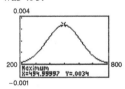

5. To find the number of days that 250 students are infected, let $y = 250$ and solve for t.

$$\frac{5000}{1 + 4999e^{-0.8t}} = y \qquad\qquad \text{Write original model.}$$

$$\frac{5000}{1 + 4999e^{-0.8t}} = 250 \qquad\qquad \text{Substitute 250 for } y.$$

$$\frac{5000}{250} = 1 + 4999e^{-0.8t} \qquad\qquad \text{Divide each side by 250 and multiply each side by } 1 + 4999e^{-0.8t}.$$

$$20 = 1 + 4999e^{-0.8t} \qquad\qquad \text{Simplify.}$$

$$19 = 4999e^{-0.8t} \qquad\qquad \text{Subtract 1 from each side.}$$

$$\frac{19}{4999} = e^{-0.8t} \qquad\qquad \text{Divide each side by 4999.}$$

$$\ln\left(\frac{19}{4999}\right) = \ln e^{-0.8t} \qquad\qquad \text{Take natural log of each side.}$$

$$\ln\left(\frac{19}{4999}\right) = -0.8t \qquad\qquad \text{Inverse Property}$$

$$-5.5726 \approx -0.8t \qquad\qquad \text{Use a calculator.}$$

$$t \approx 6.97 \qquad\qquad \text{Divide each side by } -0.8.$$

So, after about 7 days, 250 students will be infected.

Graphical Solution

To find the number of days that 250 students are infected, use a graphing utility to graph.

$$y_1 = \frac{5000}{1 + 4999e^{-0.8x}} \text{ and } y_2 = 250$$

in the same viewing window. Use the *intersect* feature of the graphing utility to find the point of intersection of the graphs.

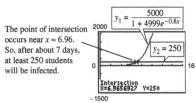

The point of intersection occurs near $x \approx 6.96$. So, after about 7 days, at least 250 students will be infected.

6. (a) Because $I_0 = 1$ and $R = 6.0$, you have the following.

$$R = \log\frac{I}{I_0}$$

$$6.0 = \log\frac{I}{1} \qquad\qquad \text{Substitute 1 for } I_0 \text{ and 6.0 for } R.$$

$$10^{6.0} = 10^{\log I} \qquad\qquad \text{Exponentiate each side (base 10).}$$

$$10^{6.0} = I \qquad\qquad \text{Inverse Property}$$

$$1{,}000{,}000 = I \qquad\qquad \text{Simplify.}$$

(b) Because $I_0 = 1$ and $R = 7.9$, you have the following.

$$7.9 = \log\frac{I}{1} \qquad\qquad \text{Substitute 1 for } I_0 \text{ and 7.9 for } R.$$

$$10^{7.9} = 10^{\log I} \qquad\qquad \text{Exponentiate each side (base 10).}$$

$$10^{7.9} = I \qquad\qquad \text{Inverse Property}$$

$$79{,}432{,}823 \approx I \qquad\qquad \text{Simplify.}$$

Chapter 6

Checkpoints for Section 6.1

1. (a) The slope of this line is $m = \frac{4}{5}$.

So, its inclination is determined from $\tan \theta = \frac{4}{5}$.

Note that $m \geq 0$. This means that

$$\theta = \arctan\left(\tfrac{4}{5}\right) \approx 0.675 \text{ radian} \approx 38.7°.$$

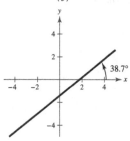

(b) The slope of the line $x + y = -1$ is $m = -1$. So, use $\tan \theta = -1$ to determine its inclination.

Note that $m \leq 0$.

This means that

$$\theta = \pi + \arctan(-1)$$

$$= \pi + \left(-\frac{\pi}{4}\right)$$

$$= \frac{3\pi}{4} \text{ radians} = 135°.$$

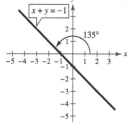

2. The two lines, $4x - 5y + 10 = 0$ and

$3x + 2y + 5 = 0$, have slopes of $m_1 = \dfrac{4}{5}$ and

$m_2 = -\dfrac{3}{2}$, respectively.

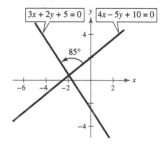

So, the tangent of the angle between the two lines is

$$\tan \theta = \left|\frac{m_2 - m_1}{1 + m_1 m_2}\right| = \left|\frac{-\dfrac{3}{2} - \dfrac{4}{5}}{1 + \left(-\dfrac{3}{2}\right)\left(\dfrac{4}{5}\right)}\right| = \left|\frac{-\dfrac{23}{10}}{-\dfrac{1}{5}}\right| = \frac{23}{2}$$

Finally, you can conclude that the angle is

$$\theta = \arctan\left(\frac{23}{2}\right) \approx 1.4841 \text{ radians} \approx 85.0°.$$

3. The general form of $y = -3x + 2$ is $3x + y - 2 = 0$. So, the distance between the point and a line is

$$d = \frac{|Ax_1 + By_1 + C|}{\sqrt{A^2 + B^2}}$$

$$= \frac{|3(5) + 1(-1) + (-2)|}{\sqrt{(3)^2 + (1)^2}}$$

$$= \frac{12}{\sqrt{10}}$$

$$\approx 3.79 \text{ units.}$$

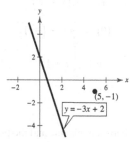

4. The general form is $3x - 5y - 2 = 0$. So, the distance between the point and a line is

$$d = \frac{|Ax_1 + By_1 + C|}{\sqrt{A^2 + B^2}}$$

$$= \frac{|3(3) + (-5)(2) + (-2)|}{\sqrt{(3)^2 + (-5)^2}}$$

$$= \frac{3}{\sqrt{34}} \approx 0.51 \text{ unit.}$$

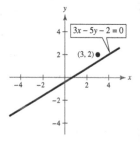

5.

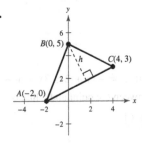

(a) To find the altitude, use the formula for the distance between line AC and the point $(0, 5)$.

The equation of line AC is obtained as follows.

Slope: $m = \dfrac{y_2 - y_1}{x_2 - x_1} = \dfrac{3 - 0}{4 - (-2)} = \dfrac{3}{6} = \dfrac{1}{2}$

Equation: $y - y_1 = m(x - x_1)$ Point-slope form

$y - (0) = \dfrac{1}{2}\left[x - (-2)\right]$ Substitute.

$y = \dfrac{1}{2}x + 1$ Slope-intercept form

$2y = x + 2$ Multiply each side by 2.

$x - 2y + 2 = 0$ General form

So, the distance between this line and the point $(0, 5)$ is

$\text{Altitude} = h = \dfrac{\left|(1)(0) + (-2)(5) + (2)\right|}{\sqrt{(1)^2 + (-2)^2}}$

$= \dfrac{8}{\sqrt{5}}$ units.

(b) Using the formula for the distance between two points, you can find the length of the base AC to be

$b = \sqrt{(x_2 - x_1)^2 + (y_2 - y_1)^2}$ Distance Formula

$= \sqrt{\left[4 - (-2)\right]^2 + (3 - 0)^2}$ Substitute.

$= \sqrt{6^2 + 3^2}$ Simplify.

$= \sqrt{45}$ Simplify.

$= 3\sqrt{5}$ units.   Simplify.

Finally, the area of the triangle is

$A = \dfrac{1}{2}bh$

$= \dfrac{1}{2}\left(3\sqrt{5}\right)\left(\dfrac{8}{\sqrt{5}}\right)$

$= 12$ square units.

Checkpoints for Section 6.2

1. The axis of the parabola is vertical, passing through $(0, 0)$ and $\left(0, \dfrac{3}{8}\right)$. The standard form is $x^2 = 4py$, where $p = \dfrac{3}{8}$.

So, the equation is $x^2 = 4\left(\dfrac{3}{8}\right)y$

$$x^2 = \dfrac{3}{2}y.$$

You can use a graphing utility to confirm this equation. To do this, graph $y_1 = \dfrac{3}{2}x^2$.

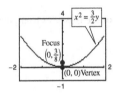

2. Because the axis of the parabola is horizontal, passing through $(2, -3)$ and $(4, -3)$, consider the equation

$$(y - k)^2 = 4p(x - h)$$

where $h = 2, k = -3$, and $p = 4 - 2 = 2$. So, the standard form is

$$\left[y - (-3)\right]^2 = 4(2)(x - 2)$$

$$(y + 3)^2 = 8(x - 2).$$

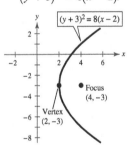

3. Convert to standard form by completing the square.

$x = \dfrac{1}{4}y^2 + \dfrac{3}{2}y + \dfrac{13}{4}$	Write orginal equation.
$4x = y^2 + 6y + 13$	Multiply each side by 4.
$4x - 13 = y^2 + 6y$	Subtract 13 from each side.
$4x - 13 + 9 = y^2 + 6y + 9$	Add 9 to each side.
$4x - 4 = y^2 + 6y + 9$	Combine like terms.
$4(x - 1) = (y + 3)^2$	Standard form

Comparing this equation with $(y - k)^2 = 4p(x - h)$, you can conclude that $h = 1, k = -3$, and $p = 1$. Because p is positive, the parabola opens to the right. So, the focus is $(h + p, k) = (1 + 1, -3) = (2, -3)$.

4. For this parabola, $p = \dfrac{1}{12}$ and the focus is $\left(0, \dfrac{1}{12}\right)$ as shown in the figure below.

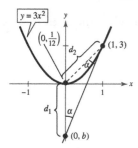

You can find the y-intercept $(0, b)$ of the tangent line by equating the lengths of the two sides of the isosceles triangle shown in the figure:

$$d_1 = \frac{1}{12} - b$$

and

$$d_2 = \sqrt{(1 - 0)^2 + \left(3 - \frac{1}{12}\right)^2} = \sqrt{1^2 + \left(\frac{35}{12}\right)^2}$$

$$= \frac{37}{12}.$$

Note that $d_1 = \dfrac{1}{12} - b$ rather than $b - \dfrac{1}{12}$. The order of subtraction for the distance is important because the distance must be positive. Setting $d_1 = d_2$ produces

$$\frac{1}{12} - b = \frac{37}{12}$$
$$b = -3.$$

So, the slope of the tangent line through $(0, -3)$ and $(1, 3)$ is

$$m = \frac{3 - (-3)}{1 - 0} = 6$$

and the equation of the tangent line in slope-intercept form is $y = 6x - 3$.

Checkpoints for Section 6.3

1. Because the foci occur at $(2, 0)$ and $(2, 6)$, the center of the ellipse is $(2, 3)$ and the distance from the center to one of the foci is $C = 3$. Because $2a = 8$, you know that $a = 4$. Now, from $c^2 = a^2 - b^2$, you have

$$b = \sqrt{a^2 - c^2} = \sqrt{4^2 - 3^2} = \sqrt{7}.$$

Because the major axis is vertical, the standard equation is as follows.

$$\frac{(y - k)^2}{a^2} + \frac{(x - h)^2}{b^2} = 1$$

$$\frac{(y - 3)^2}{4^2} + \frac{(x - 2)^2}{\left(\sqrt{7}\right)^2} = 1$$

$$\frac{(y - 3)^2}{16} + \frac{(x - 2)^2}{7} = 1$$

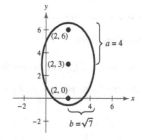

2. $x^2 + 9y^2 = 81$

$$\frac{x^2}{81} + \frac{y^2}{9} = 1$$

The center of the ellipse is $(0, 0)$. The denominator of the x^2-term is greater than the denominator of the y^2-term, so the major axis is horizontal. Because $a^2 = 81$, the endpoints of the major axis lie 9 units from the center at $(9, 0)$ and $(-9, 0)$. Because $b^2 = 9$, the minor axis is vertical and the endpoints of the minor axis lie 3 units above and below the center at $(0, 3)$ and $(0, -3)$.

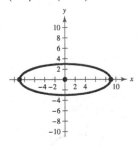

Center: $(0, 0)$

Vertices: $(-9, 0), (9, 0)$

3. Begin by writing the original equation in standard form.

$9x^2 + 4y^2 + 36x - 8y + 4 = 0$	Write original equation.
$9x^2 + 36x + 4y^2 - 8y = -4$	Group terms.
$9\left(x^2 + 4x + \boxed{}\right) + 4\left(y^2 - 2y + \boxed{}\right) = -4$	Factor out leading coefficients.
$9\left(x^2 + 4x + 4\right) + 4\left(y^2 - 2y + 1\right) = -4 + 36 + 4$	Complete the square.
$9(x + 2)^2 + 4(y - 1)^2 = 36$	Write in completed square form.
$\dfrac{(x + 2)^2}{4} + \dfrac{(y - 1)^2}{9} = 1$	Divide each side by 36.
$\dfrac{(x + 2)^2}{2^2} + \dfrac{(y - 1)^2}{3^2} = 1$	Write in standard form.

The center is $(h, k) = (-2, 1)$. Because the denominator of the y-term is $a^2 = 3^2$, the endpoints of the major axis lie 3 units above and below the center. So, the vertices are $(-2, 4)$ and $(-2, -2)$.

Similarly, because the denominator of the x-term is $b^2 = 2^2$, the endpoints of the minor axis lie 2 units to the right and left of the center at $(-4, 1)$ and $(0, 1)$.

Now, from $c^2 = a^2 - b^2$, you have $c = \sqrt{3^2 - 2^2} = \sqrt{5}$.

So, the foci of the ellipse are $\left(-2, 1 + \sqrt{5}\right)$ and $\left(-2, 1 - \sqrt{5}\right)$.

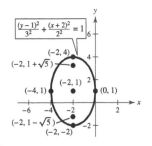

4. By completing the square, you can write the original equation in standard form.

$$5x^2 + 9y^2 + 10x - 54y + 41 = 0 \qquad \text{Write original equation.}$$

$$5x^2 + 10x + 9y^2 - 54y = -41 \qquad \text{Group terms.}$$

$$5\left(x^2 + 2x + \boxed{}\right) + 9\left(y^2 - 6y + \boxed{}\right) = -41 \qquad \text{Factor out leading coefficients.}$$

$$5\left(x^2 + 2x + 1\right) + 9\left(y^2 - 6y + 9\right) = -41 + 5 + 81$$

$$5(x + 1)^2 + 9(y - 3)^2 = 45 \qquad \text{Write in completed square form.}$$

$$\frac{(x + 1)^2}{9} + \frac{(y - 3)^2}{5} = 1 \qquad \text{Divide each side by 45.}$$

$$\frac{(x + 1)^2}{3^2} + \frac{(y - 3)^2}{\left(\sqrt{5}\right)^2} = 1 \qquad \text{Write in standard form.}$$

The major axis is horizontal,

where $h = -1$, $k = 3$, $a = 3$, $b = \sqrt{5}$ and $c = \sqrt{a^2 - b^2}$

$$= \sqrt{3^2 - \left(\sqrt{5}\right)^2} = \sqrt{4} = 2.$$

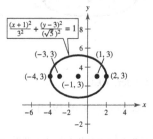

So, you have the following.

Center: $(-1, 3)$ Vertices: $(-4, 3)$ Foci: $(-3, 3)$

                                       $(2, 3)$ $(1, 3)$

5.

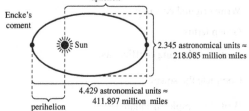

Because $2a = 411.897$ and $2b = 218.085$, you have $a \approx 205.95$ and $b \approx 109.04$

which implies that $c = \sqrt{a^2 - b^2}$

$$= \sqrt{205.95^2 - 109.04^2}$$

$$\approx 174.72.$$

So, the greatest distance, the aphelion, from the sun's center to the comet's center is

$a + c \approx 205.95 + 174.72 = 380.67$ million miles

and the least distance, the perihelion, is

$a - c = 205.95 - 174.72 = 31.23$ million miles.

Checkpoints for Section 6.4

1. By the Midpoint Formula, the center of the hyperbola occurs at the point of

$$(h, k) = \left(\frac{2 + 2}{2}, -\frac{4 + 2}{2} \right) = (2, -1).$$

Furthermore, $a = 2 - (-1) = 3$ and $c = 3 - (-1) = 4$, and it follows that $b = \sqrt{c^2 - a^2} = \sqrt{4^2 - 3^2} = \sqrt{7}$.

So, the hyperbola has a vertical transverse axis and the standard form of the equation is

$$\frac{(y - k)^2}{a^2} - \frac{(x - h)^2}{b^2} = 1$$

$$\frac{(y + 1)^2}{3^2} - \frac{(x - 2)^2}{\left(\sqrt{7} \right)^2} = 1.$$

This equation simplifies to

$$\frac{(y + 1)^2}{9} - \frac{(x - 2)^2}{7} = 1$$

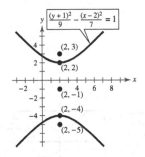

2. Algebraic Solution

Divide each side of the equation by 36, and write the equation in standard form.

$$4y^2 - 9x^2 = 36$$

$$\frac{y^2}{9} - \frac{x^2}{4} = 1$$

$$\frac{y^2}{3^2} - \frac{x^2}{2^2} = 1$$

From this, you can conclude that $a = 3$, $b = 2$, and the transverse axis is vertical. Because the center is $(0, 0)$, the vertices occur at $(0, 3)$ and $(0, -3)$. The endpoints of the conjugate axis occur at $(2, 0)$ and $(-2, 0)$. Using these four points, sketch a rectangle that is $2a = 6$ units tall and $2b = 4$ units wide.

Now from $c^2 = a^2 + b^2$, you have $c = \sqrt{a^2 + b^2} = \sqrt{3^2 + 2^2} = \sqrt{13}$. So, the foci of the hyperbola are $\left(0, \sqrt{13}\right)$ and $\left(0, -\sqrt{13}\right)$.

Finally, by drawing the asymptotes through the corners of this rectangle, you can complete the sketch. Note that the asymptotes are $y = \frac{3}{2}x$ and $y = -\frac{3}{2}x$.

Graphical Solution

Solve the equation of the hyperbola for y as follows.

$$4y^2 - 9x^2 = 36$$

$$4y^2 = 9x^2 + 36$$

$$y^2 = \frac{9x^2 + 36}{4}$$

$$y = \pm\sqrt{\frac{9x^2 + 36}{4}}$$

$$y = \pm\frac{3}{2}\sqrt{x^2 + 4}$$

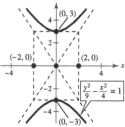

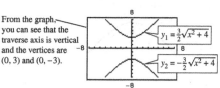

From the graph, you can see that the traverse axis is vertical and the vertices are $(0, 3)$ and $(0, -3)$.

Then use a graphing utility to graph $y_1 = \frac{3}{2}\sqrt{x^2 + 4}$ and $y_2 = -\frac{3}{2}\sqrt{x^2 + 4}$ in the same viewing window.

3.

$9x^2 - 4y^2 + 8y - 40 = 0$	Write original equation.
$9x^2 - 4y^2 + 8y = 40$	Group terms.
$9x^2 - 4\left(y^2 - 2y + \square\right) = 40$	Factor out leading coefficients.
$9x^2 - 4\left(y^2 - 2y + 1\right) = 40 - 4$	Complete the square.
$9x^2 - 4(y - 1)^2 = 36$	Write in completed square form.
$\dfrac{x^2}{4} - \dfrac{(y - 1)^2}{9} = 1$	Divide each side by 36.
$\dfrac{x^2}{2^2} - \dfrac{(y - 1)^2}{3^2} = 1$	Write in standard form.

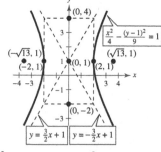

From this equation you can conclude that the hyperbola has a horizontal transverse axis centered at $(0, 1)$. The vertices are at $(-2, 1)$ and $(2, 1)$, and has a conjugate axis with endpoints $(0, 4)$ and $(0, -2)$. To sketch the hyperbola, draw a rectangle through these four points. The asymptotes are the lines passing through the corners of the rectangle.

Using $a = 2$ and $b = 3$, you can conclude that the equations of the asymptotes are $y = \frac{3}{2}x + 1$ and $y = -\frac{3}{2}x + 1$.

Finally, you can determine the foci by using the equation $c^2 = a^2 + b^2$. So, you have $c = \sqrt{2^2 + 3^2} = \sqrt{13}$, and the foci are $\left(\sqrt{13}, 1\right)$ and $\left(-\sqrt{13}, 1\right)$.

4. Using the Midpoint Formula you can determine the center is $(h, k) = \left(\dfrac{3 + 9}{2}, \dfrac{2 + 2}{2}\right) = (6, 2)$.

Furthermore, the hyperbola has a horizontal transverse axis with $a = 3$. From the original equations, you can determine the slopes of the asymptotes to be $m_1 = \dfrac{b}{a} = \dfrac{2}{3}$ and $m_2 = -\dfrac{b}{a} = -\dfrac{2}{3}$, and because $a = 3$, you can conclude that $b = 2$.

So, the standard form of the equation of the hyperbola is

$$\dfrac{(x - 6)^2}{3^2} - \dfrac{(y - 2)^2}{2^2} = 1.$$

5. Begin by representing the situation in a coordinate plane. The distance between the microphones is 1 mile, or 5280 feet. So, position the point representing microphone A 2640 units to the right of the origin and the point representing microphone B 2640 units to the left of the origin, as shown.

Assuming sound travels at 1100 feet per second, the explosion took place 4400 feet farther from B than from A. The locus of all points that are 4400 feet closer to A than to B is one branch of a hyperbola with foci at A and B. Because the hyperbola is centered at the origin and has a horizontal transverse axis, the standard form of its equation is

$$\dfrac{x^2}{a^2} - \dfrac{y^2}{b^2} = 1.$$

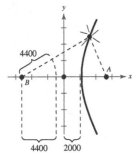

Because the foci are 2640 units from the center, $c = 2640$. Let d_A and d_B be the distances of any point on the hyperbola from the foci at A and B, respectively. From page 713, you have

$$|d_B - d_A| = 2a$$

$$|4400| = 2a \qquad \text{The points are 4400 feet closer to A than to B}$$

$$2200 = a \qquad \text{Divide each side by 2}$$

So, $b^2 = c^2 - a^2 = 2640^2 - 2200^2 = 2,129,600$ and you can conclude that the explosion occurred somewhere on the right branch of the hyperbola

$$\dfrac{x^2}{4,840,000} - \dfrac{y^2}{2,129,600} = 1.$$

6. (a) For the equation $3x^2 + 3y^2 - 6x + 6y + 5 = 0$, you have $A = C = 3$. So, the graph is a circle.

(b) For the equation $2x^2 - 4y^2 + 4x + 8y - 3 = 0$, you have $AC = 2(-4) < 0$. So, the graph is a hyperbola.

(c) For the equation $3x^2 + y^2 + 6x - 2y + 3 = 0$, you have $AC = 3(1) > 0$. So, the graph is an ellipse.

(d) For the equation $2x^2 + 4x + y - 2 = 0$, you have $AC = 2(0) = 0$. So, the graph is a parabola.

Checkpoints for Section 6.5

1. Because $A = 0$, $B = 1$, and $C = 0$, you have $\cot 2\theta = \dfrac{A - C}{B} = 0 \Rightarrow 2\theta = \dfrac{\pi}{2} \Rightarrow \theta = \dfrac{\pi}{4}$ which implies that

$$x = x'\cos\frac{\pi}{4} - y'\sin\frac{\pi}{4} \quad \text{and} \quad y = x'\sin\frac{\pi}{4} + y'\cos\frac{\pi}{4}$$

$$= x'\left(\frac{1}{\sqrt{2}}\right) - y'\left(\frac{1}{\sqrt{2}}\right) \qquad = x'\left(\frac{1}{\sqrt{2}}\right) + y'\left(\frac{1}{\sqrt{2}}\right)$$

$$= \frac{x' - y'}{\sqrt{2}} \qquad\qquad\qquad = \frac{x' + y'}{\sqrt{2}}.$$

The equation in the $x'y'$-system is obtained by substituting these expressions in the original equation.

$$xy + 6 = 0$$

$$\left(\frac{x' - y'}{\sqrt{2}}\right)\left(\frac{x' + y'}{\sqrt{2}}\right) + 6 = 0$$

$$\frac{(x')^2 - (y')^2}{2} + 6 = 0$$

$$\frac{(x')^2 - (y')^2}{2} = -6$$

$$\frac{(y')^2 - (x')^2}{12} = 1$$

$$\frac{(y')^2}{12} - \frac{(x')^2}{12} = 1$$

$$\frac{(y')^2}{\left(2\sqrt{3}\right)^2} - \frac{(x')^2}{\left(2\sqrt{3}\right)^2} = 1$$

In the $x'y'$-system this is a hyperbola centered at the origin with vertices at $\left(0, \pm 2\sqrt{3}\right)$, as shown. To find the coordinates of the vertices in the xy-system, substitute the coordinates $\left(0, \pm 2\sqrt{3}\right)$ in the equations

$$x = \frac{x' - y'}{\sqrt{2}} \quad \text{and} \quad y = \frac{x' + y'}{\sqrt{2}}.$$

This substitution yields the vertices

$$\left(\frac{0 - 2\sqrt{3}}{\sqrt{2}}, \frac{0 + 2\sqrt{3}}{\sqrt{2}}\right) = \left(-\sqrt{6}, \sqrt{6}\right) \text{ and } \left(\frac{0 - \left(-2\sqrt{3}\right)}{\sqrt{2}}, \frac{0 - 2\sqrt{3}}{\sqrt{2}}\right) = \left(\sqrt{6}, -\sqrt{6}\right)$$

in the xy-system. Note that the asymptotes of the hyperbola have equations $y' = \pm x'$ which correspond to the original x- and y-axes.

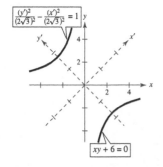

2. Because $A = 12$, $B = 16\sqrt{3}$, and $C = 28$, you have

$$\cot 2\theta = \frac{A - C}{B} = \frac{12 - 28}{16\sqrt{3}} = \frac{-16}{16\sqrt{3}} = \frac{-1}{\sqrt{3}}$$

which implies that $2\theta = \dfrac{2\pi}{3} \Rightarrow \theta = \dfrac{\pi}{3}$. The equation in the $x'y'$-system is obtained by making the substitutions

$$x = x'\cos\frac{\pi}{3} - y'\sin\frac{\pi}{3}$$

$$= x'\left(\frac{1}{2}\right) - y'\left(\frac{\sqrt{3}}{2}\right)$$

$$= \frac{x' - \sqrt{3}y'}{2}$$

and $y = x'\sin\dfrac{\pi}{3} + y'\cos\dfrac{\pi}{3}$

$$= x'\left(\frac{\sqrt{3}}{2}\right) + y'\left(\frac{1}{2}\right)$$

$$= \frac{\sqrt{3}x' + y'}{2}$$

in the original equation. So, you have

$$12x^2 + 16\sqrt{3}xy + 28y^2 - 36 = 0$$

$$12\left(\frac{x' - \sqrt{3}y'}{2}\right)^2 + 16\sqrt{3}\left(\frac{x' - \sqrt{3}y'}{2}\right)\left(\frac{\sqrt{3}x' + y'}{2}\right) + 28\left(\frac{\sqrt{3}x' + y'}{2}\right)^2 - 36 = 0$$

which simplifies to

$$36(x')^2 + 4(y')^2 - 36 = 0$$

$$36(x')^2 + 4(y')^2 = 36$$

$$\frac{(x')^2}{1} + \frac{(y')^2}{9} = 1$$

$$\frac{(x')^2}{1^2} + \frac{(y')^2}{3^2} = 1.$$

This is the equation of an ellipse centered at the origin with vertices $(0, \pm 3)$ in the $x'y'$-system as shown.

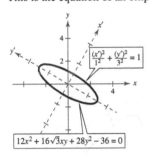

3.

Because $A = 4$, $B = 4$ and $C = 1$ you have

$$\cot 2\theta = \frac{A - C}{B} = \frac{4 - 1}{4} = \frac{3}{4}.$$

Using this information, draw a right triangle as shown in Figure 10.35. From the figure, you can see that $\cos 2\theta = \frac{3}{5}$. To find the values of $\sin \theta$ and $\cos \theta$, you can use the half-angle formulas in the forms

$$\sin \theta = \sqrt{\frac{1 - \cos 2\theta}{2}} \text{ and } \cos \theta = \sqrt{\frac{1 + \cos 2\theta}{2}}.$$

So,

$$\sin \theta = \sqrt{\frac{1 - \cos 2\theta}{2}} = \sqrt{\frac{1 - \frac{3}{5}}{2}} = \sqrt{\frac{1}{5}} = \frac{1}{\sqrt{5}}$$

$$\cos \theta = \sqrt{\frac{1 + \cos 2\theta}{2}} = \sqrt{\frac{1 + \frac{3}{5}}{2}} = \sqrt{\frac{4}{5}} = \frac{2}{\sqrt{5}}.$$

Consequently, you use the substitutions

$$x = x'\cos \theta - y'\sin \theta = x'\left(\frac{2}{\sqrt{5}}\right) - y'\left(\frac{1}{\sqrt{5}}\right) = \frac{2x' - y'}{\sqrt{5}}$$

and

$$y = x'\sin \theta + y'\cos \theta = x'\left(\frac{1}{\sqrt{5}}\right) + y'\left(\frac{2}{\sqrt{5}}\right) = \frac{x' + 2y'}{\sqrt{5}}.$$

Substituting these expressions in the original equation, you have

$$4x^2 + 4xy + y^2 - 2\sqrt{5}x + 4\sqrt{5}y - 30 = 0$$

$$4\left(\frac{2x' - y'}{\sqrt{5}}\right)^2 + 4\left(\frac{2x' - y'}{\sqrt{5}}\right)\left(\frac{x' + 2y'}{\sqrt{5}}\right) + \left(\frac{x' + 2y'}{\sqrt{5}}\right)^2 - 2\sqrt{5}\left(\frac{2x' - y'}{\sqrt{5}}\right) + 4\sqrt{5}\left(\frac{x' + 2y'}{\sqrt{5}}\right) - 30 = 0$$

which simplifies as follows.

$$5(x')^2 + 10y' - 30 = 0$$

$$5(x')^2 = -10y' + 30$$

$$(x')^2 = -2y' + 6$$

$$(x')^2 = -2(y' - 3)$$

The graph of this equation is a parabola with vertex $(0, 3)$ in the $x'y'$-system. Its axis is parallel to the y'-axis in the $x'y'$-system and because

$$\sin \theta = \frac{1}{\sqrt{5}}, \theta \approx 26.6°, \text{ as shown.}$$

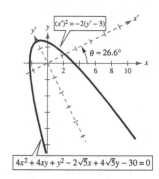

4. Because $B^2 - 4AC = (-8)^2 - 4(2)(8) = 64 - 64 = 0,$ the graph is a parabola.

$$2x^2 - 8xy + 8y^2 + 3x + 5 = 0$$

$$8y^2 - 8xy + 2x^2 + 3x + 5 = 0$$

$$8y^2 + (-8x)y + (2x^2 + 3x + 5) = 0$$

$$y = \frac{-(-8x) \pm \sqrt{(-8x)^2 - 4(8)(2x^2 + 3x + 5)}}{2(8)}$$

$$y = \frac{8x \pm \sqrt{64x^2 - 64x^2 - 96x - 160}}{16}$$

$$y = \frac{8x \pm \sqrt{-96x - 160}}{16}$$

So, $y_1 = \dfrac{8x + \sqrt{-96x - 160}}{16}$ and

$$y_2 = \frac{8x - \sqrt{-96x - 160}}{16}.$$

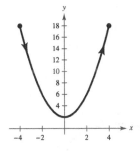

Checkpoints for Section 6.6

1. Using values of t in the specified interval, $-2 \le t \le 2,$ the parametric equations yield the points (x, y) shown.

t	x	y
-2	-4	18
-1	-2	6
0	0	2
1	2	6
2	4	18

By plotting these points in the order of increasing values of t, you obtain the curve shown. The arrows on the curve indicate its orientation as t increases from -2 to 2.

The curve starts at $(-4, 18)$ and ends at $(4, 18).$

2. Solving for t in the equation for x produces the following.

$$x = \frac{1}{\sqrt{t-1}}$$

$$x^2 = \frac{1}{t-1}$$

$$t - 1 = \frac{1}{x^2}$$

$$t = \frac{1}{x^2} + 1$$

$$t = \frac{1 + x^2}{x^2}$$

Now, substituting in the equation for y, you obtain the following rectangular equation.

$$y = \frac{\dfrac{1+x^2}{x^2} + 1}{\dfrac{1+x^2}{x^2} - 1} = \frac{\dfrac{1+x^2+x^2}{x^2}}{\dfrac{1+x^2-x^2}{x^2}} = \frac{\dfrac{2x^2+1}{x^2}}{\dfrac{1}{x^2}} = \frac{2x^2+1}{x^2} \cdot \frac{x^2}{1} = 2x^2 + 1$$

From this rectangular equation, you can recognize that the curve is a parabola that opens upward and has its vertex at $(0, 1)$.

Also, this rectangular equation is defined for all values of x. The parametric equation for x, however, is defined only when $t > 1$. This implies that you should restrict the domain of x to positive values.

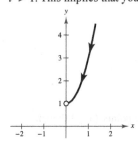

3. (a) Begin by solving for $\cos\theta$ and $\sin\theta$ in the equations

$$x = 5\cos\theta \implies \cos\theta = \frac{x}{5}$$

$$y = 3\sin\theta \implies \sin\theta = \frac{y}{3}$$

Use the Pythagorean identity $\sin^2\theta + \cos^2\theta = 1$ to form an equation involving only x and y.

$$\sin^2\theta + \cos^2\theta = 1 \qquad \text{Pythagorean identity}$$

$$\left(\frac{y}{3}\right)^2 + \left(\frac{x}{5}\right)^2 = 1 \qquad \text{Substitute } \frac{y}{3} \text{ for } \sin\theta$$

$$\text{and } \frac{x}{5} \text{ for } \cos\theta.$$

$$\frac{y^2}{9} + \frac{x^2}{25} = 1 \qquad \text{Simplify.}$$

$$\frac{x^2}{25} + \frac{y^2}{9} = 1 \qquad \text{Rectangular equation}$$

From this rectangular equation, you can see that the graph is an ellipse centered at $(0, 0)$, with horizontal major axis, vertices at $(5, 0)$ and $(-5, 0)$ and minor axis of length $2b = 6$.

Note that the elliptic curve is traced out counterclockwise, starting at $(5, 0)$, as θ increases on the interval $[0, 2\pi)$.

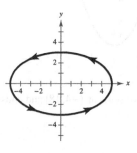

(b) Begin by solving for $\tan \theta$ and $\sec \theta$ in the equations

$x = -1 + \tan \theta \Rightarrow \tan \theta = x + 1$

$y = 2 + 2 \sec \theta \Rightarrow \sec \theta = \dfrac{y - 2}{2}.$

Use the Pythagorean identity $\tan^2 \theta + 1 = \sec^2 \theta$ to form an equation involving only x and y.

$\tan^2 \theta + 1 = \sec^2 \theta$ ⠀⠀⠀⠀Pythagorean identity

$(x + 1)^2 + 1 = \left(\dfrac{y - 2}{2}\right)^2$ ⠀⠀Substitute $x + 1$ for $\tan \theta$ and $\dfrac{y - 2}{2}$ for $\sec \theta$.

$\left(\dfrac{y - 2}{2}\right)^2 - (x + 1)^2 = 1$ ⠀⠀⠀⠀Rewrite.

$\dfrac{(y - 2)^2}{4} - (x + 1)^2 = 1$ ⠀⠀⠀⠀Rectangular equation

From this rectangular equation, you can see that the graph is a hyperbola centered at $(-1, 2)$, with a vertical transverse axis. Because $a = 2$, the vertices are $(-1, 4)$ and $(-1, 0)$. However, the restriction on θ, $\dfrac{\pi}{2} \le \theta \le \dfrac{3\pi}{2}$, corresponds to the lower branch of the hyperbola.

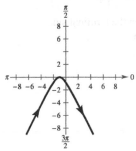

4. (a) Letting $t = x$, you obtain the parametric equations

$x = t$ and $y = x^2 + 2 = t^2 + 2.$

The curve represented by the parametric equations is shown.

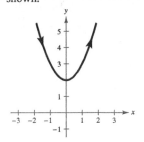

(b) Letting $t = 2 - x$, you obtain the parametric equations

$t = 2 - x \Rightarrow x = 2 - t$ and

$y = x^2 + 2 = t^2 - 4t + 6.$

The curve represented by the parametric equations is shown.

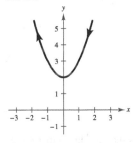

5. As the parameter, let θ be the measure of the circle's rotation, and let the point $P(x, y)$ begin at $(0, 2a)$.

When $\theta = 0$, P is at $(0, 2a)$; when $\theta = \pi$, P is at a minimum point $(\pi a, 0)$; and when $\theta = 2\pi$, P is at a maximum point $(2\pi a, 2a)$. From the figure, $\angle APC = \theta$. So,

$$\sin \theta = \sin(\angle APC) = \frac{AC}{a} = \frac{BD}{a}$$

$$\cos \theta = \cos(\angle APC) = \frac{AP}{a}$$

which implies that $BD = a \sin \theta$ and $AP = a \cos \theta$. Because the circle rolls along the x-axis, $OD = \overset{\frown}{QD} = a\theta$.

Furthermore, because $BA = DC = a$,

$$x = OD + BD = a\theta + a \sin \theta$$

$$y = BA + AP = a + a \cos \theta.$$

So, the parametric equations are $x = a(\theta + \sin \theta)$ and $y = a(1 + \cos \theta)$.

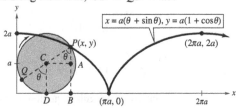

Checkpoints for Section 6.7

1. (a) The point $(r, \theta) = \left(3, \dfrac{\pi}{4}\right)$ lies three units from the pole on the terminal side of the angle $\theta = \dfrac{\pi}{4}$.

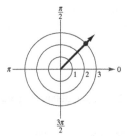

(b) The point $(r, \theta) = \left(2, -\dfrac{\pi}{3}\right)$ lies two units from the pole on the terminal side of the angle $\theta = -\dfrac{\pi}{3}$.

(c) The point $(r, \theta) = \left(2, \dfrac{5\pi}{3}\right)$ lies two units from the pole on the terminal side of the angle $\theta = \dfrac{5\pi}{3}$, which coincides with the point $\left(2, -\dfrac{\pi}{3}\right)$.

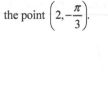

The point is shown. Three other representations are as follows.

$$\left(-1, \frac{3\pi}{4} + 2\pi\right) = \left(-1, \frac{11\pi}{4}\right) \quad \text{Add } 2\pi \text{ to } \theta.$$

$$\left(1, \frac{3\pi}{4} - \pi\right) = \left(1, -\frac{\pi}{4}\right) \quad \text{Replace } r \text{ with } -r; \text{ subtract } \pi \text{ from } \theta.$$

$$\left(1, \frac{3\pi}{4} + \pi\right) = \left(1, \frac{7\pi}{4}\right) \quad \text{Replace } r \text{ with } -r; \text{ add } \pi \text{ to } \theta.$$

$$\left(-1, \frac{3\pi}{4}\right) = \left(-1, \frac{11\pi}{4}\right) = \left(1, -\frac{\pi}{4}\right) = \left(1, \frac{7\pi}{4}\right) = \ldots$$

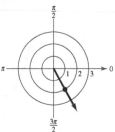

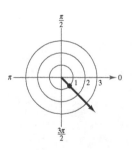

3. For the point $(r, \theta) = (2, \pi)$, you have the following.

$$x = r \cos \theta = 2 \cos \pi = (2)(-1) = -2$$
$$y = r \sin \theta = 2 \sin \pi = (2)(0) = 0$$

The rectangular coordinates are $(x, y) = (-2, 0)$.

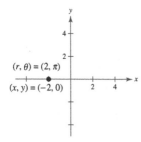

4. For the point $(x, y) = (0, 2)$, which lies on the positive y-axis, you have

$$\tan \theta = \frac{2}{0} \text{ is undefined} \Rightarrow \theta = \frac{\pi}{2}.$$

Choosing a positive value for r,

$$r = \sqrt{x^2 + y^2} = \sqrt{0^2 + 2^2} = 2.$$

So, *one* set of polar coordinates is $(r, \theta) = \left(2, \frac{\pi}{2}\right)$ as shown.

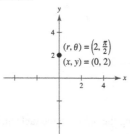

5. (a)
$$r = 7$$
$$r^2 = 49$$
$$x^2 + y^2 = 49$$

The graph consists of all points that are seven units from the pole, which is a circle centered at the origin with a radius of 7.

(b)
$$\theta = \frac{\pi}{4}$$
$$\tan \theta = \tan \frac{\pi}{4}$$
$$\frac{y}{x} = 1$$
$$y = x$$

The graph consists of all points on the line that makes an angle of $\frac{\pi}{4}$ with the polar axis and passes through the pole.

(c)
$$r = 6 \sin \theta$$
$$r^2 = 6r \sin \theta$$
$$x^2 + y^2 = 6y$$
$$x^2 + y^2 - 6y = 0$$
$$x^2 + y^2 - 6y + 9 = 9$$
$$x^2 + (y - 3)^2 = 9$$

The graph is a circle with center $(0\ 3)$ and radius 3.

Checkpoints for Section 6.8

1. The cosine function is periodic, so you can get a full range of *r*-values by considering values of θ in the interval $0 \le \theta \le 2\pi$, as shown.

θ	0	$\dfrac{\pi}{6}$	$\dfrac{\pi}{4}$	$\dfrac{\pi}{3}$	$\dfrac{\pi}{2}$	$\dfrac{2\pi}{3}$	$\dfrac{3\pi}{4}$	$\dfrac{5\pi}{6}$	π
r	6	$3\sqrt{3}$	$3\sqrt{2}$	3	0	-3	$-3\sqrt{2}$	$-3\sqrt{3}$	-6

θ	$\dfrac{7\pi}{6}$	$\dfrac{5\pi}{4}$	$\dfrac{4\pi}{3}$	$\dfrac{3\pi}{2}$	$\dfrac{5\pi}{3}$	$\dfrac{7\pi}{4}$	$\dfrac{11\pi}{6}$	2π
r	$-3\sqrt{3}$	$-3\sqrt{2}$	-3	0	3	$3\sqrt{2}$	$3\sqrt{3}$	6

By plotting these points, it appears that the graph is a circle of radius 3 whose center is at the point $(x, y) = (3, 0)$.

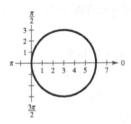

2. Replacing (r, θ) by $(r, \pi - \theta)$ produces the following.

$$r = 3 + 2\sin(\pi - \theta)$$
$$= 3 + 2(\sin\pi\cos\theta - \cos\pi\sin\theta)$$
$$= 3 + 2\left[(0)\cos\theta - (-1)\sin\theta\right]$$
$$= 3 + 2(0 + \sin\theta)$$
$$= 3 + 2\sin\theta$$

So, you can conclude that the curve is symmetric with respect to the line $\theta = \pi/2$. Plotting the points in the table and using symmetry with respect to the line $\theta = \pi/2$, you obtain the graph shown, called a dimpled limaçon.

θ	0	$\dfrac{\pi}{6}$	$\dfrac{\pi}{4}$	$\dfrac{\pi}{3}$	$\dfrac{\pi}{2}$	$\dfrac{2\pi}{3}$	$\dfrac{3\pi}{4}$	$\dfrac{5\pi}{6}$	π
r	3	4	$3 + \sqrt{2}$	$3 + \sqrt{3}$	5	$3 + \sqrt{3}$	$3 + \sqrt{2}$	4	3

θ	$\dfrac{7\pi}{6}$	$\dfrac{5\pi}{4}$	$\dfrac{4\pi}{3}$	$\dfrac{3\pi}{2}$	$\dfrac{5\pi}{3}$	$\dfrac{7\pi}{4}$	$\dfrac{11\pi}{6}$	2π
r	2	$3 - \sqrt{2}$	$3 - \sqrt{3}$	1	$3 - \sqrt{3}$	$3 - \sqrt{2}$	2	3

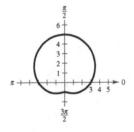

3. From the equation $r = 1 + 2 \sin \theta$, you can obtain the following:

$\quad$ *Symmetry*: With respect to the line $\theta = \dfrac{\pi}{2}$

$\quad$ *Maximum value of* $|r|$: $r = 3$ when $\theta = \dfrac{\pi}{2}$

$\quad\quad$ *Zero of r*: $r = 0$ when $\theta = \dfrac{7\pi}{6}$ and $\dfrac{11\pi}{6}$

The table shows several θ-values in the interval $[0, 2\pi]$

By plotting the corresponding points, you can sketch the graph.

θ	0	$\dfrac{\pi}{6}$	$\dfrac{\pi}{4}$	$\dfrac{\pi}{3}$	$\dfrac{\pi}{2}$	$\dfrac{2\pi}{3}$	$\dfrac{3\pi}{4}$	$\dfrac{5\pi}{6}$	π
r	1	2	$1 + \sqrt{2}$	$1 + \sqrt{3}$	3	$1 + \sqrt{3}$	$1 + \sqrt{2}$	2	1

θ	$\dfrac{7\pi}{6}$	$\dfrac{5\pi}{4}$	$\dfrac{4\pi}{3}$	$\dfrac{3\pi}{2}$	$\dfrac{5\pi}{3}$	$\dfrac{7\pi}{4}$	$\dfrac{11\pi}{6}$	2π
r	0	$1 - \sqrt{2}$	$1 - \sqrt{3}$	-1	$1 - \sqrt{3}$	$1 - \sqrt{2}$	0	1

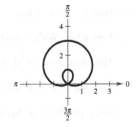

4. From the equation $r = 2 \sin 3\theta$, you can obtain the following.

$\quad$ *Symmetry*: With respect to the line $\theta = \dfrac{\pi}{2}$

$\quad$ *Maximum value of* $|r|$: $|r| = 2$ when $3\theta = \dfrac{\pi}{2}, \dfrac{3\pi}{2}, \dfrac{5\pi}{2}$

$\quad\quad$ or $\theta = \dfrac{\pi}{6}, \dfrac{\pi}{2}, \dfrac{5\pi}{6}$

$\quad$ *Zeros of r*: $r = 0$ when $3\theta = 0, \pi, 2\pi, 3\pi$

$\quad\quad$ or $\theta = 0, \dfrac{\pi}{3}, \dfrac{2\pi}{3}, \pi$

By plotting these points and using symmetry with respect to $\theta = \dfrac{\pi}{2}$, zeros, and maximum values, you obtain the graph.

θ	0	$\dfrac{\pi}{12}$	$\dfrac{\pi}{6}$	$\dfrac{\pi}{4}$	$\dfrac{\pi}{3}$	$\dfrac{5\pi}{12}$	$\dfrac{\pi}{2}$	$\dfrac{7\pi}{12}$	$\dfrac{2\pi}{3}$	$\dfrac{3\pi}{4}$	$\dfrac{5\pi}{6}$	π
r	0	$\sqrt{2}$	2	$\sqrt{2}$	0	$-\sqrt{2}$	-2	$-\sqrt{2}$	0	$\sqrt{2}$	2	0

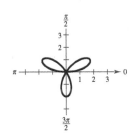

5. *Type of curve*: Rose curve with $n = 3$ petals

   Symmetry: With respect to the polar axis

   Maximum value of $|r|$: $|r| = 3$ when $\theta = 0, \dfrac{\pi}{3}, \dfrac{2\pi}{3}, \pi$

   Zeros of r: $r = 0$ when $\theta = \dfrac{\pi}{6}, \dfrac{\pi}{2}, \dfrac{5\pi}{6}$

   Using this information together with the points shown, you obtain the graph.

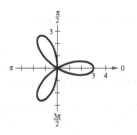

θ	0	$\dfrac{\pi}{6}$	$\dfrac{\pi}{4}$	$\dfrac{\pi}{3}$	$\dfrac{\pi}{2}$	$\dfrac{2\pi}{3}$	$\dfrac{3\pi}{4}$	$\dfrac{5\pi}{6}$	π
r	3	0	$-\dfrac{3\sqrt{3}}{2}$	-3	0	3	$\dfrac{3\sqrt{3}}{4}$	0	-3

6. *Type of curve*: Lemniscate

   Symmetry: With respect to the polar axis, the line $\theta = \dfrac{\pi}{2}$, and the pole

   Maximum value of $|r|$: $|r| = 2$ when $\theta = 0, \pi$

   Zeros of r: $r = 0$ when $\theta = \dfrac{\pi}{4}, \dfrac{3\pi}{4}$

   When $\cos 2\theta < 0$, this equation has no solution points.

   So, you can restrict the values of θ to three for which $\cos 2\theta \geq 0. \; 0 \leq \theta \leq \dfrac{\pi}{4}$ and $\dfrac{3\pi}{4} \leq \theta \leq \pi$

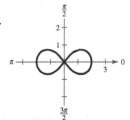

θ	0	$\dfrac{\pi}{6}$	$\dfrac{\pi}{4}$	$\dfrac{3\pi}{4}$	$\dfrac{5\pi}{6}$	π
$r = \pm 2\sqrt{\cos 2\theta}$	± 2	$\pm\dfrac{2}{\sqrt{2}}$	0	0	$\pm\dfrac{2}{\sqrt{2}}$	± 2

   Using symmetry and these points, you obtain the graph shown.

Checkpoints for Section 6.9

1. Algebraic Solution

   To identify the type of conic, rewrite the equation in the form $r = \dfrac{ep}{1 \pm e \sin\theta}$.

   $r = \dfrac{8}{2 - 3 \sin \theta}$ Write original equation.

   $r = \dfrac{4}{1 - \frac{3}{2}\sin \theta}$ Divide numerator and denominator by 2.

   Because $e = \dfrac{3}{2} > 1$, you can conclude the graph is a hyperbola.

Graphical Solution

   Use a graphing utility in *polar* mode and be sure to use a square setting as shown.

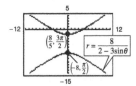

   The graph of the conic appears to be a hypberbola.

2. Dividing the numerator and denominator by 2, you have

$$r = \frac{3/2}{1 - 2 \sin \theta}.$$

Because $e = 2 > 1$, the graph is a hyperbola.

The transverse axis of the hyperbola lies on the line $\theta = \frac{\pi}{2}$, and the vertices occur at $\left(-\frac{3}{2}, \frac{\pi}{2}\right)$ and $\left(\frac{1}{2}, \frac{3\pi}{2}\right)$.

Because the length of the transverse axis is 1, you can see that $a = \frac{1}{2}$. To find b, write

$$b^2 = a^2\left(e^2 - 1\right) = \left(\frac{1}{2}\right)^2\left[(2)^2 - 1\right] = \frac{3}{4}.$$

So, $b = \frac{\sqrt{3}}{2}$. You can use a and b to determine

that the asymptotes of the hyperbola are $y = -1 \pm \frac{\sqrt{3}}{3}x$.

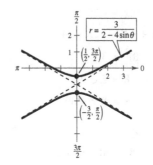

3. Because the directrix is vertical and left of the pole, use an equation of the form

$$r = \frac{ep}{1 - e \cos \theta}.$$

Moreover, because the eccentricity of a parabola is $e = 1$ and the distance between the pole and the directrix is $p = 2$, you have the equation

$$r = \frac{ep}{1 - e \cos \theta} = \frac{2}{1 - \cos \theta}.$$

4. Using a vertical major axis as shown, choose an equation of the form $r = \frac{ep}{1 + e \sin \theta}$.

Because the vertices of the ellipse occur when $\theta = \frac{\pi}{2}$ and $\theta = \frac{3\pi}{2}$, you can determine the length of the major axis to be the sum of the r-values of the vertices.

$$2a = \frac{0.847p}{1 + 0.847} + \frac{0.847p}{1 - 0.847} \approx 5.995p \approx 4.420$$

So, $p \approx 0.737$ and $ep \approx (0.847)(0.737) \approx 0.625$.

Using this value of ep in the equation, you have

$$r = \frac{ep}{1 + e \sin \theta} = \frac{0.625}{1 + 0.847 \sin \theta}$$

where r is measured in astronomical units.

To find the closest point to the sun (the focus), substitute $\theta = \frac{\pi}{2}$ into this equation.

$$r = \frac{0.625}{1 + 0.847 \sin \pi/2} = \frac{0.625}{1 + 0.847(1)}$$

$$\approx 0.340 \text{ astronomical unit}$$

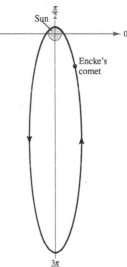

Chapter P   Practice Test Solutions

1. $5(x - 2) - 4 = 3x + 8$

$5x - 10 - 4 = 3x + 8$

$5x - 14 = 3x + 8$

$2x = 22$

$x = 11$

2. $x^2 - 16x + 25 = 0$

$x^2 - 16x = -25$

$x^2 - 16x + 64 = -25 + 64$

$(x - 8)^2 = 39$

$x - 8 = \pm\sqrt{39}$

$x = 8 \pm \sqrt{39}$

3. $|3x + 5| = 8$

$3x + 5 = -8$ OR $3x + 5 = 8$

$3x = -13$ $3x = 3$

$x = -\frac{13}{3}$ $x = 1$

4. $[x - (-3)]^2 + (y - 5)^2 = 6^2$

$(x + 3)^2 + (y - 5)^2 = 36$

5. $(2, 4)$ and $(3, -1)$

$d = \sqrt{(3 - 2)^2 + (-1 - 4)^2}$

$= \sqrt{(1)^2 + (-5)^2}$

$= \sqrt{26}$

6. $y = \frac{4}{3}x - 3$

7. $2x + 3y = 0$

$y = -\frac{2}{3}x$

$m_1 = -\frac{2}{3}$

$\perp m_2 = \frac{3}{2}$ through $(4, 1)$

$y - 1 = \frac{3}{2}(x - 4)$

$y - 1 = \frac{3}{2}x - 6$

$y = \frac{3}{2}x - 5$

8. $(5, 32)$ and $(9, 44)$

$$m = \frac{44 - 32}{9 - 5} = \frac{12}{4} = 3$$

$y - 32 = 3(x - 5)$

$y - 32 = 3x - 15$

$y = 3x + 17$

When $x = 20$, $y = 3(20) + 17$

$y = \$77.$

9. $f(x - 3) = (x - 3)^2 - 2(x - 3) + 1$

$= x^2 - 6x + 9 - 2x + 6 + 1$

$= x^2 - 8x + 16$

10. $f(3) = 12 - 11 = 1$

$\dfrac{f(x) - f(3)}{x - 3} = \dfrac{(4x - 11) - 1}{x - 3}$

$= \dfrac{4x - 12}{x - 3}$

$= \dfrac{4(x - 3)}{x - 3} = 4,\ x \neq 3$

11. $f(x) = \sqrt{36 - x^2} = \sqrt{(6 + x)(6 - x)}$

Domain: $[-6, 6]$ since $36 - x^2 \geq 0$ on this interval.

Range: $[0, 6]$

12. (a) $6x - 5y + 4 = 0$

$y = \dfrac{6x + 4}{5}$ is a function of x.

(b) $x^2 + y^2 = 9$

$y = \pm\sqrt{9 - x^2}$ is not a function of x.

(c) $y^3 = x^2 + 6$

$y = \sqrt[3]{x^2 + 6}$ is a function of x.

13. Parabola

Vertex: $(0, -5)$

Intercepts: $(0, -5), (\pm\sqrt{5}, 0)$

y-axis symmetry

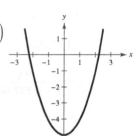

14. Intercepts: $(0, 3)$, $(-3, 0)$

x	0	1	-1	2	-2	-3	-4
y	3	4	2	5	1	0	1

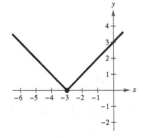

15.

x	0	1	2	3	-1	-2	-3
y	1	3	5	7	2	6	12

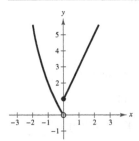

16. (a) $f(x + 2)$

Horizontal shift two units to the left

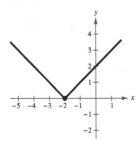

(b) $-f(x) + 2$

Reflection in the x-axis and a vertical shift two units upward

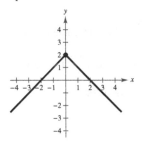

17. (a) $(g - f)(x) = g(x) - f(x)$

$$= \left(2x^2 - 5\right) - (3x + 7)$$

$$= 2x^2 - 3x - 12$$

(b) $(fg)(x) = f(x)g(x)$

$$= (3x + 7)\left(2x^2 - 5\right)$$

$$= 6x^3 + 14x^2 - 15x - 35$$

18. $f\big(g(x)\big) = f(2x + 3)$

$$= (2x + 3)^2 - 2(2x + 3) + 16$$

$$= 4x^2 + 12x + 9 - 4x - 6 + 16$$

$$= 4x^2 + 8x + 19$$

19. $f(x) = x^3 + 7$

$$y = x^3 + 7$$

$$x = y^3 + 7$$

$$x - 7 = y^3$$

$$\sqrt[3]{x - 7} = y$$

$$f^{-1}(x) = \sqrt[3]{x - 7}$$

20. (a) $f(x) = |x - 6|$ does not have an inverse.

Its graph does not pass the Horizontal Line Test.

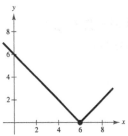

(b) $f(x) = ax + b,\ a \neq 0$ does have an inverse.

$$y = ax + b$$

$$x = ay + b$$

$$\frac{x - b}{a} = y$$

$$f^{-1}(x) = \frac{x - b}{a}$$

(c) $f(x) = x^3 - 19$ does have an inverse.

$$y = x^3 - 19$$

$$x = y^3 - 19$$

$$x + 19 = y^3$$

$$\sqrt[3]{x + 19} = y$$

$$f^{-1}(x) = \sqrt[3]{x + 19}$$

21. $f(x) = \sqrt{\dfrac{3-x}{x}}, \; 0 < x \le 3, \; y \ge 0$

$y = \sqrt{\dfrac{3-x}{x}}$

$x = \sqrt{\dfrac{3-y}{y}}$

$x^2 = \dfrac{3-y}{y}$

$x^2 y = 3 - y$

$x^2 y + y = 3$

$y(x^2 + 1) = 3$

$y = \dfrac{3}{x^2 + 1}$

$f^{-1}(x) = \dfrac{3}{x^2 + 1}, \; x \ge 0$

22. False. The slopes of 3 and $\frac{1}{3}$ are not *negative* reciprocals.

23. True. Let $y = (fg)(x)$. Then $x = (f \circ g)^{-1}(y)$. Also,

$(f \circ g)(x) = y$

$f(g(x)) = y$

$g(x) = f^{-1}(y)$

$x = g^{-1}(f^{-1}(y))$

$x = (g^{-1} \circ f^{-1})(y).$

Since $x = x$, we have $(f \circ g)^{-1}(y) = (g^{-1} \circ f^{-1})(y)$.

24. True. It must pass the Vertical Line Test to be a function and it must pass the Horizontal Line Test to have an inverse.

25. $y \approx 0.669x + 2.669$

Chapter 1 Practice Test Solutions

1. $350° = 350\left(\dfrac{\pi}{180}\right) = \dfrac{35\pi}{18}$

2. $\dfrac{5\pi}{9} = \dfrac{5\pi}{9} \cdot \dfrac{180}{\pi} = 100°$

3. $135° \, 14' \, 12'' = \left(135 + \frac{14}{60} + \frac{12}{3600}\right)°$

$\approx 135.2367°$

4. $-22.569° = -\left(22° + 0.569(60)'\right)$

$= -22° \, 34.14'$

$= -\left(22° \, 34' + 0.14(60)''\right)$

$\approx -22° \, 34' \, 8''$

5. $\cos \theta = \dfrac{2}{3}$

$x = 2, \, r = 3, \, y = \pm\sqrt{9-4} = \pm\sqrt{5}$

$\tan \theta = \dfrac{y}{x} = \pm\dfrac{\sqrt{5}}{2}$

6. $\sin \theta = 0.9063$

$\theta = \arcsin(0.9063)$

$\theta = 65° = \dfrac{13\pi}{36} \;$ or $\; \theta = 180° - 65° = 115° = \dfrac{23\pi}{36}$

7. $\tan 20° = \dfrac{35}{x}$

$x = \dfrac{35}{\tan 20°}$

≈ 96.1617

8. $\theta = \dfrac{6\pi}{5}, \, \theta$ is in Quadrant III.

Reference angle: $\dfrac{6\pi}{5} - \pi = \dfrac{\pi}{5} \;$ or $\; 36°$

9. $\csc 3.92 = \dfrac{1}{\sin 3.92} \approx -1.4242$

10. $\tan \theta = 6 = \dfrac{6}{1}, \, \theta$ lies in Quandrant III.

$y = -6, \, x = -1, \, r = \sqrt{36+1} = \sqrt{37}$, so

$\sec \theta = \dfrac{\sqrt{37}}{-1} \approx -6.0828.$

11. Period: 4π

Amplitude: 3

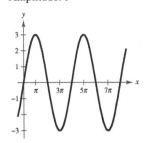

12. Period: 2π

Amplitude: 2

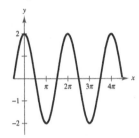

13. Period: $\dfrac{\pi}{2}$

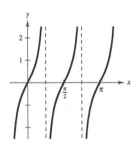

14. Period: 2π

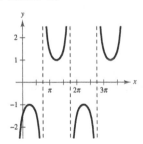

15.

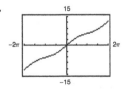

16.

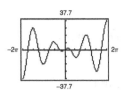

17. $\theta = \arcsin 1$

$\sin \theta = 1$

$\theta = \dfrac{\pi}{2} = 90°$

18. $\theta = \arctan(-3)$

$\tan \theta = -3$

$\theta \approx -1.249 \approx -71.565°$

19. $\sin\left(\arccos \dfrac{4}{\sqrt{35}}\right)$

$\sin \theta = \dfrac{\sqrt{19}}{\sqrt{35}} \approx 0.7368$

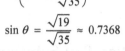

20. $\cos\left(\arcsin \dfrac{x}{4}\right)$

$\cos \theta = \dfrac{\sqrt{16 - x^2}}{4}$

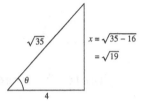

21. Given $A = 40°, c = 12$

$B = 90° - 40° = 50°$

$\sin 40° = \dfrac{a}{12}$

$a = 12 \sin 40° \approx 7.713$

$\cos 40° = \dfrac{b}{12}$

$b = 12 \cos 40° \approx 9.193$

22. Given $B = 6.84°, a = 21.3$

$A = 90° - 6.84° = 83.16°$

$\sin 83.16° = \dfrac{21.3}{c}$

$c = \dfrac{21.3}{\sin 83.16°} \approx 21.453$

$\tan 83.16° = \dfrac{21.3}{b}$

$b = \dfrac{21.3}{\tan 83.16°} \approx 2.555$

23. Given $a = 5, b = 9$

$$c = \sqrt{25 + 81} = \sqrt{106} \approx 10.296$$

$$\tan A = \frac{5}{9}$$

$$A = \arctan \frac{5}{9} \approx 29.055°$$

$$B \approx 90° - 29.055° = 60.945°$$

24. $\sin 67° = \dfrac{x}{20}$

$$x = 20 \sin 67° \approx 18.41 \text{ feet}$$

25. $\tan 5° = \dfrac{250}{x}$

$$x = \frac{250}{\tan 5°}$$

$$\approx 2857.513 \text{ feet}$$

$$\approx 0.541 \text{ mi}$$

Chapter 2   Practice Test Solutions

1. $\tan x = \dfrac{4}{11}, \sec x < 0 \Rightarrow x$ is in Quadrant III.

$$y = -4, x = -11, r = \sqrt{16 + 121} = \sqrt{137}$$

$$\sin x = -\frac{4}{\sqrt{137}} = -\frac{4\sqrt{137}}{137} \qquad \csc x = -\frac{\sqrt{137}}{4}$$

$$\cos x = -\frac{11}{\sqrt{137}} = -\frac{11\sqrt{137}}{137} \qquad \sec x = -\frac{\sqrt{137}}{11}$$

$$\tan x = \frac{4}{11} \qquad\qquad\qquad \cot x = \frac{11}{4}$$

2.
$$\frac{\sec^2 x + \csc^2 x}{\csc^2 x (1 + \tan^2 x)} = \frac{\sec^2 x + \csc^2 x}{\csc^2 x + (\csc^2 x) \tan^2 x}$$

$$= \frac{\sec^2 x + \csc^2 x}{\csc^2 x + \dfrac{1}{\sin^2 x} \cdot \dfrac{\sin^2 x}{\cos^2 x}}$$

$$= \frac{\sec^2 x + \csc^2 x}{\csc^2 x + \dfrac{1}{\cos^2 x}}$$

$$= \frac{\sec^2 x + \csc^2 x}{\csc^2 x + \sec^2 x} = 1$$

3. $\ln|\tan \theta| - \ln|\cot \theta| = \ln \left|\dfrac{\tan \theta}{\cot \theta}\right| = \ln \left|\dfrac{\sin \theta / \cos \theta}{\cos \theta / \sin \theta}\right| = \ln \left|\dfrac{\sin^2 \theta}{\cos^2 \theta}\right| = \ln|\tan^2 \theta| = 2 \ln|\tan \theta|$

4. $\cos\left(\dfrac{\pi}{2} - x\right) = \dfrac{1}{\csc x}$ is true since $\cos\left(\dfrac{\pi}{2} - x\right) = \sin x = \dfrac{1}{\csc x}$.

5. $\sin^4 x + (\sin^2 x) \cos^2 x = \sin^2 x (\sin^2 x + \cos^2 x)$

$$= \sin^2 x (1) = \sin^2 x$$

6. $(\csc x + 1)(\csc x - 1) = \csc^2 x - 1 = \cot^2 x$

7. $\dfrac{\cos^2 x}{1 - \sin x} \cdot \dfrac{1 + \sin x}{1 + \sin x} = \dfrac{\cos^2 x (1 + \sin x)}{1 - \sin^2 x} = \dfrac{\cos^2 x (1 + \sin x)}{\cos^2 x} = 1 + \sin x$

8. $\dfrac{1 + \cos \theta}{\sin \theta} + \dfrac{\sin \theta}{1 + \cos \theta} = \dfrac{(1 + \cos \theta)^2 + \sin^2 \theta}{\sin \theta (1 + \cos \theta)}$

$$= \frac{1 + 2 \cos \theta + \cos^2 \theta + \sin^2 \theta}{\sin \theta (1 + \cos \theta)} = \frac{2 + 2 \cos \theta}{\sin \theta (1 + \cos \theta)} = \frac{2}{\sin \theta} = 2 \csc \theta$$

9. $\tan^4 x + 2 \tan^2 x + 1 = \left(\tan^2 x + 1\right)^2 = \left(\sec^2 x\right)^2 = \sec^4 x$

10. (a) $\sin 105° = \sin(60° + 45°) = \sin 60° \cos 45° + \cos 60° \sin 45°$

$$= \frac{\sqrt{3}}{2} \cdot \frac{\sqrt{2}}{2} + \frac{1}{2} \cdot \frac{\sqrt{2}}{2} = \frac{\sqrt{2}}{4}\left(\sqrt{3} + 1\right)$$

(b) $\tan 15° = \tan(60° - 45°) = \dfrac{\tan 60° - \tan 45°}{1 + \tan 60° \tan 45°}$

$$= \frac{\sqrt{3} - 1}{1 + \sqrt{3}} \cdot \frac{1 - \sqrt{3}}{1 - \sqrt{3}} = \frac{2\sqrt{3} - 1 - 3}{1 - 3} = \frac{2\sqrt{3} - 4}{-2} = 2 - \sqrt{3}$$

11. $\left(\sin 42°\right) \cos 38° - \left(\cos 42°\right) \sin 38° = \sin(42° - 38°) = \sin 4°$

12. $\tan\left(\theta + \dfrac{\pi}{4}\right) = \dfrac{\tan \theta + \tan\left(\dfrac{\pi}{4}\right)}{1 - \left(\tan \theta\right) \tan\left(\dfrac{\pi}{4}\right)} = \dfrac{\tan \theta + 1}{1 - \tan \theta(1)} = \dfrac{1 + \tan \theta}{1 - \tan \theta}$

13. $\sin(\arcsin x - \arccos x) = \sin(\arcsin x) \cos(\arccos x) - \cos(\arcsin x) \sin(\arccos x)$

$$= (x)(x) - \left(\sqrt{1 - x^2}\right)\left(\sqrt{1 - x^2}\right) = x^2 - \left(1 - x^2\right) = 2x^2 - 1$$

14. (a) $\cos(120°) = \cos\left[2(60°)\right] = 2 \cos^2 60° - 1 = 2\left(\dfrac{1}{2}\right)^2 - 1 = -\dfrac{1}{2}$

(b) $\tan(300°) = \tan\left[2(150°)\right] = \dfrac{2 \tan 150°}{1 - \tan^2 150°} = \dfrac{-\dfrac{2\sqrt{3}}{3}}{1 - \left(\dfrac{1}{3}\right)} = -\sqrt{3}$

15. (a) $\sin 22.5° = \sin \dfrac{45°}{2} = \sqrt{\dfrac{1 - \cos 45°}{2}} = \sqrt{\dfrac{1 - \dfrac{\sqrt{2}}{2}}{2}} = \dfrac{\sqrt{2 - \sqrt{2}}}{2}$

(b) $\tan \dfrac{\pi}{12} = \tan \dfrac{\dfrac{\pi}{6}}{2} = \dfrac{\sin \dfrac{\pi}{6}}{1 + \cos\left(\dfrac{\pi}{6}\right)} = \dfrac{\dfrac{1}{2}}{1 + \dfrac{\sqrt{3}}{2}} = \dfrac{1}{2 + \sqrt{3}} = 2 - \sqrt{3}$

16. $\sin \theta = \dfrac{4}{5}$, θ lies in Quadrant II $\Rightarrow \cos \theta = -\dfrac{3}{5}$.

$$\cos \frac{\theta}{2} = \sqrt{\frac{1 + \cos \theta}{2}} = \sqrt{\frac{1 - \dfrac{3}{5}}{2}} = \sqrt{\frac{2}{10}} = \frac{1}{\sqrt{5}} = \frac{\sqrt{5}}{5}$$

17. $\left(\sin^2 x\right) \cos^2 x = \dfrac{1 - \cos 2x}{2} \cdot \dfrac{1 + \cos 2x}{2} = \dfrac{1}{4}\left[1 - \cos^2 2x\right] = \dfrac{1}{4}\left[1 - \dfrac{1 + \cos 4x}{2}\right]$

$$= \frac{1}{8}\left[2 - (1 + \cos 4x)\right] = \frac{1}{8}\left[1 - \cos 4x\right]$$

18. $6(\sin 5\theta) \cos 2\theta = 6\left\{\dfrac{1}{2}\left[\sin(5\theta + 2\theta) + \sin(5\theta - 2\theta)\right]\right\} = 3\left[\sin 7\theta + \sin 3\theta\right]$

19. $\sin(x + \pi) + \sin(x - \pi) = 2\left(\sin\dfrac{\left[(x + \pi) + (x - \pi)\right]}{2}\right)\cos\dfrac{\left[(x + \pi) - (x - \pi)\right]}{2}$

$$= 2\sin x \cos \pi = -2\sin x$$

20. $\dfrac{\sin 9x + \sin 5x}{\cos 9x - \cos 5x} = \dfrac{2\sin 7x \cos 2x}{-2\sin 7x \sin 2x} = -\dfrac{\cos 2x}{\sin 2x} = -\cot 2x$

21. $\frac{1}{2}\left[\sin(u + v) - \sin(u - v)\right] = \frac{1}{2}\left\{(\sin u)\cos v + (\cos u)\sin v - \left[(\sin u)\cos v - (\cos u)\sin v\right]\right\}$

$$= \frac{1}{2}\left[2(\cos u)\sin v\right] = (\cos u)\sin v$$

22. $4\sin^2 x = 1$

$\sin^2 x = \dfrac{1}{4}$

$\sin x = \pm\dfrac{1}{2}$

$\sin x = \dfrac{1}{2}$ or $\sin x = -\dfrac{1}{2}$

$x = \dfrac{\pi}{6}$ or $\dfrac{5\pi}{6}$ $x = \dfrac{7\pi}{6}$ or $\dfrac{11\pi}{6}$

23. $\tan^2 \theta + \left(\sqrt{3} - 1\right)\tan \theta - \sqrt{3} = 0$

$\left(\tan \theta - 1\right)\left(\tan \theta + \sqrt{3}\right) = 0$

$\tan \theta = 1$ or $\tan \theta = -\sqrt{3}$

$\theta = \dfrac{\pi}{4}$ or $\dfrac{5\pi}{4}$ $\theta = \dfrac{2\pi}{3}$ or $\dfrac{5\pi}{3}$

24. $\sin 2x = \cos x$

$2(\sin x)\cos x - \cos x = 0$

$\cos x(2\sin x - 1) = 0$

$\cos x = 0$ or $\sin x = \dfrac{1}{2}$

$x = \dfrac{\pi}{2}$ or $\dfrac{3\pi}{2}$ $x = \dfrac{\pi}{6}$ or $\dfrac{5\pi}{6}$

25. $\tan^2 x - 6\tan x + 4 = 0$

$\tan x = \dfrac{-(-6) \pm \sqrt{(-6)^2 - 4(1)(4)}}{2(1)}$

$\tan x = \dfrac{6 \pm \sqrt{20}}{2} = 3 \pm \sqrt{5}$

$\tan x = 3 + \sqrt{5}$ or $\tan x = 3 - \sqrt{5}$

$x \approx 1.3821$ or 4.5237 $x \approx 0.6524$ or 3.7940

Chapter 3   Practice Test Solutions

1. $C = 180° - \left(40° + 12°\right) = 128°$

$a = \sin 40°\left(\dfrac{100}{\sin 12°}\right) \approx 309.164$

$c = \sin 128°\left(\dfrac{100}{\sin 12°}\right) \approx 379.012$

2. $\sin A = 5\left(\dfrac{\sin 150°}{20}\right) = 0.125$

$A \approx 7.181°$

$B \approx 180° - \left(150° + 7.181°\right) = 22.819°$

$b = \sin 22.819°\left(\dfrac{20}{\sin 150°}\right) \approx 15.513$

3. $\text{Area} = \frac{1}{2}ab \sin C = \frac{1}{2}(3)(6)\sin 130° \approx 6.894$ square units

4. $h = b \sin A = 35 \sin 22.5° \approx 13.394$

$a = 10$

Since $a < h$ and A is acute, the triangle has no solution.

5. $\cos A = \dfrac{(53)^2 + (38)^2 - (49)^2}{2(53)(38)} \approx 0.4598$

$A \approx 62.627°$

$\cos B = \dfrac{(49)^2 + (38)^2 - (53)^2}{2(49)(38)} \approx 0.2782$

$B \approx 73.847°$

$C \approx 180° - \left(62.627° + 73.847°\right)$

$= 43.526°$

6. $c^2 = (100)^2 + (300)^2 - 2(100)(300)\cos 29°$

$\approx 47{,}522.8176$

$c \approx 218$

$\cos A = \dfrac{(300)^2 + (218)^2 - (100)^2}{2(300)(218)} \approx 0.97495$

$A \approx 12.85°$

$B \approx 180° - (12.85° + 29°) = 138.15°$

7. $s = \dfrac{a + b + c}{2} = \dfrac{4.1 + 6.8 + 5.5}{2} = 8.2$

Area $= \sqrt{s(s-a)(s-b)(s-c)}$

$= \sqrt{8.2(8.2 - 4.1)(8.2 - 6.8)(8.2 - 5.5)}$

≈ 11.273 square units

8. $x^2 = (40)^2 + (70)^2 - 2(40)(70)\cos 168°$

$\approx 11{,}977.6266$

$x \approx 190.442$ miles

9. $\mathbf{w} = 4(3\mathbf{i} + \mathbf{j}) - 7(-\mathbf{i} + 2\mathbf{j})$

$= 19\mathbf{i} - 10\mathbf{j}$

10. $\dfrac{\mathbf{v}}{\|\mathbf{v}\|} = \dfrac{5\mathbf{i} - 3\mathbf{j}}{\sqrt{25 + 9}} = \dfrac{5}{\sqrt{34}}\mathbf{i} - \dfrac{3}{\sqrt{34}}\mathbf{j}$

$= \dfrac{5\sqrt{34}}{34}\mathbf{i} - \dfrac{3\sqrt{34}}{34}\mathbf{j}$

11. $\mathbf{u} = 6\mathbf{i} + 5\mathbf{j},\ \mathbf{v} = 2\mathbf{i} - 3\mathbf{j}$

$\mathbf{u} \cdot \mathbf{v} = 6(2) + 5(-3) = -3$

$\|\mathbf{u}\| = \sqrt{61}, \qquad \|\mathbf{v}\| = \sqrt{13}$

$\cos \theta = \dfrac{-3}{\sqrt{61}\sqrt{13}}$

$\theta \approx 96.116°$

12. $4(\mathbf{i}\cos 30° + \mathbf{j}\sin 30°) = 4\left(\dfrac{\sqrt{3}}{2}\mathbf{i} + \dfrac{1}{2}\mathbf{j}\right)$

$= \langle 2\sqrt{3}, 2\rangle$

13. $\text{proj}_{\mathbf{v}}\mathbf{u} = \left(\dfrac{\mathbf{u}\cdot\mathbf{v}}{\|\mathbf{v}\|^2}\right)\mathbf{v} = \dfrac{-10}{20}\langle -2, 4\rangle = \langle 1, -2\rangle$

14. $r = \sqrt{25 + 25} = \sqrt{50} = 5\sqrt{2}$

$\tan \theta = \dfrac{-5}{5} = -1$

Because z is in Quadrant IV, $\theta = 315°$.

$z = 5\sqrt{2}(\cos 315° + i\sin 315°)$

15. $\cos 225° = -\dfrac{\sqrt{2}}{2},\ \sin 225° = -\dfrac{\sqrt{2}}{2}$

$z = 6\left(-\dfrac{\sqrt{2}}{2} - i\dfrac{\sqrt{2}}{2}\right) = -3\sqrt{2} - 3\sqrt{2}i$

16. $\left[7(\cos 23° + i\sin 23°)\right]\left[4(\cos 7° + i\sin 7°)\right] = 7(4)\left[\cos(23° + 7°) + i\sin(23° + 7°)\right]$

$= 28(\cos 30° + i\sin 30°)$

17. $\dfrac{9\left(\cos\dfrac{5\pi}{4} + i\sin\dfrac{5\pi}{4}\right)}{3(\cos\pi + i\sin\pi)} = \dfrac{9}{3}\left[\cos\left(\dfrac{5\pi}{4} - \pi\right) + i\sin\left(\dfrac{5\pi}{4} - \pi\right)\right] = 3\left(\cos\dfrac{\pi}{4} + i\sin\dfrac{\pi}{4}\right)$

18. $(2 + 2i)^8 = \left[2\sqrt{2}(\cos 45° + i\sin 45°)\right]^8 = (2\sqrt{2})^8\left[\cos(8)(45°) + i\sin(8)(45°)\right]$

$= 4096[\cos 360° + i\sin 360°] = 4096$

19. $z = 8\left(\cos\dfrac{\pi}{3} + i\sin\dfrac{\pi}{3}\right),\ n = 3$

The cube roots of z are: $\sqrt[3]{8}\left[\cos\dfrac{\left(\dfrac{\pi}{3}\right) + 2\pi k}{3} + i\sin\dfrac{\left(\dfrac{\pi}{3}\right) + 2\pi k}{3}\right],\ k = 0, 1, 2$

For $k = 0$: $\sqrt[3]{8}\left[\cos\dfrac{\dfrac{\pi}{3}}{3} + i\sin\dfrac{\dfrac{\pi}{3}}{3}\right] = 2\left(\cos\dfrac{\pi}{9} + i\sin\dfrac{\pi}{9}\right)$

For $k = 1$: $\sqrt[3]{8}\left[\cos\dfrac{\left(\dfrac{\pi}{3}\right) + 2\pi}{3} + i\sin\dfrac{\left(\dfrac{\pi}{3}\right) + 2\pi}{3}\right] = 2\left(\cos\dfrac{7\pi}{9} + i\sin\dfrac{7\pi}{9}\right)$

For $k = 2$: $\sqrt[3]{8}\left[\cos\dfrac{\left(\dfrac{\pi}{3}\right) + 4\pi}{3} + i\sin\dfrac{\left(\dfrac{\pi}{3}\right) + 4\pi}{3}\right] = 2\left(\cos\dfrac{13\pi}{9} + i\sin\dfrac{13\pi}{9}\right)$

20. $x^4 = -i = 1\left(\cos\dfrac{3\pi}{2} + i\sin\dfrac{3\pi}{2}\right)$

The fourth roots are: $\sqrt[4]{1}\left[\cos\dfrac{\left(\dfrac{3\pi}{2}\right) + 2\pi k}{4} + i\sin\dfrac{\left(\dfrac{3\pi}{2}\right) + 2\pi k}{4}\right],\ k = 0, 1, 2, 3$

For $k = 0$: $\cos\dfrac{\dfrac{3\pi}{2}}{4} + i\sin\dfrac{\dfrac{3\pi}{2}}{4} = \cos\dfrac{3\pi}{8} + i\sin\dfrac{3\pi}{8}$

For $k = 1$: $\cos\dfrac{\left(\dfrac{3\pi}{2}\right) + 2\pi}{4} + i\sin\dfrac{\left(\dfrac{3\pi}{2}\right) + 2\pi}{4} = \cos\dfrac{7\pi}{8} + i\sin\dfrac{7\pi}{8}$

For $k = 2$: $\cos\dfrac{\left(\dfrac{3\pi}{2}\right) + 4\pi}{4} + i\sin\dfrac{\left(\dfrac{3\pi}{2}\right) + 4\pi}{4} = \cos\dfrac{11\pi}{8} + i\sin\dfrac{11\pi}{8}$

For $k = 3$: $\cos\dfrac{\left(\dfrac{3\pi}{2}\right) + 6\pi}{4} + i\sin\dfrac{\left(\dfrac{3\pi}{2}\right) + 6\pi}{4} = \cos\dfrac{15\pi}{8} + i\sin\dfrac{15\pi}{8}$

Chapter 4 Practice Test Solutions

1. $4 + \sqrt{-81} - 3i^2 = 4 + 9i + 3 = 7 + 9i$

2. $\dfrac{3 + i}{5 - 4i} \cdot \dfrac{5 + 4i}{5 + 4i} = \dfrac{15 + 12i + 5i + 4i^2}{25 + 16} = \dfrac{11 + 17i}{41} = \dfrac{11}{41} + \dfrac{17}{41}i$

3. $x = \dfrac{-(-4) \pm \sqrt{(-4)^2 - 4(1)(7)}}{2(1)} = \dfrac{4 \pm \sqrt{-12}}{2} = \dfrac{4 \pm 2\sqrt{3}i}{2} = 2 \pm \sqrt{3}i$

4. False: $\sqrt{-6}\,\sqrt{-6} = \left(\sqrt{6}i\right)\left(\sqrt{6}i\right) = \sqrt{36}i^2 = 6(-1) = -6$

5. $b^2 - 4ac = (-8)^2 - 4(3)(7) = -20 < 0$

Two complex solutions

6. $x^4 + 13x^2 + 36 = 0$

$(x^2 + 4)(x^2 + 9) = 0$

$x^2 + 4 = 0 \Rightarrow x^2 = -4 \Rightarrow x = \pm 2i$

$x^2 + 9 = 0 \Rightarrow x^2 = -9 \Rightarrow x = \pm 3i$

7. $f(x) = (x - 3)\big[x - (-1 + 4i)\big]\big[x - (-1 - 4i)\big]$

$\quad = (x - 3)\big[(x + 1) - 4i\big]\big[(x + 1) + 4i\big]$

$\quad = (x - 3)\big[(x + 1)^2 - 16i^2\big]$

$\quad = (x - 3)(x^2 + 2x + 17)$

$\quad = x^3 - x^2 + 11x - 51$

8. Since $x = 4 + i$ is a zero, so is its conjugate $4 - i$.

$\big[x - (4 + i)\big]\big[x - (4 - i)\big] = \big[(x - 4) - i\big]\big[(x - 4) + i\big]$

$\qquad\qquad\qquad\qquad\qquad = (x - 4)^2 - i^2$

$\qquad\qquad\qquad\qquad\qquad = x^2 - 8x + 17$

$$
\begin{array}{r}
x - 2 \\
x^2 - 8x + 17 \overline{)\, x^3 - 10x^2 + 33x - 34} \\
\underline{x^3 - 8x^2 + 17x} \\
-2x^2 + 16x - 34 \\
\underline{-2x^2 + 16x - 34} \\
0
\end{array}
$$

Thus, $f(x) = (x^2 - 8x + 17)(x - 2)$ and the zeros of $f(x)$ are: $4 \pm i$, 2.

9. $r = \sqrt{25 + 25} = \sqrt{50} = 5\sqrt{2}$

$\tan \theta = \dfrac{-5}{5} = -1$

Since z is in Quadrant IV, $\theta - 315°$,

$z = 5\sqrt{2}(\cos 315° + i \sin 315°)$.

10. $\cos 225° = -\dfrac{\sqrt{2}}{2}$, $\sin 225° = -\dfrac{\sqrt{2}}{2}$

$z = 6\left(-\dfrac{\sqrt{2}}{2} - i\dfrac{\sqrt{2}}{2}\right) = -3\sqrt{2} - 3\sqrt{2}\,i$

11. $\big[7(\cos 23° + i \sin 23°)\big]\big[4(\cos 7° + i \sin 7°)\big] = 7(4)\big[\cos(23° + 7°) + i \sin(23° + 7°)\big]$

$\qquad\qquad\qquad\qquad\qquad\qquad\qquad\qquad\quad = 28(\cos 30° + i \sin 30°)$

$\qquad\qquad\qquad\qquad\qquad\qquad\qquad\qquad\quad = 14\sqrt{3} + 14i$

12. $\dfrac{9\left(\cos \dfrac{5\pi}{4} + i \sin \dfrac{5\pi}{4}\right)}{3(\cos \pi + i \sin \pi)} = \dfrac{9}{3}\left[\cos\left(\dfrac{5\pi}{4} - \pi\right) + i \sin\left(\dfrac{5\pi}{4} - \pi\right)\right] = 3\left(\cos \dfrac{\pi}{4} + i \sin \dfrac{\pi}{4}\right) = \dfrac{3\sqrt{2}}{2} + \dfrac{3\sqrt{2}}{2}i$

13. $(2 + 2i)^8 = \big[2\sqrt{2}(\cos 45° + i \sin 45°)\big]^8$

$\quad = \big(2\sqrt{2}\big)^8\big[\cos(8)(45°) + i \sin(8)(45°)\big]$

$\quad = 4096[\cos 360° + i \sin 360°]$

$\quad = 4096$

14. $z = 8\left(\cos\dfrac{\pi}{3} + i\sin\dfrac{\pi}{3}\right),\ n = 3$

The cube roots of z are: $\sqrt[3]{8}\left[\cos\dfrac{(\pi/3) + 2\pi k}{3} + i\sin\dfrac{(\pi/3) + 2\pi k}{3}\right],\ k = 0, 1, 2$

For $k = 0$: $\sqrt[3]{8}\left[\cos\dfrac{\pi/3}{3} + i\sin\dfrac{\pi/3}{3}\right] = 2\left(\cos\dfrac{\pi}{9} + i\sin\dfrac{\pi}{9}\right)$

For $k = 1$: $\sqrt[3]{8}\left[\cos\dfrac{(\pi/3) + 2\pi}{3} + i\sin\dfrac{(\pi/3) + 2\pi}{3}\right] = 2\left(\cos\dfrac{7\pi}{9} + i\sin\dfrac{7\pi}{9}\right)$

For $k = 2$: $\sqrt[3]{8}\left[\cos\dfrac{(\pi/3) + 4\pi}{3} + i\sin\dfrac{(\pi/3) + 4\pi}{3}\right] = 2\left(\cos\dfrac{13\pi}{9} + i\sin\dfrac{13\pi}{9}\right)$

15. $x^4 = -i = 1\left(\cos\dfrac{3\pi}{2} + i\sin\dfrac{3\pi}{2}\right)$

The fourth roots of are: $\sqrt[4]{1}\left[\cos\dfrac{(3\pi/2) + 2\pi k}{4} + i\sin\dfrac{(3\pi/2) + 2\pi k}{4}\right],\ k = 0, 1, 2, 3$

For $k = 0$: $\cos\left(\dfrac{3\pi/2}{4}\right) + i\sin\left(\dfrac{3\pi/2}{4}\right) = \cos\dfrac{3\pi}{8} + i\sin\dfrac{3\pi}{8}$

For $k = 1$: $\cos\left(\dfrac{(3\pi/2) + 2\pi}{4}\right) + i\sin\left(\dfrac{(3\pi/2) + 2\pi}{4}\right) = \cos\dfrac{7\pi}{8} + i\sin\dfrac{7\pi}{8}$

For $k = 2$: $\cos\left(\dfrac{(3\pi/2) + 4\pi}{4}\right) + i\sin\left(\dfrac{(3\pi/2) + 4\pi}{4}\right) = \cos\dfrac{11\pi}{8} + i\sin\dfrac{11\pi}{8}$

For $k = 3$: $\cos\left(\dfrac{(3\pi/2) + 6\pi}{4}\right) + i\sin\left(\dfrac{(3\pi/2) + 6\pi}{4}\right) = \cos\dfrac{15\pi}{8} + i\sin\dfrac{15\pi}{8}$

Chapter 5   Practice Test Solutions

1. $x^{3/5} = 8$

$x = 8^{5/3} = \left(\sqrt[3]{8}\right)^5 = 2^5 = 32$

2. $3^{x-1} = \dfrac{1}{81}$

$3^{x-1} = 3^{-4}$

$x - 1 = -4$

$x = -3$

3. $f(x) = 2^{-x} = \left(\dfrac{1}{2}\right)^x$

x	-2	-1	0	1	2
$f(x)$	4	2	1	$\dfrac{1}{2}$	$\dfrac{1}{4}$

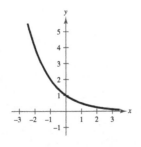

4. $g(x) = e^x + 1$

x	-2	-1	0	1	2
$g(x)$	1.14	1.37	2	3.72	8.39

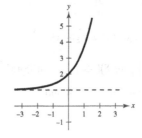

5. (a) $A = P\left(1 + \dfrac{r}{n}\right)^{nt}$

$A = 5000\left(1 + \dfrac{0.09}{12}\right)^{12(3)} \approx \$6543.23$

(b) $A = P\left(1 + \dfrac{r}{n}\right)^{nt}$

$A = 5000\left(1 + \dfrac{0.09}{4}\right)^{4(3)} \approx \$6530.25$

(c) $A = Pe^{rt}$

$A = 5000e^{(0.09)(3)} \approx \$6549.82$

6. $7^{-2} = \dfrac{1}{49}$

$\log_7 \dfrac{1}{49} = -2$

7. $x - 4 = \log_2 \dfrac{1}{64}$

$2^{x-4} = \dfrac{1}{64}$

$2^{x-4} = 2^{-6}$

$x - 4 = -6$

$x = -2$

8. $\log_b \sqrt[4]{\dfrac{8}{25}} = \dfrac{1}{4}\log_b \dfrac{8}{25}$

$= \dfrac{1}{4}[\log_b 8 - \log_b 25]$

$= \dfrac{1}{4}\left[\log_b 2^3 - \log_b 5^2\right]$

$= \dfrac{1}{4}[3\log_b 2 - 2\log_b 5]$

$= \dfrac{1}{4}\left[3(0.3562) - 2(0.8271)\right]$

$= -0.1464$

9. $5\ln x - \dfrac{1}{2}\ln y + 6\ln z = \ln x^5 - \ln \sqrt{y} + \ln z^6$

$= \ln\left(\dfrac{x^5 z^6}{\sqrt{y}}\right),\ z > 0$

10. $\log_9 28 = \dfrac{\log 28}{\log 9} \approx 1.5166$

11. $\log N = 0.6646$

$N = 10^{0.6646} \approx 4.62$

12.

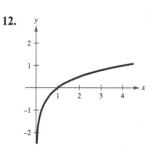

13. Domain:

$x^2 - 9 > 0$

$(x + 3)(x - 3) > 0$

$x < -3 \text{ or } x > 3$

14.

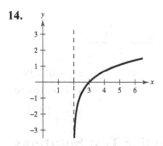

15. False. $\dfrac{\ln x}{\ln y} \neq \ln(x - y)$ because $\dfrac{\ln x}{\ln y} = \log_y x.$

16. $5^3 = 41$

$x = \log_5 41 = \dfrac{\ln 41}{\ln 5} \approx 2.3074$

17. $x - x^2 = \log_5 \dfrac{1}{25}$

$5^{x-x^2} = \dfrac{1}{25}$

$5^{x-x^2} = 5^{-2}$

$x - x^2 = -2$

$0 = x^2 - x - 2$

$0 = (x + 1)(x - 2)$

$x = -1 \text{ or } x = 2$

18. $\log_2 x + \log_2(x - 3) = 2$

$\log_2\left[x(x - 3)\right] = 2$

$x(x - 3) = 2^2$

$x^2 - 3x = 4$

$x^2 - 3x - 4 = 0$

$(x + 1)(x - 4) = 0$

$x = 4$

$x = -1 \text{ (extraneous)}$

$x = 4$ is the only solution.

19. $\dfrac{e^x + e^{-x}}{3} = 4$

$e^x(e^x + e^{-x}) = 12e^x$

$e^{2x} + 1 = 12e^x$

$e^{2x} - 12e^x + 1 = 0$

$e^x = \dfrac{12 \pm \sqrt{144 - 4}}{2}$

$e^x \approx 11.9161$ or $e^x \approx 0.0839$

$x = \ln 11.9161$ $x = \ln 0.0839$

$x \approx 2.478$ $x \approx -2.478$

20. $A = Pe^{rt}$

$12{,}000 = 6000e^{0.13t}$

$2 = e^{0.13t}$

$0.13t = \ln 2$

$t = \dfrac{\ln 2}{0.13}$

$t \approx 5.3319$ years or 5 years 4 months

Chapter 6   Practice Test Solutions

1. $3x + 4y = 12 \Rightarrow y = -\dfrac{3}{4}x + 3 \Rightarrow m_1 = -\dfrac{3}{4}$

$4x - 3y = 12 \Rightarrow y = \dfrac{4}{3}x - 4 \Rightarrow m_2 = \dfrac{4}{3}$

$\tan \theta = \left| \dfrac{(4/3) - (-3/4)}{1 + (4/3)(-3/4)} \right| = \left| \dfrac{25/12}{0} \right|$

Since $\tan \theta$ is undefined, the lines are perpendicular (note that $m_2 = -1/m_1$) and $\theta = 90°$.

2. $x_1 = 5, x_2 = -9, A = 3, B = -7, C = -21$

$d = \dfrac{|3(5) + (-7)(-9) + (-21)|}{\sqrt{3^2 + (-7)^2}} = \dfrac{57}{\sqrt{58}} \approx 7.484$

3. $x^2 - 6x - 4y + 1 = 0$

$x^2 - 6x + 9 = 4y - 1 + 9$

$(x - 3)^2 = 4y + 8$

$(x - 3)^2 = 4(1)(y + 2) \Rightarrow p = 1$

Vertex: $(3, -2)$

Focus: $(3, -1)$

Directrix: $y = -3$

4. Vertex: $(2, -5)$

Focus: $(2, -6)$

Vertical axis; opens downward with $p = -1$

$(x - h)^2 = 4p(y - k)$

$(x - 2)^2 = 4(-1)(y + 5)$

$x^2 - 4x + 4 = -4y - 20$

$x^2 - 4x + 4y + 24 = 0$

5. $x^2 + 4y^2 - 2x + 32y + 61 = 0$

$(x^2 - 2x + 1) + 4(y^2 + 8y + 16) = -61 + 1 + 64$

$(x - 1)^2 + 4(y + 4)^2 = 4$

$\dfrac{(x - 1)^2}{4} + \dfrac{(y + 4)^2}{1} = 1$

$a = 2, b = 1, c = \sqrt{3}$

Horizontal major axis

Center: $(1, -4)$

Foci: $\left(1 \pm \sqrt{3}, -4\right)$

Vertices: $(3, -4), (-1, -4)$

Eccentricity: $e = \dfrac{\sqrt{3}}{2}$

6. Vertices: $(0, \pm 6)$

Eccentricity: $e = \dfrac{1}{2}$

Center: $(0, 0)$

Vertical major axis

$a = 6, e = \dfrac{c}{a} = \dfrac{c}{6} = \dfrac{1}{2} \Rightarrow c = 3$

$b^2 = (6)^2 - (3)^2 = 27$

$\dfrac{x^2}{27} + \dfrac{y^2}{36} = 1$

7. $16y^2 - x^2 - 6x - 128y + 231 = 0$

$16(y^2 - 8y + 16) - (x^2 + 6x + 9) = -231 + 256 - 9$

$16(y - 4)^2 - (x + 3)^2 = 16$

$$\frac{(y - 4)^2}{1} - \frac{(x + 3)^2}{16} = 1$$

$a = 1, b = 4, c = \sqrt{17}$

Center: $(-3, 4)$

Vertical transverse axis

Vertices: $(-3, 5), (-3, 3)$

Foci: $\left(-3, 4 \pm \sqrt{17}\right)$

Asymptotes: $y = 4 \pm \dfrac{1}{4}(x + 3)$

8. Vertices: $(\pm 3, 2)$

Foci: $(\pm 5, 2)$

Center: $(0, 2)$

Horizontal transverse axis

$a = 3, c = 5, b = 4$

$$\frac{(x - 0)^2}{9} - \frac{(y - 2)^2}{16} = 1$$

$$\frac{x^2}{9} - \frac{(y - 2)^2}{16} = 1$$

9. $5x^2 + 2xy + 5y^2 - 10 = 0$

$A = 5, B = 2, C = 5$

$\cot 2\theta = \dfrac{5 - 5}{2} = 0$

$2\theta = \dfrac{\pi}{2} \Rightarrow \theta = \dfrac{\pi}{4}$

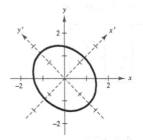

$x = x' \cos\dfrac{\pi}{4} - y' \sin\dfrac{\pi}{4}$

$= \dfrac{x' - y'}{\sqrt{2}}$

$x = x' \cos\dfrac{\pi}{4} + y' \sin\dfrac{\pi}{4}$

$= \dfrac{x' + y'}{\sqrt{2}}$

$$5\left(\frac{x' - y'}{\sqrt{2}}\right)^2 + 2\left(\frac{x' - y'}{\sqrt{2}}\right)\left(\frac{x' + y'}{\sqrt{2}}\right) + 5\left(\frac{x' + y'}{\sqrt{2}}\right)^2 - 10 = 0$$

$$\frac{5(x')^2}{2} - \frac{10x'y'}{2} + \frac{5(y')^2}{2} + (x')^2 - (y')^2 + \frac{5(x')^2}{2} + \frac{10x'y'}{2} + \frac{5(y')^2}{2} - 10 = 0$$

$$6(x')^2 + 4(y')^2 - 10 = 0$$

$$\frac{3(x')^2}{5} + \frac{2(y')^2}{5} = 1$$

$$\frac{(x')^2}{5/3} + \frac{(y')^2}{5/2} = 1$$

Ellipse centered at the origin

10. (a) $6x^2 - 2xy + y^2 = 0$

$A = 6, B = -2, C = 1$

$B^2 - 4AC = (-2)^2 - 4(6)(1) = -20 < 0$

Ellipse

(b) $x^2 + 4xy + 4y^2 - x - y + 17 = 0$

$A = 1, B = 4, C = 4$

$B^2 - 4AC = (4)^2 - 4(1)(4) = 0$

Parabola

11. Polar: $\left(\sqrt{2}, \dfrac{3\pi}{4}\right)$

$x = \sqrt{2} \cos\dfrac{3\pi}{4} = \sqrt{2}\left(-\dfrac{1}{\sqrt{2}}\right) = -1$

$y = \sqrt{2} \sin\dfrac{3\pi}{4} = \sqrt{2}\left(\dfrac{1}{\sqrt{2}}\right) = 1$

Rectangular: $(-1, 1)$

12. Rectangular: $\left(\sqrt{3}, -1\right)$

$$r = \pm\sqrt{\left(\sqrt{3}\right)^2 + (-1)^2} = \pm 2$$

$$\tan \theta = \frac{\sqrt{3}}{-1} = -\sqrt{3}$$

$$\theta = \frac{2\pi}{3} \text{ or } \theta = \frac{5\pi}{3}$$

Polar: $\left(-2, \dfrac{2\pi}{3}\right)$ or $\left(2, \dfrac{5\pi}{3}\right)$

13. Rectangular: $4x - 3y = 12$

Polar: $4r \cos \theta - 3r \sin \theta = 12$

$$r(4 \cos \theta - 3 \sin \theta) = 12$$

$$r = \frac{12}{4 \cos \theta - 3 \sin \theta}$$

14. Polar: $r = 5 \cos \theta$

$$r^2 = 5r \cos \theta$$

Rectangular: $x^2 + y^2 = 5x$

$$x^2 + y^2 - 5x = 0$$

15. $r = 1 - \cos \theta$

Cardioid

Symmetry: Polar axis

Maximum value of $|r|$: $r = 2$ when $\theta = \pi$

Zero of r: $r = 0$ when $\theta = 0$

θ	0	$\dfrac{\pi}{2}$	π	$\dfrac{3\pi}{2}$
r	0	1	2	1

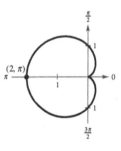

16. $r = 5 \sin 2\theta$

Rose curve with four petals

Symmetry: Polar axis, $\theta = \dfrac{\pi}{2}$, and pole

Maximum value of $|r|$: $|r| = 5$ when

$$\theta = \frac{\pi}{4}, \frac{3\pi}{4}, \frac{5\pi}{4}, \frac{7\pi}{4}$$

Zeros of r: $r = 0$ when $\theta = 0, \dfrac{\pi}{2}, \pi, \dfrac{3\pi}{2}$

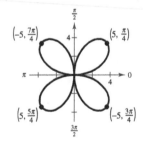

17. $r = \dfrac{3}{6 - \cos \theta}$

$$r = \frac{1/2}{1 - (1/6) \cos \theta}$$

$e = \dfrac{1}{6} < 1$, so the graph is an ellipse.

θ	0	$\dfrac{\pi}{2}$	π	$\dfrac{3\pi}{2}$
r	$\dfrac{3}{5}$	$\dfrac{1}{2}$	$\dfrac{3}{7}$	$\dfrac{1}{2}$

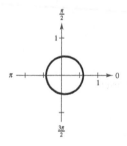

18. Parabola

Vertex: $\left(6, \dfrac{\pi}{2}\right)$

Focus: $(0, 0)$

$e = 1$

$r = \dfrac{ep}{1 + e\sin\theta}$

$r = \dfrac{p}{1 + \sin\theta}$

$6 = \dfrac{p}{1 + \sin(\pi/2)}$

$6 = \dfrac{p}{2}$

$12 = p$

$r = \dfrac{12}{1 + \sin\theta}$

19. $x = 3 - 2\sin\theta,\ y = 1 + 5\cos\theta$

$\dfrac{x - 3}{-2} = \sin\theta,\ \dfrac{y - 1}{5} = \cos\theta$

$\left(\dfrac{x - 3}{-2}\right)^2 + \left(\dfrac{y - 1}{5}\right)^2 = 1$

$\dfrac{(x - 3)^2}{4} + \dfrac{(y - 1)^2}{25} = 1$

20. $x = e^{2t},\ y = e^{4t}$

$x > 0,\ y > 0$

$y = \left(e^{2t}\right)^2 = (x)^2 = x^2,\ x > 0,\ y > 0$

PART II

Chapter Test Solutions for Chapter P

1. $-\dfrac{10}{3} < -\dfrac{5}{3}$

2. $d\left(-\dfrac{7}{4}, \dfrac{5}{4}\right) = \left|\dfrac{5}{4} - \left(-\dfrac{7}{4}\right)\right| = \left|\dfrac{12}{4}\right| = 3$

3. $(5 - x) + 0 = 5 - x$

   Additive Identity Property

4.
$$\tfrac{2}{3}(x - 1) + \tfrac{1}{4}x = 10$$
$$12\left[\tfrac{2}{3}(x - 1) + \tfrac{1}{4}x\right] = 12(10)$$
$$8(x - 1) + 3x = 120$$
$$8x - 8 + 3x = 120$$
$$11x = 128$$
$$x = \tfrac{128}{11}$$

5. $(x - 4)(x + 2) = 7$
$$x^2 - 2x - 8 = 7$$
$$x^2 - 2x - 15 = 0$$
$$(x + 3)(x - 5) = 0$$
$$x = -3 \quad \text{or} \quad x = 5$$

6. $\dfrac{x - 2}{x + 2} + \dfrac{4}{x + 2} + 4 = 0,\ x \neq -2$

$$\dfrac{x + 2}{x + 2} = -4$$

$1 \neq -4 \Rightarrow$ No solution because the variable is divided out.

7. $|3x - 1| = 7$

   $3x - 1 = 7 \quad \text{or} \quad 3x - 1 = -7$

   $\quad 3x = 8 \qquad\qquad 3x = -6$

   $\quad\ x = \tfrac{8}{3} \qquad\qquad\ x = -2$

8.

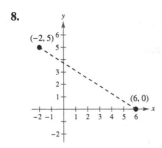

Midpoint: $\left(\dfrac{-2 + 6}{2}, \dfrac{5 + 0}{2}\right) = \left(2, \dfrac{5}{2}\right)$

Distance: $d = \sqrt{(-2 - 6)^2 + (5 - 0)^2}$
$$= \sqrt{64 + 25}$$
$$= \sqrt{89}$$

9. $y = 4 - \tfrac{3}{4}x$

   No axis or origin symmetry

   x-intercept: $\left(\tfrac{16}{3}, 0\right)$

   y-intercept: $(0, 4)$

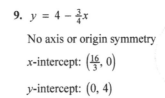
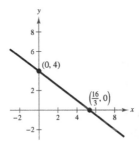

10. $y = 4 - \tfrac{3}{4}|x|$

   y-axis symmetry

   x-intercepts: $\left(\pm\tfrac{16}{3}, 0\right)$

   y-intercept: $(0, 4)$

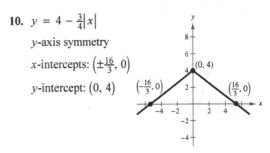

11. $y = x - x^3$

Origin symmetry

x-intercepts: $(0, 0), (1, 0), (-1, 0)$

$0 = x - x^3$

$0 = x(x + 1)(1 - x)$

$x = 0, x = \pm 1$

y-intercept: $(0, 0)$

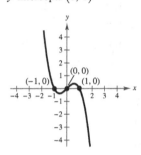

12. $(x - 3)^2 + y^2 = 9$

$(x - 3)^2 + (y - 0)^2 = 3^2$

Center: $(3, 0)$

Radius: 3

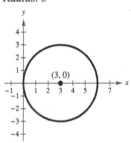

13. $(-2, 5), (1, -7)$

$m = \dfrac{-7 - 5}{1 - (-2)} = \dfrac{-12}{3} = -4$

$y - 5 = -4(x - (-2))$

$y - 5 = -4(x + 2)$

$y - 5 = -4x - 8$

$y = -4x - 3$

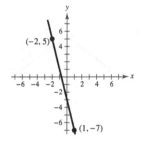

14. $(-4, -7), \left(1, \frac{4}{3}\right)$

$m = \dfrac{\frac{4}{3} - (-7)}{1 - (-4)} = \dfrac{\frac{25}{3}}{5} = \dfrac{5}{3}$

$y - (-7) = \dfrac{5}{3}(x - (-4))$

$y + 7 = \dfrac{5}{3}(x + 4)$

$y + 7 = \dfrac{5}{3}x + \dfrac{20}{3}$

$y = \dfrac{5}{3}x - \dfrac{1}{3}$

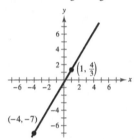

15. $5x + 2y = 3$

$2y = -5x + 3$

$y = -\frac{5}{2}x + \frac{3}{2}$

(a) Parallel line:

$m = -\frac{5}{2}$

$y - 4 = -\frac{5}{2}(x - 0)$

$y - 4 = -\frac{5}{2}x$

$y = -\frac{5}{2}x + 4$

(b) Perpendicular line:

$m = \frac{2}{5}$

$y - 4 = \frac{2}{5}(x - 0)$

$y - 4 = \frac{2}{5}x$

$y = \frac{2}{5}x + 4$

16. $f(x) = |x + 2| - 15$

(a) $f(-8) = -9$

(b) $f(14) = 1$

(c) $f(x - 6) = |x - 4| - 15$

17. $f(x) = |x + 5|$

(a)

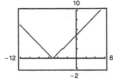

(b) Increasing on $(-5, \infty)$

    Decreasing on $(-\infty, -5)$

(c) The function is neither odd nor even.

18. $f(x) = 4x\sqrt{3 - x}$

(a)

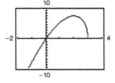

(b) Increasing on $(-\infty, 2)$

    Decreasing on $(2, 3)$

(c) The function is neither odd nor even.

19. $f(x) = 2x^6 + 5x^4 - x^2$

(a)

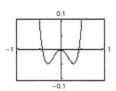

(b) Increasing on $(-0.31, 0), (0.31, \infty)$

    Decreasing on $(-\infty, -0.31), (0, 0.31)$

(c) y-axis symmetry $\Rightarrow$ the function is even.

20. $h(x) = 4[\![x]\!]$

(a) The parent function is $f(x) = [\![x]\!]$

(b) The graph of h is a vertical stretch of the graph of f.

(c)

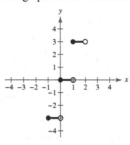

21. $h(x) = -\sqrt{x + 5} + 8$

(a) Parent function: $f(x) = \sqrt{x}$

(b) Transformation: Reflection in the x-axis,
          a horizontal shift 5 units to the left,
          and a vertical shift 8 units upward

(c)

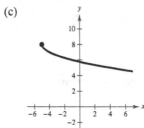

22. $h(x) = -2(x - 5)^3 + 3$

(a) Parent function: $f(x) = x^3$

(b) Transformation:
    Vertical stretch, reflection in x-axis, horizontal shift
    5 units to the right, vertical shift 3 units upward

(c)

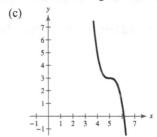

23. $f(x) = 3x^2 - 7, g(x) = -x^2 - 4x + 5$

(a) $(f + g)(x) = (3x^2 - 7) + (-x^2 - 4x + 5) = 2x^2 - 4x - 2$

(b) $(f - g)(x) = (3x^2 - 7) - (-x^2 - 4x + 5) = 4x^2 + 4x - 12$

(c) $(fg)(x) = (3x^2 - 7)(-x^2 - 4x + 5) = -3x^4 - 12x^3 + 22x^2 + 28x - 35$

(d) $\left(\dfrac{f}{g}\right)(x) = \dfrac{3x^2 - 7}{-x^2 - 4x + 5}$

(e) $(f \circ g)(x) = f(g(x)) = f(-x^2 - 4x + 5) = 3(-x^2 - 4x + 5)^2 - 7 = 3x^4 + 24x^3 + 18x^2 - 120x + 68$

(f) $(g \circ f)(x) = g(f(x)) = g(3x^2 - 7) = -(3x^2 - 7)^2 - 4(3x^2 - 7) + 5 = -9x^4 + 30x^2 - 16$

24. $f(x) = \dfrac{1}{x}, \; g(x) = 2\sqrt{x}$

(a) $(f + g)(x) = \dfrac{1}{x} + 2\sqrt{x} = \dfrac{1 + 2x^{3/2}}{x}, \; x > 0$

(b) $(f - g)(x) = \dfrac{1}{x} - 2\sqrt{x} = \dfrac{1 - 2x^{3/2}}{x}, \; x > 0$

(c) $(fg)(x) = \left(\dfrac{1}{x}\right)(2\sqrt{x}) = \dfrac{2\sqrt{x}}{x}, \; x > 0$

(d) $\left(\dfrac{f}{g}\right)(x) = \dfrac{1/x}{2\sqrt{x}} = \dfrac{1}{2x\sqrt{x}} = \dfrac{1}{2x^{3/2}}, \; x > 0$

(e) $(f \circ g)(x) = f(g(x)) = f(2\sqrt{x}) = \dfrac{1}{2\sqrt{x}} = \dfrac{\sqrt{x}}{2x}, \; x > 0$

(f) $(g \circ f)(x) = g(f(x)) = g\left(\dfrac{1}{x}\right) = 2\sqrt{\dfrac{1}{x}} = \dfrac{2}{\sqrt{x}} = \dfrac{2\sqrt{x}}{x}, \; x > 0$

25. $f(x) = x^3 + 8$

Since f is one-to-one, f has an inverse.

$$y = x^3 + 8$$
$$x = y^3 + 8$$
$$x - 8 = y^3$$
$$\sqrt[3]{x - 8} = y$$
$$f^{-1}(x) = \sqrt[3]{x - 8}$$

26. $f(x) = |x^2 - 3| + 6$

Since f is not one-to-one, f does not have an inverse.

27. $f(x) = 3x\sqrt{x} = 3x^{3/2}, \; x \geq 0$

Because f is one-to-one, f has an inverse.

$$y = 3x^{3/2}$$
$$x = 3y^{3/2}$$
$$\tfrac{1}{3}x = y^{3/2}$$
$$\left(\tfrac{1}{3}x\right)^{2/3} = y, \; x \geq 0$$
$$f^{-1}(x) = \left(\tfrac{1}{3}x\right)^{2/3}, \; x \geq 0$$

Chapter Test Solutions for Chapter 1

1. $\theta = \dfrac{5\pi}{4}$

(a)

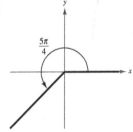

(b) $\dfrac{5\pi}{4} + 2\pi = \dfrac{13\pi}{4}$

$\dfrac{5\pi}{4} - 2\pi = -\dfrac{3\pi}{4}$

(c) $\dfrac{5\pi}{4}\left(\dfrac{180°}{\pi}\right) = 225°$

2. $\dfrac{105 \text{ km}}{\text{hr}} \times \dfrac{1 \text{ hr}}{60 \text{ min}} = 1.75 \text{ km per min}$

diameter $= 1 \text{ meter} = 0.001 \text{ km}$

radius $= \dfrac{1}{2}$ diameter $= 0.0005 \text{ km}$

Angular speed $= \dfrac{\theta}{t}$

$\qquad\qquad\quad = \dfrac{1.75}{2\pi(0.0005)} \cdot 2\pi$

$\qquad\qquad\quad = 3500 \text{ radians per minute}$

3. $130° = \dfrac{130\pi}{180} = \dfrac{13\pi}{18}$ radians

$A = \dfrac{1}{2}r^2\theta = \dfrac{1}{2}(25)^2\left(\dfrac{13\pi}{18}\right) \approx 709.04 \text{ square feet}$

4. Given $\tan \theta = \dfrac{3}{2}$.

$\text{hyp} = \sqrt{3^2 + 2^2} = \sqrt{13}$

$\sin \theta = \dfrac{\text{opp}}{\text{hyp}} = \dfrac{3}{\sqrt{13}} = \dfrac{3\sqrt{13}}{13}$

$\cos \theta = \dfrac{\text{adj}}{\text{hyp}} = \dfrac{2}{\sqrt{13}} = \dfrac{2\sqrt{13}}{13}$

$\cot \theta = \dfrac{\text{adj}}{\text{opp}} = \dfrac{2}{3}$

$\sec \theta = \dfrac{\text{hyp}}{\text{adj}} = \dfrac{\sqrt{13}}{2}$

$\csc \theta = \dfrac{\text{hyp}}{\text{opp}} = \dfrac{\sqrt{13}}{3}$

5. $x = -2, \ y = 6$

$r = \sqrt{(-2)^2 + (6)^2} = 2\sqrt{10}$

$\sin \theta = \dfrac{y}{r} = \dfrac{6}{2\sqrt{10}} = \dfrac{3}{\sqrt{10}} = \dfrac{3\sqrt{10}}{10}$

$\cos \theta = \dfrac{x}{r} = \dfrac{-2}{2\sqrt{10}} = -\dfrac{1}{\sqrt{10}} = -\dfrac{\sqrt{10}}{10}$

$\tan \theta = \dfrac{y}{x} = \dfrac{6}{-2} = -3$

$\csc \theta = \dfrac{r}{y} = \dfrac{2\sqrt{10}}{6} = \dfrac{\sqrt{10}}{3}$

$\sec \theta = \dfrac{r}{x} = \dfrac{2\sqrt{10}}{-2} = -\sqrt{10}$

$\cot \theta = \dfrac{x}{y} = \dfrac{-2}{6} = -\dfrac{1}{3}$

6. $\theta = 205°$

$\theta' = 205° - 180° = 25°$

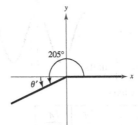

7. $\sec \theta < 0$ and $\tan \theta > 0$

$\dfrac{r}{x} < 0$ and $\dfrac{y}{x} > 0$

Quadrant III

8. $\cos \theta = -\dfrac{\sqrt{3}}{2}$

Reference angle is $30°$ and θ is in Quadrant II or III.

$\theta = 150°$ or $210°$.

9. $\cos \theta = \dfrac{3}{5}, \ \tan \theta < 0 \Rightarrow \theta$ lies in Quadrant IV.

Let $x = 3, r = 5 \Rightarrow y = -4$.

$\sin \theta = -\dfrac{4}{5}$

$\cos \theta = \dfrac{3}{5}$

$\tan \theta = -\dfrac{4}{3}$

$\csc \theta = -\dfrac{5}{4}$

$\sec \theta = \dfrac{5}{3}$

$\cot \theta = -\dfrac{3}{4}$

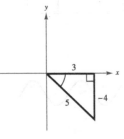

10. $\sec \theta = -\dfrac{29}{20}, \ \sin \theta > 0 \Rightarrow \theta$ lies in Quadrant II.

Let $r = 29, \ x = -20 \Rightarrow y = 21$.

$\sin \theta = \dfrac{21}{29}$

$\cos \theta = -\dfrac{20}{29}$

$\tan \theta = -\dfrac{21}{20}$

$\csc \theta = \dfrac{29}{21}$

$\cot \theta = -\dfrac{20}{21}$

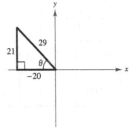

11. $g(x) = -2 \sin\left(x - \dfrac{\pi}{4}\right)$

Period: 2π

Amplitude: $|-2| = 2$

Shifted to the right by $\dfrac{\pi}{4}$ units and reflected in the x-axis.

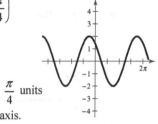

x	0	$\dfrac{\pi}{4}$	$\dfrac{3\pi}{4}$	$\dfrac{5\pi}{4}$	$\dfrac{7\pi}{4}$
y	$\sqrt{2}$	0	-2	0	2

12. $f(t) = \cos\left(t + \dfrac{\pi}{2}\right) - 1$

Period: 2π

Amplitude: $|1| = 1$

Shifted to the left by $\dfrac{\pi}{2}$ units and vertically down one unit.

t	$-\dfrac{\pi}{2}$	0	$\dfrac{\pi}{2}$	π	$\dfrac{3\pi}{2}$
y	0	-1	-2	-1	0

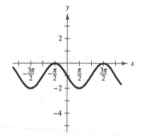

13. $f(\alpha) = \dfrac{1}{2} \tan 2\alpha$

Period: $\dfrac{\pi}{2}$

Asymptotes:

$x = -\dfrac{\pi}{4}, x = \dfrac{\pi}{4}$

α	$-\dfrac{\pi}{8}$	0	$\dfrac{\pi}{8}$
$f(\alpha)$	$-\dfrac{1}{2}$	0	$\dfrac{1}{2}$

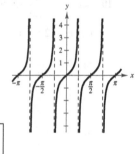

14. $y = \sin 2\pi x + 2 \cos \pi x$

Periodic: period $= 2$

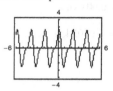

15. $y = 6x \cos(0.25x)$

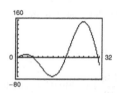

16. $f(x) = a \sin(bx + c)$

Amplitude: $2 \Rightarrow |a| = 2$

Reflected in the x-axis: $a = -2$

Period: $4\pi = \dfrac{2\pi}{b} \Rightarrow b = \dfrac{1}{2}$

Phase shift: $\dfrac{c}{b} = -\dfrac{\pi}{2} \Rightarrow c = -\dfrac{\pi}{4}$

$f(x) = -2 \sin\left(\dfrac{x}{2} - \dfrac{\pi}{4}\right)$

17. $\cot\left(\arcsin \dfrac{3}{8}\right)$

Let $y = \arcsin \dfrac{3}{8}$. Then $\sin y = \dfrac{3}{8}$ and

$\cot\left(\arcsin \dfrac{3}{8}\right) = \cot y = \dfrac{\sqrt{55}}{3}$.

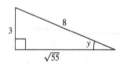

18. $f(x) = 2 \arcsin\left(\dfrac{1}{2}x\right)$

Domain: $[-2, 2]$

Range: $[-\pi, \pi]$

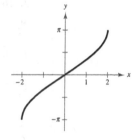

19. $\tan \theta = -\dfrac{110}{90}$

$\theta = \arctan\left(-\dfrac{110}{90}\right)$

$\theta \approx -50.7$

$\theta \approx 309.3°$

20. $d = a \cos bt$

$a = -6$

$\dfrac{2\pi}{b} = 2 \Rightarrow b = \pi$

$d = -6 \cos \pi t$

Chapter Test Solutions for Chapter 2

1. $\csc \theta = \dfrac{5}{2}$, $\tan \theta < 0$

θ is in Quadrant II.

$\sin \theta = \dfrac{1}{\csc \theta} = \dfrac{1}{\frac{5}{2}} = \dfrac{2}{5}$

$\cos \theta = -\sqrt{1 - \sin^2 \theta} = -\sqrt{1 - \left(\dfrac{2}{5}\right)^2} = -\dfrac{\sqrt{21}}{5}$

$\sec \theta = \dfrac{1}{\cos \theta} = -\dfrac{5}{\sqrt{21}} = -\dfrac{5\sqrt{21}}{21}$

$\tan \theta = \dfrac{\sin \theta}{\cos \theta} = \dfrac{\frac{2}{5}}{-\frac{\sqrt{21}}{5}} = -\dfrac{2}{\sqrt{21}} = -\dfrac{2\sqrt{21}}{21}$

$\cot \theta = \dfrac{1}{\tan \theta} = -\dfrac{\sqrt{21}}{2}$

2. $\csc^2 \beta\left(1 - \cos^2 \beta\right) = \dfrac{1}{\sin^2 \beta}\left(\sin^2 \beta\right) = 1$

3. $\dfrac{\sec^4 x - \tan^4 x}{\sec^2 x + \tan^2 x} = \dfrac{\left(\sec^2 x + \tan^2 x\right)\left(\sec^2 x - \tan^2 x\right)}{\sec^2 x + \tan^2 x}$

$= \sec^2 x - \tan^2 x = 1$

4. $\dfrac{\cos \theta}{\sin \theta} + \dfrac{\sin \theta}{\cos \theta} = \dfrac{\cos^2 \theta + \sin^2 \theta}{\sin \theta \cos \theta} = \dfrac{1}{\sin \theta \cos \theta}$

$= \csc \theta \sec \theta$

5. $\sin \theta \sec \theta = \sin \theta \dfrac{1}{\cos \theta} = \dfrac{\sin \theta}{\cos \theta} = \tan \theta$

6. $\sec^2 x \tan^2 x + \sec^2 x = \sec^2 x\left(\sec^2 x - 1\right) + \sec^2 x = \sec^4 x - \sec^2 x + \sec^2 x = \sec^4 x$

7. $\dfrac{\csc \alpha + \sec \alpha}{\sin \alpha + \cos \alpha} = \dfrac{\dfrac{1}{\sin \alpha} + \dfrac{1}{\cos \alpha}}{\sin \alpha + \cos \alpha} = \dfrac{\dfrac{\cos \alpha + \sin \alpha}{\sin \alpha \cos \alpha}}{\sin \alpha + \cos \alpha} = \dfrac{1}{\sin \alpha \cos \alpha}$

$= \dfrac{\cos^2 \alpha + \sin^2 \alpha}{\sin \alpha \cos \alpha} = \dfrac{\cos^2 \alpha}{\sin \alpha \cos \alpha} + \dfrac{\sin^2 \alpha}{\sin \alpha \cos \alpha}$

$= \dfrac{\cos \alpha}{\sin \alpha} + \dfrac{\sin \alpha}{\cos \alpha} = \cot \alpha + \tan \alpha$

8. $\tan\left(x + \dfrac{\pi}{2}\right) = \tan\left(\dfrac{\pi}{2} - (-x)\right) = \cot(-x) = -\cot x$

9. Using the power reducing formula for cosine,

$2 \cos^2 5y = 2\left(\dfrac{1 + \cos(2 \cdot 5y)}{2}\right)$

$= 1 + \cos 10y.$

So, $1 + \cos 10y = 2 \cos^2 5y.$

10. Using the double angle formula for sines,

$\dfrac{1}{2} \sin\left(2 \cdot \dfrac{\alpha}{3}\right) = \dfrac{1}{2} \cdot 2 \sin \dfrac{\alpha}{3} \cos \dfrac{\alpha}{3}$

$= \sin \dfrac{\alpha}{3} \cos \dfrac{\alpha}{3}.$

So, $\sin \dfrac{\alpha}{3} \cos \dfrac{\alpha}{3} = \dfrac{1}{2} \sin \dfrac{2\alpha}{3}.$

11. $4 \sin 3\theta \cos 2\theta = 4 \cdot \frac{1}{2}\left[\sin(3\theta + 2\theta) + \sin(3\theta - 2\theta)\right]$

$\qquad\qquad\quad = 2(\sin 5\theta + \sin \theta)$

12. $\cos\left(\theta + \dfrac{\pi}{2}\right) - \cos\left(\theta - \dfrac{\pi}{2}\right) = \cos \theta \cos \dfrac{\pi}{2} - \sin \theta \sin \dfrac{\pi}{2} - \left(\cos \theta \cos \dfrac{\pi}{2} + \sin \theta \sin \dfrac{\pi}{2}\right)$

$\qquad\qquad\qquad\qquad\qquad\quad = \cos \theta(0) - \sin \theta(1) - \cos \theta(0) - \sin \theta(1)$

$\qquad\qquad\qquad\qquad\qquad\quad = -2 \sin \theta$

13. $\tan^2 x + \tan x = 0$

$\tan x(\tan x + 1) = 0$

$\tan x = 0 \quad$ or $\quad \tan x + 1 = 0$

$\qquad x = 0, \pi \qquad\qquad \tan x = -1$

$\qquad\qquad\qquad\qquad\qquad x = \dfrac{3\pi}{4}, \dfrac{7\pi}{4}$

14. $\sin 2\alpha - \cos \alpha = 0$

$2 \sin \alpha \cos \alpha - \cos \alpha = 0$

$\cos \alpha(2 \sin \alpha - 1) = 0$

$\cos \alpha = 0 \quad$ or $\quad 2 \sin \alpha - 1 = 0$

$\alpha = \dfrac{\pi}{2}, \dfrac{3\pi}{2} \qquad \sin \alpha = \dfrac{1}{2}$

$\qquad\qquad\qquad\qquad \alpha = \dfrac{\pi}{6}, \dfrac{5\pi}{6}$

15. $4 \cos^2 x - 3 = 0$

$\cos^2 x = \dfrac{3}{4}$

$\cos x = \pm\sqrt{\dfrac{3}{4}} = \pm\dfrac{\sqrt{3}}{2}$

$x = \dfrac{\pi}{6}, \dfrac{5\pi}{6}, \dfrac{7\pi}{6}, \dfrac{11\pi}{6}$

16. $\csc^2 x - \csc x - 2 = 0$

$(\csc x - 2)(\csc x + 1) = 0$

$\csc x - 2 = 0 \quad$ or $\quad \csc x + 1 = 0$

$\csc x = 2 \qquad\qquad \csc x = -1$

$\dfrac{1}{\sin x} = 2 \qquad\qquad \dfrac{1}{\sin x} = -1$

$\sin x = \dfrac{1}{2} \qquad\qquad \sin x = -1$

$x = \dfrac{\pi}{6}, \dfrac{5\pi}{6} \qquad\qquad x = \dfrac{3\pi}{2}$

17. $5 \sin x - x = 0$

$x \approx 0, 2.596$

18. $105° = 135° - 30°$

$\cos 105° = \cos(135° - 30°)$

$\qquad\qquad = \cos 135° \cos 30° + \sin 135° \sin 30°$

$\qquad\qquad = -\cos 45° \cos 30° + \sin 45° \sin 30°$

$\qquad\qquad = \left(-\dfrac{\sqrt{2}}{2}\right)\left(\dfrac{\sqrt{3}}{2}\right) + \left(\dfrac{\sqrt{2}}{2}\right)\left(\dfrac{1}{2}\right)$

$\qquad\qquad = \dfrac{-\sqrt{6} + \sqrt{2}}{4} = \dfrac{\sqrt{2} - \sqrt{6}}{4}$

19. $x = 2, y = -5, r = \sqrt{29}$

$\sin 2u = 2 \sin u \cos u = 2\left(-\dfrac{5}{\sqrt{29}}\right)\left(\dfrac{2}{\sqrt{29}}\right) = -\dfrac{20}{29}$

$\cos 2u = \cos^2 u - \sin^2 u = \left(\dfrac{2}{\sqrt{29}}\right)^2 - \left(-\dfrac{5}{\sqrt{29}}\right)^2 = -\dfrac{21}{29}$

$\tan 2u = \dfrac{2 \tan u}{1 - \tan^2 u} = \dfrac{2\left(-\dfrac{5}{2}\right)}{1 - \left(-\dfrac{5}{2}\right)^2} = \dfrac{20}{21}$

20. Let $y_1 = 2.914 \sin(0.017t - 1.321) + 12.134$ and $y_2 = 10$.

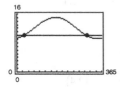

The points of intersection occur when $t \approx 30$ and $t \approx 310$.

The number of days that $D > 10$ hours is 280, from day 30 to day 310.

21.
$$28 \cos 10t + 38 = 28 \cos\left[10\left(t - \frac{\pi}{6}\right)\right] + 38$$

$$\cos 10t = \cos\left[10\left(t - \frac{\pi}{6}\right)\right]$$

$$0 = \cos\left[10\left(t - \frac{\pi}{6}\right)\right] - \cos 10t$$

$$= -2 \sin\left(\frac{10(t - (\pi/6)) + 10t}{2}\right) \sin\left(\frac{10(t - (\pi/6)) - 10t}{2}\right)$$

$$= -2 \sin\left(10t - \frac{5\pi}{6}\right) \sin\left(-\frac{5\pi}{6}\right)$$

$$= -2 \sin\left(10t - \frac{5\pi}{6}\right)\left(-\frac{1}{2}\right)$$

$$= \sin\left(10t - \frac{5\pi}{6}\right)$$

$$10t - \frac{5\pi}{6} = n\pi \text{ where } n \text{ is any integer.}$$

$$t = \frac{n\pi}{10} + \frac{\pi}{12} \text{ where } n \text{ is any integer.}$$

The first six times the two people are at the same height are: 0.26 minutes, 0.58 minutes, 0.89 minutes, 1.20 minutes, 1.52 minutes, 1.83 minutes.

Chapter Test Solutions for Chapter 3

1. The law of Cosines cannot be used.

Given: $A = 24°$, $B = 68°$, $a = 12.2$

Law of Sines: AAS

$$C = 180° - A - B = 180° - 24° - 68° = 88°$$

$$\frac{a}{\sin A} = \frac{b}{\sin B} \Rightarrow b = \frac{a}{\sin A}(\sin B)$$

$$b = \frac{12.2}{\sin 24°}(\sin 68°) \approx 27.81$$

$$\frac{a}{\sin A} = \frac{c}{\sin C} \Rightarrow c = \frac{a}{\sin A}(\sin C)$$

$$= \frac{12.2}{\sin 24°}(\sin 88°) \approx 29.98$$

2. The law of Cosines cannot be used.

Given: $B = 110°$, $C = 28°$, $a = 15.6$

Law of Sines: AAS

$$A = 180° - B - C = 180° - 110° - 28° = 42°$$

$$\frac{a}{\sin A} = \frac{b}{\sin B} \Rightarrow b = \frac{a}{\sin A}(\sin B)$$

$$= \frac{15.6}{\sin 42°}(\sin 110°) \approx 21.91$$

$$\frac{a}{\sin A} = \frac{c}{\sin C} \Rightarrow c = \frac{a}{\sin A}(\sin C)$$

$$= \frac{15.6}{\sin 42°}(\sin 28°) \approx 10.95$$

3. The law of Cosines cannot be used.

Given: $A = 24°$, $a = 11.2$, $b = 13.4$

Law of Sines: SSA

$$\frac{\sin A}{a} = \frac{\sin B}{b} \Rightarrow \sin B = b\left(\frac{\sin A}{a}\right)$$

$$\sin B = 13.4\left(\frac{\sin 24°}{11.2}\right) \approx 0.4866$$

There are two angles between $0°$ and $180°$ where $\sin \theta \neq 0.4866$, $B_1 \approx 29.12°$ and $B_2 \approx 150.88$.

For $B_1 \approx 29.12°$,

$$C_1 = 180° - 29.12° - 24° = 126.88°$$

$$\frac{c}{\sin C} = \frac{a}{\sin A} \Rightarrow c = \frac{a}{\sin A}(\sin C)$$

$$= \frac{11.2}{\sin 24°}(\sin 126.88°) \approx 22.03$$

For $B_2 \approx 150.88°$,

$$C = 180° - 150.88° - 24° = 5.12°.$$

$$\frac{c}{\sin C} = \frac{a}{\sin A} \Rightarrow c = \frac{a}{\sin A}(\sin C)$$

$$= \frac{11.2}{\sin 24°}(\sin 5.12°) \approx 2.46$$

5. The law of Cosines cannot be used.

Given: $B = 100°$, $a = 23$, $b = 15$

Law of Sines: SSA

$$\frac{\sin A}{a} = \frac{\sin B}{b} \Rightarrow \sin A = a\left(\frac{\sin B}{b}\right) = 23\left(\frac{\sin 100°}{15}\right) \approx 1.5100$$

Because there are no values of A such that $\sin A = 1.5100$, there is no possible triangle that can be formed.

6. The law of Cosines can be used.

Given: $C = 121°$, $a = 34$, $b = 55$

Law of Cosines: SAS

$$c^2 = a^2 + b^2 - 2ab \cos C$$

$$c^2 = (34)^2 + (55)^2 - 2(34)(55) \cos 121°$$

$$c^2 = 6107.2424$$

$$c \approx 78.15$$

$$\frac{\sin B}{b} = \frac{\sin C}{c} \Rightarrow \sin B = b\left(\frac{\sin C}{c}\right)$$

$$= 55\left(\frac{\sin 121°}{78.15}\right)$$

$$\approx 0.6033$$

So, $B \approx 37.11°$.

$$A = 180° - B - C = 180° - 37.11° - 121° = 21.89.$$

4. The law of Cosines can be used.

Given: $a = 6.0$, $b = 7.3$, $c = 12.4$

Law of Cosines: SSS

$$\cos C = \frac{a^2 + b^2 - c^2}{2ab}$$

$$= \frac{(6.0)^2 + (7.3)^2 - (12.4)^2}{2(6.0)(7.3)}$$

$$\approx -0.7360$$

$$c \approx 1378.39°$$

7. $a = 60, b = 70, c = 82$

$$s = \frac{a + b + c}{2} = \frac{60 + 70 + 82}{2} = 106$$

$$\text{Area} = \sqrt{s(s - a)(s - b)(s - c)} = \sqrt{106(46)(36)(24)} \approx 2052.5 \text{ square meters}$$

8. $b^2 = 370^2 + 240^2 - 2(370)(240)\cos 167°$

$\quad b \approx 606.3$ miles

$$\sin A = \frac{a \sin B}{b} = \frac{240 \sin 167°}{606.3}$$

$\quad A \approx 5.1°$

Bearing: $24° + 5.1° = 29.1°$

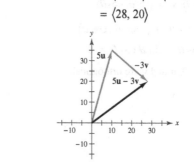

Not drawn to scale

9. Initial point: $(-3, 7)$

Terminal point: $(11, -16)$

$\mathbf{v} = \langle 11 - (-3), -16 - 7 \rangle = \langle 14, -23 \rangle$

10. $\mathbf{v} = 12\left(\dfrac{\mathbf{u}}{\|\mathbf{u}\|}\right) = 12\left(\dfrac{\langle 3, -5 \rangle}{\sqrt{3^2 + (-5)^2}}\right) = \dfrac{12}{\sqrt{34}}\langle 3, -5 \rangle$

$\quad = \dfrac{6\sqrt{34}}{17}\langle 3, -5 \rangle = \left\langle \dfrac{18\sqrt{34}}{17}, -\dfrac{30\sqrt{34}}{17} \right\rangle$

11. $\mathbf{u} = \langle 2, 7 \rangle, \mathbf{v} = \langle -6, 5 \rangle$

$\quad \mathbf{u} + \mathbf{v} = \langle 2, 7 \rangle + \langle -6, 5 \rangle = \langle -4, 12 \rangle$

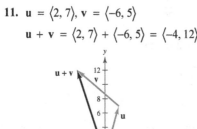

12. $\mathbf{u} = \langle 2, 7 \rangle, \mathbf{v} = \langle -6, 5 \rangle$

$\quad \mathbf{u} - \mathbf{v} = \langle 2, 7 \rangle - \langle -6, 5 \rangle = \langle 8, 2 \rangle$

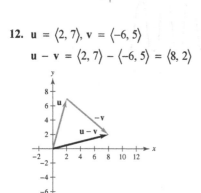

13. $\mathbf{u} = \langle 2, 7 \rangle, \mathbf{v} = \langle -6, 5 \rangle$

$\quad 5\mathbf{u} - 3\mathbf{v} = 5\langle 2, 7 \rangle - 3\langle -6, 5 \rangle$

$\qquad\qquad = \langle 10, 35 \rangle - \langle -18, 15 \rangle$

$\qquad\qquad = \langle 28, 20 \rangle$

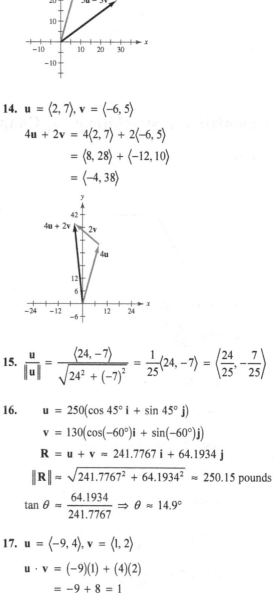

14. $\mathbf{u} = \langle 2, 7 \rangle, \mathbf{v} = \langle -6, 5 \rangle$

$\quad 4\mathbf{u} + 2\mathbf{v} = 4\langle 2, 7 \rangle + 2\langle -6, 5 \rangle$

$\qquad\qquad = \langle 8, 28 \rangle + \langle -12, 10 \rangle$

$\qquad\qquad = \langle -4, 38 \rangle$

15. $\dfrac{\mathbf{u}}{\|\mathbf{u}\|} = \dfrac{\langle 24, -7 \rangle}{\sqrt{24^2 + (-7)^2}} = \dfrac{1}{25}\langle 24, -7 \rangle = \left\langle \dfrac{24}{25}, -\dfrac{7}{25} \right\rangle$

16. $\mathbf{u} = 250(\cos 45° \, \mathbf{i} + \sin 45° \, \mathbf{j})$

$\quad \mathbf{v} = 130(\cos(-60°)\mathbf{i} + \sin(-60°)\mathbf{j})$

$\quad \mathbf{R} = \mathbf{u} + \mathbf{v} \approx 241.7767 \, \mathbf{i} + 64.1934 \, \mathbf{j}$

$\quad \|\mathbf{R}\| \approx \sqrt{241.7767^2 + 64.1934^2} \approx 250.15 \text{ pounds}$

$\quad \tan \theta \approx \dfrac{64.1934}{241.7767} \Rightarrow \theta \approx 14.9°$

17. $\mathbf{u} = \langle -9, 4 \rangle, \mathbf{v} = \langle 1, 2 \rangle$

$\quad \mathbf{u} \cdot \mathbf{v} = (-9)(1) + (4)(2)$

$\qquad\quad = -9 + 8 = 1$

18. $\mathbf{u} = \langle -1, 5 \rangle$, $\mathbf{v} = \langle 3, -2 \rangle$

$$\cos \theta = \frac{\mathbf{u} \cdot \mathbf{v}}{\|\mathbf{u}\| \|\mathbf{v}\|} = \frac{-13}{\sqrt{26}\sqrt{13}} \Rightarrow \theta = 135°$$

19. $\mathbf{u} = \langle 6, -10 \rangle$, $\mathbf{v} = \langle 5, 3 \rangle$

$\mathbf{u} \cdot \mathbf{v} = 6(5) + (-10)(3) = 0$

$\mathbf{u}$ and $\mathbf{v}$ are orthogonal.

20. $\mathbf{u} = \langle \cos\theta, -\sin\theta \rangle$, $\mathbf{v} = \langle \sin\theta, \cos\theta \rangle$

$\mathbf{u} \cdot \mathbf{v} = (\cos\theta)(\sin\theta) + (-\sin\theta)(\cos\theta)$

$\qquad = \sin\theta \cos\theta - \sin\theta \cos\theta = 0$

Because $\mathbf{u} \cdot \mathbf{v} = 0$, $\mathbf{u}$ and $\mathbf{v}$ are orthogonal.

21. $\mathbf{u} = \langle 6, 7 \rangle$, $\mathbf{v} = \langle -5, -1 \rangle$

$$\mathbf{w}_1 = \text{proj}_{\mathbf{v}}\, \mathbf{u} = \left(\frac{\mathbf{u} \cdot \mathbf{v}}{\|\mathbf{v}\|^2} \right) \mathbf{v} = -\frac{37}{26}\langle -5, -1 \rangle = \frac{37}{26}\langle 5, 1 \rangle$$

$$\mathbf{w}_2 = \mathbf{u} - \mathbf{w}_1 = \langle 6, 7 \rangle - \frac{37}{26}\langle 5, 1 \rangle$$

$$= \left\langle -\frac{29}{26}, \frac{145}{26} \right\rangle$$

$$= \frac{29}{26}\langle -1, 5 \rangle$$

$$\mathbf{u} = \mathbf{w}_1 + \mathbf{w}_2 = \frac{37}{26}\langle 5, 1 \rangle + \frac{29}{26}\langle -1, 5 \rangle$$

22. $\mathbf{F} = -500\mathbf{j}$, $\mathbf{v} = (\cos 12°)\mathbf{i} + (\sin 12°)\mathbf{j}$

$$\mathbf{w}_1 = \text{proj}_{\mathbf{v}}\, \mathbf{F} = \left(\frac{\mathbf{F} \cdot \mathbf{v}}{\|\mathbf{v}\|^2} \right) \mathbf{v} = (\mathbf{F} \cdot \mathbf{v})\mathbf{v}$$

$$= (-500 \sin 12°)\mathbf{v}$$

The magnitude of the force is $500 \sin 12° \approx 104$ pounds.

Cumulative Test Solutions for Chapters 1–3

1. (a)

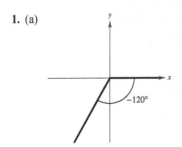

(b) $-120° + 360° = 240°$

(c) $-120\left(\dfrac{\pi}{180°} \right) = -\dfrac{2\pi}{3}$

(d) $-120° + 360° = 240°$

$\theta' = 240° - 180° = 60°$

(e) $\sin(-120°) = -\sin 60° = -\dfrac{\sqrt{3}}{2}$

$\cos(-120°) = -\cos 60° = -\dfrac{1}{2}$

$\tan(-120°) = \tan 60° = \sqrt{3}$

$\csc(-120°) = \dfrac{1}{-\sin 60°} = -\dfrac{2\sqrt{3}}{3}$

$\sec(-120°) = \dfrac{1}{-\cos 60°} = -2$

$\cot(-120°) = \dfrac{1}{\tan 60°} = \dfrac{\sqrt{3}}{3}$

2. $-1.45\left(\dfrac{180}{\pi} \right) \approx -83.079°$

3. $\tan \theta = \dfrac{y}{x} = -\dfrac{21}{20} \Rightarrow r = 29$

Because $\sin \theta < 0$, θ is in Quadrant IV $\Rightarrow x = 20$.

$\cos \theta = \dfrac{x}{r} = \dfrac{20}{29}$

4. $f(x) = 3 - 2 \sin \pi x$

Period: $\dfrac{2\pi}{\pi} = 2$

Amplitude: $|a| = |-2| = 2$

Upward shift of 3 units (reflected in x-axis prior to shift)

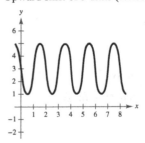

5. $g(x) = \frac{1}{2} \tan\left(x - \frac{\pi}{2}\right)$

Period: π

Asymptotes: $x = 0, x = \pi$

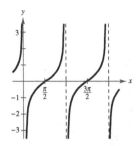

6. $h(x) = -\sec(x + \pi)$

Graph $y = -\cos(x + \pi)$ first.

Period: 2π

Amplitude: 1

Set $x + \pi = 0$ and $x + \pi = 2\pi$ for one cycle.

$x = -\pi \qquad\qquad x = \pi$

The asymptotes of $h(x)$ corresponds to the

x-intercepts of $y = -\cos(x + \pi)$.

$x + \pi = \frac{(2n + 1)\pi}{2}$

$x = \frac{(2n - 1)\pi}{2}$ where n is any integer

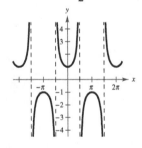

7. $h(x) = a \cos(bx + c)$

Graph is reflected in x-axis.

Amplitude: $a = -3$

Period: $2 = \frac{2\pi}{\pi} \Rightarrow b = \pi$

No phase shift: $c = 0$

$h(x) = -3 \cos(\pi x)$

8. $f(x) = \frac{x}{2}\sin x, -3\pi \le x \le 3\pi$

$-\frac{x}{2} \le f(x) \le \frac{x}{2}$

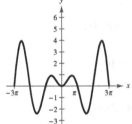

9. $\tan(\arctan 4.9) = 4.9$

10. $\tan\left(\arcsin \frac{3}{5}\right) = \frac{3}{4}$

11. $y = \arccos(2x)$

$\sin y = \sin[\arccos(2x)] = \sqrt{1 - 4x^2}$

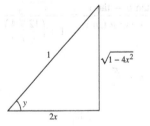

12. $\cos\left(\frac{\pi}{2} - x\right)\csc x = \sin x\left(\frac{1}{\sin x}\right) = 1$

13. $\dfrac{\sin\theta - 1}{\cos\theta} - \dfrac{\cos\theta}{\sin\theta - 1} = \dfrac{\sin\theta - 1}{\cos\theta} - \dfrac{\cos\theta(\sin\theta + 1)}{\sin^2\theta - 1}$

$= \dfrac{\sin\theta - 1}{\cos\theta} + \dfrac{\cos\theta(\sin\theta + 1)}{\cos^2\theta} = \dfrac{\sin\theta - 1}{\cos\theta} + \dfrac{\sin\theta + 1}{\cos\theta} = \dfrac{2\sin\theta}{\cos\theta} = 2\tan\theta$

14. $\cot^2\alpha(\sec^2\alpha - 1) = \cot^2\alpha\tan^2\alpha = 1$

15. $\sin(x + y) \sin(x - y) = \frac{1}{2}\Big[\cos\big(x + y - (x - y)\big) - \cos\big(x + y + x - y\big)\Big]$

$\qquad\qquad\qquad\qquad = \frac{1}{2}[\cos 2y - \cos 2x] = \frac{1}{2}\Big[1 - 2\sin^2 y - \big(1 - 2\sin^2 x\big)\Big] = \sin^2 x - \sin^2 y$

16. $\sin^2 x \cos^2 x = \left(\dfrac{1 - \cos 2x}{2}\right)\left(\dfrac{1 + \cos 2x}{2}\right)$

$\qquad\qquad = \dfrac{1}{4}(1 - \cos 2x)(1 + \cos 2x)$

$\qquad\qquad = \dfrac{1}{4}(1 - \cos^2 2x)$

$\qquad\qquad = \dfrac{1}{4}\left(1 - \dfrac{1 + \cos 4x}{2}\right)$

$\qquad\qquad = \dfrac{1}{8}(2 - (1 + \cos 4x))$

$\qquad\qquad = \dfrac{1}{8}(1 - \cos 4x)$

17. $2\cos^2 \beta - \cos \beta = 0$

$\quad \cos \beta(2\cos \beta - 1) = 0$

$\cos \beta = 0 \qquad$ or $\quad 2\cos \beta - 1 = 0$

$\beta = \dfrac{\pi}{2}, \dfrac{3\pi}{2} \qquad\qquad \cos \beta = \dfrac{1}{2}$

$\qquad\qquad\qquad\qquad \beta = \dfrac{\pi}{3}, \dfrac{5\pi}{3}$

Answer: $\dfrac{\pi}{3}, \dfrac{\pi}{2}, \dfrac{3\pi}{2}, \dfrac{5\pi}{3}$

18. $3\tan \theta - \cot \theta = 0$

$\quad 3\tan \theta - \dfrac{1}{\tan \theta} = 0$

$\quad \dfrac{3\tan^2 \theta - 1}{\tan \theta} = 0$

$\quad 3\tan^2 \theta - 1 = 0$

$\quad \tan^2 \theta = \dfrac{1}{3}$

$\quad \tan \theta = \pm\dfrac{\sqrt{3}}{3}$

$\quad \theta = \dfrac{\pi}{6}, \dfrac{5\pi}{6}, \dfrac{7\pi}{6}, \dfrac{11\pi}{6}$

19. $\sin^2 x + 2\sin x + 1 = 0$

$\quad (\sin x + 1)(\sin x + 1) = 0$

$\quad \sin x + 1 = 0$

$\quad \sin x = -1$

$\quad x = \dfrac{3\pi}{2}$

20. $\sin u = \frac{12}{13} \Rightarrow \cos u = \frac{5}{13}$ and $\tan u = \frac{12}{5}$ because u is in Quadrant I.

$\cos v = \frac{3}{5} \Rightarrow \sin v = \frac{4}{5}$ and $\tan v = \frac{4}{3}$ because v is in Quadrant I.

$\tan(u - v) = \dfrac{\tan u - \tan v}{1 + \tan u \tan v} = \dfrac{\dfrac{12}{5} - \dfrac{4}{3}}{1 + \left(\dfrac{12}{5}\right)\left(\dfrac{4}{3}\right)} = \dfrac{16}{63}$

21. $\tan u = \dfrac{1}{2}, \, 0 < u < \dfrac{\pi}{2}$

$\tan 2u = \dfrac{2\tan u}{1 - \tan^2 u} = \dfrac{2\left(\dfrac{1}{2}\right)}{1 - \left(\dfrac{1}{2}\right)^2} = \dfrac{4}{3}$

22. $\tan u = \dfrac{4}{3}, 0 < u < \dfrac{\pi}{2}$

$\tan u = \dfrac{4}{3} \Rightarrow \cos u = \dfrac{3}{5}$

$\sin \dfrac{u}{2} = \sqrt{\dfrac{1 - \cos u}{2}} = \sqrt{\dfrac{1 - \dfrac{3}{5}}{2}} = \dfrac{\sqrt{5}}{5}$

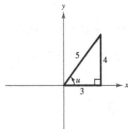

23. $5 \sin \dfrac{3\pi}{4} \cos \dfrac{7\pi}{4} = \dfrac{5}{2}\left[\sin\left(\dfrac{3\pi}{4} + \dfrac{7\pi}{4}\right) + \sin\left(\dfrac{3\pi}{4} - \dfrac{7\pi}{4}\right)\right]$

$= \dfrac{5}{2}\left[\sin \dfrac{5\pi}{2} + \sin(-\pi)\right]$

$= \dfrac{5}{2}\left(\sin \dfrac{5\pi}{2} - \sin \pi\right)$

24. $\cos 9x - \cos 7x = -2 \sin\left(\dfrac{9x + 7x}{2}\right) \sin\left(\dfrac{9x - 7x}{2}\right)$

$= -2 \sin 8x \sin x$

25. Given two sides and an angle opposite one of them, the Law of Cosines cannot be used, therefore use the Law of Sines: SSA.

$A = 30°, a = 9, b = 8$

$\dfrac{\sin B}{8} = \dfrac{\sin 30°}{9}$

$\sin B = \dfrac{8}{9}\left(\dfrac{1}{2}\right)$

$B = \arcsin\left(\dfrac{4}{9}\right)$

$B \approx 26.39°$

$C = 180° - A - B \approx 123.61°$

$\dfrac{c}{\sin 123.61°} = \dfrac{9}{\sin 30°}$

$c \approx 14.99$

26. Given two sides and the included angle, the Law of Cosines can be used, therefore use the Law of Cosines: SAS.

$A = 30°, b = 8, c = 10$

$a^2 = 8^2 + 10^2 - 2(8)(10) \cos 30°$

$a^2 \approx 25.4359$

$a \approx 5.04$

$\cos B = \dfrac{5.04^2 + 10^2 - 8^2}{2(5.04)(10)}$

$\cos B \approx 0.6091$

$B \approx 52.48°$

$C = 180° - A - B \approx 97.52°$

27. Because $C = 90°$, use right triangle ratios.

$A = 30°, C = 90°, b = 10$

$B = 180° - 30° - 90° = 60°$

$\tan 30° = \dfrac{a}{10} \Rightarrow a = 10 \tan 30° \approx 5.77$

$\cos 30° = \dfrac{10}{c} \Rightarrow c = \dfrac{10}{\cos 30°} \approx 11.55$

28. Given three sides, the Law of Cosines can be used, therefore use the Law of Cosines: SSS.

$a = 4.7, b = 8.1, c = 10.3$

$\cos C = \dfrac{a^2 + b^2 - c^2}{2ab} = \dfrac{4.7^2 + 8.1^2 - 10.3^2}{2(4.7)(8.1)} = -0.2415 \Rightarrow C \approx 103.98°$

$\sin A = \dfrac{a \sin C}{c} \approx \dfrac{4.7 \sin 103.98°}{10.3} \approx 0.4428 \Rightarrow A \approx 26.28°$

$B \approx 180° - 26.28° - 103.98° = 49.74°$

29. Given two angles and a side opposite one of them, the Law of Cosines cannot be used, therefore use the Law of Sines: AAS.

$A = 45°, B = 26°, c = 20$

$C = 180° - A - B = 180° - 45° - 26° = 109°$

$\dfrac{a}{\sin A} = \dfrac{c}{\sin C} \Rightarrow a = \dfrac{c}{\sin C}(\sin A) = \dfrac{20}{\sin 109°}(\sin 45°) \approx 14.96$

$\dfrac{b}{\sin B} = \dfrac{c}{\sin C} \Rightarrow b = \dfrac{c}{\sin C}(\sin B) = \dfrac{20}{\sin 109°}(\sin 26°) \approx 9.27$

30. Given two angles and the included side, the Law of Cosines can be used, therefore use the Law of Cosines: SAS.

$a = 1.2, b = 10, C = 80°$

$c^2 = a^2 + b^2 - 2ab \cos C$

$c^2 = (1.2)^2 + (10)^2 - 2(1.2)(10) \cos 80°$

$c^2 \approx 97.2724$

$c \approx 9.86$

$\dfrac{a}{\sin A} = \dfrac{c}{\sin C}$

$\sin A = a\left(\dfrac{\sin C}{c}\right)$

$= 1.2\left(\dfrac{\sin 80°}{9.86}\right)$

≈ 0.1199

$A \approx 6.88°$

$B \approx 180° - 80° - 6.88° = 93.12°$

31. Area $= \dfrac{1}{2}(7)(12) \sin 99° = 41.48$ in.^{2}

32. $a = 30, b = 41, c = 45$

$s = \dfrac{a + b + c}{2} = \dfrac{30 + 41 + 45}{2} = 58$

Area $= \sqrt{s(s - a)(s - b)(s - c)}$

$= \sqrt{58(28)(17)(13)}$

≈ 599.09 m^2

33. Terminal point: $(6, 10)$; Initial point: $(-1, 2)$

$\mathbf{u} = \langle 6 - (-1), 10 - 2 \rangle = \langle 7, 8 \rangle = 7\mathbf{i} + 8\mathbf{j}$

34. $\mathbf{v} = \mathbf{i} + \mathbf{j}$

$\|\mathbf{v}\| = \sqrt{1^2 + 1^2} = \sqrt{2}$

$\mathbf{u} = \dfrac{\mathbf{v}}{\|\mathbf{v}\|} = \dfrac{1}{\sqrt{2}}(\mathbf{i} + \mathbf{j}) = \dfrac{\sqrt{2}}{2}(\mathbf{i} + \mathbf{j})$

35. $\mathbf{u} = 3\mathbf{i} + 4\mathbf{j}, \mathbf{v} = \mathbf{i} - 2\mathbf{j}$

$\mathbf{u} \cdot \mathbf{v} = 3(1) + 4(-2) = -5$

36. $\mathbf{u} = \langle 8, -2 \rangle, \mathbf{v} = \langle 1, 5 \rangle$

$\mathbf{w}_1 = \text{proj}_{\mathbf{v}}\ \mathbf{u} = \left(\dfrac{\mathbf{u} \cdot \mathbf{v}}{\|\mathbf{v}\|^2}\right)\mathbf{v} = \dfrac{-2}{26}\langle 1, 5 \rangle = -\dfrac{1}{13}\langle 1, 5 \rangle$

$\mathbf{w}_2 = \mathbf{u} - \mathbf{w}_1 = \langle 8, -2 \rangle - \left\langle -\dfrac{1}{13}, -\dfrac{5}{13} \right\rangle = \left\langle \dfrac{105}{13}, -\dfrac{21}{13} \right\rangle$

$= \dfrac{21}{13}\langle 5, -1 \rangle$

$\mathbf{u} = \mathbf{w}_1 + \mathbf{w}_2 = -\dfrac{1}{13}\langle 1, 5 \rangle + \dfrac{21}{13}\langle 5, -1 \rangle$

37. Angular speed $= \dfrac{\theta}{t} = \dfrac{2\pi(63)}{1} \approx 395.8$ radians per minute

Linear speed $= \dfrac{s}{t} = \dfrac{42\pi(63)}{1} \approx 8312.7$ inches per minute

38. Area $= \dfrac{\theta r^2}{2} = \dfrac{(105°)\left(\dfrac{\pi}{180°}\right)(12)^2}{2} = 42\pi \approx 131.95$ yd^2

39. Height of smaller triangle:

$$\tan 16° \, 45' = \frac{h_1}{200}$$

$$h_1 = 200 \tan 16.75°$$

$$\approx 60.2 \text{ feet}$$

Height of larger triangle:

$$\tan 18° = \frac{h_2}{200}$$

$$h_2 = 200 \tan 18° \approx 65.0 \text{ feet}$$

Height of flag: $h_2 - h_1 = 65.0 - 60.2 \approx 5$ feet

Not drawn to scale

40. $\tan \theta = \dfrac{5}{12} \Rightarrow \theta \approx 22.6°$

41. $d = a \cos bt$

$$|a| = 4 \Rightarrow a = 4$$

$$\frac{2\pi}{b} = 8 \Rightarrow b = \frac{\pi}{4}$$

$$d = 4 \cos \frac{\pi}{4} t$$

42.

$$\mathbf{v}_1 = 500\langle \cos 60°, \sin 60° \rangle = \langle 250, 250\sqrt{3} \rangle$$

$$\mathbf{v}_2 = 50\langle \cos 30°, \sin 30° \rangle = \langle 25\sqrt{3}, 25 \rangle$$

$$\mathbf{v} = \mathbf{v}_1 + \mathbf{v}_2 = \langle 250 + 25\sqrt{3}, 250\sqrt{3} + 25 \rangle \approx \langle 293.3, 458.0 \rangle$$

$$\|\mathbf{v}\| = \sqrt{(293.3)^2 + (458.0)^2} \approx 543.9$$

$$\tan \theta = \frac{458.0}{293.3} \approx 1.56 \Rightarrow \theta \approx 57.4°$$

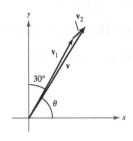

Bearing: $90° - 57.4° = 32.6°$

The plane is traveling on a bearing of $32.6°$ at 543.9 kilometers per hour.

43. $\mathbf{w} = (85)(10)\cos 60° = 425$ foot-pounds

Chapter Test Solutions for Chapter 4

1. $-5 + \sqrt{-100} = -5 + 10i$

2. $\sqrt{-16} - 2(7 + 2i) = 4i - 14 - 4i = -14$

3. $(4 + 9i)^2 = 16 + 72i + 81i^2 = -65 + 72i$

4. $(6 + \sqrt{7}i)(6 - \sqrt{7}i) = 36 - 7i^2 = 43$

5. $\dfrac{8}{1 + 2i} = \dfrac{8}{1 + 2i} \cdot \dfrac{1 - 2i}{1 - 2i} = \dfrac{8 - 16i}{1 - 4t^2} = \dfrac{8}{5} - \dfrac{16}{5}i$

6. $2x^2 - 2x + 3 = 0$

$$x = \frac{-(-2) \pm \sqrt{(-2)^2 - 4(2)(3)}}{2(2)} = \frac{2 \pm \sqrt{-20}}{4}$$

$$= \frac{2 \pm 2\sqrt{5}i}{4} = \frac{1}{2} \pm \frac{\sqrt{5}}{2}i$$

7. Since $x^5 + x^3 - x + 1 = 0$ is a fifth degree polynomial equation, it has five solutions in the complex number system.

8.
$$x^3 - 6x^2 + 5x - 30 = 0$$
$$x^2(x - 6) + 5(x - 6) = 0$$
$$(x - 6)(x^2 + 5) = 0$$
$$(x - 6)(x + \sqrt{5}i)(x - \sqrt{5}i) = 0$$

Zeros: $x = 6, \pm\sqrt{5}i$

9.
$$x^4 - 2x^2 - 24 = 0$$
$$(x^2 - 6)(x^2 + 4) = 0$$
$$(x + \sqrt{6})(x - \sqrt{6})(x + 2i)(x - 2i) = 0$$
$$x^2 - 6 = 0 \text{ or } x^2 + 4 = 0$$

Zeros: $x = \pm\sqrt{6}, \pm 2i$

10. $h(x) = x^4 - 2x^2 - 8$

Zeros: $x = \pm 2 \Rightarrow (x - 2)(x + 2) = x^2 - 4$ is a factor of $h(x)$.

$$
\begin{array}{r}
x^2 \qquad\qquad + 2 \\
x^2 + 0x - 4{\overline{\smash{\big)}\,x^4 + 0x^3 - 2x^2 + 0x - 8}} \\
\underline{x^4 + 0x^3 - 4x^2} \\
2x^2 + 0x - 8 \\
\underline{2x^2 + 0x - 8} \\
0
\end{array}
$$

Thus, $h(x) = (x^2 - 4)(x^2 + 2)$

$\qquad\quad = (x + 2)(x - 2)(x + \sqrt{2}i)(x - \sqrt{2}i).$

The zeros of $h(x)$ are $x = \pm 2, \pm\sqrt{2}i$.

11. $g(v) = 2v^3 - 11v^2 + 22v - 15$

Zero: $\frac{3}{2} \Rightarrow 2v - 3$ is a factor of $g(v)$.

$$
\begin{array}{r}
v^2 - 4v + 5 \\
2v - 3{\overline{\smash{\big)}\,2v^3 - 11v^2 + 22v - 15}} \\
\underline{2v^3 - 3v^2} \\
-8v^2 + 22v \\
\underline{-8v^2 + 12v} \\
10v - 15 \\
\underline{10v - 15} \\
0
\end{array}
$$

Thus, $g(v) = (2v - 3)(v^2 - 4v + 5)$. By the Quadratic Formula, the zeros of $v^2 - 4v + 5$ are $2 \pm i$.

The zeros of $g(v)$ are $v = \frac{3}{2}, 2 \pm 1$.

$\quad g(v) = (2v - 3)(v - 2 - i)(v - 2 + i)$

12. $f(x) = (x - 0)(x - 2)(x - 3i)(x + 3i)$

$\qquad\; = (x^2 - 2x)(x^2 + 9)$

$\qquad\; = x^4 - 2x^3 + 9x^2 - 18x$

13. $f(x) = (x - 1)(x - 1)\left[x - \left(2 + \sqrt{3}i\right)\right]\left[x - \left(2 - \sqrt{3}i\right)\right]$

$\qquad\; = (x^2 - 2x + 1)\left[(x - 2) - \sqrt{3}i\right]\left[(x - 2) + \sqrt{3}i\right]$

$\qquad\; = (x^2 - 2x + 1)\left[(x - 2)^2 - 3i^2\right]$

$\qquad\; = (x^2 - 2x + 1)(x^2 - 4x + 7)$

$\qquad\; = x^4 - 6x^3 + 16x^2 - 18x + 7$

14. No, complex zeros occur in conjugate pairs for polynomial functions with *real* coefficients. If $a + bi$ is a zero, so is $a - bi$.

15. The distance between $4 + 3i$ and $1 - i$ is

$d = \sqrt{(1 - 4)^2 + (-1 - 3)^2} = \sqrt{25} = 5$ units.

16. $z = 4 - 4i$

$|z| = \sqrt{(4)^2 + (-4)^2} = \sqrt{32} = 4\sqrt{2}$

$\tan\theta = -\dfrac{4}{4} = -1$ and is in Quadrant IV $\Rightarrow \theta = \dfrac{7\pi}{4}$

$z = 4\sqrt{2}\left(\cos\dfrac{7\pi}{4} + i\sin\dfrac{7\pi}{4}\right)$

17. $z = 6(\cos 120° + i \sin 120°)$

$\quad = 6\left(-\dfrac{1}{2} + \dfrac{\sqrt{3}}{2}i\right)$

$\quad = -3 + 3\sqrt{3}i$

18. $\left[3\left(\cos\dfrac{7\pi}{6} + i\sin\dfrac{7\pi}{6}\right)\right]^8 = 3^8\left(\cos\dfrac{28\pi}{3} + i\sin\dfrac{28\pi}{3}\right)$

$\qquad\qquad\qquad\qquad = 6561\left(-\dfrac{1}{2} - \dfrac{\sqrt{3}}{2}i\right)$

$\qquad\qquad\qquad\qquad = -\dfrac{6561}{2} - \dfrac{6561\sqrt{3}}{2}i$

19. $(3 - 3i)^6 = \left[3\sqrt{2}\left(\cos\dfrac{7\pi}{4} + i\sin\dfrac{7\pi}{4}\right)\right]^6$

$\qquad\quad = \left(3\sqrt{2}\right)^6\left(\cos\dfrac{21\pi}{2} + i\sin\dfrac{21\pi}{2}\right)$

$\qquad\quad = 5832(0 + i)$

$\qquad\quad = 5832i$

20. $z = 256(1 + 0i)$

$|z| = 256\sqrt{1^2 + (0)^2} = 256\sqrt{1} = 256$

$\tan \theta = \dfrac{0}{1} \Rightarrow \theta = 0$

$z = 256(\cos 0 + i \sin 0)$

Fourth roots of z: $\sqrt[4]{256}\left[\cos\left(\dfrac{0 + 2\pi k}{4} \right) + i \sin\left(\dfrac{0 + 2\pi k}{4} \right) \right],\ k = 0,\ 1,\ 2,\ 3$

$k = 0: 4(\cos 0 + i \sin 0)$

$k = 1: 4\left(\cos \dfrac{\pi}{2} + i \sin \dfrac{\pi}{2} \right)$

$k = 2: 4(\cos \pi + i \sin \pi)$

$k = 3: 4\left(\cos \dfrac{3\pi}{2} + i \sin \dfrac{3\pi}{2} \right)$

21. $x^3 - 27i = 0 \Rightarrow x^3 = 27i$

The solutions to the equation are the cube roots of $z = 27i = 27\left(\cos \dfrac{\pi}{2} + i \sin \dfrac{\pi}{2} \right)$.

Cube roots of z: $\sqrt[3]{27}\left[\cos\left(\dfrac{\dfrac{\pi}{2} + 2\pi k}{3} \right) + i \sin\left(\dfrac{\dfrac{\pi}{2} + 2\pi k}{3} \right) \right],\ k = 0,\ 1,\ 2$

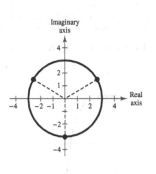

$k = 0: 3\left(\cos \dfrac{\pi}{6} + i \sin \dfrac{\pi}{6} \right)$

$k = 1: 3\left(\cos \dfrac{5\pi}{6} + i \sin \dfrac{5\pi}{6} \right)$

$k = 2: 3\left(\cos \dfrac{3\pi}{2} + i \sin \dfrac{3\pi}{2} \right)$

22.

$h = -16t^2 + 88t,\ 0 \le t \le 5.5$

$125 = -16t^2 + 88t$

$16t^2 - 88t + 125 = 0$

$t = \dfrac{88 \pm \sqrt{(-88)^2 - 4(16)(125)}}{2(16)} = \dfrac{88 \pm \sqrt{15{,}256}\,i}{32}$

Since the solutions are imaginary, it is not possible for the projectile to reach a height of 125 feet.

Chapter Test Solutions for Chapter 5

1. $0.7^{2.5} \approx 0.410$

2. $3^{-\pi} \approx 0.032$

3. $e^{-7/10} \approx 0.497$

4. $e^{3.1} \approx 22.198$

5. $f(x) = 10^{-x}$

x	-1	$-\frac{1}{2}$	0	$\frac{1}{2}$	1
$f(x)$	10	3.162	1	0.316	0.1

Horizontal asymptote: $y = 0$

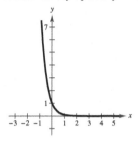

6. $f(x) = -6^{x-2}$

x	-1	0	1	2	3
$f(x)$	-0.005	-0.028	-0.167	-1	-6

Horizontal asymptote: $y = 0$

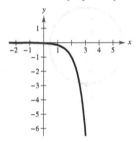

7. $f(x) = 1 - e^{2x}$

x	-1	$-\frac{1}{2}$	0	$\frac{1}{2}$	1
$f(x)$	0.865	0.632	0	-1.718	-6.389

Horizontal asymptote: $y = 1$

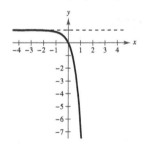

8. (a) $\log_7 7^{-0.89} = -0.89$

(b) $4.6 \ln e^2 = 4.6(2) = 9.2$

9. $f(x) = 4 + \log x$

Domain: $(0, \infty)$

x-intercept:

$$4 + \log x = 0$$
$$\log x = -4$$
$$10^{\log x} = 10^{-4}$$
$$x = 10^{-4}$$
$$\left(10^{-4}, 0\right) = (0.0001, 0)$$

Vertical asymptote: $x = 0$

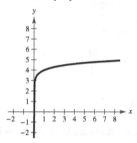

10. $f(x) = \ln(x - 4)$

Domain: $(4, \infty)$

$$\ln(x - 4) = 0$$

x-intercept: $e^{\ln(x-4)} = e^0$
$$x - 4 = 1$$
$$x = 5$$
$$(5, 0)$$

Vertical asymptote: $x = 4$

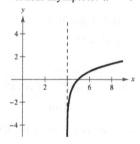

11. $f(x) = 1 + \ln(x + 6)$

Domain: $(-6, \infty)$

x-intercept:

$1 + \ln(x + 6) = 0$

$\ln(x + 6) = -1$

$e^{\ln(x+6)} = e^{-1}$

$x + 6 = e^{-1}$

$x = -6 + e^{-1}$

$\left(-6 + e^{-1}, 0\right) \approx (-5.632, 0)$

Vertical asymptote: $x = -6$

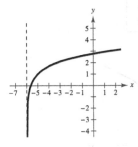

12. $\log_5 35 = \dfrac{\ln 35}{\ln 5} = \dfrac{\log 35}{\log 5} \approx 2.209$

13. $\log_{16} 0.63 = \dfrac{\log 0.63}{\log 16} \approx -0.167$

14. $\log_{3/4} 24 = \dfrac{\log 24}{\log (3/4)} \approx -11.047$

15. $\log_2 3a^4 = \log_2 3 + \log_2 a^4 = \log_2 3 + 4 \log_2 |a|$

16. $\ln \dfrac{\sqrt{x}}{7} = \ln\left(\sqrt{x}\right) - \ln 7$

$= \ln\sqrt{x} - \ln 7$

$= \ln x^{1/2} - \ln 7$

$= \dfrac{1}{2} \ln x - \ln 7$

17. $\ln \dfrac{10x^2}{y^3} = \ln\left(10x^2\right) - \ln y^3$

$= \ln 10 + \ln x^2 - \ln y^3$

$= \ln 10 + 2 \ln x - 3 \ln y$

18. $\log_3 13 + \log_3 y = \log_3 13y$

19. $4 \ln x - 4 \ln y = \ln x^4 - \ln y^4 = \ln \dfrac{x^4}{y^4}$

20. $3 \ln x - \ln(x + 3) + 2 \ln y = \ln x^3 - \ln(x + 3) + \ln y^2 = \ln \dfrac{x^3 y^2}{x + 3}$

21. $5^x = \dfrac{1}{25}$

$5^x = 5^{-2}$

$x = -2$

22. $3e^{-5x} = 132$

$e^{-5x} = 44$

$-5x = \ln 44$

$x = \dfrac{\ln 44}{-5} \approx -0.757$

23. $\dfrac{1025}{8 + e^{4x}} = 5$

$1025 = 5\left(8 + e^{4x}\right)$

$205 = 8 + e^{4x}$

$197 = e^{4x}$

$\ln 197 = 4x$

$x = \dfrac{\ln 197}{4} \approx 1.321$

24. $\ln x = \dfrac{1}{2}$

$x = e^{1/2} \approx 1.649$

25. $18 + 4 \ln x = 7$

$4 \ln x = -11$

$\ln x = -\dfrac{11}{4}$

$x = e^{-11/4} \approx 0.0639$

26. $\log x + \log(x - 15) = 2$

$\log\left[x(x - 15)\right] = 2$

$x(x - 15) = 10^2$

$x^2 - 15x - 100 = 0$

$(x - 20)(x + 5) = 0$

$x - 20 = 0 \quad \text{or} \quad x + 5 = 0$

$x = 20 \qquad\qquad x = -5$

The value $x = -5$ is extraneous. The only solution is $x = 20$.

27. $y = ae^{bt}$

$(0, 2745)$: $2745 = ae^{b(0)} \Rightarrow a = 2745$

$$y = 2745e^{bt}$$

$(9, 11{,}277)$: $\qquad 11{,}277 = 2745e^{b(9)}$

$$\frac{11{,}277}{2745} = e^{9b}$$

$$\ln\left(\frac{11{,}277}{2745}\right) = 9b$$

$$\frac{1}{9}\ln\left(\frac{11{,}277}{2745}\right) = b \Rightarrow b \approx 0.1570$$

So, $y = 2745e^{0.1570t}$.

28. $y = ae^{bt}$

$$\frac{1}{2}a = ae^{b(21.77)}$$

$$\frac{1}{2} = e^{21.77b}$$

$$\ln\left(\frac{1}{2}\right) = 21.77b$$

$$b = \frac{\ln(1/2)}{21.77} \approx -0.0318$$

$$y = ae^{-0.0318t}$$

When $t = 19$: $y = ae^{-0.0318(19)} \approx 0.55a$

So, 55% will remain after 19 years.

29. $H = 70.228 + 5.104x + 9.222 \ln x, \frac{1}{4} \le x \le 6$

(a)

x	H (cm)
$\frac{1}{4}$	58.720
$\frac{1}{2}$	66.388
1	75.332
2	86.828
3	95.671
4	103.43
5	110.59
6	117.38

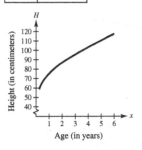

(b) Estimate: 103

When $x = 4$, $H \approx 103.43$ cm.

Chapter Test Solutions for Chapter 6

1. $4x - 7y + 6 = 0$

$$y = \tfrac{4}{7}x + \tfrac{6}{7}$$

$$\tan \theta = \tfrac{4}{7}$$

$$\theta \approx 0.5191 \text{ radian} \approx 29.7°$$

2. $3x + y = 6 \Rightarrow y = -3x + 6 \Rightarrow m_1 = -3$

$5x - 2y = -4 \Rightarrow y = \tfrac{5}{2}x + 2 \Rightarrow m_2 = \tfrac{5}{2}$

$$\tan \theta = \left|\frac{\tfrac{5}{2} - (-3)}{1 + \tfrac{5}{2}(-3)}\right| = \frac{\tfrac{11}{2}}{\tfrac{13}{2}} = \tfrac{11}{13}$$

$$\theta \approx 0.7023 \text{ radian} \approx 40.2°$$

3. $(x_1, y_1) = (2, 9)$

$y = 3x + 4 \Rightarrow 3x - y + 4 = 0 \Rightarrow A = 3, B = -1, C = 4$

$$d = \frac{|(3)(2) + (-1)(9) + 4|}{\sqrt{3^2 + (-1)^2}} = \frac{1}{\sqrt{10}} = \frac{\sqrt{10}}{10}$$

4. $y^2 - 2x + 2 = 0$

$$y^2 = 2(x - 1)$$

Parabola

Vertex: $(1, 0)$

Focus: $\left(\tfrac{3}{2}, 0\right)$

5. $x^2 - 4y^2 - 4x = 0$

$(x - 2)^2 - 4y^2 = 4$

$\dfrac{(x - 2)^2}{4} - y^2 = 1$

Hyperbola

Center: $(2, 0)$

Horizontal transverse axis

$a = 2, b = 1, c^2 = 1 + 4 = 5 \Rightarrow c = \sqrt{5}$

Vertices: $(0, 0), (4, 0)$

Foci: $\left(2 \pm \sqrt{5}, 0\right)$

Asymptotes: $y = \pm\dfrac{1}{2}(x - 2)$

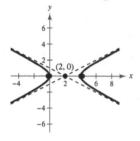

6. $9x^2 + 16y^2 + 54x - 32y - 47 = 0$

$9(x^2 + 6x + 9) + 16(y^2 - 2y + 1) = 47 + 81 + 16$

$9(x + 3)^2 + 16(y - 1)^2 = 144$

$\dfrac{(x + 3)^2}{16} + \dfrac{(y - 1)^2}{9} = 1$

Ellipse

Center: $(-3, 1)$

$a = 4, b = 3, c = \sqrt{7}$

Foci: $\left(-3 \pm \sqrt{7}, 1\right)$

Vertices: $(1, 1), (-7, 1)$

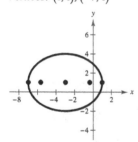

7. $2x^2 + 2y^2 - 8x - 4y + 9 = 0$

$2(x^2 - 4x + 4) + 2(y^2 - 2y + 1) = -9 + 8 + 2$

$2(x - 2)^2 + 2(y - 1)^2 = 1$

$(x - 2)^2 + (y - 1)^2 = \dfrac{1}{2}$

Circle

Center: $(2, 1)$

Radius: $\sqrt{\dfrac{1}{2}} = \dfrac{\sqrt{2}}{2} \approx 0.707$

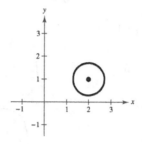

8. Parabola

Vertex: $(3, -4)$

Focus: $(6, -4)$

Horizontal axis

$p = 6 - 3 = 3$

$(y - k)^2 = 4p(x - h)$

$(y - (-4))^2 = 4(3)(x - 3)$

$(y + 4)^2 = 12(x - 3)$

9. Foci: $(0, -2)$ and $(0, 2) \Rightarrow c = 2$

Center: $(0, 0)$

Asymptotes: $y = \pm\dfrac{1}{9}x$

Vertical transverse axis

$\dfrac{a}{b} = \dfrac{1}{9} \Rightarrow b = 9a$

$c^2 = a^2 + b^2$

$4 = a^2 + (9a)^2$

$4 = 82a^2$

$\dfrac{2}{41} = a^2$

$b^2 = (9a)^2 = 81a^2 = \dfrac{162}{41}$

$\dfrac{y^2}{a^2} - \dfrac{x^2}{b^2} = 1$

$\dfrac{y^2}{2/41} - \dfrac{x^2}{162/41} = 1$

10. $xy + 1 = 0$

$A = 0, B = 1, C = 0$

$$\cot 2\theta = \frac{1-1}{1} = 0$$

$$2\theta = 90°$$

$$\theta = 45°$$

$$x = x' \cos 45° - y' \sin 45° = \frac{x' - y'}{\sqrt{2}}$$

$$y = x' \sin 45° + y' \cos 45° = \frac{x' + y'}{\sqrt{}}$$

$$\left(\frac{x' - y'}{\sqrt{2}}\right)\left(\frac{x' + y'}{\sqrt{2}}\right) + 1 = 0$$

$$\frac{(x')^2 - (y')^2}{2} + 1 = 0$$

$$\frac{(x')^2}{2} - \frac{(y')^2}{2} = -1$$

$$\frac{(y')^2}{2} - \frac{(x')^2}{2} = 1$$

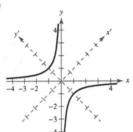

11. $x = 2 + 3\cos\theta$

$y = 2\sin\theta$

$$x = 2 + 3\cos\theta \Rightarrow \frac{x-2}{3} = \cos\theta$$

$$y = 2\sin\theta \Rightarrow \frac{y}{2} = \sin\theta$$

$$\cos^2\theta + \sin^2\theta = 1$$

$$\frac{(x-2)^2}{9} + \frac{y^2}{4} = 1$$

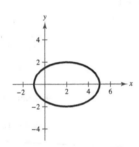

θ	0	$\pi/2$	π	$3\pi/2$
x	5	2	−1	2
y	0	2	0	−2

12. $y = 3 - x^2$

(a) $t = x \Rightarrow x = t$ and $y = 3 - t^2$

(b) $t = x + 2 \Rightarrow x = t - 2$ and $y = 3 - (t-2)^2 = 3 - (t^2 - 4t + 4) = -t^2 + 4t - 1$

13. Polar coordinates: $\left(-2, \dfrac{5\pi}{6}\right)$

$$x = -2\cos\frac{5\pi}{6} = -2\left(-\frac{\sqrt{3}}{2}\right) = \sqrt{3}$$

$$y = -2\sin\frac{5\pi}{6} = -2\left(\frac{1}{2}\right) = -1$$

Rectangular coordinates: $\left(\sqrt{3}, -1\right)$

14. Rectangular coordinates: $(2, -2)$

$$r = \pm\sqrt{2^2 + (-2)^2} = \pm\sqrt{8} = \pm 2\sqrt{2}$$

$$\tan\theta = -1 \Rightarrow \theta = \frac{3\pi}{4}, \frac{7\pi}{4}$$

Polar coordinates:

$$\left(2\sqrt{2}, \frac{7\pi}{4}\right), \left(-2\sqrt{2}, \frac{3\pi}{4}\right), \left(2\sqrt{2}, -\frac{\pi}{4}\right), \left(-2\sqrt{2}, -\frac{5\pi}{4}\right)$$

15. $x^2 + y^2 = 64$

$r^2 = 64$

$r = 8$

16. $r = \dfrac{4}{1 + \cos \theta}$

$e = 1 \Rightarrow$ Parabola

Vertex: $(2, 0)$

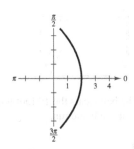

17. $r = \dfrac{4}{2 + \sin \theta}$

$= \dfrac{2}{1 + \dfrac{1}{2} \sin \theta}$

$e = \dfrac{1}{2} \Rightarrow$ Ellipse

Vertices: $\left(\dfrac{4}{3}, \dfrac{\pi}{2} \right), \left(-4, \dfrac{3\pi}{2} \right)$

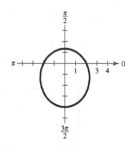

18. $r = 2 + 3 \sin \theta$

$\dfrac{a}{b} = \dfrac{2}{3} < 1$

Limaçon with inner loop

θ	0	$\dfrac{\pi}{2}$	π	$\dfrac{3\pi}{2}$
r	2	5	2	-1

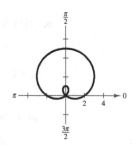

19. $r = 2 \sin 4\theta$

Rose curve $(n = 4)$ with eight petals

$|r| = 2$ when

$\theta = \dfrac{\pi}{8}, \dfrac{3\pi}{8}, \dfrac{5\pi}{8}, \dfrac{7\pi}{8}, \dfrac{9\pi}{8}, \dfrac{11\pi}{8}, \dfrac{13\pi}{8}, \dfrac{15\pi}{8}$

$r = 0$ when

$\theta = 0, \dfrac{\pi}{4}, \dfrac{\pi}{2}, \dfrac{3\pi}{4}, \pi, \dfrac{5\pi}{4}, \dfrac{3\pi}{2}, \dfrac{7\pi}{4}, 2\pi$

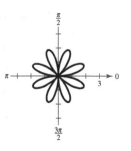

20. Ellipse $e = \dfrac{1}{4}$, focus at the pole, directrix $y = 4$

For a horizontal directrix above the pole:

$r = \dfrac{ep}{1 + e \sin \theta}$

$p = $ distance between the pole and the directrix $\Rightarrow p = 4$

So, $r = \dfrac{(1/4)(4)}{1 + (1/4) \sin \theta} = \dfrac{1}{1 + 0.25 \sin \theta}$.

21. Slope: $m = \tan 0.15 \approx 0.1511$

$\sin 0.15 = \dfrac{x}{5280 \text{ feet}}$

$x = 5280 \sin 0.15 \approx 789 \text{ feet}$

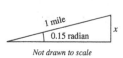

1 mile

0.15 radian

x

Not drawn to scale

22. $x = (115 \cos \theta)t$ and $y = 3 + (115 \sin \theta)t - 16t^2$

When $\theta = 30°: x = (115 \cos 30°)t$

$\qquad\qquad y = 3 + (115 \sin 30°)t - 16t^2$

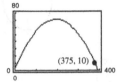

The ball hits the ground inside the ballpark, so it is not a home run.

When $\theta = 35°: x = (115 \cos 35°)t$

$\qquad\qquad y = 3 + (115 \sin 35°)t - 16t^2$

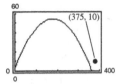

The ball clears the 10 foot fence at 375 feet, so it is a home run.

Cumulative Test Solutions for Chapters 4–6

1. $6 - \sqrt{-49} = 6 - 7i$

2. $6i - \left(2 + \sqrt{-81}\right) = 6i - (2 + 9i)$

$\qquad\qquad = 6i - 2 - 9i$

$\qquad\qquad = -2 - 3i$

3. $(5i - 2)^2 = (5i)^2 - 2(5i)(2) + 2^2$

$\qquad\qquad = 25i^2 - 20i + 4$

$\qquad\qquad = -25 - 20i + 4$

$\qquad\qquad = -21 - 20i$

4. $\left(\sqrt{3} + i\right)\left(\sqrt{3} - i\right) = \left(\sqrt{3}\right)^2 - i^2 = 3 + 1 = 4$

5. $\dfrac{8i}{10 + 2i} = \dfrac{8i}{10 + 2i} \cdot \dfrac{10 - 2i}{10 - 2i}$

$\qquad = \dfrac{80i - 16i^2}{100 - 4i^2}$

$\qquad = \dfrac{16 + 80i}{100 + 4}$

$\qquad = \dfrac{2}{13} + \dfrac{10}{13}i$

6. $f(x) = x^3 + 2x^2 + 4x + 8$

$x^3 + 2x^2 + 4x + 8 = 0$

$x^2(x + 2) + 4(x + 2) = 0$

$(x + 2)(x^2 + 4) = 0$

$(x + 2)(x + 2i)(x - 2i) = 0$

$x = -2$ or $x = \pm 2i$

7. $f(x) = x^4 + 4x^3 - 21x^2$

$x^4 + 4x^3 - 21x^2 = 0$

$x^2(x^2 + 4x - 21) = 0$

$x^2(x + 7)(x - 3) = 0$

$x = 0, x = -7,$ or $x = 3$

8. Zeros: $-6, -3,$ and $4 + \sqrt{5}i$

Because $4 + \sqrt{5}i$ is a zero, so is $4 - \sqrt{5}i$.

$f(x) = (x + 6)(x + 3)\left[x - \left(4 + \sqrt{5}i\right)\right]\left[x - \left(4 - \sqrt{5}i\right)\right]$

$\qquad = \left(x^2 + 9x + 18\right)\left[(x - 4) - \sqrt{5}i\right]\left[(x - 4) + \sqrt{5}i\right]$

$\qquad = \left(x^2 + 9x + 18\right)\left(x^2 - 8x + 21\right)$

$\qquad = x^4 + x^3 - 33x^2 + 45x + 378$

9. $r = |-2 + 2i| = \sqrt{(-2)^2 + (2)^2} = 2\sqrt{2}$

$\tan \theta = \dfrac{2}{-2} = -1$ and θ is in Quandrant II $\Rightarrow$ $\theta = \dfrac{3\pi}{4}$

Thus, $-2 + 2i = 2\sqrt{2}\left(\cos \dfrac{3\pi}{4} + i \sin \dfrac{3\pi}{4}\right).$

10. $\left[4(\cos 30° + i \sin 30°)\right]\left[6(\cos 120° + i \sin 120°)\right] = (4)(6)\left[\cos(30° + 120°) + i \sin(30° + 120°)\right]$

$$= 24(\cos 150° + i \sin 150°)$$

11.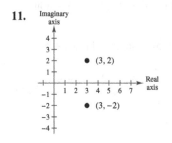

13. $\left[2\left(\cos \dfrac{2\pi}{3} + i \sin \dfrac{2\pi}{3}\right)\right]^4 = 2^4\left(\cos \dfrac{8\pi}{3} + i \sin \dfrac{8\pi}{3}\right)$

$$= 16\left(\cos \dfrac{2\pi}{3} + i \sin \dfrac{2\pi}{3}\right)$$

$$= 16\left(-\dfrac{1}{2} + \dfrac{\sqrt{3}}{2}i\right)$$

$$= -8 + 8\sqrt{3}i$$

12. $d = \sqrt{(3 - 2)^2 + \left[5 - (-1)^2\right]}$

$ = \sqrt{1 + 36}$

$ = \sqrt{37}$

14. $\left(-\sqrt{3} - i\right)^6 = \left[2\left(\cos \dfrac{7\pi}{6} + i \sin \dfrac{7\pi}{6}\right)\right]^6$

$$= 2^6(\cos 7\pi + i \sin 7\pi)$$

$$= 64(-1 + 0i)$$

$$= -64$$

15. $1 = 1(\cos 0 + i \sin 0)$

Cube roots of: $\sqrt[3]{1}\left[\cos\left(\dfrac{0 + 2\pi k}{3}\right) + i \sin\left(\dfrac{0 + 2\pi k}{3}\right)\right], \; k = 0, 1, 2$

$k = 0$: $\sqrt[3]{1}\left[\cos\left(\dfrac{0 + 2\pi(0)}{3}\right) + i \sin\left(\dfrac{0 + 2\pi(0)}{3}\right)\right] = \cos 0 + i \sin 0 = 1$

$k = 1$: $\sqrt[3]{1}\left[\cos\left(\dfrac{0 + 2\pi(1)}{3}\right) + i \sin\left(\dfrac{0 + 2\pi(1)}{3}\right)\right] = \cos \dfrac{2\pi}{3} + i \sin \dfrac{2\pi}{3} = -\dfrac{1}{2} + \dfrac{\sqrt{3}}{2}i$

$k = 2$: $\sqrt[3]{1}\left[\cos\left(\dfrac{0 + 2\pi(2)}{3}\right) + i \sin\left(\dfrac{0 + 2\pi(2)}{3}\right)\right] = \cos \dfrac{4\pi}{3} + i \sin \dfrac{4\pi}{3} = -\dfrac{1}{2} - \dfrac{\sqrt{3}}{2}i$

16. $x^4 - 81i = 0 \Rightarrow x^4 = 81i$

The solutions to the equation are the fourth roots of

$$z = 81i = 81\left(\cos\frac{\pi}{2} + i \sin\frac{\pi}{2}\right) \text{ which are:}$$

$$\sqrt[4]{81}\left[\cos\left(\frac{\frac{\pi}{2} + 2\pi k}{4}\right) + i \sin\left(\frac{\frac{\pi}{2} + 2\pi k}{4}\right)\right], k = 0, 1, 2, 3$$

$$k = 0: 3\left(\cos\frac{\pi}{8} + i \sin\frac{\pi}{8}\right)$$

$$k = 1: 3\left(\cos\frac{5\pi}{8} + i \sin\frac{5\pi}{8}\right)$$

$$k = 2: 3\left(\cos\frac{9\pi}{8} + i \sin\frac{9\pi}{8}\right)$$

$$k = 3: 3\left(\cos\frac{13\pi}{8} + i \sin\frac{13\pi}{8}\right)$$

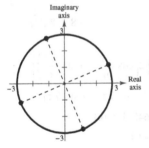

17. $f(x) = \left(\frac{2}{5}\right)^x$

$g(x) = -\left(\frac{2}{5}\right)^{-x+3}$

g is a reflection in the x-axis, a reflection in the y-axis, and a horizontal shift three units to the right of the graph of f.

18. $f(x) = 2.2^x$

$g(x) = -2.2^x + 4$

g is a reflection in the x-axis, and a vertical shift four units upward of the graph of f.

19. $\log_{10} 98 \approx 1.991$

20. $\log_{10} \frac{6}{7} \approx -0.067$

21. $\ln\sqrt{31} \approx 1.717$

22. $\ln\left(\sqrt{30} - 4\right) \approx 0.390$

23. $\log_5 4.3 = \dfrac{\log_{10} 4.3}{\log_{10} 5} = \dfrac{\ln 4.3}{\ln 5} \approx 0.906$

24. $\log_3 0.149 = \dfrac{\log_{10} 0.149}{\log_{10} 3} = \dfrac{\ln 0.149}{\ln 3} \approx -1.733$

25. $\log_{1/2} 17 = \dfrac{\log_{10} 17}{\log_{10}\left(\frac{1}{2}\right)} = \dfrac{\ln 17}{\ln\left(\frac{1}{2}\right)} \approx -4.087$

26. $\ln\left(\dfrac{x^2 - 25}{x^4}\right) = \ln\left(x^2 - 25\right) - \ln x^4$

$$= \ln(x + 5)(x - 5) - 4 \ln x$$

$$= \ln(x + 5) + \ln(x - 5) - 4 \ln x$$

27. $2 \ln x - \dfrac{1}{2} \ln(x + 5) = \ln x^2 - \ln \sqrt{x + 5}$

$$= \ln \dfrac{x^2}{\sqrt{x + 5}}, \, x > 0$$

28. $6e^{2x} = 72$

$$e^{2x} = 12$$

$$2x = \ln 12$$

$$x = \dfrac{\ln 12}{2} \approx 1.242$$

29. $4^{x-5} + 21 = 30$

$$4^{x-5} = 9$$

$$\ln 4^{x-5} = \ln 9$$

$$(x - 5) \ln 4 = \ln 9$$

$$x - 5 = \dfrac{\ln 9}{\ln 4}$$

$$x = 5 + \dfrac{\ln 9}{\ln 4}$$

$$x \approx 6.585$$

30. $\log_2 x + \log_2 5 = 6$

$$\log_2 5x = 6$$

$$5x = 2^6$$

$$x = \dfrac{64}{5} = 12.800$$

31. $\ln(4x) - \ln 2 = 8$

$$\ln\left(\dfrac{4x}{2}\right) = 8$$

$$\ln 2x = 8$$

$$2x = e^8$$

$$x = \dfrac{e^8}{2}$$

$$x \approx 1490.479$$

32. $f(x) = \dfrac{1000}{1 + 4e^{-0.2x}}$

Horizontal asymptotes: $y = 0$ and $y = 1000$

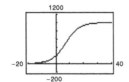

33.

$$N = 175e^{kt}$$
$$420 = 175e^{k(8)}$$
$$2.4 = e^{8k}$$
$$\ln 2.4 = 8k$$
$$\dfrac{\ln 2.4}{8} = k$$
$$k \approx 0.1094$$
$$N = 175e^{0.1094t}$$
$$350 = 175e^{0.1094t}$$
$$2 = e^{0.1094t}$$
$$\ln 2 = 0.1094t$$
$$t = \dfrac{\ln 2}{0.1094} \approx 6.3 \text{ hours to double}$$

34. $P = 20.913e^{0.0184t}$, where P is in millions and where t is the year and $t = 1$ corresponds to 2001.

$$32 = 20.913e^{0.0184t}$$
$$\dfrac{32}{20.913} = e^{0.0184t}$$
$$\ln\left(\dfrac{32}{20.913}\right) = 0.0184t$$
$$t = \dfrac{\ln(32/20.913)}{0.0184} \approx 23.11$$

According to the model, the population will reach 32 million during the year 2023.

35. $2x + y - 3 \Rightarrow y = -2x + 3 \Rightarrow m_1 = -2$

$x - 3y + 6 \Rightarrow y = \frac{1}{3}x + 2 \Rightarrow m_2 = \frac{1}{3}$

$$\tan\theta = \left|\dfrac{m_2 - m_1}{1 + m_1 m_2}\right| = \left|\dfrac{(1/3) - (-2)}{1 + (-2)(1/3)}\right| = \left|\dfrac{7/3}{1/3}\right| = 7$$

$\theta = \arctan 7 \approx 81.87°$

36. $y = 2x - 4 \Rightarrow 2x - y - 4 = 0 \Rightarrow$

$A = 2, B = -1, C = -4$

$(6, -3) \Rightarrow x_1 = 6$ and $y_1 = -3$

$$d = \dfrac{|Ax_1 + By_1 + C|}{\sqrt{A^2 + B^2}}$$
$$= \dfrac{|2(6) + (-1)(-3) + (-4)|}{\sqrt{(2)^2 + (-1)^2}}$$
$$= \dfrac{11}{\sqrt{5}}$$
$$= \dfrac{11\sqrt{5}}{5}$$

37. $9x^2 + 4y^2 - 36x + 8y + 4 = 0$

$AC > 0 \Rightarrow$ The conic is an ellipse.

$$9x^2 - 36x + 4y^2 + 8y = -4$$
$$9(x^2 - 4x + 4) + 4(y^2 + 2y + 1) = -4 + 36 + 4$$
$$9(x - 2)^2 + 4(y + 1)^2 = 36$$
$$\dfrac{(x - 2)^2}{4} + \dfrac{(y + 1)^2}{9} = 1$$

Center: $(2, -1)$

$a = 3, b = 2, c^2 = 9 - 4 = 5 \Rightarrow c = \sqrt{5}$

Vertical major axis

Vertices: $(2, -1 \pm 3) \Rightarrow (2, 2)$ and $(2, -4)$

Foci: $\left(2, -1 \pm \sqrt{5}\right)$

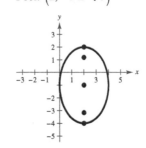

38. $4x^2 - y^2 - 4 = 0$

$AC < 0 \Rightarrow$ The conic is a hyperbola.

$4x^2 - y^2 = 4$

$\dfrac{x^2}{1} - \dfrac{y^2}{4} = 1$

Center: $(0, 0)$

$a = 1, b = 2, c^2 = 1 + 4 = 5 \Rightarrow c = \sqrt{5}$

Horizontal transverse axis

Vertices: $(\pm 1, 0)$

Foci: $\left(\pm\sqrt{5}, 0\right)$

Asymptotes: $y = \pm 2x$

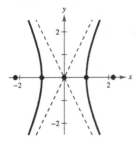

39. $x^2 + y^2 + 2x - 6y - 12 = 0$

$A = C \Rightarrow$ The conic is a circle.

$$x^2 + 2x + y^2 - 6y = 12$$
$$\left(x^2 + 2x + 1\right) + \left(y^2 - 6y + 9\right) = 12 + 1 + 9$$
$$\left(x + 1\right)^2 + \left(y - 3\right)^2 = 22$$

Center: $(-1, 3)$

Radius: $\sqrt{22}$

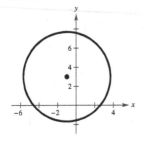

40. $y^2 + 2x + 2 = 0$

$AC = 0 \Rightarrow$ The conic is a parabola.

$y^2 = -2x - 2$

$y^2 = -2(x + 1)$

$y^2 = 4\left(-\tfrac{1}{2}\right)(x + 1)$

Vertex: $(-1, 0)$

Opens to the left since $p < 0$.

Focus: $\left(-1 - \tfrac{1}{2}, 0\right) = \left(-\tfrac{3}{2}, 0\right)$

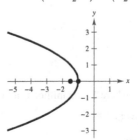

41. Ellipse

Vertices: $(0, 0)$ and $(0, 4) \Rightarrow$ Vertical major axis

Center: $(0, 2)$ and $a = 2$

Ends of minor axis: $(1, 2)$ and $(-1, 2) \Rightarrow b = 1$

Since $a^2 = b^2 + c^2 \Rightarrow 4 = 1 + c^2 \Rightarrow 3 = c^3$

Equation: $\dfrac{\left(x - 0\right)^2}{1^2} + \dfrac{\left(y - 2\right)^2}{2^2} = 1$

$$x^2 + \dfrac{\left(y - 2\right)^2}{4} = 1$$

42. Hyperbola

Foci: $(0, 0)$ and $(0, 6)$ $\Rightarrow$ Vertical transverse axis

Center: $(0, 3)$ and $c = 3$

Asymptotes: $y = \pm\dfrac{2\sqrt{5}}{5}x + 3 \Rightarrow \pm\dfrac{a}{b} = \pm\dfrac{2\sqrt{5}}{5} \Rightarrow \dfrac{a}{b} = \dfrac{2}{\sqrt{5}} \Rightarrow b = \dfrac{\sqrt{5}a}{2}$

Since $c^2 = a^2 + b^2 \Rightarrow 9 = a^2 + \left(\dfrac{\sqrt{5}a}{2}\right)^2 \Rightarrow 9 = \dfrac{9}{4}a^2,$

$a^2 = 4 \Rightarrow b^2 = \dfrac{5a^2}{4} = \dfrac{5(4)}{4} = 5.$

Equation: $\dfrac{(y-k)^2}{a^2} - \dfrac{(x-h)^2}{b^2} = 1$

$$\dfrac{(y-3)^2}{4} - \dfrac{x^2}{5} = 1$$

$$5(y-3)^2 - 4x^2 = 20$$

$$5(y^2 - 6y + 9) - 4x^2 - 20 = 0$$

$$5y^2 - 30y + 45 - 4x^2 - 20 = 0$$

$$5y^2 - 4x^2 - 30y + 25 = 0$$

43. $x^2 + xy + y^2 + 2x - 3y - 30 = 0$

(a) $\cot 2\theta = \dfrac{1-1}{1} = 0 \Rightarrow 2\theta = \dfrac{\pi}{2} \Rightarrow \theta = \dfrac{\pi}{4}$ or $45°$

(b) Since $b^2 - 4ac < 0$, the graph is an ellipse.

$x = x' \cos 45° - y' \sin 45° = \dfrac{x' - y'}{\sqrt{2}}$

$y = x' \sin 45° + y' \cos 45° = \dfrac{x' + y'}{\sqrt{2}}$

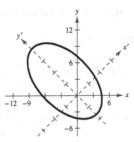

$x^2 + xy + y^2 + 2x - 3y - 30 = 0$

$\left(\dfrac{x'-y'}{\sqrt{2}}\right)^2 + \left(\dfrac{x'-y'}{\sqrt{2}}\right)\left(\dfrac{x'+y'}{\sqrt{2}}\right) + \left(\dfrac{x'+y'}{\sqrt{2}}\right)^2 + 2\left(\dfrac{x'-y'}{\sqrt{2}}\right) - 3\left(\dfrac{x'+y'}{\sqrt{2}}\right) - 30 = 0$

$\dfrac{1}{2}\left[(x')^2 - 2x'y' + (y')^2\right] + \dfrac{1}{2}\left[(x')^2 - (y')^2\right] + \dfrac{1}{2}\left[(x')^2 + 2x'y' + (y')^2\right] + \sqrt{2}(x' - y') - \dfrac{3\sqrt{2}}{2}(x' + y') - 30 = 0$

$\dfrac{3}{2}(x')^2 + \dfrac{1}{2}(y')^2 - \dfrac{\sqrt{2}}{2}x' - \dfrac{5\sqrt{2}}{2}y' - 30 = 0$

$3(x')^2 + (y')^2 - \sqrt{2}x' - 5\sqrt{2}y' - 60 = 0$

$3\left[(x')^2 - \dfrac{\sqrt{2}}{3}x' + \dfrac{2}{36}\right] + \left[(y')^2 - 5\sqrt{2}y' + \dfrac{50}{4}\right] = 60 + \dfrac{1}{6} + \dfrac{25}{2}$

$3\left(x' - \dfrac{\sqrt{2}}{6}\right)^2 + \left(y' - \dfrac{5\sqrt{2}}{2}\right)^2 = \dfrac{218}{3}$

$\dfrac{\left(x' - \dfrac{\sqrt{2}}{6}\right)^2}{\dfrac{218}{9}} + \dfrac{\left(y' - \dfrac{5\sqrt{2}}{2}\right)^2}{\dfrac{218}{3}} = 1$

To use a graphing utility, solve for y in terms of x.

$y^2 + y(x - 3) + \left(x^2 + 2x - 30\right) = 0$

$y_1 = \dfrac{-(x - 3) + \sqrt{(x - 3)^2 - 4\left(x^2 + 2x - 30\right)}}{2}$

$y_2 = \dfrac{-(x - 3) - \sqrt{(x - 3)^2 - 4\left(x^2 + 2x - 30\right)}}{2}$

44. $x = 3 + 4 \cos \theta \Rightarrow \cos \theta = \dfrac{x - 3}{4}$

$y = \sin \theta$

$\cos^2 \theta + \sin^2 \theta = 1$

$\left(\dfrac{x - 3}{4}\right)^2 + (y)^2 = 1$

$\dfrac{(x - 3)^2}{16} + y^2 = 1$

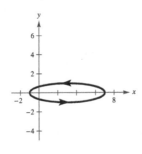

The graph is an ellipse with a horizontal major axis.

Center: $(3, 0)$

$a = 4$ and $b = 1$

Vertices: $(3 \pm 4, 0) \Rightarrow (7, 0)$ and $(-1, 0)$

45. Line through $(3, -2)$ and $(-3, 4)$

$x = x_1 + t(x_2 - x_1) = 3 - 6t$

$y = y_1 + t(y_2 - y_1) = -2 + 6t$

46.

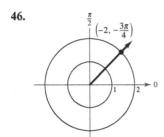

$\left(-2, \dfrac{5\pi}{4}\right), \left(2, \dfrac{\pi}{4}\right), \left(2, -\dfrac{7\pi}{4}\right)$

47. $x^2 + y^2 - 16y = 0$

$r^2 - 16r \sin \theta = 0$

$r(r - 16 \sin \theta) = 0$

$r = 16 \sin \theta$

48.

$$r = \frac{2}{4 - 5 \cos \theta}$$

$4r - 5r \cos \theta = 2$

$4(x^2 + y^2)^{1/2} - 5x = 2$

$\qquad 16(x^2 + y^2) = (5x + 2)^2 = 25x^2 + 20x + 4$

$9x^2 + 20x - 16y^2 + 4 = 0$

49. $r = \dfrac{4}{2 + \cos \theta} = \dfrac{2}{1 + (1/2) \cos \theta}$

$e = \dfrac{1}{2}$

Ellipse with a vertical directrix to the right of the pole.

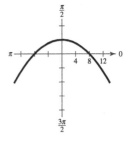

θ	0	$\dfrac{\pi}{2}$	π	$\dfrac{3\pi}{2}$
r	$\dfrac{4}{3}$	2	4	2

50. $r = \dfrac{8}{1 + \sin \theta}$

$e = 1$

Parabola with a horizontal directrix above the pole.

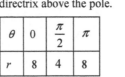

θ	0	$\dfrac{\pi}{2}$	π
r	8	4	8

51. (a) $r = 2 + 3 \sin \theta$ is a limaçon. Matches (iii).

(b) $r = 3 \sin \theta$ is a circle. Matches (i).

(c) $r = 3 \sin 2\theta$ is a rose curve. Matches (ii).